IET TELECOMMUNICATIONS SERIES 93

Antennas and propagation for 5G and beyond

Other volumes in this series

Volume 9 **Phase Noise in Signal Sources** W.P. Robins
Volume 12 **Spread Spectrum in Communications** R. Skaug and J.F. Hjelmstad
Volume 13 **Advanced Signal Processing** D.J. Creasey (Editor)
Volume 19 **Telecommunications Traffic, Tariffs and Costs** R.E. Farr
Volume 20 **An Introduction to Satellite Communications** D.I. Dalgleish
Volume 26 **Common-Channel Signalling** R.J. Manterfield
Volume 28 **Very Small Aperture Terminals (VSATs)** J.L. Everett (Editor)
Volume 29 **ATM: The broadband telecommunications solution** L.G. Cuthbert and J.C. Sapanel
Volume 31 **Data Communications and Networks, 3rd Edition** R.L. Brewster (Editor)
Volume 32 **Analogue Optical Fibre Communications** B. Wilson, Z. Ghassemlooy and I.Z. Darwazeh (Editors)
Volume 33 **Modern Personal Radio Systems** R.C.V. Macario (Editor)
Volume 34 **Digital Broadcasting** P. Dambacher
Volume 35 **Principles of Performance Engineering for Telecommunication and Information Systems** M. Ghanbari, C.J. Hughes, M.C. Sinclair and J.P. Eade
Volume 36 **Telecommunication Networks, 2nd Edition** J.E. Flood (Editor)
Volume 37 **Optical Communication Receiver Design** S.B. Alexander
Volume 38 **Satellite Communication Systems, 3rd Edition** B.G. Evans (Editor)
Volume 40 **Spread Spectrum in Mobile Communication** O. Berg, T. Berg, J.F. Hjelmstad, S. Haavik and R. Skaug
Volume 41 **World Telecommunications Economics** J.J. Wheatley
Volume 43 **Telecommunications Signalling** R.J. Manterfield
Volume 44 **Digital Signal Filtering, Analysis and Restoration** J. Jan
Volume 45 **Radio Spectrum Management, 2nd Edition** D.J. Withers
Volume 46 **Intelligent Networks: Principles and applications** J.R. Anderson
Volume 47 **Local Access Network Technologies** P. France
Volume 48 **Telecommunications Quality of Service Management** A.P. Oodan (Editor)
Volume 49 **Standard Codecs: Image compression to advanced video coding** M. Ghanbari
Volume 50 **Telecommunications Regulation** J. Buckley
Volume 51 **Security for Mobility** C. Mitchell (Editor)
Volume 52 **Understanding Telecommunications Networks** A. Valdar
Volume 53 **Video Compression Systems: From first principles to concatenated codecs** A. Bock
Volume 54 **Standard Codecs: Image compression to advanced video coding, 3rd Edition** M. Ghanbari
Volume 59 **Dynamic Ad Hoc Networks** H. Rashvand and H. Chao (Editors)
Volume 60 **Understanding Telecommunications Business** A. Valdar and I. Morfett
Volume 65 **Advances in Body-Centric Wireless Communication: Applications and state-of-the- art** Q.H. Abbasi, M.U. Rehman, K. Qaraqe and A. Alomainy (Editors)
Volume 67 **Managing the Internet of Things: Architectures, theories and applications** J. Huang and K. Hua (Editors)
Volume 68 **Advanced Relay Technologies in Next Generation Wireless Communications** I. Krikidis and G. Zheng
Volume 69 **5G Wireless Technologies** A. Alexiou (Editor)
Volume 70 **Cloud and Fog Computing in 5G Mobile Networks** E. Markakis, G. Mastorakis, C.X. Mavromoustakis and E. Pallis (Editors)
Volume 71 **Understanding Telecommunications Networks, 2nd Edition** A. Valdar
Volume 72 **Introduction to Digital Wireless Communications** H.-C. Yang
Volume 73 **Network as a Service for Next Generation Internet** Q. Duan and S. Wang (Editors)
Volume 74 **Access, Fronthaul and Backhaul Networks for 5G & Beyond** M.A. Imran, S.A.R. Zaidi and M.Z. Shakir (Editors)
Volume 76 **Trusted Communications with Physical Layer Security for 5G and Beyond** T.Q. Duong, X. Zhou and H.V. Poor (Editors)

Volume 77	**Network Design, Modelling and Performance Evaluation** Q. Vien
Volume 78	**Principles and Applications of Free Space Optical Communications** A.K. Majumdar, Z. Ghassemlooy, A.A.B. Raj (Editors)
Volume 79	**Satellite Communications in the 5G Era** S.K. Sharma, S. Chatzinotas and D. Arapoglou
Volume 80	**Transceiver and System Design for Digital Communications, 5th Edition** S.R. Bullock
Volume 81	**Applications of Machine Learning in Wireless Communications** R. He and Z. Ding (Editors)
Volume 83	**Microstrip and Printed Antenna Design, 3rd Edition** R. Bancroft
Volume 84	**Low Electromagnetic Emission Wireless Network Technologies: 5G and beyond** M.A. Imran, F. Héliot and Y.A. Sambo (Editors)
Volume 86	**Advances in Communications Satellite Systems Proceedings of the 36th International Communications Satellite Systems Conference (ICSSC-2018)** I. Otung, T. Butash and P. Garland (Editors)
Volume 89	**Information and Communication Technologies for Humanitarian Services** M.N. Islam (Editor)
Volume 95	**ISDN Applications in Education and Training** R. Mason and P.D. Bacsich

Antennas and propagation for 5G and beyond

Edited by
Qammer H. Abbasi, Syeda F. Jilani,
Akram Alomainy and Muhammed A. Imran

The Institution of Engineering and Technology

Published by The Institution of Engineering and Technology, London, United Kingdom

The Institution of Engineering and Technology is registered as a Charity in England & Wales (no. 211014) and Scotland (no. SC038698).

First published 2020

The Institution of Engineering and Technology
Michael Faraday House
Six Hills Way, Stevenage
Herts, SG1 2AY, United Kingdom

www.theiet.org

British Library Cataloguing in Publication Data
A catalogue record for this product is available from the British Library

ISBN 978-1-83953-097-5 (hardback)
ISBN 978-1-83953-098-2 (PDF)

Typeset in India by MPS Limited

Contents

About the editors

Qammer H. Abbasi is a senior lecturer and deputy head of the Communications Sensing and Imaging research group at the James Watt School of Engineering in the University of Glasgow, UK. His research interests cover the fields of 5G and beyond antenna and propagation, nano communication, wearable and implantable communication and sensors. He is one of the Investigator for Glasgow's project under Scotland 5G Centre. He is a college member of EPSRC UK and FWO Belgium, a senior member of the IEEE and endorsed as UK Exceptional talent by Royal Academy of Engineering.

Syeda F. Jilani is a research scientist at the Frontier Institute for Research in Sensor Technologies of the University of Maine, USA. Her research interests cover the areas of 5G, millimetre-wave antennas, adaptable antennas, wearable antenna technology, conformal meta-textiles, harsh-environment sensors, microfabrication and RF devices.

Akram Alomainy is a reader in antennas and applied electromagnetics, and a member of the Antennas and Electromagnetics Research Group at Queen Mary University of London, UK. He is the URSI Commission B UK Representative and Associate member of the IEEE Engineering in Medicine and Biology Society (EMBS). He is a chartered engineer, a college member of EPSRC UK and FWO Belgium, a senior member of the IEEE and a member of the IET.

Muhammad A. Imran is Dean of the University of Glasgow UESTC, a professor of communication systems at the James Watt School of Engineering and a head of the Communications Sensing and Imaging research group. He is an affiliate professor at the University of Oklahoma, USA, and a visiting professor at the 5G Innovation Centre, University of Surrey, UK. He is the Principal Investigator for Glasgow's project under Scotland 5G Centre. He is a chartered engineer, a college member of EPSRC UK and FWO Belgium, an IET fellow, a senior member of the IEEE and a senior fellow of Higher Education Academy, UK.

Chapter 1

Introduction to antennas and propagation for 5G and beyond

Qammer H. Abbasi[1], Syeda F. Jilani[2], Akram Alomainy[2] and Muhammad A. Imran[1]

In the past two decades of the inventive advancement of communication systems, the wireless technology has accomplished a tremendous growth trend to realise the ever-increasing user demands of devices and applications, leading towards a massive architecture connecting everything everywhere. It is anticipated that the fifth-generation networks, or 5G, are not just an upgrade of the 4G networks but signify an evolution of many emerging services comprising Internet of Things (IoT) and device-to-device (D2D) communication [1,2]. 5G is expected to visualise an ultra-fast, ultra-flexible, efficient and a unified cloud-native core network interconnecting several heterogeneous networks under its umbrella to facilitate communication diversity [3]. High capacity, spectrum efficiency, low latency, low energy consumption, reliability, security and faster connectivity and high mobility are key features to achieve a real 5G wireless experience.

The design architecture of 5G wireless systems, although in its earlier stages of development, suggests high level of feasibility at the users' end not just in the traditional applications but leads to a multitude of smart devices beneficial for many innovative personal, clinical and healthcare, vehicular and industrial automation applications [4,5]. A proposed architecture of the future 5G wireless network is presented in Figure 1.1, which describes the feasibility of interconnection among various emerging technologies like massive multiple-input–multiple-output (MIMO) network, cognitive radio network, and mobile and static small-cell networks [5]. In addition, the role of network function virtualisation cloud in the next-generation cellular network infrastructure is also demonstrated. The proposed model has also taken into account advanced features such as IoT, D2D communication and small-cell access points [5]. Similar ideas about the 5G architecture have also been proposed at multiple 5G innovative research forums, which can serve as good platforms for developing future 5G standardisation network.

[1]James Watt School of Engineering, University of Glasgow, Glasgow, United Kingdom
[2]School of Electronic Engineering and Computer Science, Queen Mary University of London, London, United Kingdom

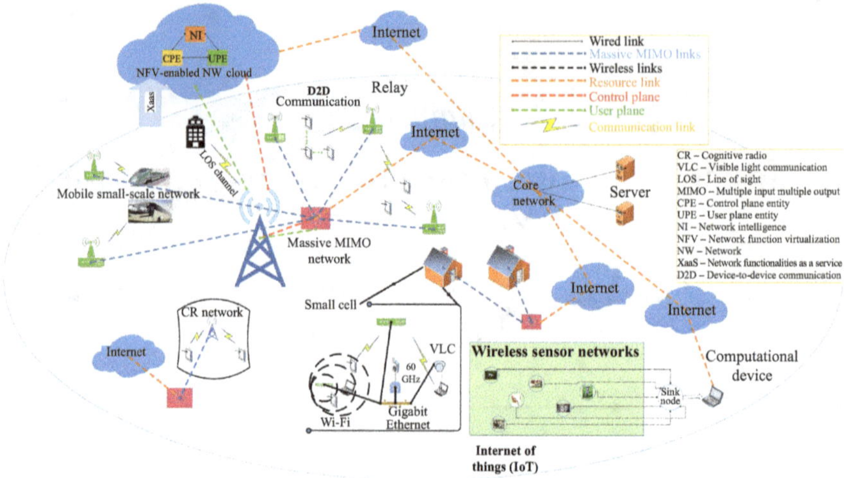

Figure 1.1 A proposed architecture for the 5G networks. © 2015 IEEE. Reprinted, with permission, from [5]

Currently deployed 4G wireless spectrum comprises a frequency bandwidth ranging from 300 MHz to 3 GHz [6]. The selection of a suitable frequency band for cellular networks is primarily dependent on its signal penetration ability, atmospheric attenuations, fading and the compatibility of the antenna size with a cell phone or wearable device. In order to increase the capacity of an allocated frequency spectrum for efficient utilisation, numerous complicated techniques and smart algorithms are in practice to maximise the throughput. In addition, adaptive techniques such as MIMO antennas, reconfigurable antennas and phased arrays have been developed to establish faster communication links [7–9]. It has been established that the current spectrum for wireless networks has theoretically reached its maximum system utilisation, and further upgrade towards 5G needs an allocation of other suitable frequency bands. In contrast with the current low-frequency wireless systems, the millimetre-wave (mm-wave) band presents a wide available spectral resource capable of supporting high data throughput in 5G systems and beyond [10].

The mm-wave spectrum offers exceptional advantages over the current wireless bands such as shorter wavelengths for compact antenna form factor, available bandwidth to allow significant enhancement in channel width, wideband spread-spectrum capability for reduced multipath and clutter and availability of certain high-attenuation bands for highly secure short-distance radio communication [11]. Despite the great potential of mm-waves, reluctance towards mm-wave use for wireless networks in past was mainly due to many challenges concerning its poor penetration power in urban areas responsible for low network coverage, high atmospheric attenuation and signal fading especially in bad weather conditions. The feasibility of mm-wave cellular communication requires careful verification and involves the dense deployment of much smaller pico- and femtocells to reduce the coverage area to avoid increasing signal attenuation with distance.

It is important to comprehend the 5G demands for smartphones and other wireless devices and exploit potential solutions to develop diversified antennas capable of supporting highly efficient and ultra-fast signal transmission and reception at mm-wave frequencies. A great deal of research work is ongoing to provide several feasible antenna designs that can fulfil 5G demands while addressing design parameters with critical importance like cellular handset effects, radiation coverage, antenna gain, hardware implementation and cost [12]. Several prospective antenna configurations have been reported in the past few years, including metamaterial-based antennas [13], three-dimensional (3D) printed antennas [14], on-chip antennas [15], wideband antennas [16,17], reflect-array antennas [18], planar phased arrays [19] and MIMO topologies [20]. Numerous efforts have been made to contribute to the design of mm-wave wireless architecture constituting cellular antennas as the core element, and promising enablers for the 5G technology transition such as ultra-dense networks, advanced inter-cell interference coordination schemes, signal processing schemes of massive MIMO, phased arrays and beamforming networks and performance modelling of wireless backhaul/fronthaul and associated impact on network provisioning for 5G and beyond [21]. This book has specifically designed to comprehend the demands of 5G wireless networks and the recent advancement in developing feasible solutions for the 5G antennas and wireless propagation.

1.1 Scope of the 5G networks

Future 5G network standards are predicted to resolve the current capacity issues and offer suitability for short-range communication to attain minimal attenuation, the capability of handling the path loss limitations and connectivity to multiple users with low latency. Unused mm-wave bands are anticipated as an attractive solution to deal with bandwidth scarcity issues in the 5G wireless applications. The declaration of prospective 5G frequency bands by the international regulatory authorities has provided much-needed clarity towards the future research objectives, especially the development of mm-wave-based 5G architecture. The frequency bands of 28 GHz or above have a potential to accommodate future 5G systems, where high capacity and fast speed are the key objectives to achieve [22].

It is noted that the antenna is the centralised unit of cellular networks, and the advancement of 5G is greatly influenced by the performance of the antenna integrated in the system, including the handheld devices and the base stations. The antenna performance is highly significant in analysing the reliability and efficiency of the wireless systems. The implementation of reliable 5G antenna solutions encompasses several critical aspects regarding antenna design geometry, fabrication and performance characterisation in terms of desired gain and bandwidth based on the requirements of a novel short-range architecture consisting of small cells. The mm-wave spectrum comprises shorter wavelengths that lead to smaller antenna dimensions and demand a high degree of fabrication accuracy. Many novel techniques of antenna fabrication have been introduced to deliver a fast, precise and cost-effective bulk fabrication, for instance, 3D printed antennas [23]. The evolution of 5G is expected to bring much more versatility and flexibility in terms of wearable gadgets and implantable antennas; hence, it is important to look for the scope of integrated on-chip antennas with desired performance attributes.

The propagation issues related with mm-waves, especially the increasing losses with increase in the distance to the target, can be avoided by using high-gain antenna arrays for the 5G communication. Array antennas are regarded as reasonable candidates to compensate the path loss issues for highly dense short-range communication. Two-dimensional (2D) antenna arrays with large electrical apertures can provide narrow beamwidth resulting in highly directive beams and high-gain characteristics essential for 5G operations. Large electrical aperture size at mm-waves does not affect the physical profile of the antenna due to short wavelengths. Massive MIMO antenna arrays, reflect arrays and 2D phased arrays with cognitive operations have been suggested for 5G wireless networks due to their compact integration and diversified operation [24,25]. The smaller sizes of mm-wave antennas compared to sub-3-GHz antennas result in a highly compact and lightweight integration of numerous antenna elements in a MIMO assembly and phased arrays, where 3D printing technique can further reduce the weight of array panels.

1.2 Standardisation and spectrum allocation for 5G

The spectrum transition towards mm-waves has numerous advantages such as shorter wavelengths to allow reduced antenna form factor, unused bandwidth for channel-width enhancement, wideband spread-spectrum capability for reduced multipath and clutter and access to some high-attenuation bands for a secure point-to-point communication link [26,27]. The chief considerations while developing 5G standardisation framework are to mitigate the capacity-related issues, suitability for high-speed short-range communication, capability of dealing with the path loss limitations, security, better connectivity and lower latency. A suitable spectrum allocation with minimal propagation losses and reduced attenuations is of great interest to the worldwide regularity authorities. Several architecture models have been presented recommending mm-waves as a future of 5G wireless networks [28]. The statistical studies have demonstrated lower atmospheric absorptions at 28 and 38 GHz and potential suitability for 5G systems [29]. In the US Federal Communications Commission (FCC) meeting in July 2016, a new set of rules to facilitate 5G progress was approved with the objectives to simplify the innovative process of 5G realisation without further delays. The FCC recommended release of nearly 11 GHz of the mm-wave spectrum for flexible, mobile and fixed-use wireless broadband that comprises bandwidth of 7 from 64 to 71 GHz for unlicensed spectrum and 3.85 GHz, including 27.5–28.35, 37–38.6 and 38.6–40 GHz for licensed spectrum [30]. Many US companies like Verizon and AT&T have publicised their 5G testing plans, and others are on the way to compete in future.

Another major international regulation authority is the Office of Communications (Ofcom), United Kingdom which has published several reports on 5G progress so far. For instance, the report in 2017 covered the research and development of 5G in United Kingdom and an overview of the anticipated 26-GHz band as the pioneer mm-wave-based spectrum for 5G in Europe [31]. It also emphasised greatly on providing access to all or a part of 26-GHz band for 5G applications at earliest. The Radio Spectrum Committee agreed on the prospective road map to develop favourable technical circumstances at the spectrum between 24.25 and 27.5 GHz with a bandwidth of 3.25 GHz, which can facilitate the 5G implementation in Europe. In the second Ofcom report, the goals for spectrum access and desired features of 5G wireless systems were addressed. The report also described that the vision of future cellular networks is to be more flexible and adaptive, expected to bring innovation and create new markets for the spectrum access and frequency reuse, and also to combine emerging technologies under a centralised and unified network [32].

1.3 Millimetre-wave networks: limitations and mitigation

Focused research and standardisation for building the framework of 5G have been addressed the corresponding challenges from the radiofrequency (RF) perspective while employing advanced technologies, such as densified cell deployment, cost of mm-wave massive MIMO, beamformers and phased arrays, security, energy efficiency and latency issues involved and wireless backhaul/fronthaul and associated impact on network provisioning. It has been discussed that mm-wave communication is highly vulnerable towards the effects of weather attenuation, rain droplets and absorptions due to water vapours and oxygen molecules. An estimate states that a heavy rainfall of 25 mm/h causes attenuation of 7 dB/km at 28 GHz; however, if the coverage cell area is restricted to a radius of 200 m, the rain attenuation interprets to about 1.4 dB [33,34]. The proposed solution to resolve these issues is to reduce the cell size of 200 m or less (pico- or femtocells), as the attenuations become more critical with the distance. Heavy rainfall is not obvious in most parts of the world, and radio links to minimise the attenuations due to rain effect can be established [34].

Another concern is that the deployment of highly dense picocells can also multiply the number of installed devices at repeatable distances and ultimately increases the system's cost. While deciding about the technology considerations for 5G, it is necessary to consider the installation expense of the new infrastructure at mm-waves as well as its energy efficiency. Due to shorter wavelengths, mm-wave antennas are smaller and require more power for data transmission and reception as compared to current cellular antennas. While moving towards picocells, ultra-dense installation of a larger number of antennas will ultimately increase the hardware cost, as sophisticated fabrication strategies such as highly integrated on-chip antennas are required for such compact structures. Efficient manufacturing processes for mass production are essential for mm-wave devices to make the overall system cost-effective. In addition to this, the shorter wavelength leads to devices more sensitive to edge diffraction as compared to sub-3-GHz devices [26]. Also, integrating an adaptive control especially in massive MIMO and phased arrays usually requires switching components, and commercially available mm-wave switching devices are either expensive or still immature and need more advancement.

One important concern is much higher free-space path loss for the same communication distance, as it is increased by 3 dB with twice the transmission frequency. The line-of-sight (LOS) link should be established where required to overcome path loss and to attain high speed at much cheaper rates. Use of LOS communication can facilitate high spectral efficiency, low latency, error-free signal detection and faster connectivity. Along with these arrangements, there is still a need for smart and efficient mm-wave 5G antenna design with desirable features of adaptivity, reliability and cognitive control. Sophisticated antenna designs are required to deal with the severe attenuations and path loss and enable effective communication links. High-gain antenna arrays should be utilised to overcome signal fading due to bad weather conditions in both LOS and non-LOS scenarios. Highly directional phased arrays and steerable beamformer antennas with insignificant shadowing or multi-path components are desirable if beam steering can provide a solution to deal with the path loss issues. However, the instalment of outdoor phased array panels could be prone to mechanical motions due to wind or storms that could change the placement of antenna arrays in the order of hundreds of millimetre-wavelengths and can cause severe alignment issues especially in the case of narrow beam antenna arrays. In addition, many ground obstacles could also be responsible for causing losses in wireless links between the base stations and access points [35,36].

Both radio latency and data throughput are of critical significance to inspect the performance of wireless networks. Latency refers to the transit time in signal transmission and determines the transmission speed, while data throughput estimates the number of packets being sent and relies on the available bandwidth to maximise the amount of data being transmitted instantaneously. The ultra-dense and heavy traffic cells should be connected to the core network utilising backhaul and usually come with high demands regarding capacity, latency, availability, energy and cost-efficiency [21]. With interactive applications like online gaming, video and multimedia ones, high latency can degrade the system performance. The huge demand for such applications is estimated in future, and 5G is expected to deliver ultra-low latency. Deploying highly efficient signal processing schemes would essentially increase the system performance. Moreover, the integration of smaller simplified buffers in a radio link will reduce the latency as well as the system's cost. Low latency can also enhance the energy efficiency of the wireless network, for instance, fast sleep and active state transitions of a cellular phone, a short active time with high data rates and lower sleep-mode power consumption will increase the energy efficiency of the device. Battery consumption of device is also a significant aspect of improvement. 5G is expected to deliver an efficient air-interface scheme, which does not require continuous transmissions towards every carrier for detection yet permits transmitter and receiver to be switched off even at the shortest interval of zero traffic for an energy-efficient performance [37].

1.4 Antennas and propagation for 5G and beyond

This book presents a collective overview of the research progress made so far in building the organisational architecture of the 5G networks and beyond, standardisation and regulations, advanced antennas and radio propagation techniques to realise a diversified and smart wireless communication system capable of delivering a seamless user experience. Antennas being the core part of the wireless networks have been greatly focused by the 5G research platforms. Novel techniques have been developed to bring the 5G antenna performance on the advanced level, and many of these have been discussed in the next chapters. For instance, a metamaterial is anticipated to offer a great potential especially in the antenna design due to the strong electromagnetic wave manipulation characteristics as well as the control. Metamaterial-based antennas allow many attractive features such as miniaturisation, bandwidth broadening and gain enhancement that can contribute significantly to the advanced 5G systems [38]. Wideband antennas are desired in 5G communication to enhance the channel width for larger throughput and high data rates. Moreover, spatial diversity at the antenna front ends can be substantially improved by deploying wideband antennas in a MIMO configuration for simultaneous multiple-channel communication.

Recently, another extraordinary approach which has drawn a great attention in the design of wireless transceivers is a fully integrated system-on-chip (SoC) technique due to its multitude of valuable advantages. A few of these include lesser complexity, low cost, miniaturisation/compactness and low power consumption for the integrated wireless modules. The SoC-based wireless system suggests tight monolithic integration of the antenna along with the RF, analogue and digital circuitry on the same substrate therefore avoiding lossy interconnections. The reduced wavelength at mm-waves allows for the antenna to shrink its size making it feasible for an on-chip integration. The antenna of such a system is referred to as an antenna-on-chip which offers several desired advantages over conventional off-chip antennas. For instance, (i) a finer impedance matching control between the antenna and the RF front-end modules and allows for conjugate matching techniques to exhibit better system optimisation, (ii) lower power requirements for the system, (iii) compact integrated circuits with less complex packaging and (iv) design flexibility especially in terms of an antenna design [39].

One significant challenge in using mm-wave is overcoming the high free-space path loss, which severely attenuates the signal strength through the blockages or long transmission distances. Compact massive MIMO antennas are regarded as a smart solution as a higher gain is achieved with more radiating elements to compensate mm-wave path loss. Furthermore, specialised high-gain antenna arrays can be designed equipped with a phasing network where the signal phase for each antenna can be shifted by the phase shifters. These phased arrays comprise a highly directive radiation pattern and provided with a controlled mechanism of phase shifting to achieve beamforming. The beamforming gain is proportional to the antenna array size and dimensions, which is an advantage due the compactness of the mm-wave massive MIMO [40]. In terms of 5G system perspectives, MIMO techniques such as beamforming can substantially improve the transmission reliability and provide higher data rates. The increase in the number of antenna elements within a compact area due to mm-waves at the transceiver is expected to deliver higher diversity and multiplexing gains, and the channel matrix tends to have favourable conditions. Analogue, digital or hybrid beamformers can be implemented based on the system requirements [41]. In addition, a reflect array comprises an array of antennas illuminated by a feed and enabled to create a focused beam with adjusted phasing. Reflect array operates like a phased array antenna but without any power divider or additional phase shifters and combines certain advantages of both reflector antennas and phased arrays such as simpler design and low power consumption that make it economically efficient choice in 5G systems [42,43]. These aspects have been comprehensively discussed in later chapters.

1.5 Conclusion

It is anticipated that mm-wave frequencies will be used in 5G communication systems and beyond to facilitate the higher data throughput for the future wireless devices. The availability of unused bands at mm-waves will increase system capacity, deliver higher data rates and capable of accommodating exponential rise in the number of users, devices and services. Besides many advantages, the mm-wave spectrum experiences severe path loss, high sensitivity to physical objects and edge diffraction issues, leading to smaller cell deployment and complex network architectures. Realising a network framework of highly dense cells with short-range radio links is an unprecedented challenge and involves several aspects such as efficient antenna designs and integrations, feasible and cost-effective fabrication techniques and reliable channel modelling techniques. It is recommended that the 5G antenna should offer a wideband operation to support high-speed communication, high gain and beamforming capabilities to overcome attenuations, and preferably flexible, compact and cost-effective. Many novel antenna topologies embedded with advanced materials, phase shifters, massive arrays and techniques for efficient wireless propagation as key milestones to complete the transition from lower frequencies towards 5G networks and beyond are thoroughly discussed in this book.

References

[1] Fang D, Qian Y, and Hu RQ. Security for 5G mobile wireless networks. IEEE Access. 2018;6:4850–74.

[2] Andrews JG, Buzzi S, Choi W, *et al.* What will 5G be?. IEEE J. Selected Areas Commun. 2014;32(6):1065–82.

[3] Qiao J, Shen X, Mark J, Shen Q, He Y, and Lei L. Enabling device-to-device communications in millimeter-wave 5G cellular networks. IEEE Commun. Mag. 2015;53(1):209–15.

[4] GSA. The Road to 5G: Drivers, Applications, Requirements and Technical Development, GSA, Washington, DC, 2015.

[5] Gupta A, and Jha RK. A survey of 5G network: Architecture and emerging technologies. IEEE Access. 2015;3:1206–32.

[6] Dahlman E, Mildh G, Parkvall S, *et al.* 5G wireless access: Requirements and realization. IEEE Commun. Magazine Mag. 2014;52(12):42–7.

[7] Jilani SF, Rahimian A, Alfadhl Y, and Alomainy A. Flexible and low-profile inkjet-printed frequency-reconfigurable millimeter-wave MIMO antenna for 5G applications. Flex. Print. Electron. 2018;3(3):035003.

[8] Agarwal P, Ali SN, and Heo D. Reconfigurable phased-array design techniques for 5G and beyond communications. IEEE Int. Symp. Radio-Frequency Integration Tech. (RFIT). 2017;53–5.

[9] Rebeiz GM, Kim S-Y, Inac O, *et al.* Millimeter-wave large-scale phased-arrays for 5G systems. IEEE MTT-S Int. Microw. Symp. 2015;1–3.

[10] Rappaport TS, Sun S, Mayzus R, *et al.* Millimeter wave mobile communications for 5G cellular: It will work!. IEEE Access. 2013;1:335–49.

[11] Samimi MK, and Rappaport TS. 3-D millimeter-wave statistical channel model for 5G wireless system design. IEEE Trans. Microw. Theory Technol. 2016;64(7):2207–25.

[12] Rangan S, Rappaport TS, and Erkip E. Millimeter-wave cellular wireless networks: Potentials and challenges. Proc. IEEE. 2014;102(3):366–85.

[13] Jiang H, Si L, Hu W, and Lv X. A symmetrical dual-beam bowtie antenna with gain enhancement using metamaterial for 5G MIMO applications. IEEE Photonics J. 2019;11(1):1–9.

[14] Alkaraki S, Andy AS, Gao Y, *et al.* Compact and low-cost 3-D printed antennas metalized using spray-coating technology for 5G mm-wave communication systems. IEEE Antennas Wireless Propag. Lett. 2018;17 (11):2051–55.

[15] Cheema HM, and Shamim A. The last barrier: On-chip antennas. IEEE Microw. Mag. 2013;14(1):79–91.

[16] Ur-Rehman M, Kalsoom T, Malik NA, *et al.* A wearable antenna for mmWave IoT applications. 2018 IEEE Int. Symp. Antennas Propag. & USNC/URSI National Radio Science Meeting. 2018;1211–12.

[17] Jilani SF, and Alomainy A. A multiband millimetre-wave two-dimensional array based on enhanced Franklin antenna for 5G wireless systems. IEEE Antennas Wireless Propag. Lett. 2017;16:2983–86.
[18] Chou H, and Liu JW. Synthesis and characteristic evaluation of convex metallic reflectarray antennas to radiate relatively orthogonal multibeams. IEEE Trans. Antennas Propag. 2018;66(8):4008–16.
[19] Ibrahim MI, Ahmed MG, El-Nozahi M, Safwat AME, and El-Hennawy H. Design and performance analysis of a miniature, dual-frequency, millimeter wave linear phased array antenna. IEEE Trans. Antennas Propag. 2017;65,(12):7029–37.
[20] Van Chien T., and Björnson E. (2017) Massive MIMO communications. In: Xiang W., Zheng K., and Shen X. (eds) 5G Mobile Communications. Springer, Cham
[21] Jaber M, Imran MA, Tafazolli R, and Tukmanov A. 5G backhaul challenges and emerging research directions: A survey. IEEE Access. 2016;4:1743–66.
[22] Sulyman AI, Alwarafy A, MacCartney GR, Rappaport TS, and Alsanie A. Directional radio propagation path loss models for millimeter-wave wireless networks in the 28-, 60-, and 73-GHz bands. IEEE Trans. Wireless Commun. 2016;15(10):6939–47.
[23] Alkaraki S, Gao Y, Torrico MOM, Stremsdoerfer S, Gayets E, and Parini C. Performance comparison of simple and low cost metallization techniques for 3D printed antennas at 10 GHz and 30 GHz. IEEE Access. 2018;6:64261–9.
[24] Patcharamaneepakorn P, Wu S, Wang C-X, *et al.* Spectral, energy, and economic efficiency of 5G multicell massive MIMO systems with generalized spatial modulation. IEEE Trans. Veh. Technol. 2016;65(12):9715–31.
[25] Syrytsin I, Zhang S, Pedersen GF, and Morris AS. Compact quad-mode planar phased array with wideband for 5G mobile terminals. IEEE Trans. Antennas Propag. 2018;66(9):4648–57.
[26] Bhartia P, and Bahl IJ. Millimeter Wave Engineering and Applications. New York, NY:John Wiley & Sons Inc. 1984.
[27] Rappaport TS, Murdock JN, and Gutierrez F. State of the art in 60 GHz integrated circuits systems for wireless communications. Proc. IEEE. 2011;99(8):1390–436.
[28] Baykas T, Sum C-S, Lan Z, *et al.* IEEE 802.15.3c: The first IEEE wireless standard for data rates over 1 Gb/s. IEEE Commun. Mag. 2011;49(7):114–21.
[29] Rappaport TS, Xing Y, MacCartney GR, Molisch AF, Mellios E, and Zhanget J. Overview of millimeter wave communications for fifth-generation (5G) wireless networks—with a focus on propagation models. IEEE Trans. Antennas Propag. 2017;65(12):6213–30.
[30] Federal Communications Commission. Use of spectrum bands above 24 GHz for mobile radio services; Proposed Rule. Fed Regist. 2016;81(164):58269–308.
[31] https://www.ofcom.org.uk/__data/assets/pdf_file/0021/97023/5G-update-08022017.pdf (accessed 10/03/2017).

[32] https://www.ofcom.org.uk/__data/assets/pdf_file/0014/104702/5G-spectrum-access-at-26-GHz.pdf (accessed 17/09/2017).
[33] Wells J. Faster than fiber: The future of multi-G/s wireless. IEEE Microw. Mag. 2009;10(3):104–12.
[34] Zhao Q, and Li J. Rain attenuation in millimeter wave ranges. IEEE Int. Symp. Antennas, Propag. EM Theory. 2006;1–4.
[35] Hur S, Kim T, Love DJ, Krogmeier JV, Thomas TA, and Ghosh A. Millimeter wave beamforming for wireless backhaul and access in small cell networks. IEEE Trans. Commun. 2013;61(10):4391–403.
[36] Bai T, Vaze R, and Heath RW. Analysis of blockage effects on urban cellular networks. IEEE Trans. Wireless Commun. 2014;13(9):5070–83.
[37] Bouras C, and Diles G. Energy efficiency in sleep mode for 5G femtocells. 2017 Wireless Days. 2017;143–5.
[38] Zhu S, Liu H, and Wen P. A new method for achieving miniaturization and gain enhancement of Vivaldi antenna array based on anisotropic metasurface. IEEE Trans. Antennas Propag. 2019;67(3):1952–6.
[39] Karim R, Iftikhar A, Ijaz B, and Mabrouk IB. The potentials, challenges, and future directions of on-chip-antennas for emerging wireless applications—A comprehensive survey. IEEE Access. 2019;7:173897–934.
[40] Cho YJ, Suk G, Kim B, Kim DK, and Chae C. RF lens-embedded antenna array for mmwave MIMO: Design and performance. IEEE Commun. Mag. 2018;56(7):42–8.
[41] Payami S, Ghoraishi M, and Dianati M. Hybrid beamforming for large antenna arrays with phase shifter selection. IEEE Trans. Wireless Commun. 2016;15(11):7258–71.
[42] Dahri MH, Jamaluddin MH, Abbasi MI, and Kamarudin MR. A review of wideband reflectarray antennas for 5G communication systems. IEEE Access. 2017;5:17803–15.
[43] Huang J, and Encinar JA. Reflectarray antennas. Piscataway, NJ/New York: IEEE Press/Wiley, 2008.

Chapter 2

Antennas for 5G: state-of-the-art and open challenges

Sajid M. Asif[1], Adnan Iftikhar[2], Muhammad S. Khan[3], Muhammad Usman[4], Raed A. Abd-Alhameed[5] and Richard J. Langley[1]

2.1 Introduction

5G technology provides users with a significantly better experience and is a key enabler of massive connectivity between people-to-people and people-to-machines as well as machines-to-machines. The low-latency transmission promised by 5G opens up the possibilities for implementing remote healthcare, self-driving cars, and virtual reality/augmented reality, as depicted in Figure 2.1. 5G technology includes fully digital beamforming, multiple access technologies, and massive multiple-input–multiple-output (MIMO) to deliver higher data rates, higher bandwidth, and lower latency. Antenna along with other microwave systems plays an important role in achieving key features of 5G technology [1]. Unlike a single base station antenna array technology, in 1990, the modern antenna system has evolved from a passive to an active antenna system with integrated radios and digital baseband processing units, as shown in Figure 2.2.

With the release of 5G Phase 1 in 2019, currently deployed 5G systems have non-stand alone (NSA) architecture in which devices have dual connectivity for 4G and 5G links. However, release 16 of 5G Phase 2 greatly focuses on the development of SA architecture where pure 5G link shall be utilized for both data and control. Also, carrier aggregation shall be combined with millimeter-wave (mm-wave) and sub-6 GHz. Both NSA and SA have opted advanced antenna system (AAS) for delivering promising features of 5G. The 5G AAS exploits beamforming

[1]Department of Electronic and Electrical Engineering, The University of Sheffield, Sheffield, United Kingdom
[2]Department of Electrical Engineering, COMSATS University Islamabad, Islamabad, Pakistan
[3]Department of Information Engineering, University of Padova, Padova, Italy
[4]School of Electrical and Electronic Engineering, University College Dublin, Dublin, Ireland
[5]Department of Biomedical and Electronics Engineering, University of Bradford, Bradford, United Kingdom

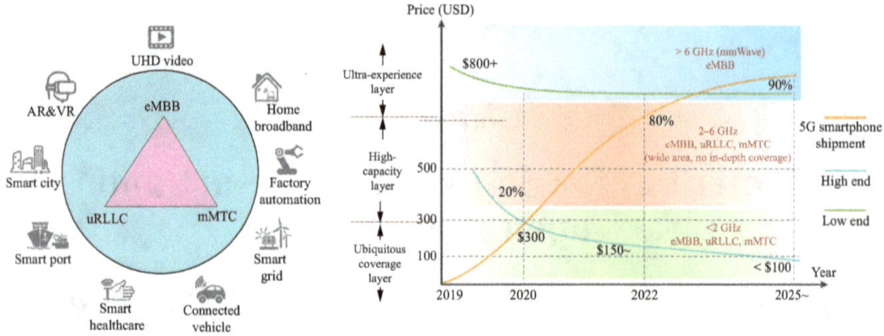

Figure 2.1 Service models of the 5G (left) and the 5G triple-layer structure of target networks (right) (eMBB, enhanced mobile broadband; mMTC, massive machine-type communications; uRLLC, ultra-reliable low-latency communication)

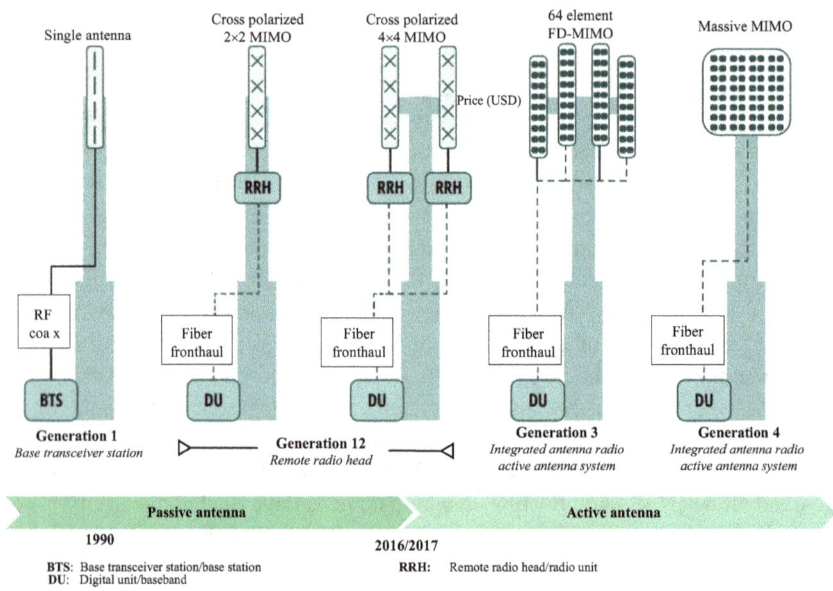

Figure 2.2 The evolution of antennas from passive antennas to active antenna system

in control and broadcast channels as compared to 5G in which beamforming was offered only in data channels [2]. The beamforming capabilities in AAS may comprise active and passive antenna counts of 4T4R (4 transmitters and 4 receivers), 8T8R, 16T16R, 32T32R, 64T64R, and 128T128R. For channel reciprocity in time division duplexing (TDD) and beamforming, 8T8R passive antennas have been utilized. On the other hand, 16T16R, 32T32R, 64T64R, and 128T128R active antennas with radio integration are normally practiced. To enable high beamforming for coverage enhancement in challenging propagation environments at mm-wave bands, AAS may include antenna elements greater than 100. The increased performance demand from the end user has been resulting in the continued growth of network traffic, hence pushing the RAN (radio access network) to not only deliver improved coverage but also more capacity and throughput [3]. The data usage has been increasing much faster than the corresponding revenue of the MNO (mobile network operators), which is pushing the MNOs to evolve the RAN so that cost reduction per bit is increased, in addition to the increase in the end-user performance. It is argued that AAS provides the right platform for the telecom operators to achieve these objectives. Overall, such a system requires technological advancements in the integration of baseband, radio, and antennas as well as cost reduction in the digital processing to achieve MIMO and beamforming.

With 5G NR (New Radio), new frequency bands are available above 6 GHz. More specifically, the allocation of 28 GHz (and 39 GHz) in the mm-wave bands allows very large operational bandwidths (order of GHz) to be used, which support the capacity need. In comparison to the lower frequency bands (1–3 GHz carrier frequencies), radio propagation in mm-wave-frequency bands is significantly challenging due to increased path loss. However, this challenge can be mitigated by using AAS that enables coverage through the beamforming gain. For the lower frequency bands (sub-6 GHz), capacity enhancement is more important; however, for frequencies above 6 GHz, coverage enhancement is more important because of issues with propagation and path loss conditions.

2.2 Key features of 5G antennas

5G antennas must support adaptive beam coverage as per application scenarios and user equipment (UE) distribution. For beam management and precise coverage in relevant area with significant interference suppression in irrelevant areas, 5G antennas must function with RAN. Therefore, 5G antennas should be flexible to support all band configurations and adaptive beam management [4,5]. In addition, to meet 5G traffic capacity and mobility management for achieving high data rates, 5G antenna must support

- sub-6 GHz frequency band and mm-wave 5G allocated frequencies (e.g., 28 and 39 GHz) and
- adaptive beamforming should be concentrated in the desired area to minimize interference and achieve peak data rates.

For high spectral efficiency, power efficiency, and connection density requirements, 5G antennas must support the following:

- Multiuser (MU) beamforming. This shall allow adaptive beam to share both time and frequency resources, which shall increase overall system efficiency. However, interference caused by resource sharing should be minimized by introducing null steering control in AAS.

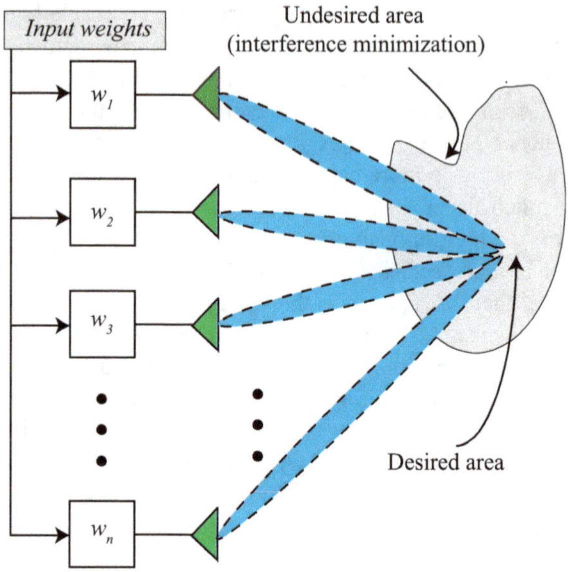

Figure 2.3 An illustration of the beamforming principle for 5G antennas

- Adaptive beamforming should enable precise pattern control in AAS to concentrate radiated energy in desired area to increase power efficiency.
- Antenna components and RF connectors must have minimized losses so that the same radiation energy can ensure deep and wide coverage to enhance connection density.

To achieve maximum radiated signal reference power and signal-to-interference-plus-noise ratio, beamforming technology capable of having narrow beams of appropriate widths and directivity is highly desirable in 5G antennas. A beamforming principle in 5G antennas showing sweeping of narrow beam across the desired area without leaving any coverage holes with minimal overlap is sketched in Figure 2.3.

Furthermore, to achieve high directed beam over a wide angle and beam steering, large antenna arrays are subdivided into sub arrays, and two dedicated chains per sub array are applied. This scenario is depicted in Figure 2.4, where antennas are configured in rectangular shape and two antennas are included in a sub-array. More number of antenna sub arrays would result in a narrower and directed beam.

The 5G antenna technology should also support MU beamforming toward multiple UE having accurate null steering to other UEs. A possible scenario of MU beamforming is sketched in Figure 2.5, where four directed beams are accurately serving four users in respective coverage area of base stations. Also, null-steering phenomenon that must be a feature of 5G antenna technology is depicted in Figure 2.5. The MU beamforming is the key to accurate coverage of 5G broadcast and traffic beams, and high-precision beamforming along with adaptive null steering is considered as a mandatory feature for 5G antenna arrays. Other than MU

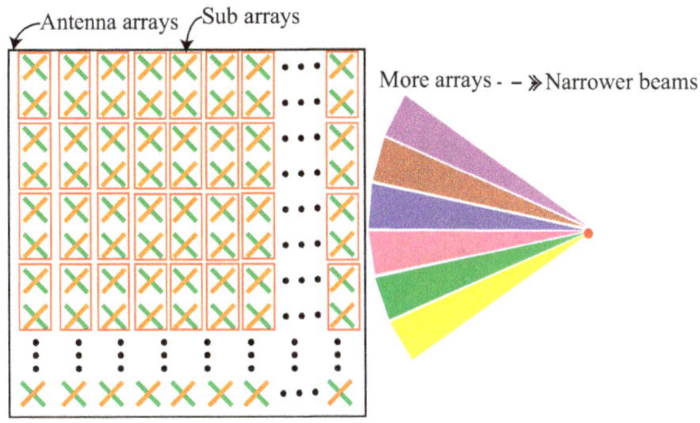

Figure 2.4 Antenna array and sub arrays schematic in 5G showing narrower and directed beamforming

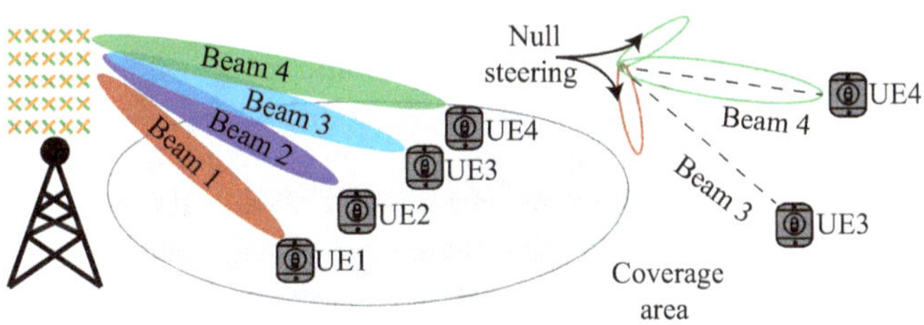

Figure 2.5 MU beamforming and null steering for accurate coverage in 5G

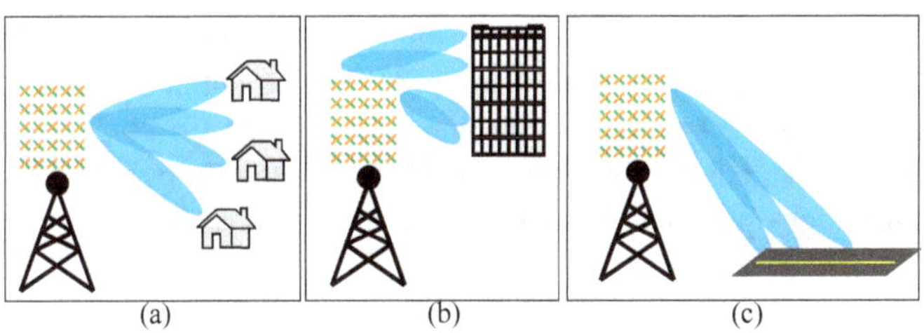

Figure 2.6 Beam visibility for coverage scenarios: (a) houses or low-rise coverage; (b) vertical beam for high-rise coverage; and (c) narrow and directed beam for roads and pathways

beamforming and adaptive null steering in 5G antennas, beam visibility support, and network operation center adjustment should also be deliverable features of 5G antenna. The network should automatically identify scenarios and beam configuration through artificial-intelligence-based algorithms integrated with 5G antennas. An example is illustrated in Figure 2.6, where beam visibility is shown in different scenarios such as houses, high-rise buildings, and roads. It can be observed from Figure 2.6 that directed beams are necessary for roads, whereas wide vertical beams are required for tall buildings. The antenna array architectures for 5G must also support azimuth and elevation beamwidth adjustment flexibility controlled by the network operation center. This would help in the efficiency improvement of the 5G network. An illustration of beam adjustment is depicted in Figure 2.7. To avoid instabilities in the 5G network, 5G antennas must also support intelligent channel shutdown by appropriately turning ON and OFF weights of the antenna elements. This would further help in improving network coverage based in UE density.

It can be revealed from required features of 5G antennas that adaptive beamforming, MU beamforming, and null steering are the most important characteristics that should be present in the 5G antenna technology.

2.3 Massive MIMO antenna technology

Massive MIMO antenna technology solutions increase the capacity and coverage while also improving energy efficiency. The continued growth of mobile broadband (MBB) services has forced the operators to maximize the efficiency of their valuable spectrum resources to increase the network capacity. Massive MIMO techniques are capable of addressing such issues. While the conventional 2×2 MIMO configuration can double the capacity of basic antenna by using two transmit and two receive antenna elements, but in massive MIMO antenna systems, a much larger number of antenna elements are used simultaneously to transmit and receive streams of signals controlled by advanced software to create a much higher network capacity (such systems are deployed in Nokia Bell Labs) [6].

Table 2.1 shows the speed enhancement of wireless networks over the years starting from single-input–single-output systems, single user (SU), and MU MIMO networks [7]. MU-MIMO systems already provide substantial benefits over earlier systems. Massive MIMO targets to further improve this (to 20 Gbps and more) using hundreds of antennas exploiting the advancements in parallel digital signal processing and high-speed electronics. Additional antennas help the signal energy

Figure 2.7 Beam adjustment in 5G antenna array

Table 2.1 Wireless network's speed evolution over the years

Wireless network generation	**2G**	**3G**	**4G**	**5G**
Technology	SISO[a]	SU-SISO[b]	MU-MIMO[c]	Massive MIMO
Year	2002	2009	2012	2020
Data rate (peak)	250 kbps	3 Mbps	100 Mbps	20 Gbps

[a]SISO, single-input–single-output.
[b]SU-SISO, single user-single-input–single-output.
[c]MU-MIMO, multiuser-multiple-input–multiple-output.

of transmission and reception to focus into very small regions of space, hence, bringing enormous improvements in throughput and energy efficiency. Beamforming and spatial multiplexing are the two important concepts that can be dynamically combined in massive MIMO systems—enabling a large number of antenna elements to focus their energy into much smaller region of space. An antenna array system capable of achieving this is called massive MIMO. While massive MIMO is primarily employed at base stations, 5G user terminals may realize basic beamforming techniques.

2.3.1 Antenna array topology

Depending on the geometry pattern, there exist various types of massive MIMO antenna systems. In this chapter, only the linear and planar typologies are discussed. In linear arrays, the scanning of the beam is possible only in the elevation or azimuth; however, in planar array, the beam can be scanned along both the elevation and the azimuth planes. Planar antenna arrays are capable of achieving high gain with lower sidelobes than the linear arrays by using more antennas in an array [8]. The geometry of the planar massive MIMO antenna arrays is designed according to the required radiation pattern. In this section, we will briefly discreet three different types of planar antenna arrays for massive MIMO: the uniform rectangular planar array (URPA), the hexagonal planar array (HPA), and the circular planar array (CPA). Here, the term "uniform" means that the input weight parameters are all unity but depending on the requirement, these parameters can be modified. The URPA is the simplest configuration offering two-dimensional element displacements, so the geometry pattern in this case is a two-dimensional matrix within which antennas are placed. Basically, an URPA is a two-dimensional antenna array, in which elements along the x-axis and y-axis are placed with equal spacing depending on the operating wavelengths. If the number of antenna elements placed in the x-axis is denoted by M and number of elements placed in y-direction is denoted by N, then the total number of elements is $N \times M$. Figure 2.8(a) shows the URPA configuration of antenna array.

Next, an HPA MIMO configuration usually comprises M hexagonal rings, in which each ring has $6m$ elements, where m is the mth ring of the configuration. The distribution of antennas in the HPA is also uniform on the hexagonal side as

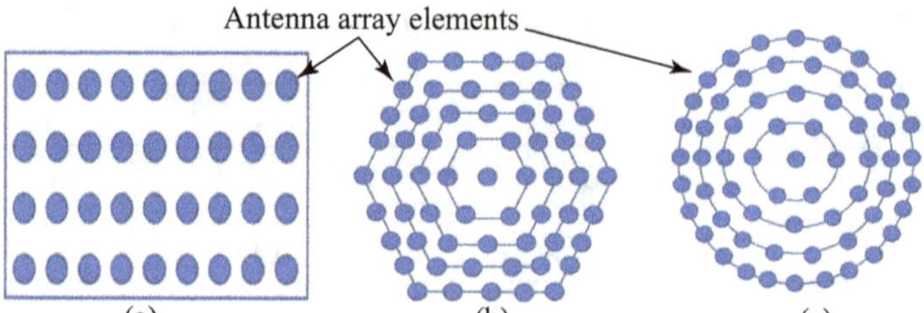

Figure 2.8 Antenna array configuration in massive MIMO systems [8]: (a) uniform rectangular planar array (URPA) configuration; (b) hexagonal planar array (HPA) configuration, and (c) circular planar array (CPA) configuration

illustrated in Figure 2.8(b). The final configuration is the CPA, which is similar to an HPA, but the hexagonal rings are replaced by circular rings. CPA configuration is illustrated in Figure 2.8(c). It consists of certain number of circular rings with different radius, and antennas are located on the circumference of each ring. CPA is a special case of HPA, so the antenna array pattern is similar to HPA.

2.3.2 Single user (SU)-MIMO and multiple user (MU)-MIMO

SU-MIMO is a multichannel technology for wireless communication that assigns the bandwidth of a wireless access point to a single device. SU-MIMO is in contrast with MU-MIMO. SU-MIMO was implemented with the 802.11n wireless standard in 2007. SU-MIMO allows multiple data streams to be transmitted to or received among Wi-Fi devices. A device that uses SU-MIMO will transmit several data streams using several antennas on the device to send data to a single end point device such as a phone, laptop, or tablet. SU-MIMO technologies will also make use of spatial multiplexing (MIMO function) or beamforming to direct and improve signal strengths.

A potential shortcoming of using SU-MIMO in comparison to another approach such as MU-MIMO is only being able to send data to one device at a time. In SU-MIMO, both transmitting and receiving antennas should support MIMO to operate. Moreover, both the devices need multiple antennas that may be a burden for smaller devices.

MU-MIMO is another technology intended for wireless communication between devices. MU-MIMO is generally used to support environments where multiple consumers access the same wireless network at one time. MU-MIMO helps relieve possible overcrowding caused by multiple devices trying to connect by creating multiple connections to a device at the same time. Compared to SU-MIMO, where data is transmitted to a single device at a time, MU-MIMO can transmit data to multiple devices at the same time. Also in SU-MIMO, multiplexing

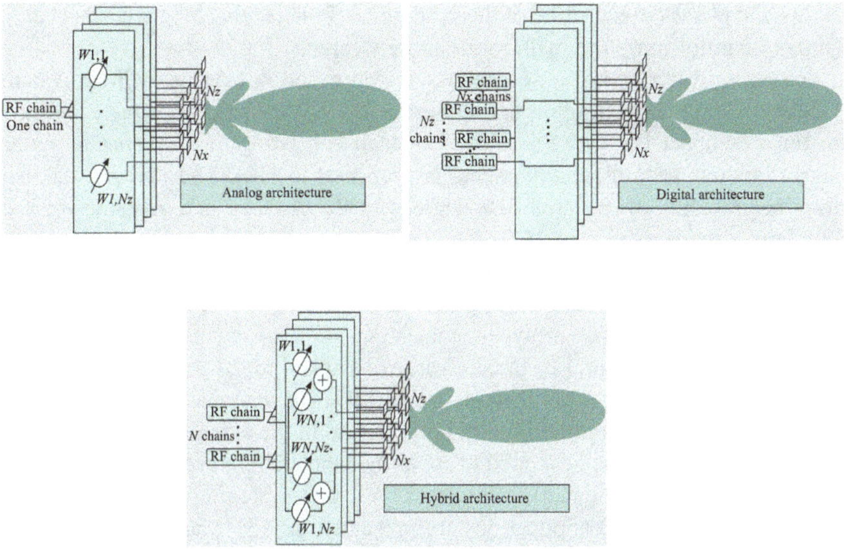

Figure 2.9 Three possible beamforming architectures for 5G mm-wave antenna systems [9]

gain is limited by a number of transmitting and receiving; however, in MU-MIMO, multiplexing gain will scale with the number of transmit antennas used.

2.3.3 Beamforming antennas in 5G massive MIMO

Beamforming in massive MIMO improves the end-user experience by considerably increasing network capacity and coverage while also reducing interference. This is due to the increase in transmission efficiency. Highly focused beams deliver a stronger signal at the receiver device event at higher distance. This also increases capacity in the congested areas, where adding another frequency is not possible. A large number of antenna elements with steerable ports allow beamforming in massive MIMO systems, which means that signals and data are directed to the device instead of broadcasting across the entire cell. This helps to reduce the radio interference across the cell. Three possible beamforming architectures (analog, digital, and hybrid) for 5G mm-wave systems are illustrated in Figure 2.9, the details of which can be found in [9].

Beamforming is a technique that allows focusing the signal from multiple antennas into one strong beam, reducing energy in sidelobes at the transmitter end. At the receiver, beamforming refers to a spatial multiplexing that combines the received signals from a certain direction, and rejecting the signals from any other direction, and considering them as interference. The direction control of the signal is done by altering the weights of the signal phase amplitudes of multiple antennas. Beamforming techniques are categorized into fixed beamformers and adaptive beamformers. In fixed beamforming, the signals can be combined using fixed weights and phases irrespective of the properties of the received signal; however, in adaptive beamforming, the main lobe can be steered.

Recently, different types of antennas are proposed in massive MIMO systems for beamforming depending on the requirements. A 4×4 antenna array configuration for a compact BS with 36 subsectors each consisting of 16 planar antipodal linearly tapered slot antenna elements is proposed in [10]. The antenna has half power beamwidths of 5.3 and 10.7 degrees in the *E*-plane and *H*-plane, respectively, and a realized gain of 25.6 dBi.

A dual-polarized (horizontal and vertical) antenna array with 144 ports operating at 3.7 GHz is proposed in [11]. The antenna array has 18 sub-arrays fed by power splitters; each sub-array is consisting of four elements as shown in Figure 2.10(a) and (b). In [12], three layers are stacked together in orthohexagonal rings to make an antenna array for massive MIMO. The antennas are dual polarized with a gain of 16.6 dBi and HPBW of 12.5 degrees in the azimuth plane. A total of 18 steerable beams can be generated to cover the whole circumference of hexagonal. Multimode antennas designed to operate as a UWB system covering 6–8.5 GHz band for massive MIMO applications are presented in [13]. Massive MIMO using 32 active antennas to increase the cell average throughput gain is presented in [14]. Dolph–Chebyshev approach to synthesize the array pattern of uniform circular array with desired sidelobe has been presented in [15].

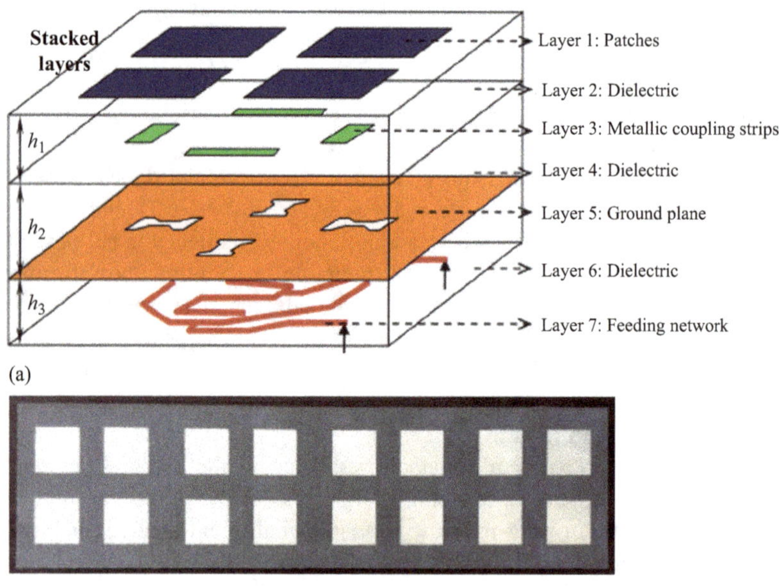

Figure 2.10 (a) Prospective view of the proposed antenna array with layers and (b) sub array with four antenna units (adapted from [11])

2.3.4 5G MIMO antenna for mobile devices

High data rates and allocation of new band for 5G have attracted the researchers to design new antennas that can overcome the current challenges in mobile communication. Over the last decade, researchers have proposed numerous designs for 5G communication in single and MIMO configuration, but the focus here will be to highlight the features of MIMO antenna designs for 5G communication. Printed antennas are more suitable among the antenna used for MIMO applications due to their light weight, low cost, and easily being integrated to small terminal devices [16–18]. However, placing multiple antennas in the limited space is also challenging [19,20]. According to cellular communication requirement, compact, wideband, and highly isolated antennas have been developed recently [21–29].

A compact four-port MIMO antenna system with four inverted-L (IL)-shaped monopole on the edges of non-ground portion of 9×9 mm^2 with substrate FR-4 and ground plane of dimensions of 136×68 mm^2 is proposed in [16]. The design operates in C-band at 3.5 GHz for future 5G with mutual coupling (less than -15 dB) and impedance bandwidth of 200 MHz. The elements are placed orthogonally to exploit the polarization and pattern diversity, and compactness is achieved by using an IL-shape shortened stripe and an IL-shape slot etched in the ground plane to increase the electrical length of the IL-shape monopole.

In [22], a new eight-port dual-polarized MIMO antenna array operating at 3.6 GHz is proposed for 5G mobile phones. The design comprises four pairs of compact dual-polarized self-complementary slot-patch antennas fed by a pair of independent coupled feeding lines. The radiation elements are placed on the corners of 75×150 mm^2 mobile phone PCB board. The miniature design with dual polarization offers suitable isolation, impedance matching, and high-gain radiation patterns.

Eight loop antenna design is proposed in a limited space for 5G communication operating at 3.6 GHz [23]. During the simulation, the mobile surroundings such as phone battery, liquid crystal display (LCD) screens, and electronic components are taken into account. Three and five loop antennas are located on the top of the FR4 board in a stacking manner of two sides. The design operates with mutual coupling (<-15 dB), over the impedance bandwidth of 200 MHz for $|S11|\leq6$ dB. Also, the envelope correlation coefficient (ECC) value of any two elements is less than 0.2 with an efficiency of more than 40% for single antenna. Another eight-element design consists of a novel balanced open slot is proposed in [25] as shown in Figure 2.11. The novel design ensures good impedance matching ($S11\leq-10$ dB), high isolation (>17.5 dB), high total efficiency (>6%), and low ECC (<0.05). Orthogonal polarization is utilized to reduce the mutual coupling in [27]. A quarter wave-mode-substrate-integrated wave-guide antenna is combined with two open-end slots to design a three-antenna tri-polarization block operating in the 3.5-GHz band within a small volume of $17\times17\times6$ mm^3 as shown in Figure 2.12. By integrating four such tri-polarization blocks, a 12-antenna MIMO array is then designed for smartphones.

The proposed array measures good antenna performances with return loss better than 10 dB, isolation higher than 12.5 dB, and antenna efficiencies higher

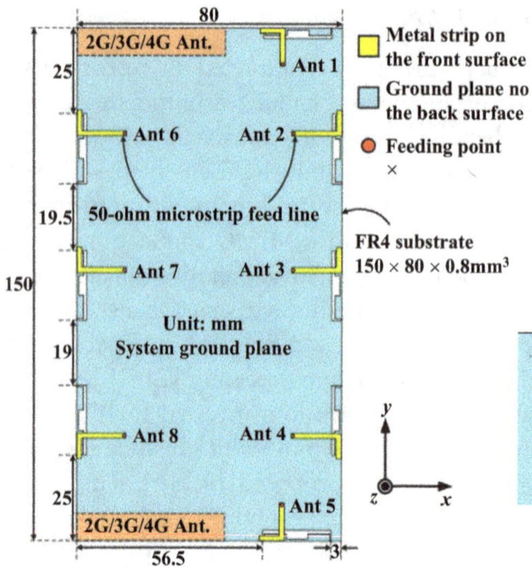

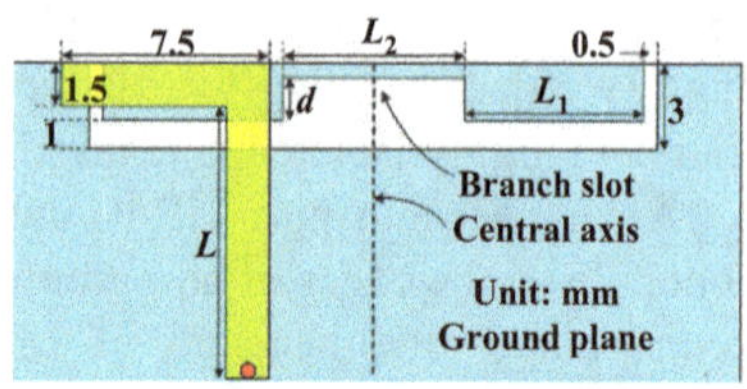

Figure 2.11 Geometry along with dimensions of the proposed MIMO antenna array: (a) perspective view of the full antenna array and (b) detailed structure of the balanced open slot antenna element (Ant 1) [25]

than 50%. The calculated channel capacity of the 12-antenna array is about 57 bps/Hz in a 12×12 MIMO system with 20-dB signal-to-noise ratio, which proves that the proposed array using tri-polarization technique is a good choice for future 5G terminals.

2.4 State-of-the-art phased arrays

Phased arrays are a standard architecture to achieve beamforming, which is an essential requirement for the realization of 5G. In this section, we provide some examples of the state-of-the-art phased arrays that showcase the developments in this technology for 5G and also for applications in other areas such as satellite communication and automotive markets.

The picture shown in Figure 2.13(a) is the antenna technology from Alcan Systems [30] that combines the latest LCD technology with microwave LC and antenna design. Alcan's phased array system is modular design and comprises a radiator, a feed network, and an LC phase shifter. The phase shifter stack is manufactured using conventional LCD processes, which allows for any aperture size and also allowing large-scale fabrication possible. The LC phase shifter stack comprises two glass sheets with LC material filled inside. Alcan has realized an LC layer thickness of a few microns enabling lower loss and more compact size. In comparison to other phased array approaches, waveguide or patch antenna with

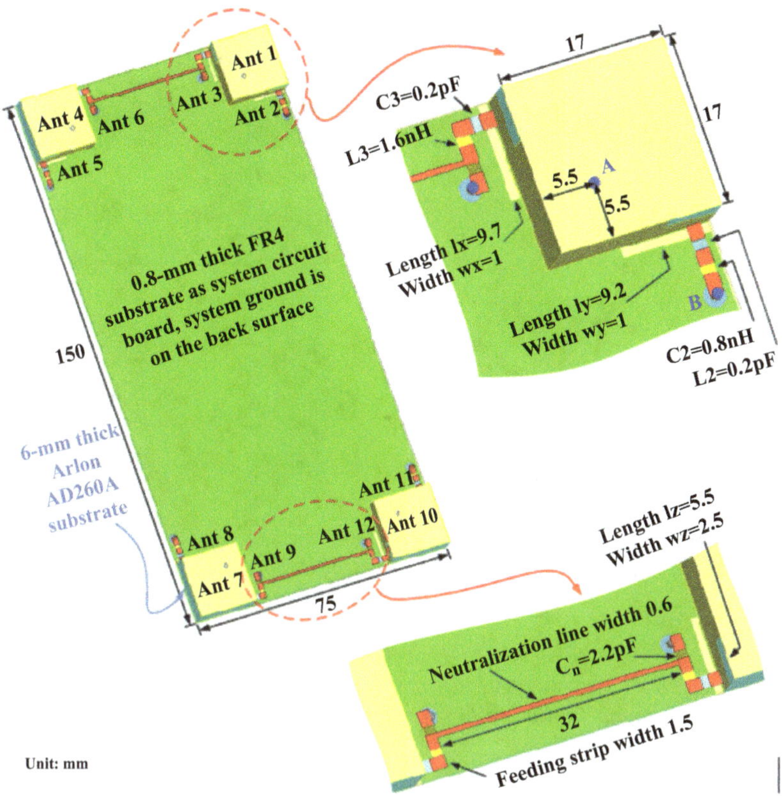

Figure 2.12 Layout of the proposed tri-polarized 12-antenna array for 5G MIMO devices and detailed structure of the tri-polarization block and neutralization line [27]

highly integrated Si devices, the power consumption of this technology is at less than half.

Anokiwave [34] also has been one of the early manufacturers of mm-wave chips for 5G and satellite communications. The chip designed and manufactured by Anokiwave has advanced features such as accurate operation of temperature without calibration using their ZERO-CAL technology. Also, specific sector of arrays can be turned on and off to save the energy (when not needed) using their Kinetic Green functionality. The 5G product portfolio of Anokiwave supports most of the mm-wave frequency bands, i.e., 28 and 39 GHz and has been reportedly used by Ball Aerospace [35] to achieve beamforming applications.

The antenna solution shown in Figure 2.13(b) has been proposed and developed by Gapwaves [31], which is 28 GHz active antenna with integrated chipset and analog beamforming capabilities. Based on the artificial-magnetic-conductor-based technology, this prototype has showed high antenna efficiency with a very

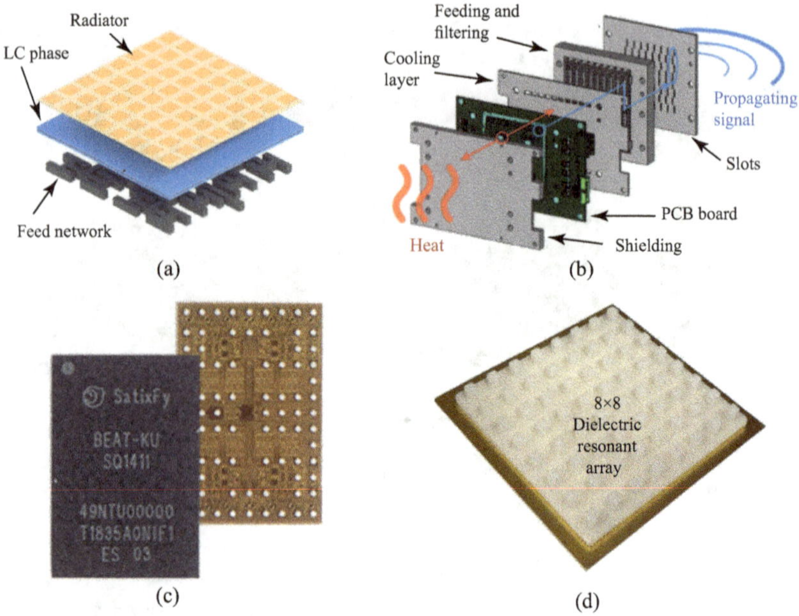

Figure 2.13 Pictures of various state-of-the-art 5G phased arrays technologies by (a) Alcan Systems [30], (b) Gapwaves [31], (c) SatiXfy [32], and (d) Antenna Company [33]

low loss (ten times less than microstrip line and three times less than SIW and approx. the same as the rectangular waveguides) as well as excellent thermal performance.

Figure 2.13(c) shows Beat RFIC chip (ASIC), developed by SatixFy, that is designed to interface directly to the phased array antenna elements [32]. The first 256-element phased array, fully digital, for Ku-band was introduced by SatixFy. The antenna array can be used as a building block for a larger array or as a stand-alone IoT terminal. The antenna is a single-board design with a shared aperture antenna (Rx and Tx), operating from 11 to 12 GHz for Rx and 13.75 to 14.5 GHz for Tx. The 256-element electronically steered multibeam array antenna consists of eight primes daisy-chained and 64 Ku-band beats. The antenna has great capabilities as it is able to concurrently point, track, and manage multiple beams with multiple polarizations.

The industry's first dielectric resonator antenna for mm-wave 5G has been designed by the Antenna Company [33]. The 8×8 (64-element) DRA shown in Figure 2.13(d) has wideband operation and can achieve the entire frequency range from 24 to 30 GHz. The competitive advantage of this mm-wave array is reduced scanning losses; lower sidelobes; and greater scan angle when compared to the traditional patch antenna array. With an overall 40 dBm of peak equivalent isotropically radiated power, this is a scalable design that can support the frequency bands of 37–40 GHz in the mm-wave frequency spectrum.

The aforementioned technologies are just a few examples of the products that the vendors are developing to fully realize the true potential for 5G technology and their applications. It would be a remiss to not mention the works in this area by Qualcomm Technologies, Qorvo, Sivers IMA and Blu Wireless, TMYTEK, Pabellon, Pivotal Commware, NXP, Analog Devices, Metawave Corporation, Kymeta, JMA Wireless, and Movandi among others.

2.5 5G antenna challenges

2.5.1 Active and passive antenna systems

With regards to the deployment of the 5G NR, the active antenna array systems have a great part to play due to their advantages such as their ability to enable MU-MIMO and beamforming, which significantly improve the spectral efficiency as well as the capacity. It should be mentioned that 5G is an upgrade and hence has to be layered on top of the existing networks (2G, 3G, and 4G) using the existing infrastructure (antenna towers and power delivery systems), which may be up to 10 years old and thus may not support the current state-of-the-art equipment being deployed. The deployment of 5G on the existing network, consisting of macro cells, poses a greater challenge as each sector may already have several antennas on them. Some additional considerations are (1) if the additional weight of the antennas as well as wind loading will be supported by the existing infrastructure; (2) if obtaining the planning permission for additional antennas will be secured; and (3) if the installation of the additional antennas will require the vendor to renegotiate the leasehold agreements, which may result in increase of cost.

In the presence of any aforementioned issues, the passive antenna technology can be considered as an alternative technology as the latest passive antennas are capable of supporting multiband and multi-radio access terminal systems. A modern high-end antenna can now support up to 30 jumper ports, compared to only 10 in 2010. By using only two columns (low and high band) of the linear array columns (wideband), an operator can support all the cellular frequencies ranging from 700 to 2,600 MHz. This enables the combination of all the current cellular technologies on a cellular site into three passive antennas, each per sector, and still supports the RET (Remote Electrical Tilt) for each frequency band. Moreover, these antennas support the 2T2R or 4T4R cellular architecture and can continue to deliver the desired diversity or MIMO capability. Passive antennas play a direct role in the rollout of the 5G NR in the sub-6 GHz (3.5 GHz) band. Active antennas do promise potential capacity and spectral efficiency; however, they do have some drawbacks, such as

- economic power delivery to the top of antenna tower;
- will antenna placement limit the RF coverage?

- will the UE's radiation pattern (across the cell) enable complete spatial diversity gain?
- sharing of antenna site with other vendors and if the site is capable of supporting more than one active antenna (per sector).

The deployment of active antennas to achieve capacity gains should be justifiable as it would increase the overall cost. To achieve an improvement of an additional 15%–20% of spectral efficiency, the network operators can save the cost by deploying passive antennas capable of beamforming along with an 8T8R receiver unit and achieve considerable capacity and other 5G benefits.

2.5.2 Antenna characterization and measurements

The 5G wireless technology is driving a much greater need for new integrated circuit packages and modules that can be used in smartphones as well as other systems. Qualcomm, Huawei, Samsung, MediaTek, and Advanced Semiconductor Engineering are some of the vendors active in developing the RF modules and IC packages for the 5G mm-waves. These devices are expected to be equipped with a multitude of new technologies, including antenna in package and phased array antennas. Unlike the current 4G smartphones and devices, where the RF components are part of a separate RF front-end module and not integrated with the antennas as single module, the 5G technology will combine the antenna and the RF chip into a single unit called antenna-in-package. By bringing antenna and the RF chip closer, the signal quality is improved and also losses are minimized [36]. Furthermore, 5G mm-wave systems use phased-array antenna to electronically steer the radiation pattern in multiple directions using the beamforming methods. The key challenges are the realization of such components at extremely high frequencies, mm-waves, and also the associated characterization and measurements before such devices are considered fit for purpose.

More specifically, the measurement or characterization of mm-wave antennas faces the following key challenges:

- Sensitivity of the measurement setup is very critical as small misalignment can cause phase and amplitude errors.
- The small form factor of the chip as well as the high integration level limits the probing of signals except the input and output signals only.
- When measuring the antenna under test, wafer probes limit the scanning area for NF (near-field) measurements (due to close proximity) and also result in unwanted reflections in the far-field measurements [37].
- The parasitic radiation of the wafer probe superimposes with the antenna's radiation and result in distortion of the results.
- Any movement and bending of the measurement cables (RF/coax) result in changing their electrical attributes and subsequently influence the overall measurement signals.
- The fabrication of various components and standard antennas (probes) is very challenging at mm-wave frequencies.

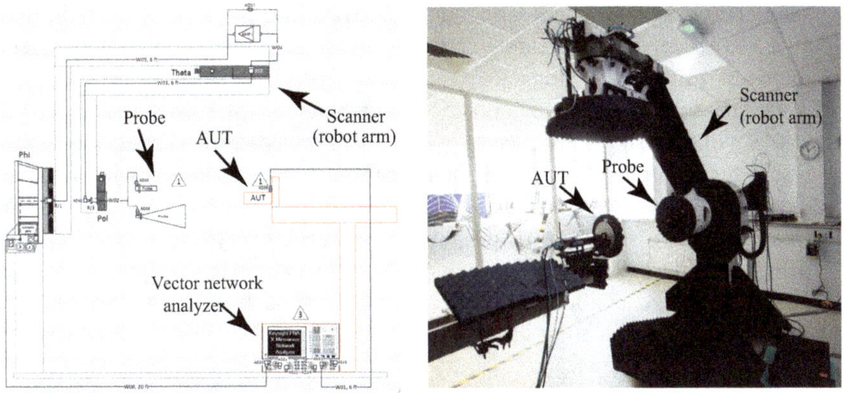

Figure 2.14 A spherical antenna measurement system. A schematic (left) and picture of the measurement setup (right) (courtesy of the University of Sheffield, UK)

- The conventional measurement methods (employed at low frequencies) such as the Wheeler cap method are generally unstable across multiple frequencies and are challenging to realize for chips due to their relatively smaller dimensions.

Besides the testing of integrated antennas (on-chip), measurement facilities capable of performing measurements at mm-wave frequencies are required for 5G systems. Such facilities are not only expensive but required dedicated personnel with a greater understanding of the measurement procedures. As shown in Figure 2.14, the University of Sheffield (UK) has a state-of-the-art antenna measurement facility, which is equipped with a Keysight's 4-port vector network analyzer (N5225A), frequency extending units (from VDI (Virginia Diodes Inc.)), and a spherical antenna measurement system (from NSI-MI technologies) capable of antenna measurements in the region of mm-wave frequencies (up to 110 GHz) [38]. It is important to mention that for a complete system level performance, 5G antennas are required to be tested in a 5G OTA (over-the-air) test-bed, which has its own challenges and limitations (topic covered in a separate chapter in this book) [39].

2.5.3 Challenges with massive MIMO antenna systems

The challenges of massive MIMO are summarized next:

- *Calibration:* Due to the increased number of antenna elements in massive MIMO, the beamforming antennas that are not calibrated accurately will experience undesired emissions in undesired directions. Examples may include the squinting of the beam in one direction.

- *Mutual coupling:* The mutual coupling between the antenna elements in massive MIMO can be challenging and results in energy loss and therefore affects the overall maximum range.
- *Nongeometric arrays:* Due to practical design constraint, some massive MIMO arrays will have to be designed in irregular or nongeometric shapes, which may result in grating lobes and hence will result in dissipating energy in unwanted directions.
- *Complexity:* In massive MIMO antenna systems, everything from design to manufacturing and calibration to deployment is extremely complex and hence presents a huge challenge. The complexity of such systems requires new design concepts as well as novel test approaches, which significantly increases the cost.
- *Data bottleneck:* The large amount of data transmission between the massive MIMO antenna systems may, potentially, create a data bottleneck (as part of the centralized RAN), hence, requiring sufficient fiber bandwidth.

2.6 Conclusion

The applications and technologies of 5G are promising to transform the MBB to enhanced MBB for ubiquitous extremely fast experience and facilitate massive machine-to-machine communication as well as allow network applications that require very high reliability and low latency. Antennas—a key component in the 3G/4G networks—have been evolving into a critical element in the 5G networks and are required to be configurable and controllable as well as manageable. Overall, 5G antennas are required to support the following key features: (1) adaptability to 5G network transition; (2) flexible coordination with other equipment; and (3) intelligent network applications. Besides, 5G will use massive MIMO antenna systems with much large number of antennas for data transmission and achieve greater capacity. This is also important for beamforming purposes that will enable focusing specific beams to the desired user terminals—an essential requirement for emerging 5G applications. While there are several technological challenges in the realization of cost-effective mm-wave 5G antenna systems, the future is promising and offers huge opportunities. The integration of ICs with the antenna elements of arrays is the key to realize efficient arrays at the mm-wave frequencies.

References

[1] 5G Americas Advanced Antenna Systems for 5G. 5G Americas; 2019.
[2] Hong W, Jiang ZH, Yu C, *et al.* Multibeam Antenna Technologies for 5G Wireless Communications. IEEE Transactions on Antennas and Propagation. 2017;65(12):6231–6249.
[3] Ericsson Ericsson Mobility Report. Ericsson, Sweden; 2018.

[4] Liu D, Hong W, Rappaport TS, *et al.* What will 5G Antennas and Propagation Be? IEEE Transactions on Antennas and Propagation. 2017;65 (12):6205–6212.
[5] Huawei Technologies Co. Ltd New 5G, New Antenna. Huawei Technologies Co. Ltd; 2019.
[6] Larsson EG, Edfors O, and Tufvesson F. Massive MIMO for Next Generation Wireless Systems. IEEE Communications Magazine. 2014;52(2):186–195.
[7] Hassan N, and Fernando X. Massive MIMO Wireless Networks: An Overview. Electronics. 2017;6(3):63.
[8] Hassan N, and Fernando X. Massive MIMO Wireless Networks: An Overview. Electronics. 2017;6(3). Available from: https://www.mdpi.com/ 2079-9292/ 6/3/63.
[9] Sun S, Rappaport TS, Heath RW, *et al.* MIMO for Millimeter-Wave Wireless Communications: Beamforming, Spatial Multiplexing, or Both? IEEE Communications Magazine. 2014;52(12):110–121.
[10] Tan W, Assimonis SD, Matthaiou M, *et al.* Analysis of Different Planar Antenna Arrays for mmWave Massive MIMO Systems. In: 2017 IEEE 85th Vehicular Technology Conference (VTC Spring); 2017. p. 1–5.
[11] Ma R, Gao Y, Cuthbert L, *et al.* Antipodal Linearly Tapered Slot Antenna Array for Millimeter-Wave Base Station in Massive MIMO Systems. In: 2014 IEEE Antennas and Propagation Society International Symposium (APSURSI); 2014. p. 1121–1122.
[12] Gao Y, Ma R, Wang Y, *et al.* Stacked Patch Antenna With Dual-Polarization and Low Mutual Coupling for Massive MIMO. IEEE Transactions on Antennas and Propagation. 2016;64(10):4544–4549.
[13] Taluja PS, and Hughes BL. Diversity Limits of Compact Broadband Multi-Antenna Systems. IEEE Journal on Selected Areas in Communications. 2013;31(2):326–337.
[14] Doose N, and Hoeher PA. Massive MIMO Ultra-Wideband Communications Using Multi-Mode Antennas. In: SCC 2015; 10th International ITG Conference on Systems, Communications and Coding; 2015. p. 1–6.
[15] Nam Y, Ng BL, Sayana K, *et al.* Full-Dimension MIMO (FD-MIMO) for Next Generation Cellular Technology. IEEE Communications Magazine. 2013;51(6):172–179.
[16] Lau BK, and Leung YH. A Dolph-Chebyshev Approach to the Synthesis of Array Patterns for Uniform Circular Arrays. In: 2000 IEEE International Symposium on Circuits and Systems (ISCAS). vol. 1; 2000. p. 124–127.
[17] Khan MS, Capobianco A, Asif S, *et al.* A 4 Element Compact Ultra-Wideband MIMO Antenna Array. In: 2015 IEEE International Symposium on Antennas and Propagation USNC/URSI National Radio Science Meeting; 2015. p. 2305–2306.
[18] Khan MS, Rigobello F, Ijaz B, *et al.* Compact 3-D Eight Elements UWB MIMO Array. Microwave and Optical Technology Letters. 2018;60 (8):1967–1971. Available from: https://onlinelibrary.wiley.com/doi/abs/10. 1002/mop.31286.

[19] Khan SM, Iftikhar A, Asif SM, *et al.* A Compact Four Elements UWB MIMO Antenna With On-Demand WLAN Rejection. Microwave and Optical Technology Letters. 2016;58(2):270–276. Available from: https://onlinelibrary.wiley.com/doi/abs/10.1002/mop.29546.

[20] Khan MS, Capobianco AD, Iftikhar A, *et al.* A Compact Dual Polarized Ultrawideband Multiple-Input-Multiple-Output Antenna. Microwave and Optical Technology Letters. 2016;58(1):163–166. Available from: https://onlinelibrary.wiley.com/doi/abs/10.1002/mop.29518.

[21] Khan MS, Capobianco A, Naqvi A, *et al.* Planar, Compact Ultra-Wideband Polarisation Diversity Antenna Array. IET Microwaves, Antennas Propagation. 2015;9(15):1761–1768.

[22] Abdullah M, Ban Y, Kang K, *et al.* Compact 4-Port MIMO Antenna System for 5G Mobile Terminal. In: 2017 International Applied Computational Electromagnetics Society Symposium - Italy (ACES); 2017. p. 1–2.

[23] Parchin NO, Al-Yasir YIA, Noras JM, *et al.* Dual-Polarized MIMO Antenna Array Design Using Miniaturized Self-Complementary Structures for 5G Smartphone Applications. In: 2019 13th European Conference on Antennas and Propagation (EuCAP); 2019. p. 1–4.

[24] Rao L, and Tsai C. 8-Loop Antenna Array in the 5 Inches Size Smartphone for 5G Communication the 3.4 GHz-3.6 GHz Band MIMO Operation. In: 2018 Progress in Electromagnetics Research Symposium (PIERS-Toyama); 2018. p. 1995–1999.

[25] Abdullah M, Ban YL, Kang K, *et al.* Eight-Element Antenna Array At 3.5 GHz for MIMO Wireless Application. Progress in Electromagnetics Research C. 2017;78:209–216.

[26] Li Y, Sim C, Luo Y, *et al.* High-Isolation 3.5 GHz Eight-Antenna MIMO Array Using Balanced Open-Slot Antenna Element for 5G Smartphones. IEEE Transactions on Antennas and Propagation. 2019;67(6):3820–3830.

[27] Sun L, Feng H, Li Y, *et al.* Compact 5G MIMO Mobile Phone Antennas With Tightly Arranged Orthogonal-Mode Pairs. IEEE Transactions on Antennas and Propagation. 2018;66(11):6364–6369.

[28] Li M, Ban Y, Xu Z, *et al.* Tri-Polarized 12-Antenna MIMO Array for Future 5G Smartphone Applications. IEEE Access. 2018;6:6160–6170.

[29] Zhao A, and Ren Z. Size Reduction of Self-Isolated MIMO Antenna System for 5G Mobile Phone Applications. IEEE Antennas and Wireless Propagation Letters. 2019;18(1):152–156.

[30] ALCAN Systems GmbH, Darmstadt, Germany. Available from: https://www.alcansystems.com/.

[31] Gapwaves AB Gothenburg, Sweden. Available from: https://www.gapwaves.com/.

[32] SatixFy, UK. Available from: https://www.satixfy.com/beat/.

[33] The Antenna Company International N.V., the Netherlands. Available from: https://www.antennacompany.com/.

[34] Anokiwave, Austin, TX, USA. Available from: https://www.anokiwave.com/5g/.

[35] Ball Aerospace and Technologies, Boulder, Colorado, USA. Available from: https://www.ball.com/aerospace.

[36] Boehm L, Boegelsack F, Hitzler M, *et al.* The Challenges of Measuring Integrated Antennas at Millimeter-Wave Frequencies [Measurements Corner]. IEEE Antennas and Propagation Magazine. 2017;59(4):84–92.

[37] Boehm L, Hitzler M, Roos F, *et al.* Probe Influence on Integrated Antenna Measurements at Frequencies Above 100 GHz. In: 2016 46th European Microwave Conference (EuMC); 2016. p. 552–555.

[38] The University of Sheffield Millimeter-Wave Antenna Measurement Facility. The University of Sheffield, UK. Available from: https://mmwave.group.shef.ac.uk/.

[39] Microwave Journal 5G Over The Air Testing (OTA) Challenges and Solutions. Microwave Journal, UK; April 2019.

Chapter 3

Metamaterial antennas for 5G and beyond

Muhammad S. Rabbani[1], *James Churm*[1] *and Alexandros Feresidis*[1]

Metamaterials are composite structures that interact with waves in an engineered way to produce effects that are not possible with conventional materials. Optical and microwave metamaterials have gained attention for their ability to produce a negative refractive index and for their potential stealth applications along with other 'non-natural' properties. Metamaterials are of particular interest for the microwave, mm-wave and THz bands as they can produce new functionalities in low-loss, low-profile and tuneable systems without some of the limitations of conventional approaches.

This chapter will outline the requirements for 5G communication systems and demonstrate how metamaterials are well placed to provide innovative solutions for the next generation and beyond. It will go onto describe aspects of metamaterial components and their tunability and showcase an example of an mm-wave antenna design based on metamaterial technology. The chapter will then go onto describe fabrication challenges of metamaterial systems, and how these can be mitigated.

3.1 Channels and antenna requirements for 5G and beyond

Since the late twentieth century, a substantial amount of research has been conducted on channel measurements at frequencies from hundreds of MHzs to tens of GHzs for cellular mobile communication networks and has resulted in reliable channel propagation models for various scenarios of indoor and outdoor wireless communication system. These channels and propagation studies determine the front-end characteristics of a radio link to deliver the required data throughput in the operating environment. An ever-increasing demand for high-throughput mobile communication systems has resulted in the use of evermore complex modulation techniques in an attempt to achieve the highest possible spectral efficiency for a wireless channel. The current mobile communication systems operating at the

[1]Department of Electronic, Electrical and Systems Engineering, University of Birmingham, Birmingham, UK

designated frequency bands below 3 GHz, such as 3G, 4G, 4G/LTE, have delivered data rates up to hundreds of Mbits/s and have reached the Shannon channel capacity limit. Since 2010, efforts have started to utilize sub-6 GHz and mm-wave frequency channels (<100 GHz), promising to deliver a data throughput in the range of Gbits/s under a 5G mobile network standard.

Initial sub-6 GHz band 5G communication is projected to be functional in several countries across the world by 2020. The mm-wave version of the 5G system, mm-wave-5G, is envisaged to make use of broader spectral bandwidth (BW) to obtain multi-Gbits/s data rate over the wireless link and is yet to be commercially implemented. Mm-Wave-5G communication systems are envisioned to be an essential part of modern smart cities, smart highways and smart healthcare infrastructures, providing a fast data rate, low latency, high reliability, greater connectivity and vast coverage. In this regard, several channel measurements have been reported, particularly at frequency bands around 28, 38, 60 and 73 GHz, for mm-wave-5G wireless systems [1–5]. These measurements used the fundamental propagation modelling techniques, with essential modifications, to build new channel models for mm-wave frequencies.

The driven channel models specify the antenna features of mm-wave-5G systems for various use cases, i.e. mobile station (MS), based station (BS), back-haul and indoor/outdoor radios links and draw a reliable prediction for the deliverable throughput in line-of-sight (LOS) and non-line-of-sight (NLOS) links. It has been revealed through the reported channel measurements that the cellular system at mm-wave frequencies gives better throughput and coverage than the conventional cellular networks at lower cm-wave frequencies (<3 GHz), provided a highly directional RF link is established, with beam steering capabilities, exploiting the benefits of analogue and digital beamforming techniques in condensed cellular network architecture [4].

In this section, the key findings of the reported mm-wave channel measurements are illustrated in order to figure out the antenna specifications required for various front-ends in the mm-wave-5G network. Some fundamental work proposed in the literature on mm-wave antenna design for 5G applications is highlighted and a future direction of the metamaterial-based antenna design approach is presented for 5G and beyond communications systems.

3.1.1 Channel measurements and capacity estimation

Various research groups and industry players across the world have reported mm-wave channel measurements and throughput capacity estimations by employing different types of antenna systems [1–21]. In these studies, the applicability of the conventional low-frequency path-loss models has been examined for mm-wave applications and the essential modifications have been proposed for their suitability for a 5G network operating at various mm-wave frequency bands.

In [1], the channel measurements have been performed at 28 and 38 GHz in New York City and Austin to develop new path-loss models through the modifications of the existing Stanford University Interim (SUI) and free-space (FS) path-loss models used for the conventional microwave cellular networks. The original SUI model is given as

$$PL_{SUI}(d)(\text{dB}) = PL(d_0) + 10n\log_{10}\left(\frac{d}{d_0}\right) + X_{fc} + X_{RX} + X_{\sigma} \quad (3.1)$$

where

$$PL(d_0) = 20\log_{10}\left(\frac{4\pi d_0}{\lambda}\right)$$

$$n = a - b \cdot h_{TX} + \frac{c}{h_{TX}}$$

$$X_{fc} = 6\log_{10}\left(\frac{f_{\text{GHz}}}{2}\right), \quad f_c > 2\ \text{GHz}$$

$$X_{RX} = -10.8 \cdot \log\left(\frac{h_{RX}}{2}\right)$$

where λ is the carrier wavelength in metres, $PL(d_0)$ denotes the FS path loss in decibels at a close-in reference distance d_0. X_{fc} and X_{RX} denote the correction factors for frequency and receiver (RX) height, respectively, and X_σ is the typical lognormal random shadowing variable with 0 dB mean and standard deviation, s, in decibels such that $8.2 < s < 10.6$ dB, and f_{GHz} is the carrier frequency in GHz. h_{TX} and h_{RX} denote the transmitter (TX) and RX antenna heights in metres, and $a = 4.6$, $b = 0.0075$ and $c = 12.6$. In order to develop a modified path-loss model at 28 and 38 GHz, channel measurements have been made by using the basic transceiver parameters given in Table 3.1.

Based on the experimental results, the SUI path-loss model is modified for NLOS and LOS mm-wave cellular networks as follows:

$$PL_{\text{mod},NLOS}(\text{dB}) = \alpha_{NLOS} \times (PL_{SUI}(d) - PL_{SUI}(d_0)) + PL(d_0) + X_\sigma \text{ for NLOS} \quad (3.2)$$

$$PL_{\text{mod},LOS}(\text{dB}) = \alpha_{LOS} \times (PL_{FS}(d) - PL_{FS}(d_0)) + PL(d_0) + X_\sigma \text{ for LOS} \quad (3.3)$$

Table 3.1 Transceiver parameters used for channel measurements in [1]

Transmission power (dBm)	**Antenna gain (dBi) BS, *MS***		**HPBW (°) BS, *MS***		**Receiver sensitivity (dBm)**	**Cell radius (m)**	**BW (GHz)**
43	at 28 GHz 25, 15	at 38 GHz 25, 13.2	at 28 GHz 10.9, 8.8	at 38 GHz 7.8, 49.5	−75	200	1

where α_{NLOS} and α_{LOS} are the mean slope correction factors obtained directly from the NLOS and LOS empirical results, and PL_{FS} is the path loss obtained directly from the Friis formula. The measured data and modified path-loss models are further used to estimate the channel capacity gain C, as in the following equation:

$$C = BW \cdot \log_2(1 + CIR) \tag{3.4}$$

where BW denotes the signal bandwidth and CIR is carrier to noise ratio at various distances from a serving BS, which in the case of tri-sector BS, is given as

$$CIR(\text{dB}) = 10 \log_{10}\left(\frac{P_{rec}(d)}{2P_{rec}(D-d)}\right) \tag{3.5}$$

where D is the BS-to-BS distance and d is the distance between BS and MS.

It has been shown in [1] that the 28- and 38-GHz channel coverage within a 200-m radius cell is in fact better than the conventional lower frequency cellular systems in both LOS and NLOS environments and the channel capacity gain is enhanced by about 20 times when compared to that of the conventional lower frequency based cellular systems. Moreover, it has been presented in [6] that the channel capacity can be further increased by employing a larger number of small cells and massive MIMO systems according to the following relation:

$$C_{total} = \sum_i \sum_j C \tag{3.6}$$

where i and j denote the number of small cells and number of MIMO RF chains, respectively. A similar study and conclusion on 28- and 38-GHz channel measurements has also been reported in [2] proving that an overall steerable antenna gain of about 37 dBi (25 dBi at BS and 12 dBi at MS) can be sufficient for SNR >10 dB to establish an NLOS RF link within a 200-m radius cell. In [3], both indoor and outdoor propagation measurement campaigns were conducted at 28 and 73 GHz, respectively, in Downtown Manhattan, New York. The system specifications used for these measurements are given in Table 3.2.

The measured data has been compared with five different existing channel models, and their accuracy has been compared for both indoor and outdoor environments. It has been shown in [3] that the measured path-loss exponent (PLE)

Table 3.2 Transceiver parameters used for channel measurements in [3]

Measurement	Antenna gain (dBi) at 28 GHz, at 73 GHz	Cell radius (m)	Bandwidth (MHz)	Beam width (°)	TX height (m)
Outdoor	24.5, 27	200	800	7, (10.9, 8.6)	7–17, 1.5–2. 4 for backhaul
Indoor	15, 20	<50	800	28.8, 15	2.5, 1.5

(which determines paths-loss performance) for a LOS link at 28 and 73 GHz agreed with that theoretically predicted for the outdoor environment (i.e. two). On the other hand, the PLE values for both LON and NLOS in the indoor environment are lower than those measured for outdoor environment due to the waveguiding effect of the walls. However, the overall measured PLE values are similar to those used for conventional cellular networks which proved to be the feasibility of the mm-wave channels for 5G systems.

In [4], it has been described that in outdoor environments, for both LOS and NLOS situations, the path-loss coefficient at mm-wave channels are in the range of 1.6–2.1 and 2.5–5, respectively, which are typically the same as observed at conventional lower frequencies. There is often no strong frequency dependency beyond the f_2 dependency of FS path loss. However, for indoor environments, i.e. office and hotspots such as lobbies, malls, train stations, the PLE is of the order of 2–3 in NLOS and 1.2–2 in LOS, which are similar to those observed at lower frequencies but with a higher outage probability due to shadowing by persons blocking the LOS as well as highly absorbing steel and concrete walls.

The frequency dependence of the path loss is more pronounced in indoor environments than in outdoor; f_k with $k = 2{:}5$. Based on the propagation difference between indoor and outdoor environments and the fact that the building materials allow very small mm-wave signal penetration, it has been proposed in [1,5,6] that the indoor and outdoor data traffic may be separated in the 5G cellular network and additional BS repeaters may be installed on each building/hotspot to provide a reliable coverage. Based on these mm-wave channels measurements, it is concluded that the PLE at the mentioned mm-wave frequencies (~2 for LOS and ~4–5 for NLOS) are similar to the ones for conventional cellular networks. However, the BS cell coverage may be confined within a 200-m radius with BS and MS TX power in the range of 43 and 23 dBm, respectively. A LOS link may be established up to around 1.5 km distance. Moreover, since polarization is well preserved in an mm-wave, a LOS link, two orthogonal polarizations may be used to support two simultaneous data streams for data-rate enhancement in a LOS link [8]. The delay spread in indoor environments is about 5 ns for LOS and 10–50 ns for NLOS channels; however, a large range of delay spread measurements is reported in the literature for outdoor environments [4]. It is estimated that a data rate of a few Gbits/s may be realized at the MS, and that of up to 12 Gbits/s may be achieved for a fixed point-to-point LOS radio link through a massive MIMO scheme. In order to deliver the required throughput, antenna design is going to be crucial for the front-end of each 5G use case.

3.1.2 Antenna design considerations

It is clear that mm-wave-5G antennas with different characteristics are required for various front-end scenarios, depending on the link type, coverage and implementation constraints [18–27]. A MIMO transmission scheme along with hybrid beam steering (i.e. a combination of digital and analogue techniques) will be preferably employed at the BS front-end for enhanced data capacity and RF link reliability. The mm-wave-5G BS may employ highly directional linearly polarized antenna arrays of a very large electrical size such as 256–1,024 antenna units, where each antenna unit may be a sub-array of multiple-antenna elements. The overall antenna array physical dimensions at mm-wave will still remain comparable to the antenna systems operating at conventional lower frequency bands due to the smaller form factor at mm-wave. It has also been proposed that a BS may deploy tri-sector, 4-sector or even 6-sector for full 360° coverage and hence the half-power beam width (HPBW) range will be specified accordingly.

In the case of separate indoor and outdoor data traffic through repeaters, high-gain antennas may be required to communicate with BS and low-gain wide beam antennas to cover the repeaters surrounding service areas. On the other hand, an MS may use compact, wider beam, polarization-independent and integrated antennas to provide for the space and power limitations and the wider angular coverage in high mobility environments. A planar type antenna structure may be preferred for MS front-end.

The required beam steering range of the antenna mainly depends on the signal angular dispersion which, in most of the cases, has been observed to be on the order of 10° at BS and in the range 30°–70° at MS in an outdoor environment [4]. This offers the possibility of beam switching or beam steering to enhance the radio link robustness when the main direction beam is blocked by a moving obstacle. Another important consideration to be made when designing a hybrid beam steering antenna system at mm-wave frequencies is the separation, *D*, between the MIMO antennas which should be [28]

$$D = \sqrt{\frac{R \cdot \lambda_0}{N}} \tag{3.7}$$

where *R* is the distance between the TX and RX, and *N* is the number of MIMO elements at the RX.

Based on the studies mentioned above, the following antenna system specifications are desired for mm-wave 5G applications:

- Multiple single-band antennas, or preferably a multi-band single antenna, may be required to operate at sub-6 GHz as well as mm-wave bands, i.e. 28, 38, 60 or 73 GHz.
- An antenna operating BW of up to 1 GHz may be required to deliver multi-Gbits/s throughput in cellular environment. The BW may exceed 5 GHz at 73 GHz central frequency for backhaul applications, for even higher throughput.
- An antenna gain in the range of 18–28 dBi (~7°–10° HPBW) and 8–15 dBi (~50°–60° HPBW) may be required at BS and MS, respectively, to compensate for the additional path losses. A total antenna gain of 37 dBi is projected to be sufficient in the link budget for a reliable mm-wave radio link.
- A continuous antenna beam steering of up to 10° and 70°, or multi-beam antenna, may be required to sustain the radio link in a dynamic LOS and NLOS

environment, respectively. The beam searching and switching time is suggested to be <45 ms [5].
- At the MS, beam switching may be applied between a small number of antenna elements for two-dimensional (2D) 360° coverage.

Finally, with regards to antenna performance and implementation low complexity and high reliability are the other criteria to be considered when a specific technology is proposed.

3.1.3 Reported antenna designs for 5G cellular systems

The design of high-gain beam-steerable antennas exhibiting the aforementioned characteristics at mm-wave frequencies is particularly challenging. State-of-the-art beam steering antenna technologies, i.e. mechanical beam steering, beam switching, electronic beam scanning, etc. are constrained by several factors, including upper frequency limits, loss performance, cost, complexity, scanning time and resolution. Conventional phased array techniques employed for beam scanning are applicable only at lower mm-wave frequencies, due to poor performance of phase shifters at higher frequencies. Similarly, beam switching methods using PIN diodes for high-gain antennas are limited to lower mm-wave frequencies due to overall incurred losses. Several techniques have been reported in the literature to design beam steering antenna elements for various front-ends of mm-wave-5G systems [5,29–39]. In [29], four different types of low-gain (<8 dBi) antenna designs, i.e. (i) dual-band L-shaped slot patch antenna at 28 and 38 GHz, (ii) inverted-F antenna at 28 GHz, (iii) T-shaped patch antenna at Ka-band and (iv) 2 × 2 MIMO antenna array at 29 GHz, have been discussed for 5G applications. In [30], a 10-element ultra-wideband mm-wave dipole antenna array is proposed for 5G cellular devices such as laptops, tablets, mobile phones, etc. at 22–33 GHz. The reported antennas in [29–31] showed up to 40° beam steering range.

In [32], a U-shaped patch antenna based on multilayer substrate integrated waveguide (SIW) technology is designed at 73 GHz for 5G systems. Advanced integrated antenna beamforming techniques have been described in [33]. Active linear waveguide antenna array and multi-beam reflector lens antenna have been presented in [34]. Beam switching phased arrays are presented in [35,36] for three-dimensional (3D) coverage in 5G mobile terminals. However, the real-life implementation and coverage performance of these reported antennas have not been fully studied in both MS and BS environments. A physically tested promising antenna technique employing hybrid beamforming has been proposed in [5] and [37] for 5G systems. The antennas are 18 dBi gain arrays of uniformly distributed multiple planar components and exhibit 3D beam steering with angular width where multiple-antenna units are employed to support MIMO operations. The lose performance of the switching and phase shifting elements and the antenna size (6×3 cm^2 for one antenna unit) may restrict these antennas only to the BS applications at lower mm-wave frequencies.

In [38], an SIW antenna is designed at 24.5 GHz with high gain (25 dBi) and ±45° beam scanning. This antenna deployed numerous switching elements along the extended feeding network which limited the upper operating frequency and degraded the antenna efficiency down to 20% at wider scanning angles. Furthermore, the beam switching is obtained at certain defined angles.

Based on the above, it is vital to develop new low-loss antenna beam steering technologies to provide a robust solution for 5G BS and MS applications. Metamaterials are emerging artificial electromagnetic materials and are anticipated to provide promising solutions for antenna applications [39] especially for mm-wave-5G systems and beyond. In recent years, metamaterials have produced promising tuneable antenna systems at mm-wave frequencies. The next sections of this chapter will elaborate some of those techniques to realize such efficient antennas systems incorporated by metamaterial technologies.

3.2 Metamaterial surfaces (metasurfaces)

Given the demanding specifications required by future communication systems, it is necessary to develop new technologies and components that are capable of providing the required performance. Over the years various metamaterial surfaces, also termed metasurfaces, have been proposed for the development of novel phase shifters [40–42], filters [43,44], absorbers [44,45], polarization converters [46,47], etc. These surfaces have often been studied and designed as frequency selective surfaces [48,49], band-gap devices [50], partially reflective surfaces (PRSs) [51,52], holographic surfaces [53,54], etc., each of which can be used in metamaterial-based systems for particular applications.

A good example of a metamaterial surface that can be made into a useful component for mm-wave systems is a high-impedance surface (HIS) [55]. A HIS consists of a passive periodic array of metallic elements over a ground plane and reflects incident waves with zero phase shift for a particular frequency. The HIS is generally used as an artificial/metamaterial ground for low-profile antennas, as with a controllable reflection phase it is possible to reduce the separation between a ground plane and an antenna. A typical HIS can be seen in Figure 3.1, where an incident EM wave passes through the PRS, travels across the dielectric cavity,

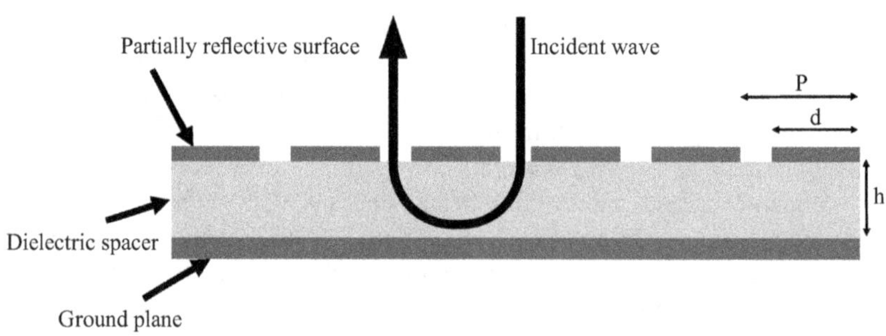

Figure 3.1 A schematic of a high impedance surface where the reflection coefficient is dictated by the structure of the layers

reflects off the ground plane and through the cavity and surface again. This gives a reflection coefficient of the HIS to be as in the following equation:

$$\varphi_{HIS} = 2\varphi_T - \frac{4\pi}{\lambda} \cdot h - \pi \tag{3.8}$$

where φ_T is the transmission coefficient of the PRS, and h is the height of the dielectric cavity. This shows that a HIS can be made tuneable if either the transmission properties of the PRS or the electric height of the dielectric can be altered.

An example of a dynamically phase shifting surface can be produced if the ground plane is separated from the periodic surface by air, and if the separation between ground plane and PRS is mechanically altered. In a recent study [56] two periodic unit cells were designed for a phase shifting surface at the low mm-wave band using this technique.

The unit cell designs can be seen in Figure 3.2 with the square loop designed to give a sharp phase shifting response and the square patch designed to maximize operational BW.

Each of the unit cells were simulated on a 0.8 mm thickness substrate ($\varepsilon_r = 2.2$, $\tan\delta = 0.0009$) with normal plane wave incidence and a varying separation between the ground and the PRS from 0 to 0.5 mm. Due to the unit cells being optimized to be extremely sensitive to the cavity height, each unit cell provided large phase shifts – with the loop unit cell providing over 200° controllable phase shift between 33 and 47 GHz and the square patch providing the same performance between 32 and 59 GHz. With further optimization and the introduction of extra resonant elements, it is possible to get phase shifters with arbitrary phase control.

Due to the nature of the tuning mechanism here, which is based on the simple mechanical movement of the ground plane, there are no active or lossy elements present in the EM system which leads to an extremely low-loss performance that is unparalleled at these frequencies, i.e. <0.35 dB for a theoretical loop element and <0.18 dB for the theoretical square element. In our experience, it is possible to obtain measured losses of less than 2 dB or even 1 dB at these bands, depending on the metasurface design and the highest achievable phase shift.

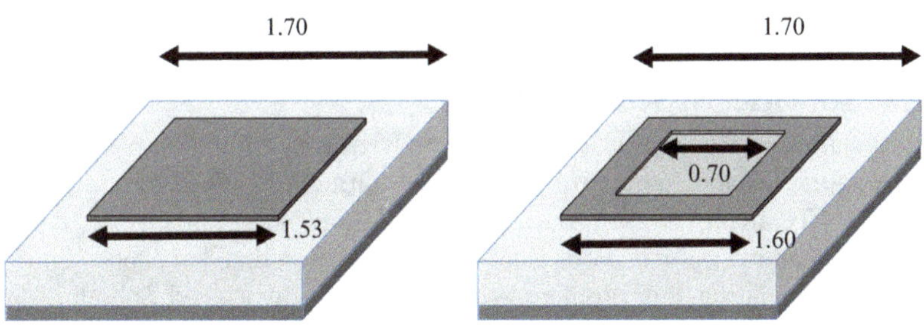

Figure 3.2 A square loop and square patch unit cell design for a phase shifting surface

The mechanical motion of the ground plane can be achieved using various techniques that will be discussed in the next section. Similar phase shifting results can be achieved by altering the transmission coefficient of the PRS which can be achieved by incorporating active/tuning elements into the unit cell design; however, these systems are generally more lossy due to the introduction of these reconfigurable elements (this is also covered in the next section).

The above phase shifting surface gives an example of what a simple reconfigurable surface can do. If complexity is added either within the 2D lattice, by adding multiple layers, or by combining multiple components together, it is possible to make highly effective antenna components such as reflect arrays, transmit arrays, leaky wave antenna (LWA) and holographic surfaces. If low-loss tunability can be incorporated into these kinds of components then metamaterials begin to be one of the most promising avenues for exploration when considering the demands of future communication and sensing systems.

3.3 Tunability in metamaterial systems

Tuneable components are required in modern communication systems with frequency reuse requiring high directivity beam scanning antennas and the market pressure to make antennas multifunctional, multiband or frequency scanning. Conventionally, diodes are the tuning element of choice, either in the form of switchable elements such as PIN diodes or varactors for variable capacitance tuning. However, as modern communication systems move into ever higher frequencies, the losses and parasitic effects experienced by diode systems dramatically increase, and when moving into the mm-wave band, these effects become significant enough that alternative tuning technologies are required.

As frequency increases, a traditional varactor will become high loss. This is due to the resistance and parasitic elements within the diode beginning to dominate the characteristic response.

There are however several emerging methods that offer the ability to providing tunability in mm-wave systems. These include diode architecture modification, dynamic dielectric materials and use of macro- and micro-electromechanical devices. Such technologies are addressed in the following sections.

3.3.1 Alternative tuning technologies

3.3.1.1 Air-bridged Schottky diodes

In traditional diodes, a significant factor in the frequency limitation problem comes from the parasitic capacitance from pad-to-pad through the dielectric on which the diode sits. Teratech, a spin-out company from the Rutherford Appleton Laboratories, and Virginia Diodes are two companies that have put significant efforts into alleviating this problem with the use of air-bridged Schottky diodes [57]. With air-bridged Schottky diodes, the substrate between the pads is split. This creates a parallel pad-to-pad capacitance which reduces the overall pad-to-pad capacitance and increases the operational frequencies of the diode. A gold finger protrudes over the gap to make the air bridge. This makes contact with the semiconductor to create the Schottky diode which is tuneable at frequencies above 60 GHz in a similar manner to a varactor diode, as under reverse bias the Schottky junction capacitance can be varied and controlled [58].

The air-bridged Schottky diodes provided by Teratech Ltd are rated between 75 GHz and 1 THz; however, as these diodes are capable of capacitance changes of between 13 and 15 pF, we have recently demonstrated that they could theoretically be used for lower frequency systems in the mm-wave, such as 60 GHz [59].

Air-bridged Schottky diodes, although not low loss, can provide viable tuneable components at mm-wave frequencies. They can be used in conventional antenna systems, however if they can be fabric junction resistance, may prove to be extremely powerful in metamaterial systems.

3.3.1.2 Tuneable dielectrics

Liquid crystals are commonly used in display technology where a voltage-controlled variation in liquid alignment can selectively block light transmittance through a surface. However, the same alignment changes can create a variation in dielectric permittivity which can be utilized in microwave and mm-wave systems [60,61].

One of the main criticisms of liquid crystal technology is that losses and tuning speed are proportional, with some of the lower loss liquid crystals taking as long as a minute to switch state, with the fastest being extremely high loss.

Ferroelectric materials can also be operated in a similar way to liquid crystals [62]. Traditionally ferroelectric materials have been seen as lossy; however, it may be possible for thin-film ferroelectric materials to provide highly effective metamaterial systems [63].

3.3.1.3 Micro-actuation

The properties of certain metamaterials can be optimized so that they are extremely sensitive to a given parameter. Often this could be a disadvantage, as it increases the reliance on fabrication tolerances; however, it can also be used as a tuning mechanism, as outlined by the example of the HIS in the previous section, where a dimension of a metamaterial can be physically varied using micro-actuators. In terms of speed and ease of integration, these micro-actuators can be either piezoelectric actuators or made from electroactive polymers (EAPs).

Piezoelectric actuators (PEAs) are layers of stacked ceramic plates, interleaved with electrodes, that when a voltage is applied across the plates expand in a given direction. These actuators typically exhibit an expansion of tens of microns; however, if placed in a flexure lever assembly, this can be amplified up to 1 mm with microsecond response times and nanometre positional resolution. These actuators can have inbuilt strain gauges that can be used for negative feedback control to oppose any positional deviation due to varying temperature, humidity, vibration, etc.

These actuators can be used, for example, to make a variable phase shifting metasurface, by taking the HIS in Figure 3.1 and attaching the ground plane to a

PEA to provide a variable cavity height. In this case, the HIS has been optimized to make its phase response very sensitive to the separation of the dielectric from the ground plane. This created a quasi-optical phase shifter that has no active elements, and it therefore extremely low loss, and is scalable across the mm-wave band and beyond.

Another method of achieving actuation on this scale is with the use of EAPs. An EAP is a polymer that alters its shape when a voltage is applied [64]. The classic example of an EAP is PDMS, a silicone that once a high voltage is applied across it will shrink in-line with the electric field, thus providing a controllable sagging of the material, which if combined with a pre-strain can cause movement in any plane.

EAPs have been shown to produce effective phase shifting in a co-planar waveguide (CPW) device [65] where a loading strip is suspended over a CPW using an EAP. As voltage is applied at either side of the EAP, the selective strain reduction this provides causes the loading strip to move into the opposite direction, creating a significant transmission phase shift. This kind of actuation could be implemented in a metamaterial system.

3.3.1.4 Other systems

There are several other tuning mechanisms that could be employed in tuneable metamaterial systems including the following:

- *Nano/micro-electromechanical systems* (MEMS) that provide micro-actuation in response to a voltage change. MEMS can have issues in metamaterials first in terms of reliability in both operation and fabrication [66] and second in terms of element density, as larger displacements require larger electrostatic actuators to move active elements. This can limit how closely unit cells can be packed.
- *Phase change materials* such as vanadium dioxide and germanium/antimony/tellurium that change their physical properties in response to temperature change [67]. These materials can be quite lossy and suffer from thermal tuning which can be slow to respond and requires excessive regulation to survive general changes.
- *Materials with tuneable mobile carriers* such as graphene which can be structurally changed to achieve semiconductor-like performance and indium tin oxide which can have properties tailored to metamaterial performance [68].

Each of these has its own merits but generally suffers from a lack of maturity or integration issues.

It is likely that a variety of technologies will be used in future communication systems depending on the specific requirements of the system.

Metamaterials also offer added complexity when it comes to tuning. It is possible with simple phase shifting surfaces to have a tuning material distributed across the metasurface – for example having a tuneable dielectric, such as liquid crystal or ferroelectric, as a substrate, or having air-bridged Schottky diodes embedded in every unit cell that tune to the same control signal. However, certain metamaterials, such as reflect and transmit arrays along with holographic surfaces, become much more possible and powerful if the phase of each element (or block of elements) is individually addressable. In this way, it is possible to have complex beam scanning, or multi-beam systems working off one surface. However, for many of the technologies, this can introduce biasing issues, where biasing lines can interfere with the electromagnetic performance of the surface. With both careful design of the metasurfaces and appropriate biasing technologies, it should be possible to avoid bias line issues being prohibitive to function.

3.4 Leaky-wave antenna and stacked metasurfaces

Once the metamaterial components and tuning mechanisms are chosen, it is possible to design practical systems for mm-wave. For example, a metasurface-based beam scanning antenna for future communication systems can be developed that is scalable across the entire mm-wave band. Figure 3.3 illustrates a typical Fabry–Perot type LWA [52], comprising PRS plane, which is a passive periodic surface, at the top and a metallic reflecting plane at the bottom with an exiting element in between the two planes. The electromagnetic field radiating from the feeding source reflects back and forth within the cavity formed by the two planes and partially leaks out from the PRS as it propagates along the cavity until no EM field is left. The total EM leakage from the PRS forms the overall antenna radiation. This type of LWA radiates to its maximum when the resonant cavity height h is set to be [52]

$$h = \frac{n\lambda}{2} + \frac{\lambda}{4\pi}(\varphi_{PRS} - \pi) \quad \text{where } n = 0, 1, 2, \ldots \tag{3.9}$$

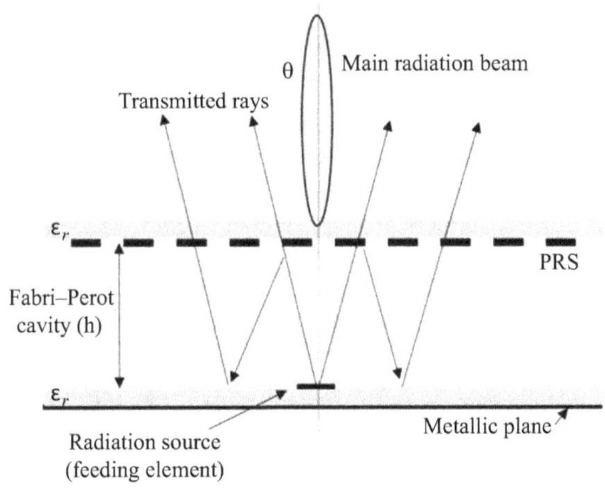

Figure 3.3 Leaky-wave antenna structure

where φ_{PRS} is the phase of the reflection coefficients of PRS and λ is the operating wavelength [69]. Approximate equations for the antenna directivity (*D*), *BW* and radiation pattern (*P*) have been presented as follows:

$$D(\text{dBi}) = 10\log\left[\frac{1+|\Gamma|}{1-|\Gamma|}\right] \tag{3.10}$$

$$BW(\%) = \frac{\lambda}{2\pi h}\left[\frac{1-|\Gamma|}{\sqrt{|\Gamma|}}\right] \tag{3.11}$$

$$P = \frac{1-\Gamma^2}{1+\Gamma^2-\cos\left(\varphi_{PRS}-\pi-\frac{4\pi}{\lambda}\cdot h\cdot\cos(\theta)\right)}\cdot F(\theta) \tag{3.12}$$

where $F(\theta)$ is the radiation pattern of the feeding element and Γ is the magnitude of the reflection coefficients of the PRS [52,53,69–71].

According to (3.9)–(3.11), the LWAs operating frequency, radiation patterns, main lobe direction, directivity and BW can be obtained as desired by designing the PRS and HIS of appropriate reflection coefficient magnitude and phase response. If a HIS is employed as an artificial ground plane instead of a simple metallic ground for this LWA structure then (3.9) and (3.12) are slightly modified as follows [70,71]:

$$h = \frac{n\lambda}{2} + \frac{\lambda}{4\pi}(\varphi_{PRS}-\varphi_{HIS}) \tag{3.13}$$

$$P = \frac{1-\Gamma^2}{1+\Gamma^2-\cos\left(\varphi_{PRS}-\varphi_{HIS}-(4\pi/\lambda)\cdot h\cdot\cos(\theta)\right)}\cdot F(\theta) \tag{3.14}$$

where φ_{HIS} is the reflection coefficient phase of the HIS. In this case, the reflection coefficients phase of HIS also contributes to determine the antenna radiation performance, i.e. HPBW, directivity and beam angle. Equation (3.14) shows that the antenna radiation beam may be steered by tuning one of four parameters in its denominator, i.e. by tuning (i) φ_{PRS}, (ii) φ_{HIS}, (iii) cavity height *h*, or (iv) λ (corresponding to the operating frequency). Recent work in metamaterial engineering shows that the phase responses of reflection coefficients of periodic metasurfaces, here functioning as PRS or HIS in LWA, can be effectively tuned through various electrical, mechanical or electro-mechanical techniques, as expressed in previous sections. Tuning of the mechanical displacement *h* and frequency calibration can also be used to achieve antenna beam steering.

Here, two example LWA designs with beam steering features are presented; one based on tuneable HIS, and the other based on frequency scanning which is a fundamental property of LWAs. The HIS tuning is obtained through electro-mechanical method by employing piezoelectric material, as discussed in previous sections. The HIS antenna is designed at 38 GHz and the frequency scanning antenna is designed for the 60 GHz band (57–66 GHz). The antennas overall performance features are set according to the antenna requirements for 5G BS front-end presented in Section 3.1.

3.4.1 Tuneable HIS-based LWA design

A schematic of the proposed LWA design, PRS and HIS unit cells, and the feeding dipole element is presented in Figure 3.4(a)–(c), respectively. The leaky-wave cavity is excited through a printed circuit dipole element, placed at one end of the cavity and backed by a metallic reflector. The periodic layer of HIS is suspended over a metal ground plane at an adjustable displacement, h_2. The periodic layer of the HIS is mechanically fixed to the rest of the antenna structure, whereas the metallic plane is mounted on a PEA so that h_2 could be varied when a voltage is applied across the PEA terminals. The change in relative displacement between the HIS periodic layer and ground plane causes the phase response of the HIS to shift, which is used to control the antenna's main radiation beam angle.

A BALUN type half-wavelength printed dipole is used as a feeding element, printed on a low-loss substrate (RT/Duroid5880) with $\varepsilon_r = 2.2$, thickness $h_4 = 0.127$ mm and copper cladding of 17.5 μm, as shown in Figure 3.4(c). Half of the dipole length is etched on the top of the thin substrate and the other half is etched on the bottom. The antenna is connected to a standard 50 Ω input through a balanced microstrip line with a tapered ground, as explained in [72]. The feeding antenna dimensions are as follows: $L_d = 3$ mm, $W_d = 0.5$ mm, $L_T = 4$ mm, $W_T = 0.3$ mm, $W_{T1} = 0.5$ mm, with a PCB area of 6 mm × 6 mm. A metallic reflector is placed at a distance $a = 0.9$ mm from the dipole. Figure 3.5(a)–(c)

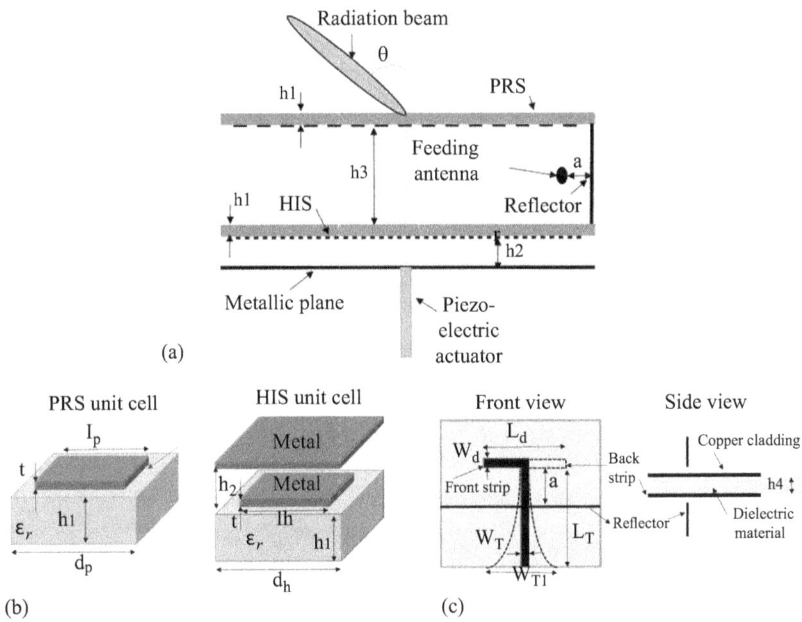

Figure 3.4 Structure of the proposed LWA: (a) overall antenna schematic, (b) PRS and HIS unit cells and (c) the feeding dipole element (dimensions not to scale)

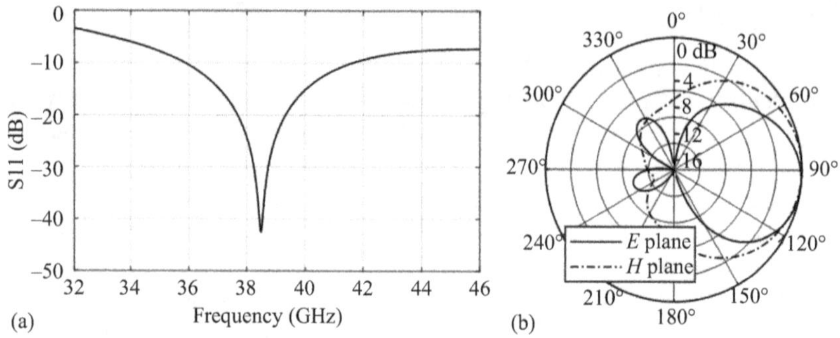

Figure 3.5 Simulation results of the feeding antenna: (a) S11 and (b) E and H plane radiation patterns at 38 GHz

presents the simulated S11, far-field radiation patterns (FRP) at 38 GHz in E and H planes, respectively. As seen from Figure 3.5, the dipole antenna -10 dB S11 BW, directivity and efficiency are about 6 GHz (i.e. 15% fractional BW), 7.44 dBi and 95%, respectively. The BW spans between 36 and 42 GHz which is well above the desired 1 GHz BW for 5G applications.

The PRS and HIS consist of metallic square patch elements, etched onto the same substrate material (RT/Duroid 5880) as the feed antenna, but with a substrate thickness, h_1, of 0.787 mm and a copper cladding of 35 μm. The PRS is designed to have a high-reflection coefficient (Γ) in order to obtain high antenna directivity ($D > 20$ dBi) with a reasonable BW ($BW > 1$ GHz). The HIS is designed to achieve a highly tuneable reflection phase, φ_{HIS}, response, sensitive to the separation between HIS periodic layer and ground plane, h_2. The HIS reflection coefficient phase can be approximated by

$$\varphi_{HIS} = 2\varphi_T - \frac{4\pi}{\lambda} \cdot h_2 - \pi \tag{3.15}$$

where φ_T is the phase of transmission coefficient of HIS periodic layer and λ is the operating wavelength [71]. Thus, φ_{HIS} is a function of h_2 and i, while φ_T depends on the geometry of the periodic unit cell. Based on these criteria, both PRS and HIS unit cells are designed and optimized in the CST Microwave Studio. The PRS unit cell dimensions were optimized to $lp = 3$ mm and $dp = 3.3$ mm. The optimized HIS unit cell dimensions are $lh = 1.53$ mm and $dh = 1.83$ mm. The separation, h_2, is varied from 40 to 400 μm to achieve the desired HIS reflected phase shift, φ_{HIS}.

Figure 3.6 shows the simulated magnitude response of the PRS reflection coefficients, and Figure 3.7(a) and (b) depicts the simulated magnitude and unwrapped phase response of the HIS reflection coefficients, respectively, for various displacements h_2. The simulated φ_{PRS} value is $-170°$. This way, using (3.10)–(3.13), it is predicted that the resonant cavity height, $h_3 = 3.45$ mm (later optimized to 3.5 mm), the directivity enhancement, i.e. excluding the contribution of the primary feed, is $D = 15.1$ dB and $BW = 22\%$.

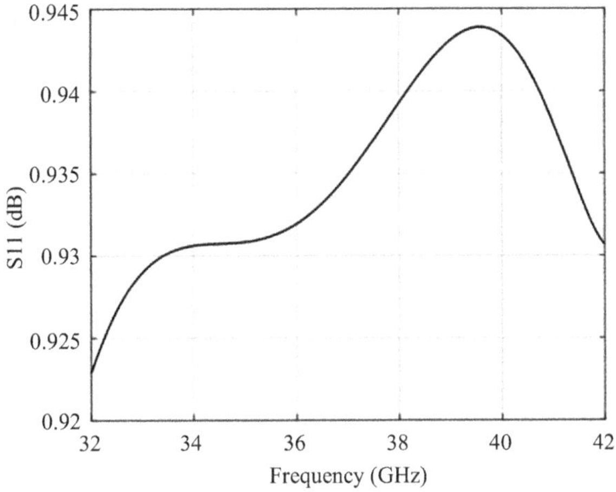

Figure 3.6 Magnitude response of PRS reflection coefficients

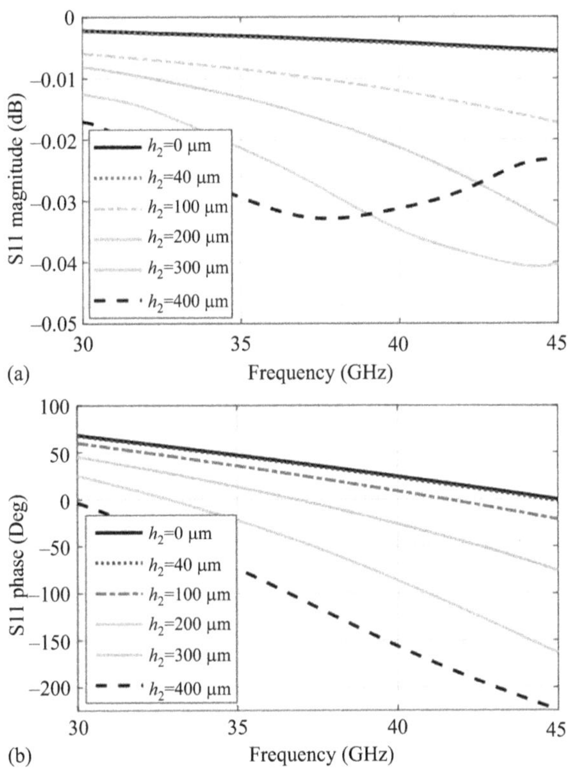

Figure 3.7 Simulation results of (a) magnitude and (b) phase response of HIS reflection coefficients

Figure 3.7 shows that the change in separation from 0 to 40 μm does not make a significant difference in the HIS S11 response. However, separations between 40 and 400 μm ($\Delta h_2 = 360$ μm) yields a phase shift of about $\Delta\varphi_{HIS} = 156°$ at 38 GHz whilst exhibiting negligible loss of about 0.03 dB.

The radiation pattern of the employed half-wavelength dipole antenna along θ having a metallic reflector at distance, a, is approximated as

$$F(\theta) = \cos\left(\theta + \frac{\pi}{2}\right) \cdot \cos\left[\frac{\beta}{2} \cdot a \cdot \cos\left(\theta + \frac{\pi}{2} + \pi\right)\right] \tag{3.16}$$

where $\beta = 2\pi/\lambda$ is the wave number [71]. The FRP and beam scanning range of the proposed LWA is predicted by substituting $\Gamma = 0.94$ and $\varphi_{PRS} = -170°$ from Figure 3.6, and $\Delta\varphi_{HIS} = 156°$ for $\Delta h_2 = 360$ μm from Figure 3.7. Figure 3.8 demonstrates the expected FRP and beam steering performance of the LWA at 38 GHz from (3.14). It can be noticed from Figure 3.8 that the estimated beam scanning range of the presented LWA is about 34° (from 22° to 58°) at 38 GHz with the HPBW reducing from lower to high scanning angles.

The LWA structure was modelled and simulated in the CST Microwave Studio. The PRS and HIS layers comprise 9 × 27 and 16 × 49 unit cells, respectively. Figure 3.9(a) and (b) shows the simulated antenna S11 and gain response, for separation, h_2, from 40 to 400 μm. The simulated S11 magnitude remains below −10 dB over the range of displacements across a 1-GHz BW between 37 and 38 GHz. The antenna gain remains above 19.5 dBi with a 1 GHz operating BW up to a separation of 300 μm; it then drops sharply due to a shift in gain at lower frequencies. The maximum gain at 37 GHz is 23.78 dBi for $h_2 = 200$ μm. The antenna simulation efficiency is as high as 81%. The normalized simulated *H*-plane FRP and beam steering performance of the LWA at 36 GHz is shown in Figure 3.9 (c) where the maximum beam scanning response is 33° (i.e. from 10° to 43°). The FRP becomes distorted at 38 GHz due to the shift in peak gain to lower frequencies.

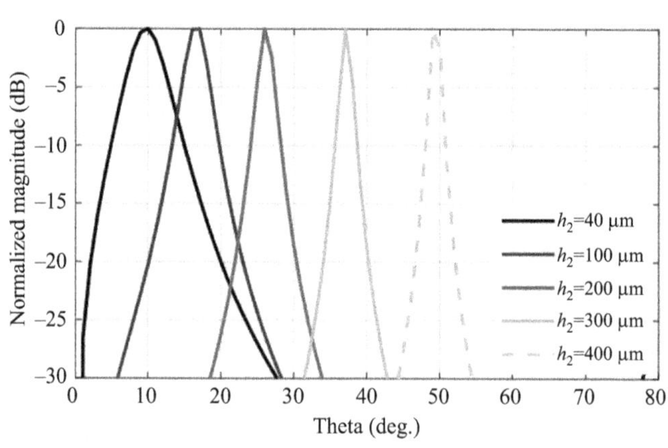

Figure 3.8 Predicted beam steering performance of the LWA at 38 GHz

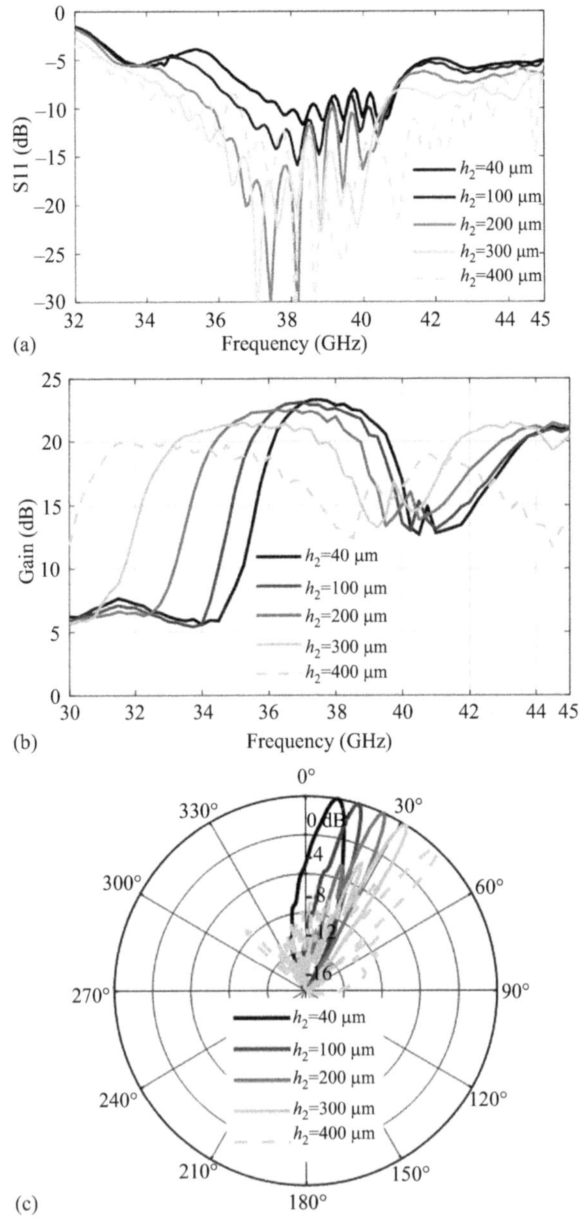

Figure 3.9 LWA simulation results: (a) S11, (b) gain response for various displacements and (c) normalized LWA H-plane FRP simulation results and beam steering performance at 36 GHz

The LWA exhibits a beam squint of between 3° and 6° within a 1-GHz operating BW, having larger beam squint at lower separation. It is found from the simulated results that the beam squint improves down to 2° when the operating BW is reduced to 500 MHz around 37 GHz, making this the best operational band for 5G.

3.4.2 Frequency scanning LWA antenna

A frequency scanning based beam steerable LWA structure designed for 60 GHz band frequencies is presented in Figure 3.10(a). The antenna consists of a PRS made of square metallic elements with patch width 1.5 mm and periodicity 1.8 mm. The PRS is printed on a standard PCB substrate and placed at a distance h_1 from a ground plane. Both the PRS and the ground layers are etched on a low loss RT/Duroid5880 substrate with thickness $h = 0.254$ mm and copper cladding thickness $t = 17.5$ μm. A two-element series-fed microstrip antenna array with gain above 10 dBi is employed to excite the LWA from a side and is etched on the top side of the substrate used for the ground plane. The PRS is designed to have reasonably high reflection coefficients ($\Gamma < 0.80$) to boost the antenna directivity ($D > 18$) for a wide range of frequencies, i.e. from 60 to 66 GHz, within 60 GHz-band.

The feeding antenna is an extended width (W), series-fed microstrip patch array of two elements, specially designed for mm-wave applications to improved fabrication tolerance and high efficiency. The design procedure of the feeding antenna is described in detail in [73]. The feeding antenna's patch length

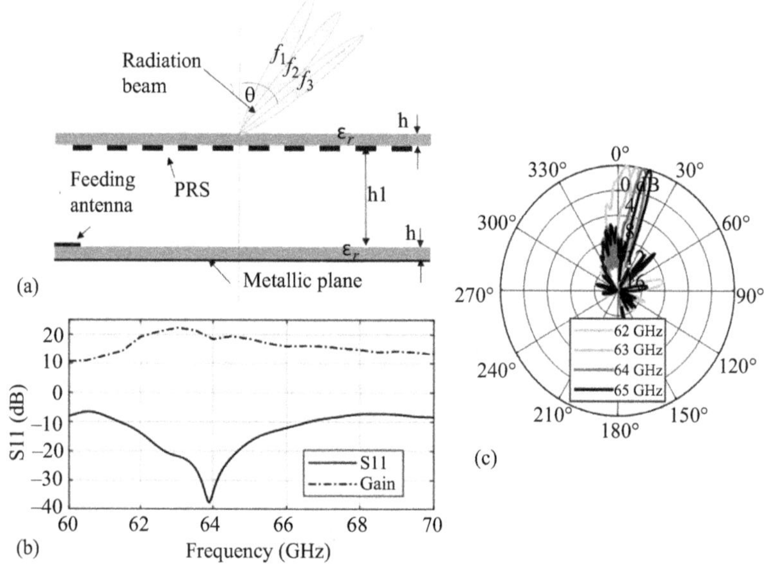

Figure 3.10 Frequency-scanning-based beam steering LWA and simulation results: (a) LWA geometry, (b) S11 and gain response, and (c) beam scanning in E-filed pattern for various frequencies

$L = 1.26$ mm, patch width $W = 5.23$ mm, transmission line width $WT = 1.1$ mm transmission line length $LT1 = 4$ mm and inter-patch separation $LT2 = 1.8$ mm. The overall PRS plane size is about 3 × 5 cm. The simulated magnitude and phase of the reflection coefficients of the PRS unit cell are $|\Gamma| = 0.87$ and $\varphi_{PRS} = 156°$, respectively. The cavity height, h_1, is computed from (3.9) choosing $n = 3$ and then it is optimized to be 7.1 mm for maximum directivity. It is predicted from (3.10) that the directivity of the feeding antenna (13.66 dBi) can be enhanced by 8.86 dB, giving a total maximum directivity of 22.52 dBi for the proposed LWA.

Figure 3.10(b) and (c) illustrates, respectively, the simulated S11 and gain versus frequency response, and FRPs in the E-plane at various frequencies within the operating band of the frequency scanning LWA. The antenna BW is about 5.18 GHz (8.22% of the central frequency) covering 61.4–66.28 GHz. The maximum gain is 22.58 dBi at 63 GHz, and it remains >18 dBi for the operating frequencies. It is clear from Figure 3.10(c) that the antenna has a beam scanning range of about 12° from 62 to 65 GHz with 3 dB scanning loss. The HPBW varies from 9.6° to 4.6° between 62 and 65 GHz. The antenna efficiency remains as high as 95.5%. If an arbitrary BW of 500 MHz is assigned to a channel operation, the beam squint will only be about 2°–3° within the channel. The beam scanning range of this LWA may be further increased by introducing a frequency sensitive static HIS instead of the simple metallic ground at the bottom. However, such HIS may reduce the usable BW of this type of LWA.

3.5 Millimetre-wave metasurface fabrication

Metamaterials are made up of electrically small structures arranged in highly precise ways to manipulate their electrical properties. Typically a metasurface will have unit cells with geometries that are a fraction of a wavelength, which when considering the mm-wave band gives a minimum feature and/or gap size of simple geometries in the order of a few hundreds or even tens of microns. Thus mm-wave metamaterials represent a transition in the technology available for fabrication, with standard PCB technologies being able to fabricate the lower end of the band, but with results that progressively get worse beyond 60 GHz.

The limitations in accuracy of standard PCB technologies also limit the achievable complexity of an individual unit cell. It is often a requirement in metamaterial design to have highly intricate unit cell designs, or unit cells that have geometries that differ from each other in subtle ways. This dramatically increases the need for fabrication accuracy. This is common in non-periodic metamaterials, which might give rise to designs such as reflectarrays or holographic surfaces [74]. The intricately coupled and interrelated nature of metamaterial design increases tolerance requirements, especially in highly resonant or sensitive structures. Therefore, in order to get the most capability out of metamaterials, new materials and processes need to be considered as standard fabrication methods.

Common PCB technologies can get down to a minimum track width or feature size of around 100 μm. A typical PCB process would include the PCB design being transferred to a silk screen that can be used to protect any wanted copper from a wet etch with a copper corrosive acid. The limitations from this kind of process are due to the divergence of light sources used.

Another issue with standard PCB processes is composition and uniformity of materials, in both the metal and the dielectric. Commercially available, high quality, high-frequency PCB laminates often use PTFE, woven-glass and/or ceramic composites, with the woven-glass providing structural support. The inhomogeneity of the dielectric in these materials can offer problems in mm-wave, highly resonant or closely coupled surfaces due to small changes in dielectric constant across the length of a unit cell that cannot be taken into account during the design stage. This also affects the surface roughness of the dielectric which can offer problems in certain systems. Thin PTFE only laminates are also available that can work well with these systems; however, these are flexible substrates, which is not always desirable, especially in systems where extreme flatness with respect to other layers in a metamaterial stack is required.

Many laminates come with electrodeposited (ED) copper, as this is a cost-efficient way of copper-cladding substrates; however, ED copper produces dendritic structures on the back side of the copper where the substrate is non-uniform and this dramatically increases the surface roughness of the backside of the copper. This is often not a problem in even high-frequency RF systems; however, in certain metamaterials, for example a HIS, the reflection from the back side of the copper is integral to the functionality of the surface, and the difference between ED copper and rolled copper (which produces a flatter copper backing) can be as high as 3 dB, as demonstrated by measurements of a 60-GHz HIS in our lab.

With all the above, it is certainly possible to make certain metamaterial systems using conventional technologies at the lower end of the mm-wave band. However, the current technologies are limiting and prohibit certain functionalities that can be designed into metamaterial systems, even if the optimum combination of commercial PCB technologies is used. So with complex and high-frequency metasurfaces, it is necessary to search for alternative manufacturing solutions, such as those used in microfabrication.

3.5.1 Microfabrication for metamaterials

Although standard PCB technology presents problems for micron-scale devices, the semiconductor industry have been capable of realizing such dimensions for over 30 years with the assistance of microlithography techniques. Current microfabrication in the semiconductor industry is capable of lithography of structures of several nanometres [75]. However, the materials used in semiconductor fabrication are often antagonistic to the requirements of antenna systems. It is possible, however, to combine some of the technologies used at the micro-scale with materials that are more suited to communication systems, including polymers such as polyimide ($\varepsilon_r = 3.24$, $\tan\delta = 0.031$), SU8 ($\varepsilon_r = 2.89$, $\tan\delta = 0.120$), PDMS, PMMA ($\varepsilon_r = 2.22$, $\tan\delta = 0.009$) and common substrates, including quartz and high-purity silicon, depending on the desired application [76].

There are three major driving factors for the suitability of microfabrication techniques to metamaterials fabrication: the absolute minimum feature size achievable with a given technology or material, the tolerance achievable by a given technique, and the flatness of a given material once processed in relation to other components in the system. For the scales and tolerances required for metamaterials in the mm-wave and THz, it is not necessary to utilize state-of-the-art silicon foundry level equipment to produce the required structures. It is possible using well-established photolithography techniques to create metamaterial devices, at scale, above the 100 GHz mark, which is ideal for next generation communication systems [77].

A standard, non-flexible, quasi-optical metamaterial could be made with any of the aforementioned materials. This would typically include the physical vapour deposition of a metallic ground layer on a standard silicon wafer. A suitable dielectric polymer is then spin coated onto the ground plane, cured and covered again with a photoresist, which undergoes photolithography resulting in the negative of desired pattern being outlined on the surface of the dielectric. Another metal layer is then deposited on top of this resist. The device is then submerged in a solvent which lifts off the undesired metal to produce a metamaterial structure. These techniques have been used in recent years to produce split ring resonator structures in [78,79], and flexible materials for metasurfaces have been reviewed in [80].

3.5.2 Other fabrication

Alongside photolithographic materials, some metamaterial systems can be designed so as to be fabricated by other means. For example a 300-GHz Bullseye was recently produced and tested using CNC milling onto a planar metal face, using a split-block method to mill the waveguide feed [81]. With readily available CNC milling, it is possible to get end mills down to a diameter of 0.2 mm, meaning that with careful design and the right kind of system, it is possible to create alternative metamaterial systems using this kind of technique.

Modern desktop 3D printers are also capable of similar resolutions as CNC machines, with some models boasting an accuracy of 200 μm for a Fuse Deposition Modelling print. With advances in 3D printing and other more accurate techniques such as stereolithography printing and material jetting, it may also be possible to make high-frequency structures using these methods [82,83]. New materials are being developed that have ceramic-loaded materials that can be 3D printed, so that dielectric constants can be varied across the face of a surface or through a 3D printed block [84]. This enables structures like polarization converting metamaterials to be designed, and as the materials available and dimensional accuracies become more applicable to high-frequency RF systems, it is likely that low-volume and highly bespoke metamaterials can be made for commercial applications using 3D printing techniques [85].

3.6 Beyond 5G

The next generation of communication systems beyond 5G, i.e. 6G and THz-Coms, will place even higher demands on performance features such as high data capacity, better connectivity, low latency, high reliability, high integrity and more security. The 6G communication systems will be based on the smart living concept where people, things, institutions, processes and sub-systems will be connected under uniformed platforms. The ultra-high definition video streaming, automotive transport and Internet of Things are projected to feed a huge amounts of data into wireless networks. Similarly, the emerging use of space communication systems and networks through ground-to-satellite, inter-satellite and intra-satellite links will also generate an extensive amount of data for the future wireless networks. This way the next-generation communication system is expected to face a dramatic increase in wireless data traffic in the coming decades. In order to accommodate the enormous data traffic, the operating spectral frequencies of the wireless front-end may have to shift towards higher mm-wave or even to lower terahertz bands, i.e. <1 THz.

Due to the versatile nature of applications in 6G, the antenna characteristics will vary depending on the application scenarios. In general, multi-band, highly directional, beam steerable, high efficiency, multifunctional and tuneable integrated antennas are expected to be commonly required for most of the wireless systems operating at frequencies from 0.1 to 1 THz in 6G. The antenna systems may contain some sort of self-intelligence to adapt their performance according to requirements in dynamic environments. Design of such new multifunctional smart antenna systems at frequencies above 100 GHz along with their fabrication and characterization may remain active research areas for coming decades.

Development of new antenna structures/materials, ultra-fast electromagnetic computing systems and advance micro- and nanofabrication techniques will be playing key roles to realize 6G wireless systems. Fortunately, multi-port vector network analysers operating at frequencies in a Terahertz range compatible with waveguide input/out port standards are currently available for research institutions. Automated anechoic chambers have also been developed by several commercial companies for antenna characterization up to low terahertz frequencies. The willingness of both private and government organizations to invest in research and development in this area is also encouraging due to increasing demand and growing market size of the wireless communications industry.

The recent exploitation of metamaterials, such as PRS, FSS, artificial magnetic conductor, HIS, electromagnetic band-gap structures and materials with negative or near-zero parameters, and their applications in antenna systems show that this technology has a high potential for future antenna developments, especially at mm-wave and terahertz bands. Metamaterials have also shown their wide applications in low-frequency antenna designs for size miniaturization, high efficiency, high gain, frequency tuning, beam steering, polarization conversion, antenna-filter integration and so on. As metamaterial structures are scalable, the proposed metamaterial designs may be scaled up to mm-wave and terahertz frequencies. However, new metamaterial designs are desired for multi-band multifunctional tuneable antennas. In this regard, the particular challenge may be to achieve an independent tuning of each band/function of the antenna without affecting the other functions – one of the significant challenges in the development of materials and tuning technologies that best utilize the benefits of metamaterial systems. Once standard methods have been built for the design and fabrication of metamaterials, it is highly likely that they will play a large part in the communication and sensing infrastructure of tomorrow.

References

[1] Sulyman, A.I., Nassar, A.T., Samimi, M.K., MacCartney, G.R., Rappaport, T.S. and Alsanie, A., 2014. Radio propagation path loss models for 5G cellular networks in the 28 GHz and 38 GHz millimeter-wave bands. *IEEE Communications Magazine*, 52(9), pp. 78–86.

[2] Rappaport, T.S., Sun, S., Mayzus, R., *et al.*, 2013. Millimeter wave mobile communications for 5G cellular: It will work!. *IEEE Access*, 1, pp. 335–349.

[3] Sun, S., MacCartney, G.R. and Rappaport, T.S., 2016, April. Millimeter-wave distance-dependent large-scale propagation measurements and path loss models for outdoor and indoor 5G systems. In *Antennas and Propagation (EuCAP), 2016 10th European Conference on* (pp. 1–5). Davos: IEEE.

[4] Shafi, M., Molisch, A.F., Smith, P.J., *et al.*, 2017. 5G: a tutorial overview of standards, trials, challenges, deployment and practice. *IEEE Journal on Selected Areas in Communications*, 35, pp. 1201–1221.

[5] Roh, W., Seol, J.Y., Park, J., *et al.*, 2014. Millimeter-wave beamforming as an enabling technology for 5G cellular communications: theoretical feasibility and prototype results. *IEEE Communications Magazine*, 52(2), pp. 106–113.

[6] Wang, C.X., Haider, F., Gao, X., *et al.*, 2014. Cellular architecture and key technologies for 5G wireless communication networks. *IEEE Communications Magazine*, 52(2), pp. 122–130.

[7] Kim, T., Park, J., Seol, J.Y., Jeong, S., Cho, J. and Roh, W., 2013, December. Tens of Gbps support with mmwave beamforming systems for next generation communications. In *Global Communications Conference (GLOBECOM), 2013 IEEE* (pp. 3685—3690). Atlanta, GA: IEEE.

[8] Cudak, M., Kovarik, T., Thomas, T.A., Ghosh, A., Kishiyama, Y. and Nakamura, T., 2014, December. Experimental mm wave 5G cellular system. In *Globecom Workshops (GC Wkshps), 2014* (pp. 377–381). Austin, TX: IEEE.

[9] Sacchi, C., Rahman, T.F., Bartolomei, N., Morosi, S., Mazzinghi, A. and Ciabini, F., 2016, December. Design and assessment of a CE-OFDM-based mm-wave 5G communication system. In *Globecom Workshops (GC Wkshps), 2016 IEEE* (pp. 1–7). Washington, DC: IEEE.

[10] Vook, F.W., Ghosh, A. and Thomas, T.A., 2014, June. MIMO and beamforming solutions for 5G technology. In *Microwave Symposium (IMS), 2014 IEEE MTT-S International* (pp. 1–4). Tampa, FL: IEEE.

[11] Hur, S., Baek, S., Kim, B., *et al.*, 2016. Proposal on millimeter-wave channel modeling for 5G cellular system. *IEEE Journal of Selected Topics in Signal Processing*, 10(3), pp. 454–469.

[12] Medbo, J., Kyosti, P., Kusume, K., *et al.*, 2016. Radio propagation modeling for 5G mobile and wireless communications. *IEEE Communications Magazine*, 54(6), pp. 144–151.

[13] Sulyman, A.I., Alwarafy, A., MacCartney, G.R., Rappaport, T.S. and Alsanie, A., 2016. Directional radio propagation path loss models for millimeter-wave wireless networks in the 28-, 60-, and 73-GHz bands. *IEEE Transactions on Wireless Communications*, 15(10), pp. 6939–6947.

[14] Haneda, K., Zhang, J., Tan, L., *et al.*, 2016, May. 5G 3GPP-like channel models for outdoor urban microcellular and macrocellular environments. In *Vehicular Technology Conference (VTC Spring), 2016 IEEE 83rd* (pp. 1–7). Nanjing: IEEE.

[15] Sun, S., Rappaport, T.S., Thomas, T.A., *et al.*, 2016. Investigation of prediction accuracy, sensitivity, and parameter stability of large-scale propagation path loss models for 5G wireless communications. *IEEE Transactions on Vehicular Technology*, 65(5), pp. 2843–2860.

[16] Jaeckel, S., Peter, M., Sakaguchi, K., Keusgen, W. and Medbo, J., 2016, May. 5G channel models in mm-wave frequency bands. In *European Wireless 2016; 22th European Wireless Conference; Proceedings of* (pp. 1–6). Oulu: VDE.

[17] Sun, S., Rappaport, T.S., Rangan, S., *et al.*, 2016, May. Propagation path loss models for 5G urban micro-and macro-cellular scenarios. In *Vehicular Technology Conference (VTC Spring), 2016 IEEE 83rd* (pp. 1–6). Nanjing: IEEE.

[18] Lee, J., Liang, J., Kim, M.D., Park, J.J., Park, B. and Chung, H.K., 2016. Measurement-based propagation channel characteristics for millimeter-wave 5G Giga communication systems. *ETRI Journal*, 38(6), pp. 1031–1041.

[19] Huang, J., Wang, C.X., Feng, R., Sun, J., Zhang, W. and Yang, Y., 2017. Multi-frequency mmwave massive MIMO channel measurements and characterization for 5G wireless communication systems. *IEEE Journal on Selected Areas in Communications*, 35, pp. 1591–1605.

[20] Zhai, M., Li, J., Liang, Y., Li, T., Gui, G. and Li, F., 2016, October. A novel coupling mode based 3D MIMO channel modeling and capacity analysis for 5G. In *Ubiquitous Wireless Broadband (ICUWB), 2016 IEEE International Conference on* (pp. 1–4). Nanjing: IEEE.

[21] Samimi, M.K., Sun, S. and Rappaport, T.S., 2016, April. MIMO channel modeling and capacity analysis for 5G millimeter-wave wireless systems. In *Antennas and Propagation (EuCAP), 2016 10th European Conference on* (pp. 1–5). Davos: IEEE.

[22] Tuovinen, T., Tervo, N. and Pärssinen, A., 2016, October. RF system requirement analysis and simulation methods towards 5G radios using massive MIMO. In *Microwave Conference (EuMC), 2016 46th European* (pp. 142–45). London: IEEE.

[23] Gavrilovska, L., Rakovic, V. and Atanasovski, V., 2016. Visions towards 5G: technical requirements and potential enablers. *Wireless Personal Communications*, 87(3), pp. 731–757.

[24] Dupleich, D., Luo, J., Haefner, S., Mueller, R., Schneider, C. and Thomae, R., 2016. A hybrid polarimetric wide-band beam-former architecture for 5G mm-wave communications. In *Smart Antennas (WSA 2016); Proceedings of*

the 20th International ITG Workshop on Smart Antennas (pp. 1–8). Munich: VDE.

[25] Ala-Laurinaho, J., Aurinsalo, J., Karttunen, A., *et al.*, 2016. 2-D beam-steerable integrated lens antenna system for 5G $ E $-band access and backhaul. *IEEE Transactions on Microwave Theory and Techniques*, 64(7), pp. 2244–2255.

[26] Psychoudakis, D., Zhou, H., Biglarbegian, B., Henige, T. and Aryanfar, F., 2016, May. Mobile station radio frequency unit for 5G communications at 28GHz. In *Microwave Symposium (IMS), 2016 IEEE MTT-S International* (pp. 1–3). San Francisco, CA: IEEE.

[27] Niknejad, A.M., Thyagarajan, S., Alon, E., Wang, Y. and Hull, C., 2015, September. A circuit designer's guide to 5G mm-wave. In *Custom Integrated Circuits Conference (CICC), 2015 IEEE* (pp. 1–8). San Jose, CA: IEEE.

[28] Khalid, N. and Akan, O.B., 2016. Experimental throughput analysis of low-THz MIMO communication channel in 5G wireless networks. *IEEE Wireless Communications Letters*, 5(6), pp. 616–619.

[29] El Shorbagy, M., Shubair, R.M., AlHajri, M.I. and Mallat, N.K., 2016, November. On the design of millimetre-wave antennas for 5G. In *Microwave Symposium (MMS), 2016 16th Mediterranean* (pp. 1–4). Busan: IEEE.

[30] Ojaroudiparchin, N., Shen, M. and Pedersen, G.F., 2016. UWB mm-wave antenna array with quasi omnidirectional beams for 5G handheld devices. In *16th IEEE International Conference on Ubiquitous Wireless Broadband (ICUWB 2016)*. Nanjing: IEEE.

[31] Curtis, J., Zhou, H., Hisayasu, P., Sarkar, A. and Aryanfar, F., 2016, January. MM-wave radio, a key enabler of 5G communication. In *Silicon Monolithic Integrated Circuits in RF Systems (SiRF), 2016 IEEE 16th Topical Meeting on* (pp. 1–3). Austin, TX: IEEE.

[32] Zhai, W., Miraftab, V., Repeta, M., Wessel, D. and Tong, W., 2016, May. Dual-band millimeter-wave interleaved antenna array exploiting low-cost PCB technology for high speed 5G communication. In *Microwave Symposium (IMS), 2016 IEEE MTT-S International* (pp. 1–4). San Francisco, CA: IEEE.

[33] Cao, Z., Ma, Q., Smolders, A.B., *et al.*, 2016. Advanced integration techniques on broadband millimeter-wave beam steering for 5G wireless networks and beyond. *IEEE Journal of Quantum Electronics*, 52(1), pp. 1–20.

[34] Sun, C.C. and Chou, H.T., 2016, August. Summary and progress of MM-wave antenna technologies for 5G application. In *Radio-Frequency Integration Technology (RFIT), 2016 IEEE International Symposium on* (pp. 1–3). Taipei: IEEE.

[35] Cai, M., He, C., Li, X. and Yang, G., 2016, August. Design of mm-wave phased array in mobile terminal for 5G mobile system. In *Progress in Electromagnetic Research Symposium (PIERS)* (pp. 2456–2458). Shanghai: IEEE.

[36] Zhang, S., Chen, X., Syrytsin, I. and Pedersen, G.F., 2017. A planar switchable 3D-coverage phased array antenna and its user effects for 28 GHz mobile terminal applications. *IEEE Transactions on Antennas and Propagation*, 65, pp. 6413–6421.

[37] Jang, J., Chung, M., Hwang, S.C., *et al.*, 2016. Smart small cell with hybrid beamforming for 5G: Theoretical feasibility and prototype results. *IEEE Wireless Communications*, 23, pp. 124–131.

[38] Milosevic, V., Jokanovic, B., Boric-Lubecke, O. and Lubecke, V., 2016, November. Key millimeter wave technologies for 5G. In *Telecommunications Forum (TELFOR), 2016 24th* (p. 1). Belgrade: IEEE.

[39] Mavridou, M., 2015. Tunable microwave and millimetre-wave metamaterial structures and applications (Doctoral dissertation, University of Birmingham).

[40] Zhu, B., Zhao, J. and Feng, Y., 2013. Active impedance metasurface with full 360° reflection phase tuning. *Scientific Reports*, 3, pp. 3059. doi:10.1038/srep03059.

[41] Mavridou, M., Konstantinidis, K. and Feresidis, A.P., 2016. Continuously tunable mm-wave high impedance surface. *IEEE Antennas and Wireless Propagation Letters*, 15, pp. 1390–1393.

[42] Antoniades, M.A. and Eleftheriades, G.V., 2003. Compact linear lead/lag metamaterial phase shifters for broadband applications. *IEEE Antennas and Wireless Propagation Letters*, 2, pp. 103–106.

[43] Baena, J.D., Bonache, J., Martín, F., *et al.*, 2005. Equivalent-circuit models for split-ring resonators and complementary split-ring resonators coupled to planar transmission lines. *IEEE Transactions on Microwave Theory and Techniques*, 53(4), pp. 1451–1461.

[44] Costa, F. and Monorchio, A., 2012. A frequency selective radome with wideband absorbing properties. *IEEE Transactions on Antennas and Propagation*, 60(6), pp. 2740–2747.

[45] Costa, F., Monorchio, A. and Manara, G., 2010. Analysis and design of ultra thin electromagnetic absorbers comprising resistively loaded high impedance surfaces. *IEEE Transactions on Antennas and Propagation*, 58(5), pp. 1551–1558.

[46] Gao, X., Han, X., Cao, W., Li, H.O., Ma, H.F. and Cui, T.J., 2015. Ultrawideband and high-efficiency linear polarization converter based on double V-shaped metasurface. *IEEE Transactions on Antennas and Propagation*, 63(8), pp. 3522–3530.

[47] Liu, X., Zhang, J., Li, W., *et al.*, 2016. Three-band polarization converter based on reflective metasurface. *IEEE Antennas and Wireless Propagation Letters*, 16, pp. 924–927.

[48] Sarabandi, K. and Behdad, N., 2007. A frequency selective surface with miniaturized elements. *IEEE Transactions on Antennas and Propagation*, 55 (5), pp. 1239–1245.

[49] Goussetis, G. and Feresidis, A.P., 2010. Perturbed frequency selective surfaces for multiband high impedance surfaces. *IET Microwaves, Antennas & Propagation*, 4(8), pp. 1105–1110.

[50] Goussetis, G., Feresidis, A.P. and Vardaxoglou, J.C., 2006. Tailoring the AMC and EBG characteristics of periodic metallic arrays printed on

grounded dielectric substrate. *IEEE Transactions on Antennas and Propagation*, 54(1), pp. 82–89.

[51] Feresidis, A.P., Goussetis, G., Wang, S. and Vardaxoglou, J.C., 2005. Artificial magnetic conductor surfaces and their application to low-profile high-gain planar antennas. *IEEE Transactions on Antennas and Propagation*, 53(1), pp. 209–215.

[52] Feresidis, A.P. and Vardaxoglou, J.C., 2001. High gain planar antenna using optimised partially reflective surfaces. *IEE Proceedings – Microwaves, Antennas and Propagation*, 148(6), pp. 345–350.

[53] Pandi, S., Balanis, C.A. and Birtcher, C.R., 2015. Design of scalar impedance holographic metasurfaces for antenna beam formation with desired polarization. *IEEE Transactions on Antennas and Propagation*, 63(7), pp. 3016–3024.

[54] Johnson, M.C., Brunton, S.L., Kundtz, N.B. and Kutz, J.N., 2015. Sidelobe canceling for reconfigurable holographic metamaterial antenna. *IEEE Transactions on Antennas and Propagation*, 63(4), pp. 1881–1886.

[55] Sievenpiper, D.F., 1999. High-impedance electromagnetic surfaces (PhD diss., University of California),

[56] Churm, J., Rabbani, M.S. and Feresidis, A., 2018. Design of millimetre wave phase shifting periodic metasurfaces. In *12th European Conference on Antennas and Propagation (EuCAP 2018)*, London (pp. 1–4).

[57] Alderman, B., Henry, M., Sanghera, H., *et al.*, 2011. Schottky diode technology at Rutherford Appleton Laboratory. In *2011 IEEE International Conference on Microwave Technology & Computational Electromagnetics*, Beijing (pp. 4–6).

[58] Crowe, T.W., Bishop, W.L., Porterfield, D.W., Hesler, J.L. and Weikle, R. M., 2005. Opening the terahertz window with integrated diode circuits. *IEEE Journal of Solid-State Circuits*, 40(10), pp. 2104–2110.

[59] Vassos, E., Churm, J. and Feresidis, A., 2019. Design of millimetre wave phase shifting element using air-bridged Schottky diodes. In *2019 13th European Conference on Antennas and Propagation (EuCAP)*, Krakow, Poland (pp. 1–3).

[60] Hu, W., Ismail, M.Y., Cahill, R., *et al.*, 2006. Phase control of reflectarray patches using liquid crystal substrate. In *2006 First European Conference on Antennas and Propagation*, Nice (pp. 1–6).

[61] Doumanis, E., Dickie, R., Baine, P., *et al.*, 2013. Nematic liquid crystals for reconfigurable millimeter wavelength antenna technology. In *2013 7th European Conference on Antennas and Propagation (EuCAP)*, Gothenburg (pp. 1791–1792).

[62] Ozbay, E., Aydin, K., Butun, S., Kolodziejak, K. and Pawlak, D., 2007. Ferroelectric based tuneable SRR based metamaterial for microwave -applications. In *2007 European Microwave Conference*, Munich (pp. 497–499).

[63] Sazegar, M., Zheng, Y., Kohler, C., *et al.*, 2012. Beam steering transmitarray using tunable frequency selective surface with integrated ferroelectric

varactors. *IEEE Transactions on Antennas and Propagation*, 60(12), pp. 5690–5699.

[64] Lotz, P., Matysek, M. and Schlaak, H.F., 2011. Fabrication and application of miniaturized dielectric elastomer stack actuators. *IEEE/ASME Transactions on Mechatronics*, 16(1), pp. 58–66.

[65] Araromi, O.A., Romano, P., Rosset, S., Perruisseau-Carrier, J., and Shea, H.R., March 2014. A tunable millimeter-wave phase shifter driven by dielectric elastomer actuators. In *Proc. SPIE 9056, Electroactive Polymer Actuators and Devices (EAPAD) 2014*, 90562M.

[66] van Spengen, W.M., 2003. MEMS reliability from a failure mechanisms perspective. *Microelectronics Reliability*, 43(7), pp. 1049–1060.

[67] Hashemi, M., Yang, S., Wang, T., *et al.*, 2016. Electronically-controlled beam-steering through vanadium dioxide metasurfaces. *Scientific Reports*, 6, pp. 35439.

[68] Zhang, C. and Cheng, Q., 2017. Optically transparent metamaterial for broadband millimeter wave absorption. In *2017 10th UK-Europe-China Workshop on Millimetre Waves and Terahertz Technologies (UCMMT)*, Liverpool (pp. 1–4).

[69] Rabbani, M.S. and Ghafouri-Shiraz, H., 2017. Dual layer partially reflective surface antennas based on extended size unit cells for 60 GHz band WLAN/WPAN. *IET Microwaves, Antennas & Propagation*, 12, pp. 789–795.

[70] Mavridou, M., 2015. Tunable microwave and millimetre-wave metamaterial structures and applications (PhD diss., University of Birmingham).

[71] Mavridou, M. and Feresidis, A.P., 2016. Dynamically reconfigurable high impedance and frequency selective metasurfaces using piezoelectric actuators. *IEEE Transactions on Antennas and Propagation*, 64(12), p. 5190.

[72] Bancroft, R., 2009. “Microstrip and printed antenna design,” Raleigh, NC, SciTech Pub.

[73] Rabbani, M.S. and Ghafouri-Shiraz, H., 2017. Ultra-wide patch antenna array design at 60 GHz band for remote vital sign monitoring with Doppler radar principle. *Journal of Infrared, Millimetre and Terahertz Wave*, 38(5), pp. 548–566.

[74] Trampuz, C., Hajian, M. and Ligthart, L.P., 2006. Design, analysis and measurements of reflected phased array microstrip antennas at Ka-band, using hollow phasing. In *2006 European Radar Conference*, Manchester (pp. 57–60).

[75] Choi, K., Sagong, H.C., Kang, W., *et al.*, 2019. Enhanced reliability of 7 nm process technology featuring EUV. In 2019 *Symposium on VLSI Technology*, Kyoto, Japan. IEEE Transactions on Electron Devices, 66(12), pp. 5399–5403.

[76] Walia, S., Shah, C.M., Gutruf, P., *et al.*, 2015. Flexible metasurfaces and metamaterials: a review of materials and fabrication processes at micro-and nano-scales. *Applied Physics Reviews*, 2(1), p. 011303.

[77] Moser, H.O., Casse, B.D.F., Wilhelmi, O. and Saw, B.T., 2005. Terahertz response of a microfabricated rod–split-ring-resonator electromagnetic metamaterial. *Physical Review Letters*, 94(6), p. 063901.

[78] Quan, B., Xu, X. and Xia, X., *et al.*, 2006. Microfabrication and properties of the meta-materials. *Microelectronic Engineering.* 83(4–9), pp. 1364–1367.

[79] Xia, X., Yang, H. and Sun, Y., *et al.* 2008. Fabrication of terahertz metamaterials using S1813/LOR stack by lift-off. *Microelectronic Engineering*, 85(5–6), pp. 1433–1436.

[80] Walia, S., Shah, C.M. and Gutruf, P., 2015. Flexible metasurfaces and metamaterials: a review of materials and fabrication processes at micro- and nano-scales. *Applied Physics Reviews*. 2(1), pp. 011303.

[81] Kampouridou, D. and Feresidis, A., 2017. A new THz corrugated-ground antenna. In *Loughborough Antennas & Propagation Conference (LAPC 2017)*, Loughborough (pp. 1–2).

[82] Zhang, B., Chen, W. and Wu, Y., *et al.*, 2017. Review of 3D printed millimeter-wave and terahertz passive devices. *International Journal of Antennas and Propagation*, 2017.

[83] Sadeqi, A., Rezaei Nejad, H., Owyeung, R.E., *et al.*, 2019. Three dimensional printing of metamaterial embedded geometrical optics (MEGO). *Microsystems & Nanoengineering*, 5(16). doi:10.1038/s41378-019-0053-6.

[84] Castles, F., Isakov, D., Lui, A., *et al.*, 2016. Microwave dielectric characterisation of 3D-printed $BaTiO_3$/ABS polymer composites. *Scientific Reports*, 6, pp. 22714.

[85] Zhang, B., Guo, Y., Zirath, H. and Zhang, Y.P., 2017. Investigation on 3-D-printing technologies for millimeter-wave and terahertz applications. *Proceedings of the IEEE*, 105(4), pp. 723–736.

Chapter 4

3D-printed millimetre-wave antennas with spray-coated metalization

Shaker Alkaraki[1], James Kelly[1] and Yue Gao[2]

4.1 Metallic corrugated plate antenna fed using rectangular waveguide

4.1.1 Introduction

This chapter is concerned with corrugated plate antennas. The corrugated plate antennas, presented here, consist of a resonant slot formed within a metallic plate. The resonant slot is surrounded by a rectangular cavity and corrugations. Subwavelength slots surrounded by periodic corrugations have been widely investigated to deliver enhanced transmission at both optical and microwave bands [1–4]. Inspired by the phenomenon observed at optical frequencies, highly directive single-band and dual-band-corrugated plate antennas were first realized and proposed by Beruete *et al.* in [5–7]. This was achieved by scaling the optical structures to frequencies in the microwave frequency range. Generally speaking, the antennas consist of resonant slot-surrounded corrugations that are formed within a metallic slab. The slot is fed using a standard rectangular waveguide. The corrugations can be either straight or circular in shape.

Figure 4.1 shows corrugated plate antenna incorporating straight corrugations on either side of the slot. The antenna with six corrugations has a simulated peak gain of 15.4 dBi [6]. The corrugations improve the gain of the antenna by 8.9 dB compared with the flat plate alone. The side lobe level (SLL) is ≈ -10 dB. A resonance is excited in the corrugations when the electric field (*E*-field) is polarization perpendicular to the width of the corrugations. The gain improvement is observed when the period of the corrugations ($d \approx \lambda$, where λ is the wavelength at the operating frequency). Under this situation, the waves are re-emitted in phase $\lambda - d$. Hence, the gain improvement can be attributed to a combination of power re-emission as well as resonance within the corrugations. To deliver good resonance and as reported in [6], the corrugations should have a depth of $\approx ((2m + 1)\lambda)/4$, where m is an integer.

[1]School of Electronic Engineering and Computer Science, Queen Mary University of London, London, United Kingdom

[2]Institute for Communication Systems, University of Surrey, Guildford, United Kingdom

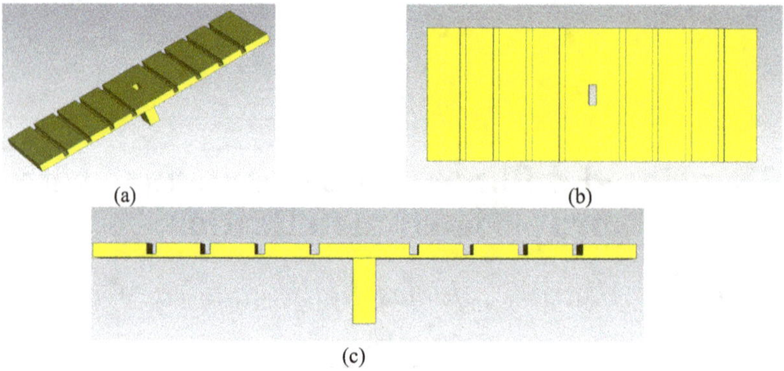

Figure 4.1 Corrugated plate antenna: (a) perspective view, (b) top view and (c) front view

The antenna with straight corrugations in [6] has overall dimensions of 320 mm × 80 mm which corresponds to $17\lambda \times 4.3\lambda$ at 16 GHz with an overall aperture efficiency of 5.5%. The antenna yields dual-band operation. The first resonant frequency of the antenna depends solely on slot length (SL) Figure 4.1(b). The slot resonates at the $SL \approx \lambda/2$. The second resonant frequency depends on the thickness of the metal slab. The reason for this is that the slot is used to couple electromagnetic (EM) power (incident TE_{10} wave) from the waveguide to free-space. However, the antenna gain is improved, by the use of corrugations, only at the second resonant frequency (16.5 GHz) as the dimensions of the corrugations are optimized to maximize the directivity at this frequency. The antenna has a gain of ~5 dBi at the first resonant frequency of 13 GHz. The reason for this is that the radiation within this band is due to the central slot and is not supported by the corrugations which are optimized to support radiation at 16.5 GHz.

A smaller version of the straight corrugated antenna was developed in [8]. The antenna has dimensions of 60 mm × 24 mm ($2.6\lambda \times 1\lambda$) with a maximum gain of 10 dBi and SLL of −10 dBi at 13 GHz. The two corrugations were filled by dielectric material which increases the antenna gain by further 1.3 dBi. The antenna aperture efficiency is 29% when the corrugations are loaded with air, and it is 40% when the corrugations are loaded by dielectric material [8].

4.1.2 Novel corrugated plate antenna operating at 28.5 GHz

This section introduces a small single-layer-corrugated plate antenna operating at 28.5 GHz. Figure 4.2(a) illustrates the geometry of the antenna. Figure 4.2(b) shows a fabricated prototype. The antenna consists of a resonant slot which is formed within an aluminium plate. The resonant slot is surrounded by a rectangular cavity and two corrugations [9]. The EM energy is transferred from the source to the antenna via a length of WR-28 rectangular waveguide with dimensions of $a \times b = 7.112$ mm × 3.556 mm. Energy, from the waveguide, is coupled through the slot which polarizes the E-field in the x-direction. The H-plane of the

(a) (b) (c)

Figure 4.2 Geometry and fabricated antenna prototype: (a) geometry of the antenna, (b) top view of the fabricated prototype and (c) the feeding waveguide

Table 4.1 The dimensions of the proposed antenna

Parameters	**L**	**W**	**T**	**SL**	**SW**	**L1**	**W1**	**K**	**C**	**D**	**X**
Dimensions (mm)	20	11	6.4	5.7	2.6	10.8	10.6	3.2	1.6	2.1	2.6
$\lambda = 10.58$ mm	1.90	1.05	0.61	0.54	0.25	1.03	1.00	0.30	0.15	0.20	0.25

waveguide is aligned along the direction of *SL*. The dimensions of the antennas are given in Table 4.1. The antenna dimensions were optimized to maximize the antenna's gain and aperture efficiency at 28.5 GHz. The optimal dimensions are given by the following equations [9]:

$$W1 \approx L1 \approx \lambda \tag{4.1}$$

$$SL \approx \frac{\lambda}{2} \tag{4.2}$$

$$X \approx \frac{\lambda}{4} \tag{4.3}$$

$$D \approx \frac{\lambda}{5} \tag{4.4}$$

$$C \approx \frac{\lambda}{6} \tag{4.5}$$

4.1.3 Radiation mechanism and operation principles

The EM energy is coupled into the antenna through the resonant slot. The slot radiates part of the energy and the remainder is converted to EM waves that are polarized in the *x*-direction. These waves propagate along the surface of the metal and ultimately are lost and leaked at metal edges. Hence, the antenna with slot only has a simulated gain of ≈6 dBi at resonance frequency. Furthermore, adding the cavity and the corrugations on the antenna surface increases the gain of the antenna by 7.2 dBi. Specifically, the corrugations increase the gain of the antenna by 3 dBi. Hence, the high gain and high aperture efficiency of the antenna can be attributed to the combination of radiation from the slot, cavity and corrugations as shown in power flow profile, surface current and E-field plots in Figures 4.3 and 4.4.

The radiation from the cavity can be attributed to the strong excitation of TM_{12} mode inside the cavity as shown in Figures 4.5–4.7. Specifically, the following E-field and magnetic field components exist in the cavity and around its edges: E_x, E_y, E_z, H_x, and H_y. Inside the cavity E_z, H_x, and H_y are dominant and H_z is completely absent, which indicates that the dominant mode inside the cavity is a TM mode (Figure 4.7). When the cavity length $L1$ is equivalent to the operating wavelength (i.e. $L1 \approx \lambda$) and when the cavity depth (K) is in the range $\lambda/3 \leq K \leq \lambda/4$ the antenna yields peak directivity, as shown in Figure 4.8(a) and (b). The TE_{11} mode is excited inside the narrow corrugations. Hence, the directivity of the antenna is sensitive to the depth of the corrugations (D) and the maximum directivity is achieved when $D \approx \lambda/5$, as shown in Figure 4.8. Furthermore, Figure 4.8 shows that the directivity of antenna is more sensitive to the length of the cavity ($L1$) and to the depth of the corrugations (D) than the depth of the cavity (K) and the width of the corrugations (C). This is due to the excitation of TM mode inside the cavity which has an opposite effect to the TE mode excitation inside the corrugations.

Please note that the antenna incorporates three resonant structures, namely the slot, cavity and corrugations. The position of the resonant frequency is mainly controlled by the length of the slot (SL), length of the cavity ($L1$) and, to a lesser extent, by the depth of the corrugations (D). In addition, Figure 4.9 shows that the location of the antenna resonance is inversely proportional to the SL and to the cavity length, while it is directly proportional to the depth of the corrugations. Also Figure 4.9 indicates that the slot resonance is dominant over the cavity resonance,

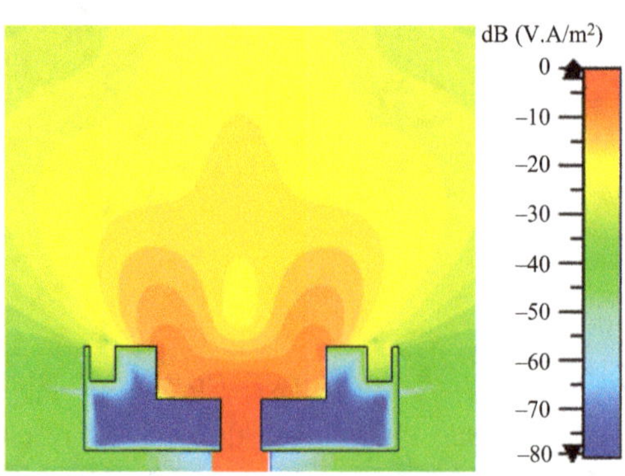

Figure 4.3 The power flow profiles at (x–z) cutting plane for the antenna at 28 GHz

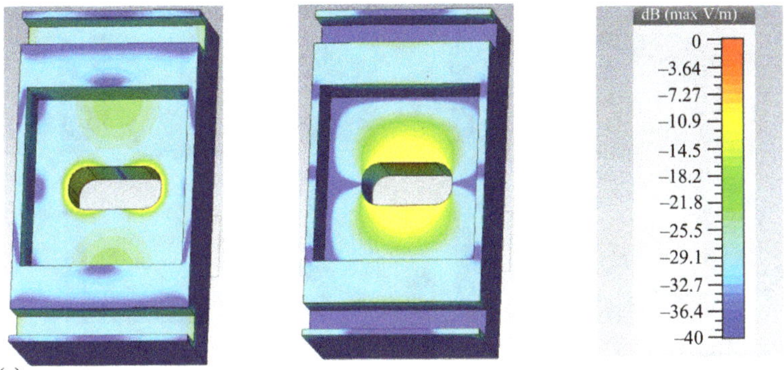

(a) (b)

Figure 4.4 Electric field (E-field) and surface current for antenna at 28.5 GHz: (a) surface current and (b) E-field

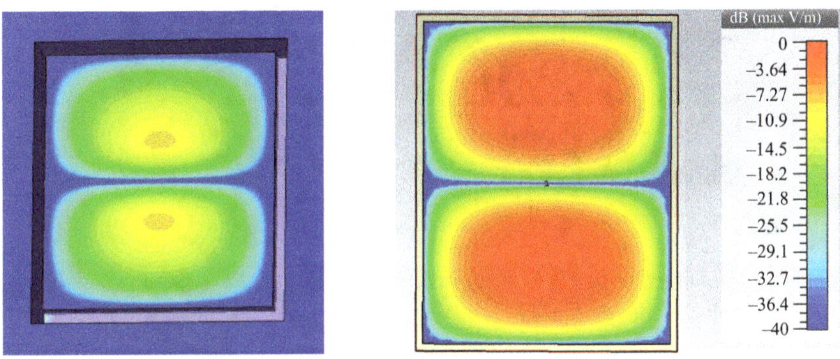

(a) (b)

Figure 4.5 TM_{21} mode inside the rectangular cavity and a rectangular waveguide at 28.5 GHz: (a) E_z on a cutting plane inside the cavity and (b) E_z inside a rectangular waveguide

while the resonance associated with the corrugations is the least dominant among three.

Increasing the number of periodic corrugations on the surface of the structure will further improve the overall antenna gain but reduce the aperture efficiency of the antenna. The reason for this is that the incremental increase in the overall size of the structure is given by the following expression: $P + D \approx \lambda$. We have studied the relationship between the size, gain, aperture efficiency and number of corrugations, and the results are shown in Table 4.2. The antenna incorporating a cavity and one corrugation has a simulated aperture efficiency (A_e) of 83% which is 51% more than the antenna without cavity. The aperture efficiency drops to 45% when two corrugations are employed and it continues to drop gradually, as more corrugations

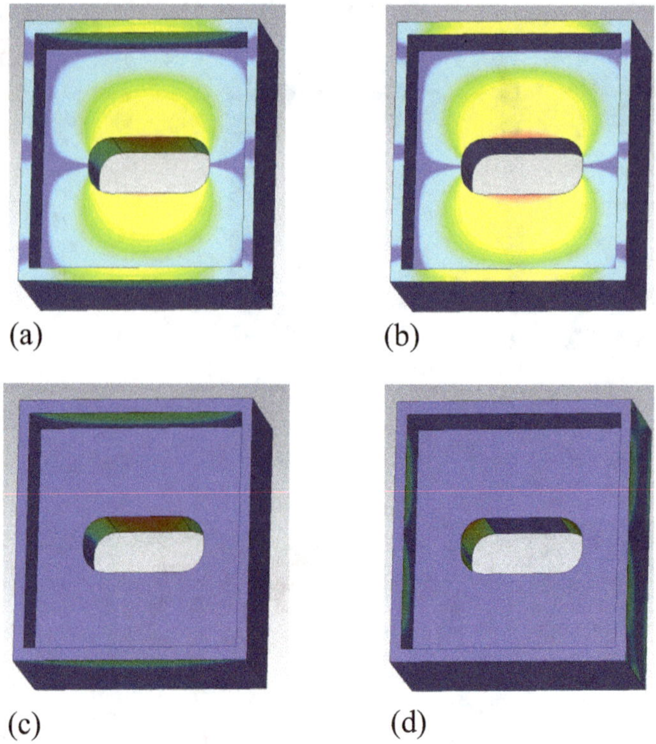

Figure 4.6 The electric field (E-field) inside the cavity: (a) absolute value, (b) E_z, (c) E_x and (d) E_y

are added, until it reaches a value of 18.8% when nine corrugations are employed. The aperture efficiency (A_e) of the antenna is inversely proportional to the number of corrugations. The antenna gain is directly proportional to the number of corrugations irrespective of whether the cavity is present or not. The antenna has a maximum gain of 16.4 dBi. This corresponds to the case where the cavity is employed alongside nine corrugations. The gain drops to 14.7 dBi when the cavity is removed.

4.1.4 Measured results at 28.5 GHz

The proposed antenna operates at 28.5 GHz with a measured −10 dB impedance bandwidth of 9.8% (i.e. an absolute bandwidth of 2.8 GHz), as shown in Figure 4.10. The gain of the antenna as well as the far-field radiation patterns were measured using a planar near-field system and the gain is calculated using gain comparison method described in [10]. Good agreement is observed between the measured and simulated radiation patterns at 28.5 GHz, as shown in Figure 4.11. In the *E*-plane, the antenna has a measured HPBW of 26.5° with a simulated SLL of −9.5 dB. In the *H*-plane, the antenna has a measured HPBW of 46° with a

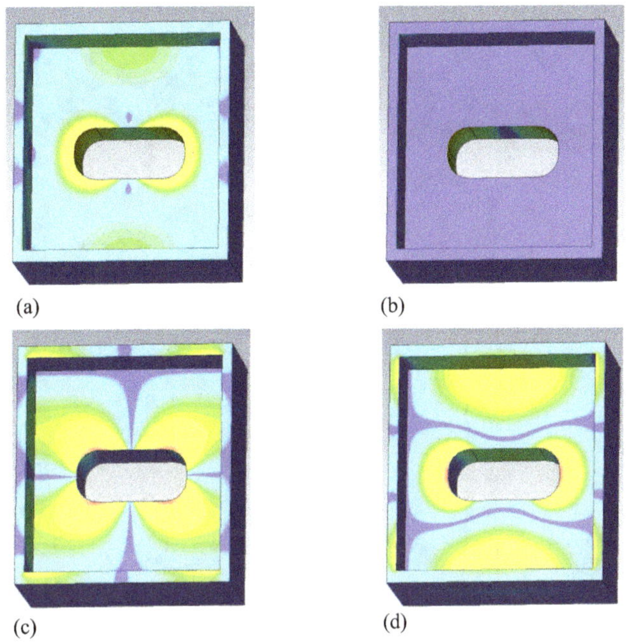

Figure 4.7 The magnetic field (H-field) inside the cavity: (a) absolute value, (b) H_z, (c) H_x and (d) H_y

simulated SLL of −23 dB. Due to the limitation of the planar near-field test system, the pattern measurements were conducted over the range $\pm\varnothing = \pm 60°$. The antenna exhibits a low level of measured cross polar radiation in both the *E*-plane and *H*-plane, as shown in Figure 4.11. Over the entire frequency operating bandwidth, and in both *E*-plane and *H*-plane, the antenna radiates in the boresight direction with a low SLL and low cross polar level. The simulated front-to-back (*F*/*B*) ratio for the proposed antenna is 20.1 dB.

The measured gain of the antenna is 11.7 dBi at 27.5 GHz, and it rises gradually to a maximum of 12.9 dBi at 28.6 GHz and then it drops gradually to 11.5 dBi at 30.3 GHz. Generally, it has been noticed that the antenna-measured gain is 0.3–1.1 dBi lower than the simulated one and can be attributed to fabrication tolerances as well as cable and WR-28 adapter losses. In addition, the antenna maintains an average measured aperture efficiency of 67% over its entire operating frequency bandwidth. The antenna's measured aperture efficiency is 64% at 27.5 GHz, it rises to a peak of 78% at 28.6 GHz and then gradually drops to 50% at 30.3 GHz, as shown in Figure 4.12. The simulated radiation efficiency of the antenna is 98% at 28.5 GHz. The antenna has a radiation efficiency higher than 90% over its entire operating frequency bandwidth. The high radiation efficiency can be attributed to the efficient feeding mechanism as well as the fact that the antenna purely consists of a metallic structure which means that it only suffers from conduction losses.

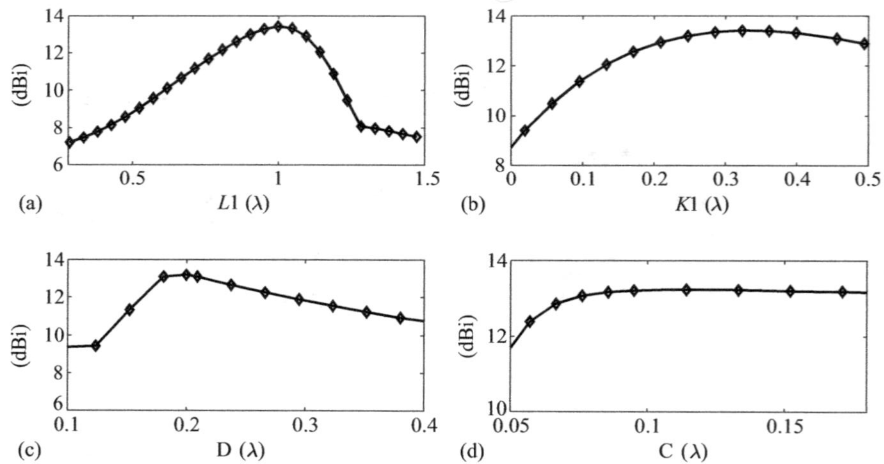

Figure 4.8 *The relationship between the cavity and corrugation dimensions and directivity of both antennas: (a) effect of L*1*, (b) effect of K, (c) effect of D and (d) effect of C*

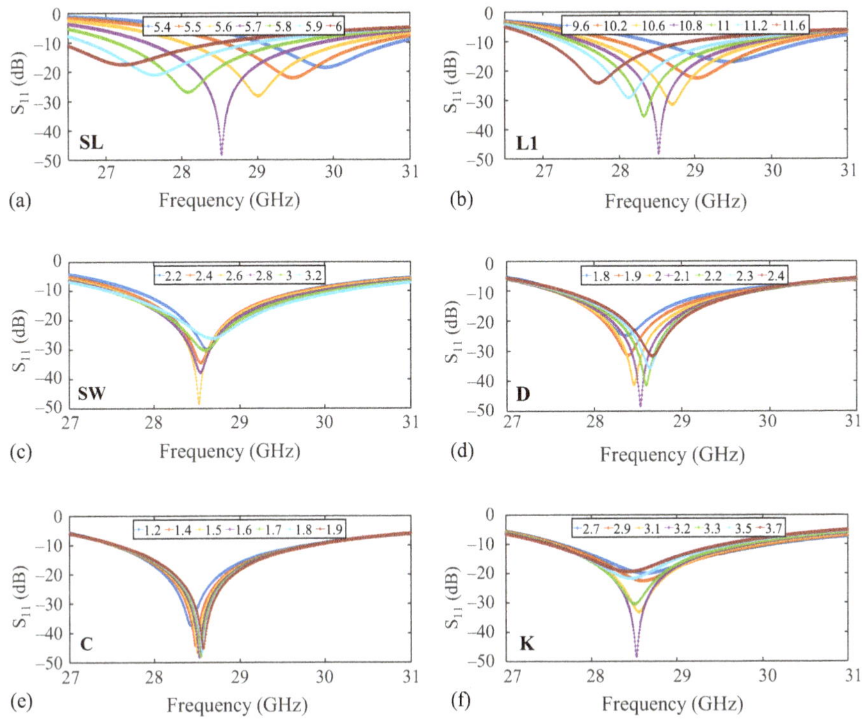

Figure 4.9 The effect of the dimensions of Ka-band antenna on the S_{11}: (a) SL, (b) L1, (c) D, (d) C, (e) SW and (f) K (all dimensions are in mm)

Table 4.2 The relationship between the number of corrugations, simulated gain and simulated aperture efficiency of the antenna at 28.6 GHz

No. of corrugations	Size	Without cavity		With cavity	
		Gain (dBi)	A_e (%)	Gain	A_e (%)
1	20 mm × 11 mm 1.9λ × 1.05λ	9	32	13.2	83
2	42 mm × 11 mm 1.9λ × 1.05λ	11.3	25.6	13.8	45
3	64 mm × 11 mm 4λ × 1.05λ	13	24.9	14.2	32.8
4	80 mm × 11 mm 7.6λ × 1.05λ	13.7	23.3	14.6	28.7
5	103 mm × 11 mm 1.9λ × 1.05λ	14.3	20.7	15	24.3
6	120 mm × 11 mm 9.8λ × 1.05λ	14.4	18.1	15.4	23
7	140 mm × 11 mm 13.3λ × 1.05λ	14.5	16.2	15.8	21.6
8	160 mm × 11 mm 15.2λ × 1.05λ	14.6	14.4	16.1	20.5
9	182 mm × 11 mm 17.3λ × 1.05λ	14.7	12.9	16.3	18.8
10	204 mm × 11 mm 19.4λ × 1.05λ	14.7	11.5	16.3	16.7

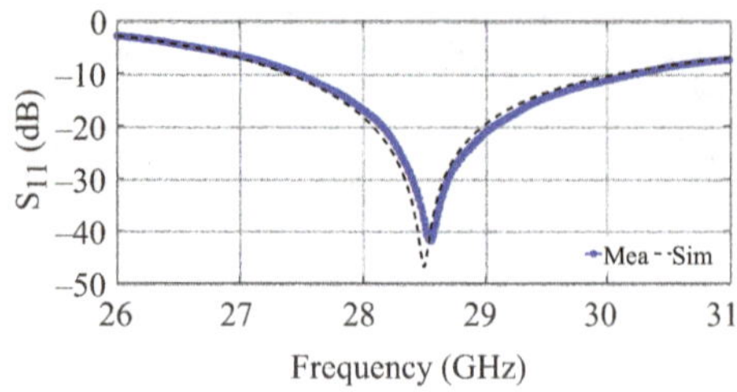

Figure 4.10 The simulated and measured S_{11} of the proposed antenna

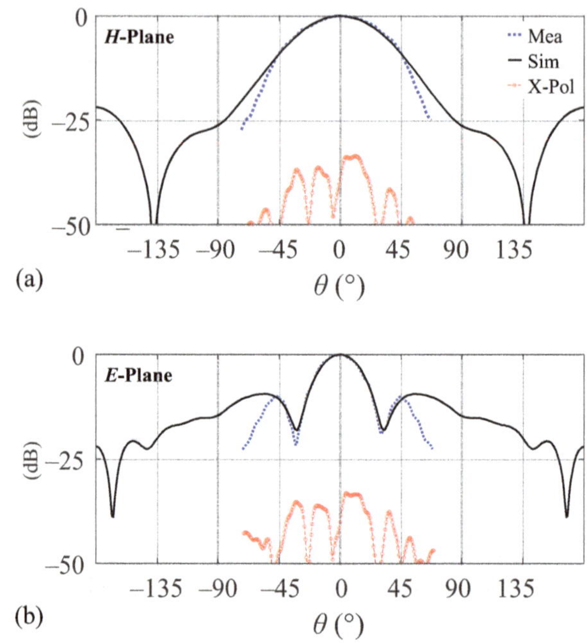

Figure 4.11 Simulated and measured radiation patterns for the Ka-band antenna at 28.5 GHz: (a) H-plane and (b) E-plane

4.2 Metallization techniques for 3D-printed antennas

4.2.1 Introduction

In this section of the chapter, we introduce a procedure for the additive manufacture of a 3D antenna. The technique involves 3D printing and metalization. We show how those techniques can be used to produce prototypes of the novel corrugated plate antenna introduced in the previous section.

Additive manufacturing or 3D printing is a powerful technology that enables low cost and rapid prototyping of a wide variety of items. The technology has been

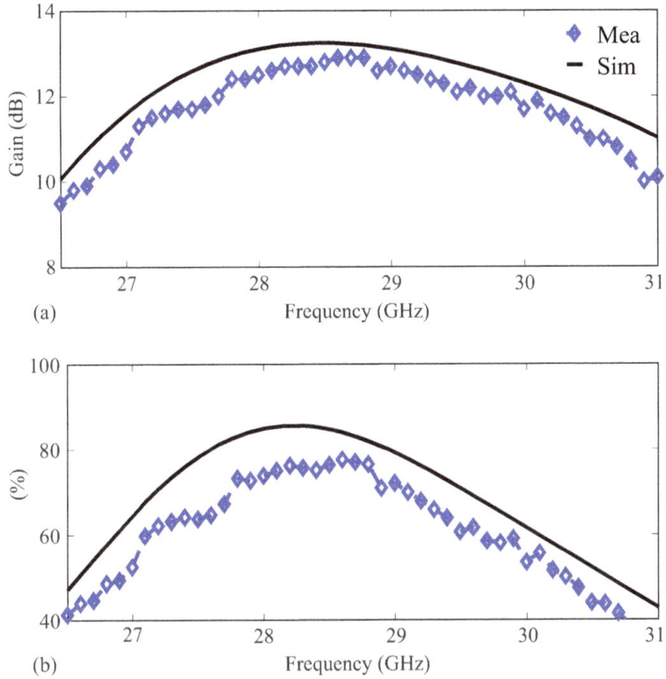

Figure 4.12 The gain and efficiencies of the Ka-band antenna at 28.5 GHz: (a) measured and simulated gain and (b) simulated and measured aperture efficiency

widely used to fabricate prototype antenna designs operating at various frequencies, including those within the microwave [11–17], mm-wave [18–20] and THz bands [21,22]. In this section, we introduce a technique for the additive manufacturing of a 3D antenna. The first step in the process is to produce a plastic/dielectric prototype using the 3D printer. The second and final step is to metallize (i.e. plate) the dielectric prototype [11]. Theoretically, the metallized prototype should deliver the same level of performance as its solid metal counterpart. This assumes that the antenna has been metallized with a smooth layer that is thicker than the skin depth at the desired operating frequency. For example, the skin depth for metals such as copper, gold, aluminium and silver is ≈ 0.4 μm at 30 GHz. The most common, reliable, high-performance metallization technique for use with 3D-printed objects is electroless metallic plating also known as electrochemical or autocatalytic plating. However, electroless plating remains a complex and expensive process as it involves a large number of processing steps and it involves using expensive chemicals such palladium in the activation process. Palladium is needed to catalyse the surface of the 3D-printed structure to initiate the deposition of nickel or copper. Moreover, electroless plating suffers from additional disadvantages such as the difficulty of plating large objects. Additionally, the chemicals have limited shelf life, are toxic and are damaging to the environment [23–26]. Later in this section, we introduce a range of alternative metallization techniques which do not suffer from these limitations.

4.2.2 Performance of metallization techniques at 30 GHz

A variant of the corrugated plate antenna was designed to operate at Ka-band. This antenna was used to evaluate the additive manufacturing process proposed in this section. The process is described in detail next. The Ka-band antenna resonates at 30 GHz. The antenna has a similar structure and operating principle to that of the aluminium antenna described previously. Figure 4.13 illustrates the geometry of the proposed antenna. The proposed antenna consists of central resonant slot surrounded by a rectangular cavity and two corrugations. The slot is fed using standard WR-28 waveguide. The H-plane of the waveguide is aligned with the width of the slot (SW). For this reason, the E-field is polarized in the x-direction. The parameters of the antennas are provided in Table 4.3 and they are described as following: antenna width W, antenna length L, antenna thickness T, corrugation depth D, corrugation width C, rectangular cavity depth K, rectangular cavity width $W1$, distance between corrugations and cavity X and rectangular cavity length $L1$.

Four copies of this antenna were produced using an Objet30 3D printer, see Figure 4.14. Each of these copies has nominally the same dimensions but was metalized using a different technique. The techniques are discussed next. By measuring the performance of the different antennas, we are able to evaluate the different metalization techniques. Ant-1 was metalized using Jet Metal process (JMT), Ant-2 was metalized using the RS EMI (electromagnetic interference)/RFI (radio frequency interference) conductive paint, Ant-3 was metalized using the nickel screening compound, Ant-4 was metalized using Electrolube silver conductive paint (SCP). The additive manufacturing processes are discussed in Sections 4.2.3 and 4.2.4. The metalization process is discussed in Section 4.2.5.

4.2.3 3D printer

We chose to use an Objet30 3D-printer to print the antennas, see Figure 4.15. This printer was developed by Stratasys and employs, the rapid PolyJet printing process. The printer incorporates two or more jetting heads (one set to produce the model

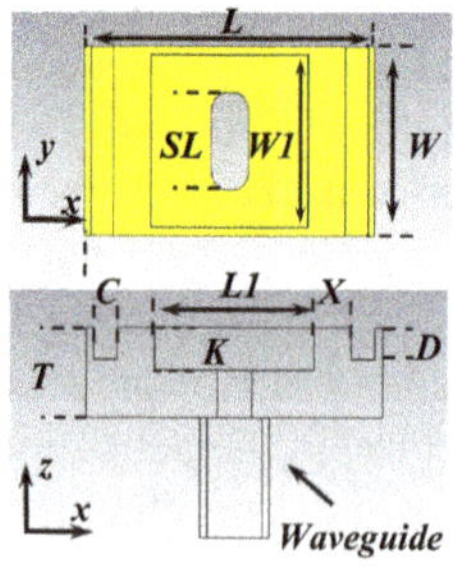

Figure 4.13 Geometry of the proposed 3D printed Ka-band antenna

Table 4.3 Dimensions of the Ka-band antenna

	L	**W**	**T**	**SL**	**SW**	**L1**	**W1**	**K**	**C**	**D**
Dimensions (mm) $\lambda = 10.58$ mm	20(2λ)	11(1.1λ)	6.1(0.61λ)	5.4(0.54λ)	3(0.3λ)	10(1.00λ)	10(1.00λ)	3.1(0.31λ)	1.7(0.17λ)	2(0.20λ)

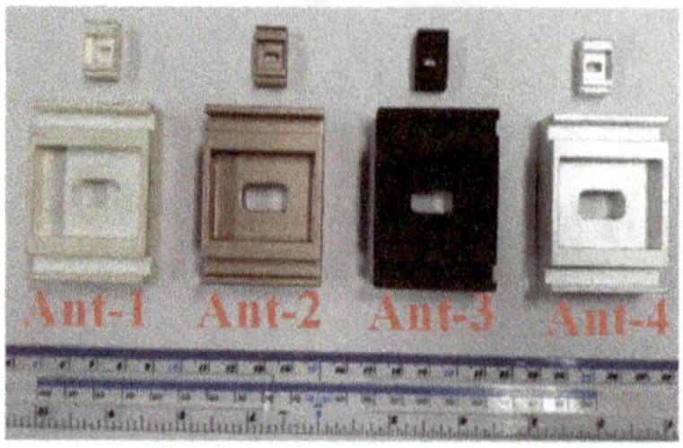

Figure 4.14 Fabricated prototype of the Ka-band antenna

Figure 4.15 Objet30 3D printer

and the other to produce the supporting material). These jetting heads deposit the outline of the part, layer-by-layer. The printer uses photopolymer materials, which get cured nearly instantly by a ultraviolet lamp within the printer, creating a solid, plastic-like model whose dimensions are accurate. The minimum layer thickness is 16 μm and the accuracy is 100 μm.

The support material is a gel-like substance, which is easily washed away after printing the final prototype. The models printed using this technique have a smooth finish and are suitable for further processing, such as sanding, painting, drilling, or tapping.

4.2.4 Metallization techniques

4.2.4.1 Jet Metal Technologies process

Jet Metal® is a new 'direct plating' technique that can be performed at room temperature and ambient pressure, see Figure 4.16. It was developed to alleviate some

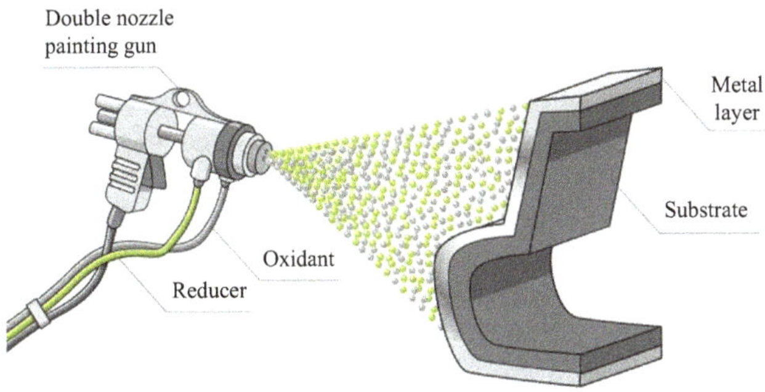

Figure 4.16 Principle of the Jet Metal metallization process

of the disadvantages associated with electroless plating. The novelty of this plating technique is that the supply of material (metallic ions Ag^+) and the energy are precisely controlled. The method does not reply on the use of hazardous chemicals, like palladium, that are carcinogenic, mutagenic, or toxic. The metal plating process involves continuous and simultaneously spraying of two solutions, using compressed air and a double nozzle paint-spraying gun. The reducing and oxidizing agents are sprayed simultaneously onto the surface of the object to be plated. This object could be either conducting or a non-conducting. This film is adsorbed into the surface and contains the active components.

The deposition speed for the JMT process is up to 22 μm h^{-1}. The JMT's spraying process allows control over the supply of chemical species and thus the silver thickness deposited. Control over the thickness of the Ag layer during the JMT process is achieved by manipulating the curing time as well as the speed of the reaction, both of which are well-defined functions of the temperature. For this reason, in order to obtain a requested Ag layer thickness, the spraying time is measured. Off-line this deposited Ag layer thickness can be verified via measuring the resistivity of the plated layer using a four-point probe and then using a lookup table to determine the corresponding layer thickness graphs via X-ray.

The conductivity at direct current of plated layers produced using the various different metallization techniques was determined by measuring the sheet resistance R_s using four-point probe. R_s is measured in $\Omega/\square$ and it is expressed as

$$R_s = \frac{\rho}{t} \tag{4.6}$$

where ρ is the resistivity (Ω) and t is the thickness of the paint, and the conductivity is $\sigma = 1/\rho = \text{Sm}^{-1}$. Hence, the measured conductivity of a 2.5-μm-thick silver film is ~4.40×10^7 Sm^{-1}.

4.2.4.2 RS EMI/RFI conductive paint

An EMI/RFI aerosol of silver coated copper compound from RS components. The compound is able to serve as a Faraday cage to protect electronic equipment thus providing RFI/EMI shielding/screening to reduce EMI. The paint exhibits low resistance, even when applied as a thin layer. The paint has several advantages; as it adheres to a wide variety of objects and it has excellent ageing performance. The paint has bronze colour and it delivers an R_s in the range of 0.3–0.7 $\Omega/\square$ which corresponds to a conductivity ranging from 0.29×10^5 to 0.67×10^5 Sm^{-1}. The paint density is 1.05 gml^{-1} and the 400 ml aerosol covers an area of 1.2 m^2. The aerosol is fully conductive 5 min after application and it delivers maximum conductivity within 24 h after application [27].

4.2.4.3 Nickel screening compound

We also evaluated a nickel-based compound. The compound is an electrically conductive coating based on nickel powder contained within a thermoplastic resin. It is an efficient EMI/RFI screening coating that adheres to a wide variety of objects. The coating becomes touch dry 5 min after application and achieves maximum conductivity after 24 h. The paint has bronze colour and it delivers an R_s ranging between 0.3 and 0.7 $\Omega/\square$ which corresponds to a conductivity ranging from 0.29×10^5 to $0.67 \times 10^5 S.m^{-1}$. The paint has dark grey colour and a density of 1.17 gml^{-1}. A 400 ml aerosol can cover an area of 1.5 m^2 [28].

4.2.4.4 Silver conductive paint

A SCP is used. The paint consists of 45% silver suspended within a solvent blend. The SCP is manufactured by Electrolube Ltd in 3-g-bottle. The paint provides a thin, smooth, adherent, flexible film with high electrical conductivity. The SCP can be applied to a variety of different objects, including plastics, paper, wood, textiles, glass, ceramics and metals. It can be applied by brushing, or spraying and it dries quickly at room temperature. Traditionally, the applications of SCP include repairing/replacing tracks within a printed circuit board (PCB) design. It can also be used to mark recording tape or to screening plastic housings against high-frequency fields. Finally, it can be used to make connections to non-solderable surfaces. SCP has a density of 1.44 gml^{-1} and the surface resistance ranges from 0.01 to 0.03 $\Omega/\square$ assuming that it has been applied at in the proportion from 0.6 to $2g/100$ cm^2 [29].

4.2.5 Metallization procedure

The Ka-band antenna which was metallized using JMT is referred to as Ant-1. The technique was used to apply a thin and a smooth silver film to the 3D-printed antennas. First, the 3D-printed prototypes were cleaned using Isopropyl alcohol in order to remove grease and stains. After cleaning, both antennas were flamed in order to achieve good wettability over the entire surface and to increase the adhesion of the silver coating. Then, the prototypes were fixed into a holder that could be rotated in order to ensure a homogeneous coating. The silver was then deposited onto the surface of the antenna through continuous and simultaneous spraying of oxidant agent (metallic ions Ag^+) and reducing agent. This was achieved using a double nozzle paint spray gun powered by compressed air. The process is conducted at room temperature. Hence, both antennas are coated with a film that has a thickness of 2.5 μm by using two spray guns simultaneously for 13 min. This enables the thickness of the silver layer to be easily controlled by controlling the spraying time.

Ant-2 was metallized using the RS EMI/RFI conductive paint which is available in an aerosol and has a bronze colour. Two coats of the paint were applied manually to the surface of the antennas. The process was performed at room temperature. The aerosol is conductive 5 min after application and it delivers maximum conductivity 24 h after application. No pre-treatment or post-processing steps were performed. The coverage area of 1.3 m^2/400 ml was sufficient to metallize more than 20 antenna prototypes of the type proposed.

The prototype corresponding to the third method for metallizing antennas is Ant-3. Ant-3 was created by coating the 3D-printed prototypes using the nickel screening compound. Three coats have been manually sprayed to the surface of the proposed antennas and the finished prototypes have a dark grey colour as shown in Figure 4.14. No pre-treatment or post-processing steps were applied before/after spraying the prototypes. The aerosol has a coverage area of 1.5 m^2/400 ml which is sufficient to metallize more than 20 pieces of the proposed prototypes.

The antennas which are metallized using the fourth metallization technique are referred to as Ant-4. Ant-4 was painted using Electrolube SCP. The paint is applied gently to the surface of the antennas using a fine paint brush. Only one coat was applied and the painted prototypes left to dry for 10 min at room temperature. No pre-treatment or post-processing steps were applied before/after painting the prototypes. The finished prototypes are shown in Figure 4.14. Finally, less than 2 g of silver were used to metallize both prototypes.

The full cost comparison of the metallization material used in the fabrication of the proposed antennas is shown in Table 4.4.

4.2.6 Operating principles

The operating principles of the Ka-band antenna are identical to those of the 28 GHz aluminium antenna, discussed earlier. The EM energy is coupled into the antenna via a length of waveguide. Energy enters the resonant cavity via the resonant slot. The resonance frequency is inversely proportional to the cavity length ($L1$) and SL. The antenna resonates at the frequency for which $SL \approx \lambda/2$. The slot width (SW) controls the frequency-operating bandwidth. The antenna yields high gain and high directivity

*Table 4.4 Cost comparison of the proposed metallization techniques**

Antenna	Ant-1	Ant-2	Ant-3	Ant-4
Cost	<3 USD	<5 USD	<5 USD	<8 USD

*The cost was calculated based on the cost of raw material/paint used to metallize both prototypes of the proposed antennas. The cost excludes any cost of labour and equipment involved in the metallization process.

in spite of its small physical size. The reason for this is that radiation is emitted from the slot, corrugations and rectangular cavity. For instance, the X-band antenna has a peak directivity of 5.3 dBi due to radiation from the slot alone. The addition of the cavity increases the directivity by 4.4–9.7 dBi, while the final directivity is increased by another 3.3–13 dBi, after adding the corrugations. The radiation from the cavity is due to the strong excitation of the TM_{21} mode inside the cavity. Maximum directivity is achieved for a cavity length $(L1) = \lambda$ and when the cavity depth (K) is in range $\lambda/3 \leq K \leq \lambda/4$, as shown in Figure 4.18. The corrugations excite a travelling wave, which propagates along the surface of the antenna. Peak boresight radiation is

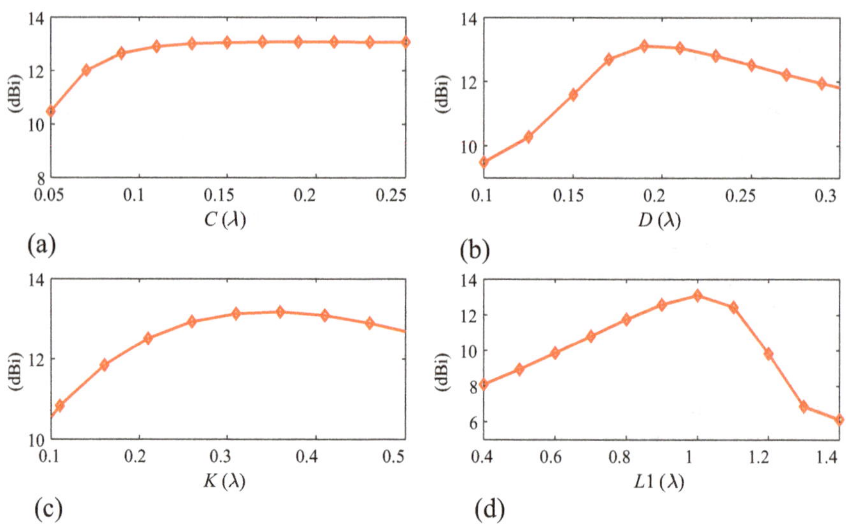

Figure 4.17 The relationship between the parameters of proposed antennas and their directivity: (a) C, (b) D, (c) K and (d) L1

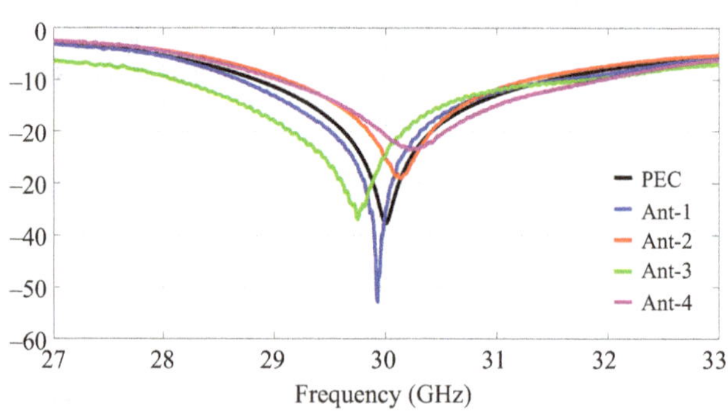

Figure 4.18 The simulated and the measured S_{11} of the proposed Ka-band antennas

obtained when the dimensions of the corrugations are optimized, as shown in Figure 4.17(b) and (c).

4.2.7 Measured results at 30 GHz

Figures 4.18–4.20 have shown the simulated and measured S_{11}, gain and radiation patterns of the Ka-band antennas. The full comparison between the figures-of-merits of the measured antennas with the simulated perfect electric conductor (PEC) antenna is shown in Table 4.5. At 30 GHz, Ant-1 has a measured gain of 12.75 dBi. This gain value is 0.25–0.4 dBi lower than that of the PEC antenna as shown in Figure 4.19. Ant-2 has a peak gain of 12.3 dBi at 30 GHz. This gain value is 0.7–1.7 dBi lower than that of the PEC antenna. Ant-4 has a gain of 12.6 dBi at 30 GHz. This gain value is 0.4–1.1 dBi lower than that of the PEC antenna. Finally, Ant-3 has a gain of 9.8 dBi at 30 GHz. This gain value is 3–3.5 dBi lower than that of the PEC antenna.

4.2.8 Discussion and analysis

Three main factors affect the performance of the proposed 3D-printed antennas. The first is the accuracy and the resolution of the 3D printer, the second is the conductivity and thickness of the metallic layer and the third is the surface roughness of the metallic layer [11,26]. The accuracy of the employed 3D printer is 0.1mm which corresponds to $\lambda/100$ at 30 GHz. The fabrication tolerances therefore have a moderate effect on the performance of the Ka-band antennas. In fact, the frequency-operating bandwidth and resonant frequency of the Ka-band antennas are both sensitive to even 50 µm variation in the slot dimensions, as shown in Figure 4.21, especially the SL and the radius of the bend at corners of the slot.

The conductivity of each of the different metallic layers was determined by measuring the sheet resistance R_s using a four-point probe. This enabled us to model the metalization and examine its effect on the performance of the antenna.

The effect of conductivity on the gain and reflection coefficient of the antenna was studied. It is observed that the changes in the conductivity affect the bandwidth and position of the resonant frequency, as shown in Figure 4.22. Lower

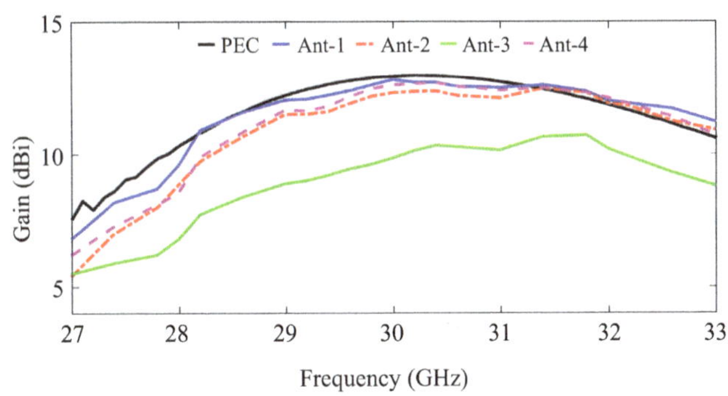

Figure 4.19 The simulated and measured gain of the Ka-band antennas

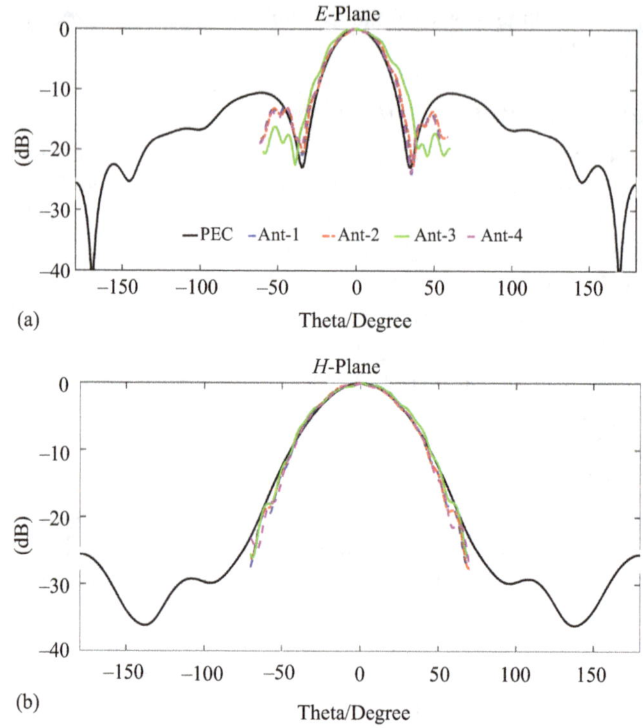

Figure 4.20 The simulated and measured radiation patterns of the Ka-band antennas: (a) E-plane and (b) H-plane

Table 4.5 The effect of variation in conductivity and paint thickness on the gain performance of the proposed antennas

	Thickness (μm)	***Sheet resistance*** R_s ($m\Omega/\square$)	***Conductivity*** (Sm^{-1})	***Gain simulated 30 GHz***
PEC	NA	0	Infinity	13
Ant-1	2.5	9	4.40×10^7	12.95
Ant-2	50–80	323–510	0.25×10^5 to 0.39×10^5	12.2–12.45
Ant-3	100–200	800–1,200	0.04×10^5 to 0.12×10^5	11.2 –11.8
Ant-4	30–45	13–27	0.82×10^6 to 2.60×10^6	12.75–12.9

conductivity values have the effect of reducing operating frequency, the bandwidth of the antennas and shifting the resonance frequency towards lower values as shown in Figure 4.22. Moreover, lower conductivity increases the conduction and gain losses as summarized in Table 4.5.

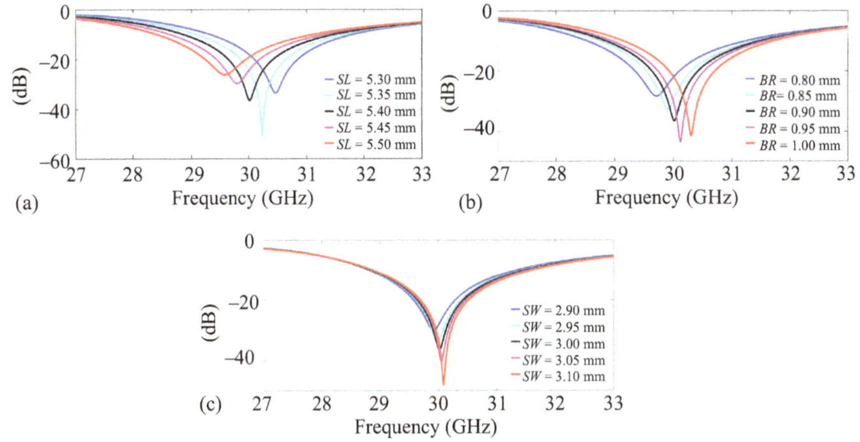

Figure 4.21 The effect of 50 μm variation in the slot dimensions on the S_{11} of the Ka-band PEC antenna: (a) slot length (SL), (b) slot width (SW) and (c) radius of the bend at slot corner (BR)

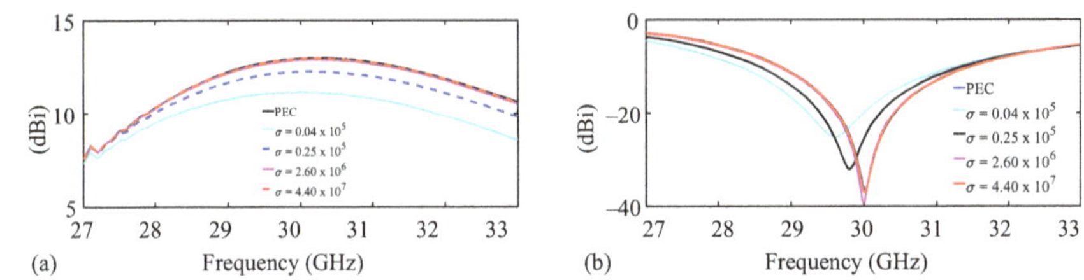

Figure 4.22 The effect of the change in conductivity on the performance of the Ka-band antenna: (a) S_{11} and (b) gain

One of the disadvantages of the methods used to metallize Ant-2, Ant-3 and Ant-4 is that it is not entirely possible to accurately control the thickness of the paint layer uniformly across the surface of the antenna, especially in the interior sides of the slot and corrugations. The thickness of the paint depends mainly on the number of coats applied and, in the case of Ant-2 and Ant-3, more importantly on duration of the spraying process. This contributes to variation in the conductivity across the antennas' surface resulting in a limited additional losses that cannot be predicted during the simulation process. For example, the thickness of the two coats of copper spray used for Ant-2 is in the range of 50–80 μm and the thickness of three coats of nickel spray used for Ant-4 is in the range of 100–200 μm. This explains the shift in the resonance frequency of Ant-2 and Ant-4 at higher frequencies as the thickness of the paint reduced the physical size of the slot resulting in a shift of the resonance to higher frequencies.

The root mean square (RMS) of surface roughness of each of the different metallic layers was measured along a 5 mm long path using a profilometer and the results are shown in Figure 4.23. The RMS surface roughness of Ant-1, Ant-2 and Ant-4 are 1.82, 8.71 and 4.62 μm, respectively. This is unlikely to contribute significantly to the gain losses at Ka-band frequencies as it is much lower than the measured surface roughness of 25.9 μm (RMS) which contributed to no losses in the horn gain at 15 GHz [12]. However, the surface roughness of Ant-3 is 41.4 (RMS) which contributes to gain losses as it is comparable to the RMS value of 39.7 μm reported in [12] which contributed to gain loss of 0.7 dB at 15 GHz. Furthermore, the nickel paint used to metallize Ant-3 has poor adhesive properties. This resulted in a number of few cracks appearing on the surface of the 3D-printed antennas, after application. These cracks justify the additional gain losses and the variation in the S_{11} performance of Ant-3 at Ka-band.

4.3 Compact 3D-printed antenna

A small corrugated antenna that is fed using standard microstrip line and mini-SMP (sub miniature push-on) connector is developed. The proposed antenna consists of two layers. The first layer is a feeding layer and the second layer is the radiating structure which is similar to the corrugated antenna previously discussed in precious sections. The EM energy is coupled to the antenna surface using a microstrip line with a special feeding structure designed to accommodate the PE 44489 mini-SMP connector as shown in Figure 4.24. The feeding structure for the proposed antenna consists of a transmission line, vias, mini-SMP ground plane/pad and Rogers RO4003C substrate. RO4003C is hydrocarbon ceramic laminate which has a dielectric constant of 3.38 and thickness of $h = 0.508$ mm. The second layer is 3D-printed using Objet30 printer that has a resolution of 100 μm and metalized using 2.5 μm silver layer using JMT metallization technique as shown in Figure 4.25. The dimensions of the antennas are as follows: width of antenna $W = 11$ mm, length of antenna $L = 19.6$ mm, length of slot $SL = 6.5$ mm, width of the slot $SW = 3$ mm, top cavity width $W1 = 10$ mm, top cavity length $L1 = 10.8$ mm, top cavity depth $Cd = 0.7$ mm, thickness $t = 3$ mm, width of

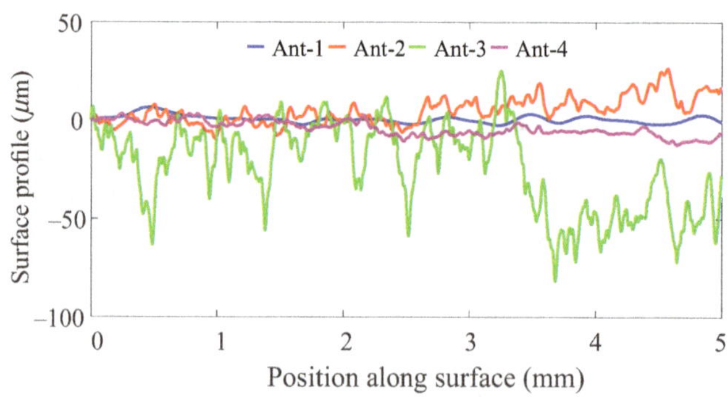

Figure 4.23 Surface profile measurements of the proposed 3D-printed antennas

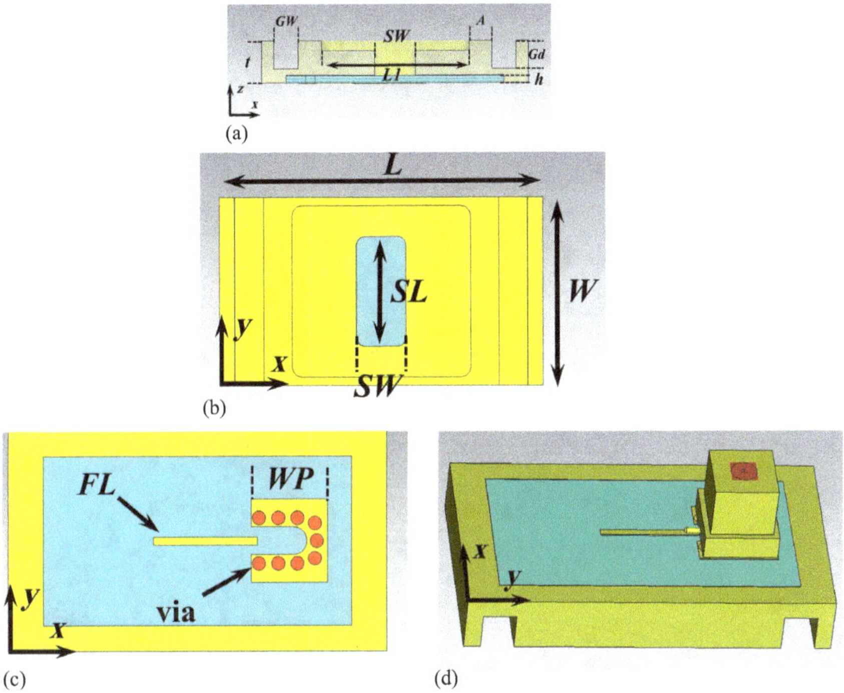

Figure 4.24 The schematic of the proposed antenna: (a) cross section of front view, (b) top view, (c) bottom view and (d) perspective view with the mini-SMP connector

corrugations/grooves $GW = 1.8$ mm, depth of corrugations $Gd = 2$ mm, substrate dimensions 16 mm × 8.5 mm, radius of via $VR = 0.7$ mm, feed line width $FW = 0.4$ mm, length of feed line $FL = 5.4$ mm, distance between cavity and corrugation $A = 1.7$ mm, mini-SMP pad $WP \times WP = 4.2$ mm × 4.2 mm.

4.3.1 Operating principles

For the fixed feed line length *FL*, the antenna resonance is inversely proportional to the *SL* dimension, and the bandwidth is directly proportional to the antenna width *SW* as shown in Figure 4.26. Furthermore, the depth of the corrugations and the dimensions of the top cavity affect the resonance frequency in similar manner to the antenna that is fed using the waveguide due to the resonance of TM_{21} and TE_{11} modes.

The antenna has a peak directivity of 9.72 dBi at 28 GHz due to the combined radiation from the slot, top-layer cavity both corrugations as the power flow figures show in Figure 4.27. The antenna has a directivity of 5 dBi due to radiation from the slot and the directivity is increased to 8.75 dBi with the corrugations and to 9.72 dBi with the top cavity. The antenna dimensions are optimized to deliver the maximum directivity and gain performance at 28 GHz. The directivity of the

Figure 4.25 The prototype of the proposed antenna

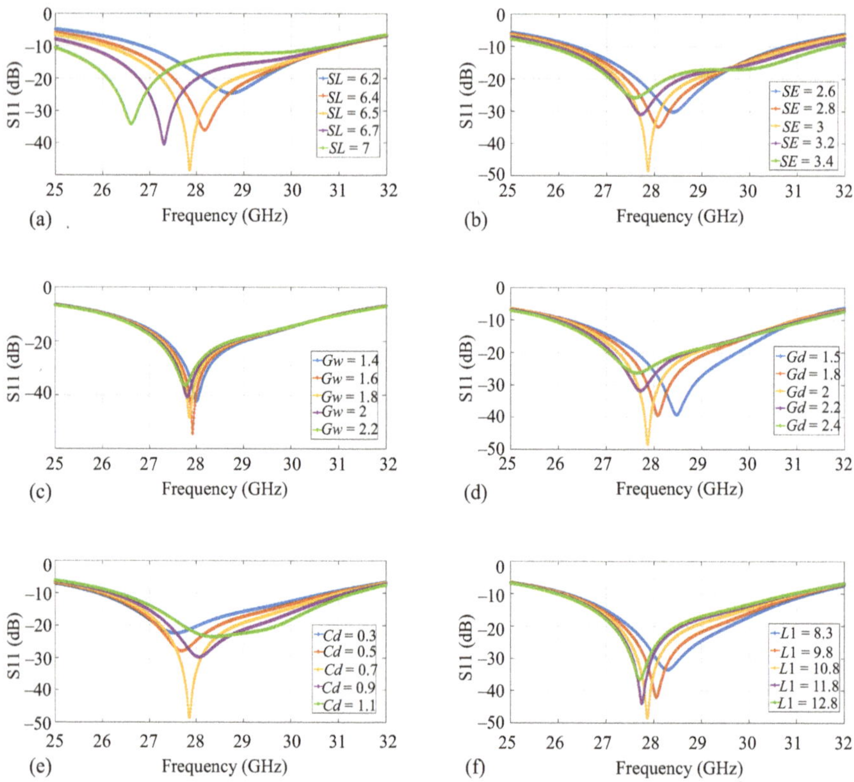

Figure 4.26 The effect of the antenna parameters on S_{11}: (a) SL, (b) SW, (c) Gw, (d) Gd, (e) Cd and (f) L1 (all dimensions are in mm)

antenna is mainly sensitive to the dimensions of the corrugations and the cavity and marginally sensitive to the dimensions of the slot as shown in Figure 4.28.

The antenna resonates at 27.9 GHz with a simulated bandwidth of 4.9 GHz ranges from 26.1 to 31 GHz as shown in Figure 4.29. The peak-simulated gain of the antenna is 9.6 dBi at 28 GHz and the antenna has a simulated gain of 8 dBi at 26 GHz rises to 9.17 dBi at 27 GHz and approaching the peak at 28 GHz. Then, the gain drops to 9 dBi at 29.1 GHz and to 8 dBi at 29.9 GHz and to 5.8 dBi at 31 GHz as shown in Figure 4.30. The antenna measured results agrees with the simulated results as the measured antenna has a -10 dB bandwidth of 5.5 GHz ranges from 25 to 29.5 GHz. The peak measured gain of the antenna is shown in Figure 4.29 and the antenna has a gain of 8.8 dBi at 28 GHz. The antenna has a measured gain of 8.3 dBi at 26 GHz, rises to 8.5 dBi at 27 GHz and to and 8.8 dBi at 28 GHz. Then the measured gain drops to 8 dBi at 29 GHz and to 6 dBi at 30 GHz. The measured gain continues to drop to 5.7 dBi at 31 GHz and to 3.5 dBi at 32 GHz.

The simulated and measured radiation patterns of the antenna are shown in Figures 4.31 and 4.33. At 28 GHz, the antenna has an HPBW of 36.7° in the *E*-plane with SLL of -9.2 dBi and the HPBW in the *H*-plane is 56° with SLL of -11 dBi. The *F*/*B* of the antenna is 6.5 dB. The antenna has a radiation in the boresight with good SLL, HPBW performance over a wide bandwidth ranges between 26 and 30 GHz as measured 2D radiation patterns shown in Figure 4.32.

4.3.2 Fabrication tolerances and antenna performance

The discrepancies between the measurement and simulated results of the proposed antenna are mainly due to the fabrication tolerances and connector soldering process. Hence, the effects of the antenna fabrication tolerances on the antenna performance are studied. The antenna consists of two major parts which are the feeding structure and the 3D-printed radiation structure. The fabrication tolerances of the 3D-printed antenna are due to accuracy of the 3D printer which is

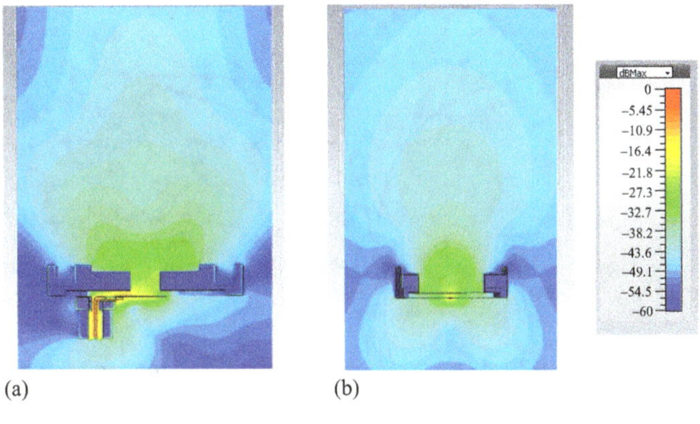

Figure 4.27 Power flow at 28 GHz: (a) x–z plane and (b) y–z plane

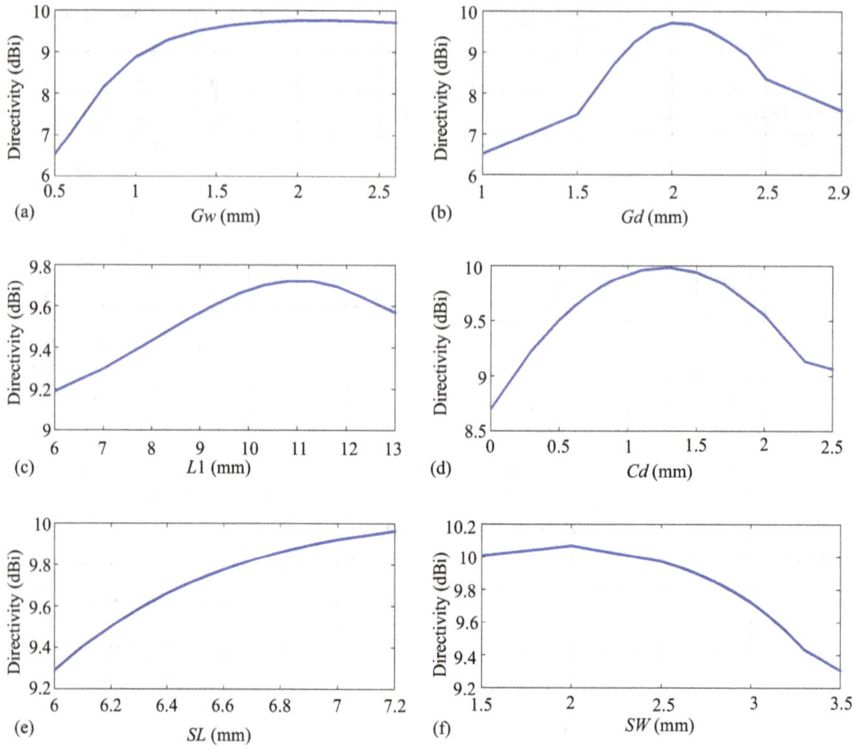

Figure 4.28 The effect of antenna parameters on the directivity of the antenna at 28 GHz: (a) Gw, (b) Gd, (c) L1, (d) Cd, (e) SL and (f) SW

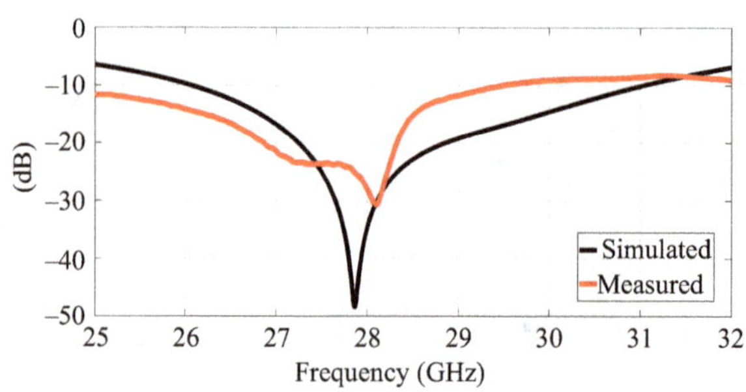

Figure 4.29 The S_{11} performance of the antenna

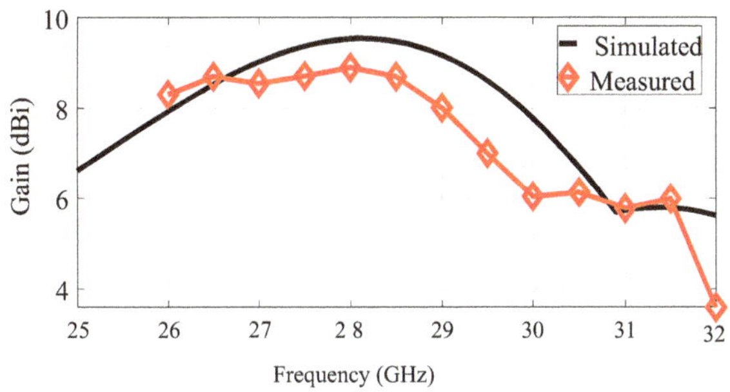

Figure 4.30 The gain of the proposed antenna

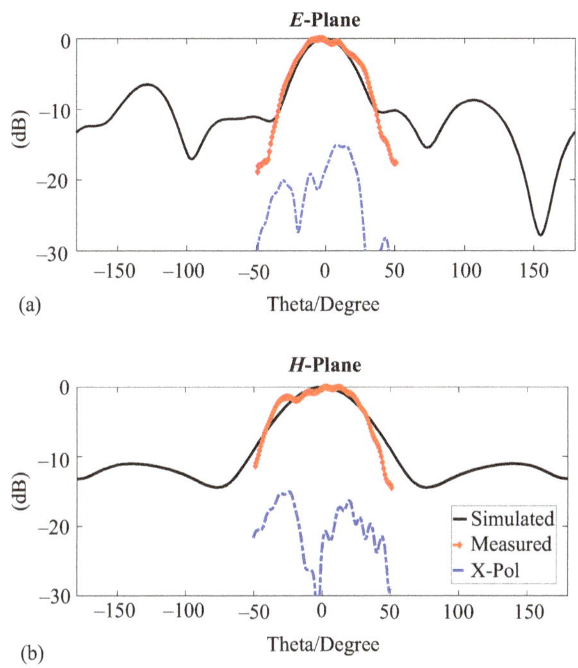

Figure 4.31 The normalized radiation pattern of the proposed antenna: (a) E-plane and (b) H-plane

100 μm = 0.1 mm and due to the traditional PCB fabrication process used to fabricate the feeding layer which is 50–100 μm.

4.3.2.1 Effect of 3D printer tolerances on the antenna performance

Simulation results show that a tolerances of 0.1 mm in the 3D-printed antenna has a very limited effect (± 0.2 dBi) on the directivity performance of the antenna, while

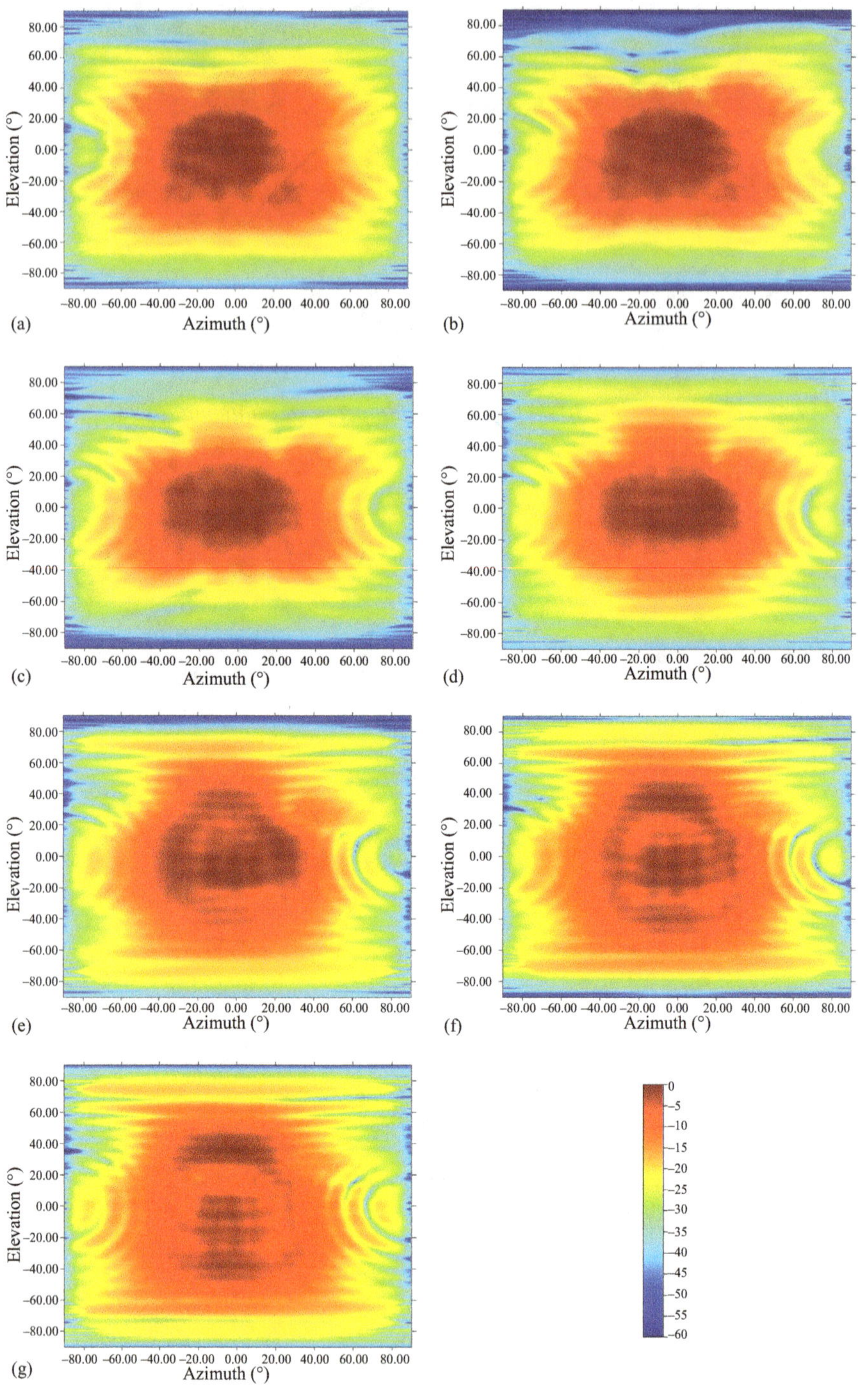

Figure 4.32 The measured 2D radiation pattern of the proposed antenna using planar NSI utilizing the measurement setup described in Appendix B: (a) 26 GHz, (b) 27 GHz, (c) 28 GHz, (d) 29 GHz, (e) 30 GHz, (f) 31 GHz and (g) 32 GHz

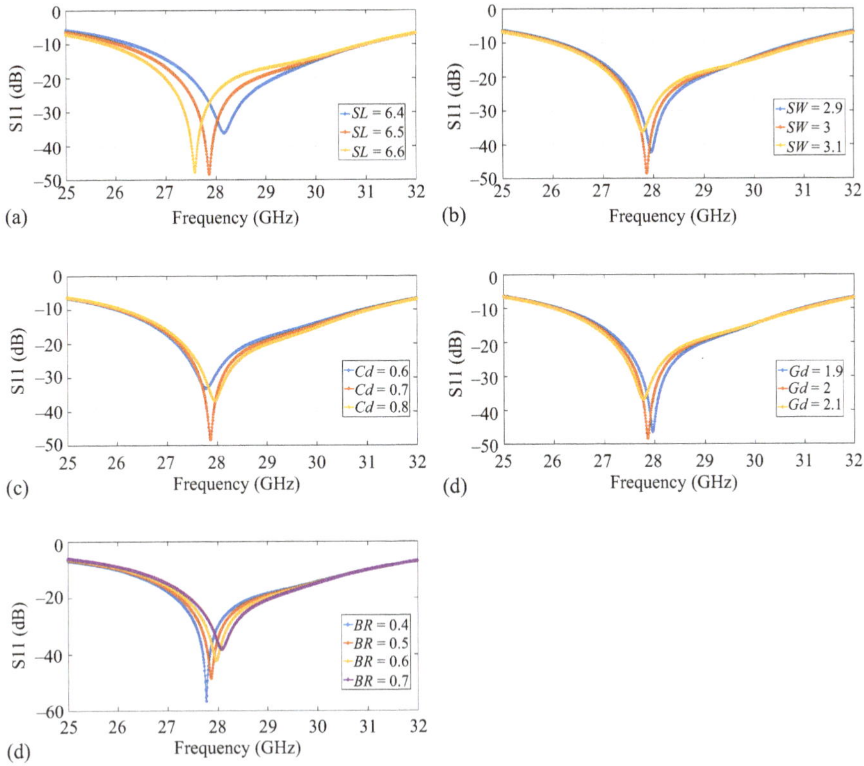

Figure 4.33 The effect of 3D printer tolerance of 0.1 mm in main antenna dimensions on the antenna performance: (a) SL, (b) SW, (c) Cd, (d) Gd and (e) BR (all dimensions are in mm)

it has a minor effect on the S_{11} performance of the antenna. In more details, the antenna S_{11} is affected mainly by the slot dimensions, especially the *SL* and the radius of the bend at slot corner *BR* resulting in a maximum shift of $\approx \pm 250$ MHz in the position of the resonance frequency as shown in Figure 4.33. The other parameters which have much less effect on S_{11} ($\approx \pm 150$ MHz) compared to *SL* and *BR* are the slot width *SW*, cavity depth *Cd* and corrugation depth *Gd* as shown in Figure 4.33.

4.3.2.2 Effect of feeding layer tolerances on the antenna performance

The S_{11} performance of the antenna is sensitive to the dimensions of the feeding line which are feeding line width *FW* and *FL* as shown in Figure 4.34. *FW* is affecting the matching between the mini-SMP connector tip and the transmission line and *FL* affects the matching and coupling of EM energy from the feeding layer to the surface of the 3D-printed antenna via the slot.

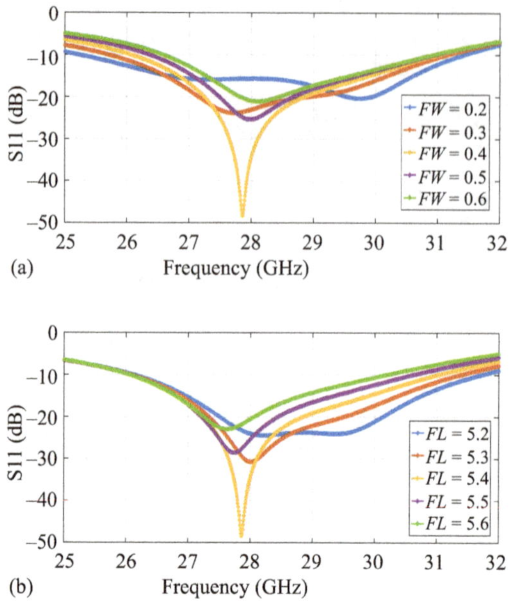

Figure 4.34 The effect of 0.1 mm variation in the feed line on S_{11} of the antenna. (a) FW and (b) FL (all dimensions are in mm)

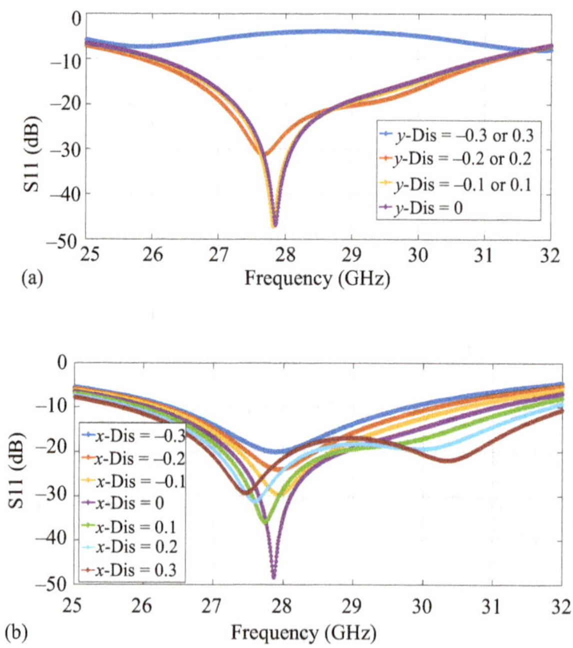

Figure 4.35 The effect of displacement in the mini-SMP connector: (a) horizontal displacement (y-axis direction), (b) vertical displacement (x-axis direction) (all dimensions are in mm)

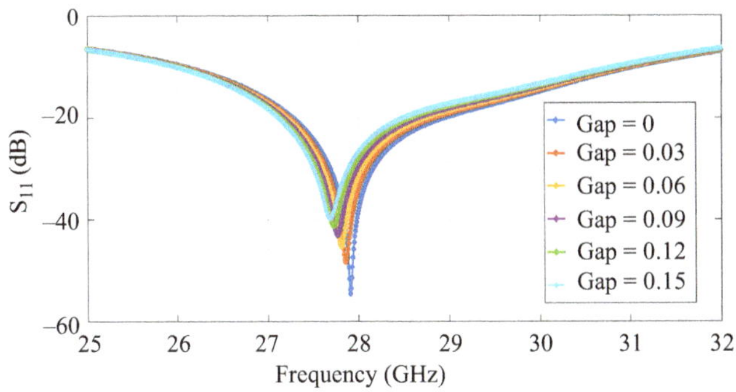

Figure 4.36 The effect of a potential air gap between the feeding layer and the 3D-printed structure (all dimensions are in mm)

4.3.2.3 Effect of connector displacement

Another fabrication error that affects the S_{11} performance of the antenna is any vertical or horizontal displacement in the position of the connector while soldering it to the transmission line and the mini-SMP pad as shown in Figure 4.35. This misalignment and displacement could be in the vertical direction (x-axis) or in the horizontal direction (y-axis). The connector tip which is soldered to the transmission line has a diameter of 0.3 mm and it is 0.8 mm long, while the transmission line has a width of $FW = 0.4$ mm. Hence, it is possible to have a misalignment during soldering, especially vertical misalignment in the x-axis.

4.3.2.4 Effect of air gap

Another source of error that causes a limited effect on the performance of the antenna is the existence of air gap between the feeding layer and the 3D-printed structure once they are assembled. This could cause a limited variation in the S_{11} performance as shown in Figure 4.36.

References

[1] M. J. Lockyear, A. P. Hibbins, J. R. Sambles, and C. R. Lawrence, "Surface-topography-induced transmission and directivity of microwave radiation through a subwavelength circular metal aperture," Applied Physics Letter, vol. 84, no. 12, pp. 2040–2042, 2004.
[2] M. J. Lockyear, A. P. Hibbins, J. R. Sambles, and C. R. Lawrence, "Enhanced microwave transmission through a single subwavelength aperture surrounded by concentric grooves," Journal of Optics A: Pure and Applied Optics, vol. 7, no. 2, pp. S153–S158, 2005.
[3] M. Beruete, I. Campillo, J. S. Dolado, *et al.*, "Enhanced microwave transmission and beaming using a subwavelength slot in corrugated plate," IEEE Antennas and Wireless Propagation Letters, vol. 3, pp. 328–331, 2004.

[4] F. J. García-Vidal, H. J. Lezec, T. W. Ebbesen, and L. Martín-Moreno, "Multiple paths to enhance optical transmission through a single subwavelength slit," Physics Review Letters, vol. 90, no. 21, pp. 213901-1–213901-4, 2003.

[5] M. Beruete, I. Campillo, J. S. Dolado, *et al.*, "Low-profile corrugated feeder antenna," IEEE Antennas Wireless Propagation Letters, vol. 4, pp. 378–380, 2005.

[6] M. Beruete, I. Campillo, J. S. Dolado, *et al.*, "Dual-band low-profile corrugated feeder antenna," IEEE Transactions on Antennas and Propagation, vol. 54, no. 2, pp. 340–350, 2006.

[7] M. Beruete, I. Campillo, J. S. Dolado, *et al.*, "Very low-profile "bull's eye" feeder antenna," IEEE Antennas Wireless Propagation Letters, vol. 4, pp. 365–368, 2005.

[8] M. Beruete, I. Campillo, J. E. Rodrigues-Seco, E. Perea, F. Falcone, and M. Sorolla, "Very low profile and dielectric loaded feeder antenna," Antennas Wireless Propagation Letters, vol. 6, pp. 544–548, 2007.

[9] S. Alkaraki, Y. Gao, and C. Parini, "High aperture efficient slot antenna surrounded by the cavity and narrow corrugations at Ka-band and Ku-band," IET Microwaves, Antennas & Propagation, vol. 12, no. 12, pp. 1926–1931,1933, 2018.

[10] C. A. Balanis, "Antenna measurement", in Antenna Theory, 3rd ed. Hoboken, NJ: John Wiley and Sons, 2005, ch. 17, sec. 4.2, pp. 1033–1034.

[11] G. P. Le Sage, "3D printed waveguide slot array antennas," IEEE Access, vol. 4, pp. 1258–1265, 2016.

[12] C. R. Garcia, R. C. Rumpf, H. H. Tsang, and J. H. Barton, "Effects of extreme surface roughness on 3D printed horn antenna," Electronics Letters, vol. 49, no. 12, pp. 734–736, 2013.

[13] M. Rizwan, M. W. A. Khan, L. Sydänheimo, J. Virkki, and L. Ukkonen, "Flexible and stretchable brush-painted wearable antenna on a three-dimensional (3-D) printed substrate," IEEE Antennas and Wireless Propagation Letters, vol. 16, pp. 3108–3112, 2017.

[14] G. L. Huang, S. G. Zhou, and T. Yuan, "Development of a wideband and high-efficiency waveguide-based compact antenna radiator with binder-jetting technique," IEEE Transactions on Components, Packaging and Manufacturing Technology, vol. 7, no. 2, pp. 254–260, 2017.

[15] S. Y. Jun, B. Sanz-Izquierdo, E. A. Parker, D. Bird, and A. McClelland, "Manufacturing considerations in the 3-D printing of fractal antennas," IEEE Transactions on Components, Packaging and Manufacturing Technology, vol. 7, no. 11, pp. 1891–1898, 2017.

[16] J. C. S. Chieh, B. Dick, S. Loui, and J. D. Rockway, "Development of a Ku-band corrugated conical horn using 3-D print technology," IEEE Antennas and Wireless Propagation Letters, vol. 13, pp. 201–204, 2014.

[17] B. Zhang, Z. Zhan, Y. Cao, *et al.*, "2 metallic 3-D printed antennas for millimeter- and submillimeter wave applications," IEEE Transactions on Terahertz Science and Technology, vol. 6, no. 4, pp. 592–600, 2016.

[18] E. A. Rojas-Nastrucci, J. T. Nussbaum, N. B. Crane, and T. M. Weller, "Ka-band characterization of binder jetting for 3-D printing of metallic

rectangular waveguide circuits and antennas," IEEE Transactions on Microwave Theory and Techniques, vol. 65, no. 9, pp. 3099–3108, 2017.

[19] E. A. Rojas-Nastrucci, J. T. Nussbaum, N. B. Crane, and T. M. Weller, "Ka-band characterization of binder jetting for 3-D printing of metallic rectangular waveguide circuits and antennas," IEEE Transactions on Microwave Theory and Techniques, vol. 65, no. 9, pp. 3099–3108, 2017.

[20] U. Beaskoetxea, S. Maci, M. Navarro-Cía, and M. Beruete, "3-D-printed 96 GHz bull's-eye antenna with off-axis beaming," IEEE Transactions on Antennas and Propagation, vol. 65, no. 1, pp. 17–25, 2017.

[21] B. Zhang, W. Chen, Y. Wu, K. Ding, and R. Li, "Review of 3D printed millimeter-wave and terahertz passive devices," International Journal of Antennas and Propagation, vol. 2017, 2017.

[22] B. Zhang, Y. X. Guo, H. Zirath, and Y. P. Zhang, "Investigation on 3-D-printing technologies for millimeter-wave and terahertz applications," Proceedings of the IEEE, vol. 105, no. 4, pp. 723–736, 2017.

[23] A. Jammes, M. Petisme, and K. Staelens, "Metallization and selective metallization of silver using spraying," Surface Coating Technology, vol. 322, pp. 560–563, 2017.

[24] G. Stremsdoerfer, Techniques de l'ingénieur, JetMetal: Procédé innovant de dépôts métalliques ou d'alliages (Reference IN204, 10/09/2012), https://www.techniques-ingenieur.fr (accessed: 22/05/2020).

[25] Principles of Jet Metal Technology, https://www.jetmetal-tech.com/technologies?lang=en (accessed: 22/05/2020).

[26] S. Alkaraki, Y. Gao, M. O. Munoz Torrico, S. Stremsdoerfer, E. Gayets, and C. Parini, "Performance comparison of simple and low cost metallization techniques for 3D printed antennas at 10 GHz and 30 GHz," IEEE Access, vol. 6, pp. 64261–64269, 2018.

[27] S. Alkaraki, A. S. Andy, Y. Gao, *et al.*, "Compact and low-cost 3-D printed antennas metalized using spray-coating technology for 5G mm-wave communication systems," IEEE Antennas and Wireless Propagation Letters, vol. 17, no. 11, pp. 2051–2055, 2018.

[28] Nickel screening compound data sheet, https://uk.rs-online.com/web/p/shielding-aerosols/7259987/ (accessed: 22/05/2020).

[29] Silver conductive paint data sheet, https://electrolube.com/product/scpsilver-conductive-paint/ (accessed: 22/05/2020).

Chapter 5

Multiband millimetre-wave antennas for 5G and beyond

Masood Ur Rehman[1] *and Qammer H. Abbasi*[1]

The demand for wireless communications has increased by multiple folds in recent years due to advent of wearable sensors and Internet of Things. Body-centric networks (BCNs) for fifth generation (5G) and beyond systems, which make the use of in-body, on-body and near-body communication devices to gather, process and transmit vital information, have seen a massive growth as a result. These networks have a wide variety of applications ranging from healthcare to infotainment and navigation to sports [1–3]. Their popularity is such that estimates suggest a demand of 187.2 million wearable units annually by 2020 [4,5]. This along with high drive for increased network capacity and data rates has put a great amount of pressure on the present fourth-generation wireless communication systems. The wireless technology is, therefore, aiming to make its 5G operational by 2020, while modalities for beyond 5G networks are already in proposition (where millimetre-wave frequency bands of 28/38 GHz are potential candidates for the cellular communications, while 60 GHz band is spotted as the front-running contender for the BCNs) [1,6,7]. Millimetre-wave is attracting much interest as it offers a number of benefits, including smaller and lightweight components, wide bandwidth, high data rate and low interference to name a few [8].

Advent of new wireless technologies requires harmonious integration of multiple services that have different operating frequencies in a miniaturised wearable device impelled often by commercialisation and user acceptability factors. For example, coverage of different cellular bands, availability of near-field communications, provision of navigation and positioning and inclusion of Wi-Fi require a wearable device to operate at various LTE-A and 5G frequencies along with Wi-Fi, NFC, Bluetooth and GPS. Maintaining the overall size of the device within certain limitations while supporting these many wireless technologies is a discernible challenge in 5G and beyond networks. Enclosure of multiple antennas in these devices makes them bulkier. Therefore, using a single antenna operating at multiple frequencies is a good substitute. This chapter highlights

[1]James Watt School of Engineering, University of Glasgow, Glasgow, UK

basic requirements for these multiband antennas besides key design challenges. A millimetre-wave antenna is studied as the test case to understand their working and advantages.

5.1 Fundamentals of multiband antennas

Rapid growth of wireless technologies has on the one end raised the popularity of multiservice wearable devices but added small form factor, compact size, low cost and lightweight to the requirements on the other end in order to guarantee the ease of mobility and flexibility of use [9]. It has complicated the design of these devices multifold and in turn the antenna design which is an integral part of such devices. Advancements in miniaturisation of RF components have made smaller wireless devices realisable, but maintaining a high performance of the antennas with similar scale of reduction in operational frequencies, size, weight and cost is a tough nut to crack.

In this multiservice scenario, the use of one antenna per service is not a viable solution as it would not only increase the weight and cost of the device but will also result in a degraded performance due to mutual coupling between the different antennas. Use of a single antenna that can support multiple services is therefore a practicable consideration [10–17].

An antenna that simultaneously operates at more than one frequency is called a *multiband* antenna. For example, a single quad-band antenna can be used to provide coverage to GSM800/900/1800/1900 bands by resonating at 824–960-MHz and 1.71–1.99-GHz frequencies.

5.1.1 Multiband techniques

Multiband operation of the antennas can be achieved in many ways. Two basic techniques are as follows:

- use of higher order resonances and
- use of multiple resonant structures.

5.1.1.1 Using higher order resonances

A multiband operation can be realised by using higher order resonances. Various basic antenna structures, including monopoles, dipoles, patches, slots and dielectric resonators, exhibit this phenomenon. Modifications in these structures can result in a multiband operation. For example, a quarter-wavelength monopole has a resonant frequency f_1 when the electric field is minimum and current is maximum at the feed point. The structure would exhibit a similar profile of the electric field and current if its length is increased to $3\lambda/4$. The resonant frequency for the lengthened monopole would be $f_2 = 3f_1$ [18,19]. Similarly, patch structures with slots would offer multiple current paths resulting in multiple resonances. Realisation of higher order resonant modes for monopoles and slotted patch is illustrated in Figure 5.1.

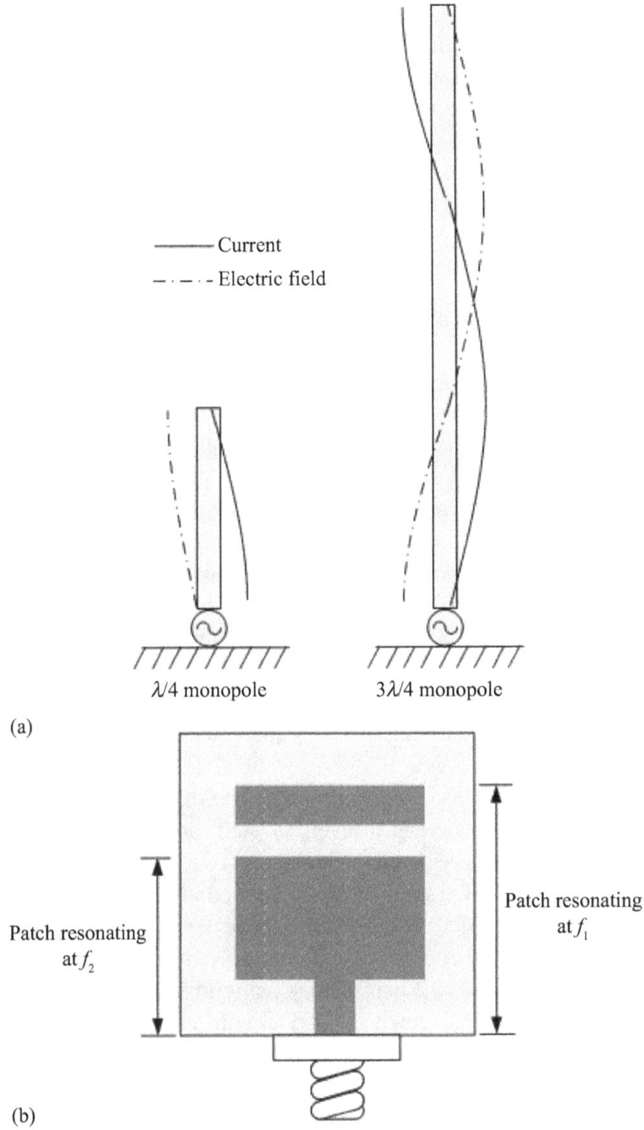

Figure 5.1 Use of higher order resonances to generate multiple resonant operation of multiband antenna: (a) monopoles with varying lengths and (b) patch with slots

5.1.1.2 Using multiple resonant structures

Another popular method to realise the multiband operation is the use of multiple resonant structures having different lengths and located in close proximity [18,19]. If we take the example of two monopoles, the element with larger length will resonate at a lower frequency (f_1) while the shorter element will resonate at a higher frequency ($f_2 = 3f_1$). Similar principal can be applied by using other basic structures, including loops, helices, rings, spirals, meander lines and patches, to contain the size and mutual coupling levels. Excitation of these structures can be achieved through a centrally fed or end-fire arrangement using a single feed. Another possibility is to excite through proximity coupling with one element being fed directly, while others act as the parasites being located in the near field of it. The two feeding methods are given in Figure 5.2.

5.2 Multiband antennas for millimetre-wave 5G and beyond networks

Efficient design and implementation of 5G and beyond BCNs depend highly on the working of antenna. Design of such antennas is very challenging due to added requirements of having low profile, lightweight, low power consumption and flexibility on top of typical specifications of impedance matching, bandwidth, radiation pattern, efficiency and gain [20,21]. Another aspect that needs careful considerations is impact of human body presence on the performance of these antennas as the human user is an integral part of BCNs [3]. It is a well-established fact that the presence of the human body in the vicinity of antennas introduces losses by detuning frequency and distorting radiation pattern as a result of electromagnetic absorptions in human body tissues and reflections/scattering from the body surface [3,22,23]. These effects are well characterised at the lower operational frequencies, but millimetre-wave frequencies need in-depth investigations [1,3, 24–26]. It is also important to design an antenna with higher gain at millimetre-wave frequencies to subdue increased levels of path loss [12].

Design of millimetre-wave antennas has received good degree of interest from the researchers. A number of studies have considered antenna design at 28, 38 and 60-GHz frequencies [27–36]. Patch antennas have become a popular choice for BCN devices for 5G and beyond systems due to ease of design, simple integration and cost-efficiency [37]. A 2 × 2 patch array using low-temperature cofired ceramic (LTCC) technology is presented in [27]. Single element consists of two circular patches with diagonal perturbations. A planar parallel stripline feeding network has been used to feed the array covering frequencies of 30–40 GHz with a maximum gain of 9.3 dBi. A dual-feed square loop antenna is proposed by Li *et al.* in [28]. The antenna operates at 38.5 GHz and offers a peak gain of 2.9 dBi. A scaled version to operate at 73.5 GHz is also discussed by the authors. A 3 × 3 series-fed patch array for 28-GHz band is presented in [29]. The patches are connected to each other in series, and 2D symmetric formation using high-impedance microstrip lines provides a gain of 14.9 dBi. Aliakbari *et al.* have proposed a single-feed circular polarised antenna for dual-band operation at 28 and 38-GHz bands [30]. The antenna makes the use of a composite structure combining two L-shaped slots etched on a square patch and achieves a peak gain of 4.5 dBi at 38 GHz. A proximity-fed elliptical patch etched non-concentrically in an elliptical slot located

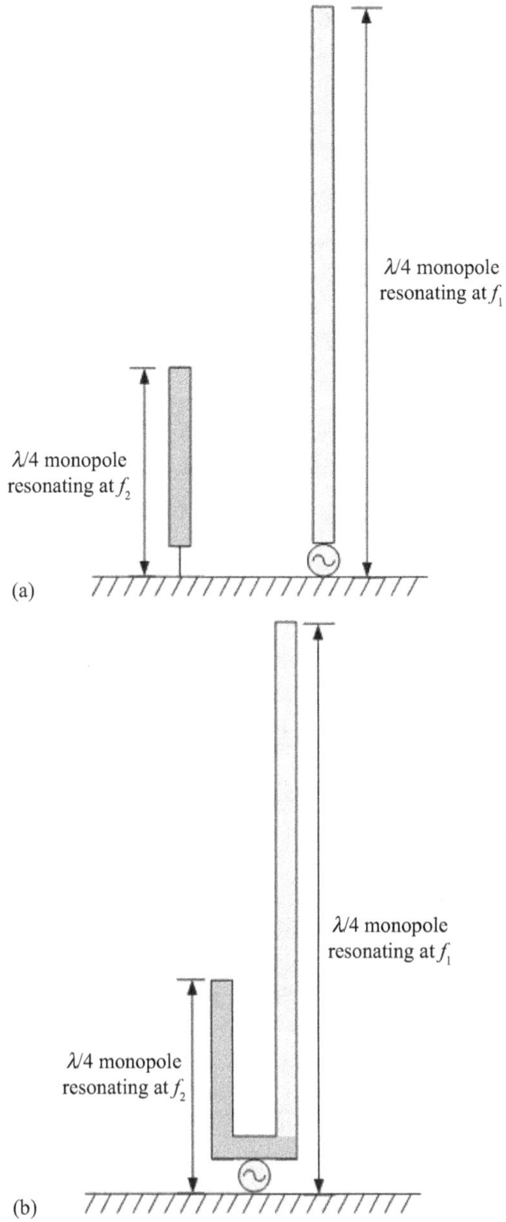

Figure 5.2 Use of multiple resonant structures to generate multiple resonant operation of multiband antenna through single feed and proximity feed: two monopoles with varying lengths using (a) single feed and (b) proximity coupling feed

at the antenna ground plane is used for dual-band operation at 28/38 GHz in [31]. An L-shaped slot is etched off in the feed line to create separation between the two bands. The antenna has a peak gain of 4.4 dBi.

A four-element linear array antenna working at 28/38-GHz bands fed via 1 × 4 Wilkinson power divider is discussed in [32]. The antenna employs a slotted substrate integrated waveguide (SIW) array structure and electromagnetic band gap ground plane to attain a peak gain of 11.9 dBi at 28 GHz. Xie *et al.* have designed a dielectric resonator antenna array with an 8-way Wilkinson power divider as the feed network [33]. The unit cell has a semi-cylindrical dielectric loaded on a rectangular patch. The array works at 28 GHz with a maximum gain of 15 dBi. A disc-like antenna having an electromagnetic coupling between a circular disc and the feeding pin is considered in [35]. This antenna is using substrate integrated horn for the feeding. Lu *et al.* have used three-stage ladder-shaped directors to enhance the directivity by 2.4–3.1 dB and bandwidth by 30% of a four-arm quasi-Yagi antenna [34]. A patch antenna on a flexible printed circuit board (FPCB) for 60-GHz BCNs' operation is discussed in [36]. This antenna employs an air gap between two layers of the FPCB to achieve a bandwidth of 9.8 GHz and offers a gain of 10.6 dBi in off-body (free-space) configuration.

These multiband antennas employ either complex geometries, array structures or require difficult feeding arrangements. Moreover, very few of these studies consider the antenna performance in wearable scenarios limiting their applicability in BCN applications. Continuous growth of the wireless personal area networks and wearables in 5G and beyond systems, however, necessitates novel solutions.

5.3 Design of multiband millimetre-wave antenna for 5G and beyond: a case study

Design case of a multiband low-profile and efficient antenna for BCNs operating at millimetre-wave frequencies for 5G and beyond networks is studied in detail here. To maintain small form factor while achieving a multi-frequency operation, the discussed antenna employs a simple slotted patch geometry fed through a microstrip line to cover three popular bands of 28, 38 and 61 GHz. The performance of the antenna in off-body, on-body and in-body scenarios is investigated through numerical analysis. The simulated results are verified through measurements. Efficient antenna performance in terms of impedance bandwidth and radiation characteristics in all three scenarios makes it a potentially viable solution for tri-band millimetre-wave BCNs.

5.3.1 Concept and topology

Requirements of low profile, simple geometry, ease of integration and low cost for the BCN antennas make a patch-type structure as one of the most feasible solutions. Such antennas usually have a ground plane that helps to minimise backward radiations towards the human user. On the other hand, substrate thickness provides a gap between the radiating element and the human user helping to slightly reduce the degrading effects on impedance matching and antenna efficiency [36]. A patch geometry combining a square patch and L- and F-shaped slots positioned close to edges of the patch is, therefore, used as a radiator in the discussed antenna, hence named L- and F-slot antenna (LFSA). This arrangement results in inductive and capacitive effects which generate the multi-frequency operation.

The antenna uses Rogers RT/duroid 5880 printed circuit board with a thickness of 0.254 mm. Substrate has a relative permittivity of 2.2 and tan $\sigma = 0.0009$. RT/duroid 5880 is suitable for high-frequency operation because of low dielectric loss minimising losses and dispersion at these frequencies [38]. The antenna is fed through a 50-Ω microstrip feed line, and overall size of the antenna is $5.1 \times 5 \times 0.254$ mm^3. The antenna is modelled and simulated using CST Microwave Studio® that employs finite integration technique to solve Maxwell's equations [39]. Geometry and dimensions of the modelled antenna are shown in Figure 5.3.

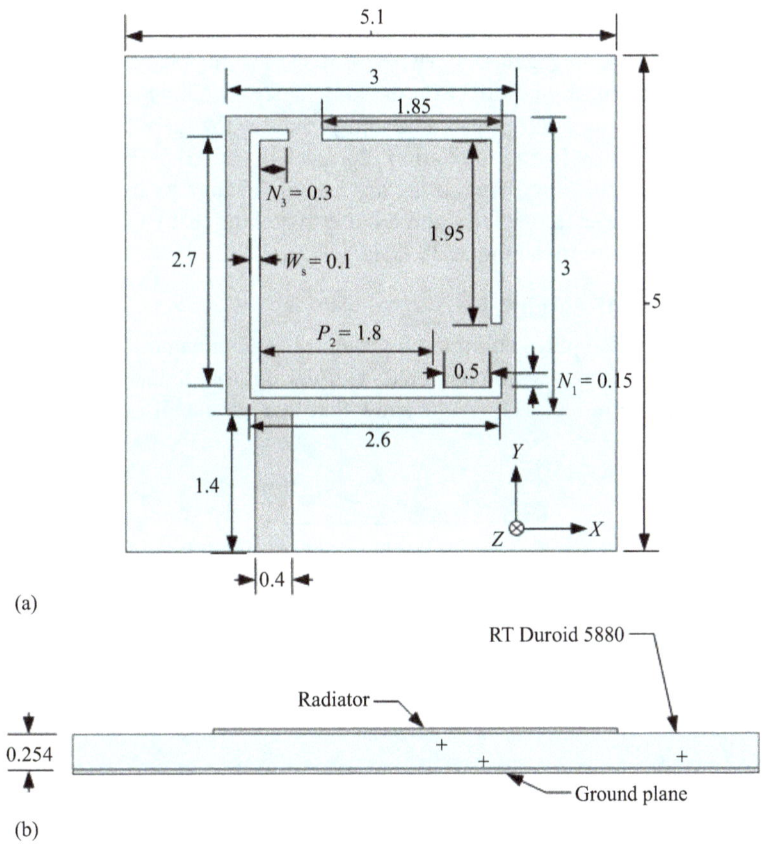

Figure 5.3 Geometry and dimensions of the tri-band LFSA (all units are in mm): (a) front and (b) side view [40]

Addition of the L- and F-shaped slots to the square patch together with off-centred microstrip feed improves impedance matching while reducing the overall size of the patch. The three structures also generate resonances at the three operating bands. A major portion of the antenna radiation comes from the slots. The current is mostly concentric on the edges of the patch and slots as evident from Figure 5.4. It also shows that the microstrip line provides feed balancing.

5.3.2 Parametric study

Optimisation of various structural parameters is necessary to realise an antenna structure having good impedance matching and radiation characteristics. Effects of three key elements, including width of the slots, length of notch 1 and length of notch 3 (refer to Figure 5.3) on the performance of the LFSA, are investigated to better understand the antenna-operating principle. Optimised values are taken as a reference.

5.3.2.1 Width of the slots (W_s)

The width of L- and F-shaped slots is one of the key design elements. Its impact on the impedance matching and bandwidth in terms of reflection coefficient (S_{11}) is studied and presented in Figure 5.5. These results show that change in the width of the slots has a very profound effect on the impedance matching and bandwidth of the LFSA. Narrower slots ($W_s = 0.05$ mm) reduce the gap between 28 and 38-GHz bands while losing operation completely at 61 GHz. Wider slots ($W_s = 0.2$ mm), on the other hand, bring an opposite change by increasing the gap between the two lower frequencies and a better matching at 61 GHz. Wider slots, however, tend to loose matching at 38 GHz.

5.3.2.2 Length of notch 1 (N_1)

Length of notch 1 also plays an important role in the antenna performance. Effects of changing this parameter are compared with the optimal value of $N_1 = 0.15$ in terms of S_{11} in Figure 5.6. The results show that this parameter is vital for the tri-

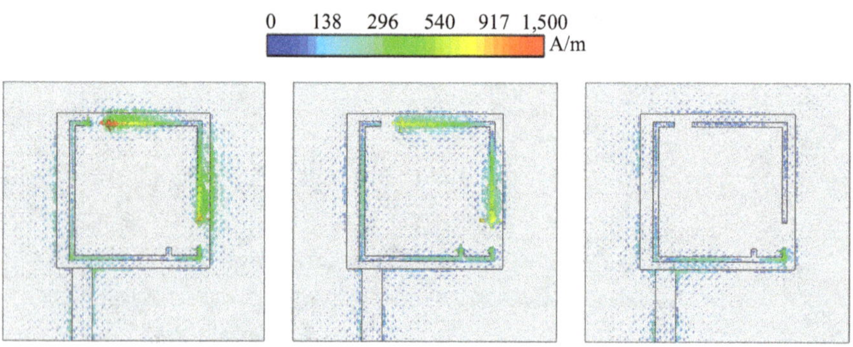

(a) (b) (c)

Figure 5.4 Surface current distribution on the LFSA at (a) 28 GHz, (b) 38 GHz and (c) 61 GHz

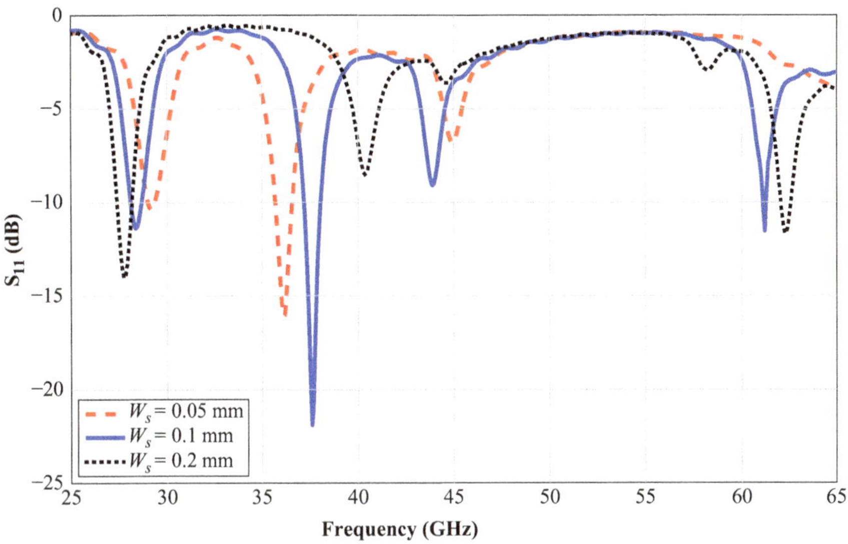

Figure 5.5 Effects of width of the slots on the performance of the LFSA [40]

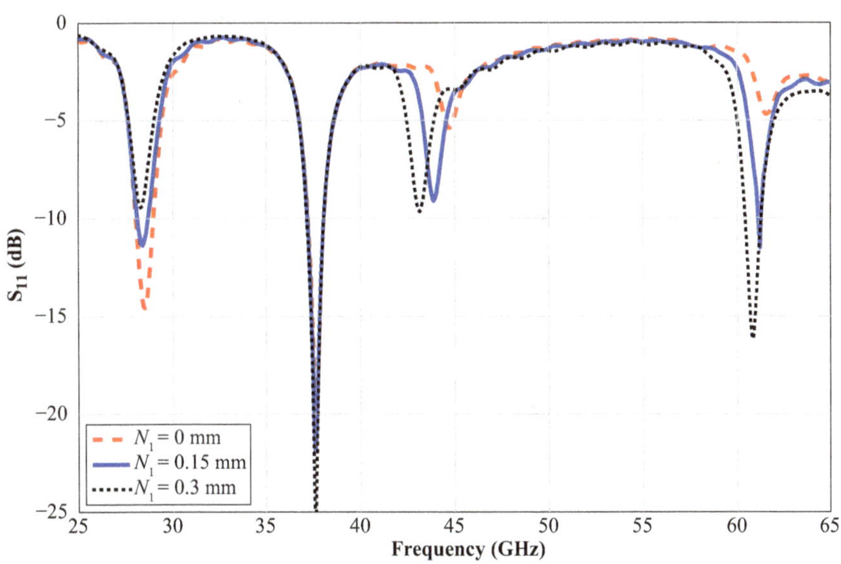

Figure 5.6 Effects of length of the notch 1 on the performance of the LFSA [40]

band operation of the LFSA. Any change in N_1 not only varies the bandwidth by shifting the resonance but also affects the impedance matching. A shorter notch improves the matching at 28 GHz with slight upward shift of the resonance and increases in the bandwidth at the expense of -10 dB at the two higher bands. On the contrary, a larger notch improves the matching but undergoes a downward resonance shift that is more significant at 61-GHz band.

5.3.2.3 Length of notch 3 (N_3)

Influence of the length of notch 3 on the LFSA performance is depicted through reflection coefficient response in Figure 5.7. Role of this parameter is also important in the matching and bandwidth but does not impact the tri-band operation as significantly as notch 1. A reduced length of (N_3) improves the matching at the three bands but with an upward shift in the resonance at 61 GHz. A larger notch 3 looses the matching at 28 and 61 GHz but maintains the performance at the middle band.

5.3.3 Antenna performance analysis

Impedance and radiation characteristics of the millimetre-wave LFSA are studied in off-body (free-space) and wearable configurations for 5G and beyond BCN applications throughout the range of the operating frequencies in terms of reflection coefficient response, radiation patterns, peak gain and radiation efficiency. Free-space results are verified through measurements of fabricated prototype of the antenna. These measurements are taken in an anechoic chamber using Anritsu's MS4647B vector network analyser.

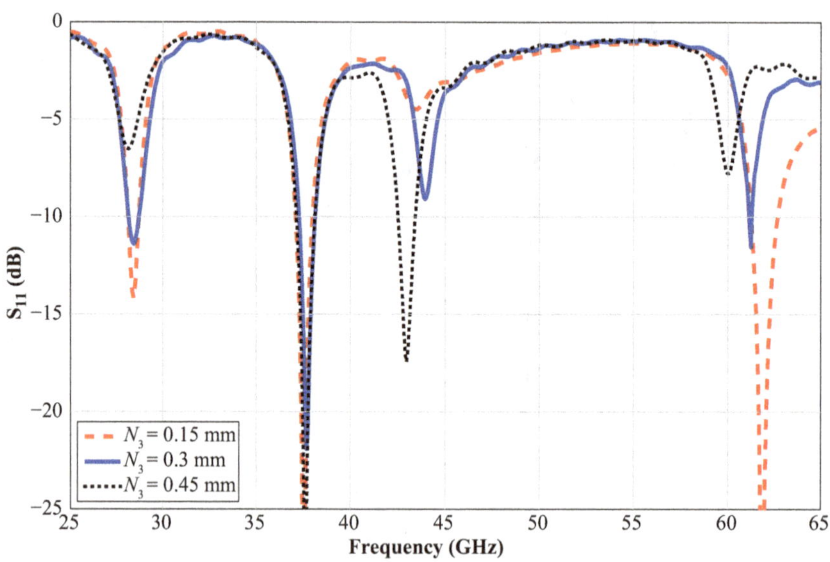

Figure 5.7 Effects of length of the notch 3 on the performance of the LFSA [40]

5.3.3.1 Off-body scenarios

This section discusses the antenna results being observed in off-body (free-space) scenarios.

The simulated and measured reflection coefficient response of the LFSA is given in Figure 5.8. The results show that the antenna achieves a good impedance matching at the three bands for the tri-band mmWave operation. It also exhibits good −10 dB bandwidth of 840, 370 and 900 MHz at 28, 38 and 61 GHz, respectively. A good agreement between simulation and measurement is observed. Discrepancies are associated with fabrication imperfections and spurious radiations from coaxial cable and connector.

Figure 5.9 compares the measured and simulated antenna radiation patterns at $\phi = 0°$ and $\phi = 90°$ planes at the three operating frequencies. A good radiation coverage with nearly omnidirectional pattern can be observed from these results. Higher frequency band brings slight distortions in $\phi = 90°$ pattern as expected. The simulation results agree well with the measurements.

Comparison of the simulated and measured peak gain and radiation efficiency of the LFSA at the three millimetre-wave bands of interest is presented in Figure 5.10. The antenna achieves maximum gain of 7.4 dBi in simulation and 7.22 dBi in measurement at 38 GHz. Minimum gain is being observed at 61 GHz with values of 6.32 dBi in simulation and 6.5 dBi in measurement.

A similar trend is noted for the antenna efficiency. Efficiency values are 89%, 92% and 87% in simulation at 28, 38 and 61 GHz, respectively. Measured efficiency values appear to be 86%, 91% and 85% at the three bands, respectively. A good agreement between the measured and simulated values is also evident.

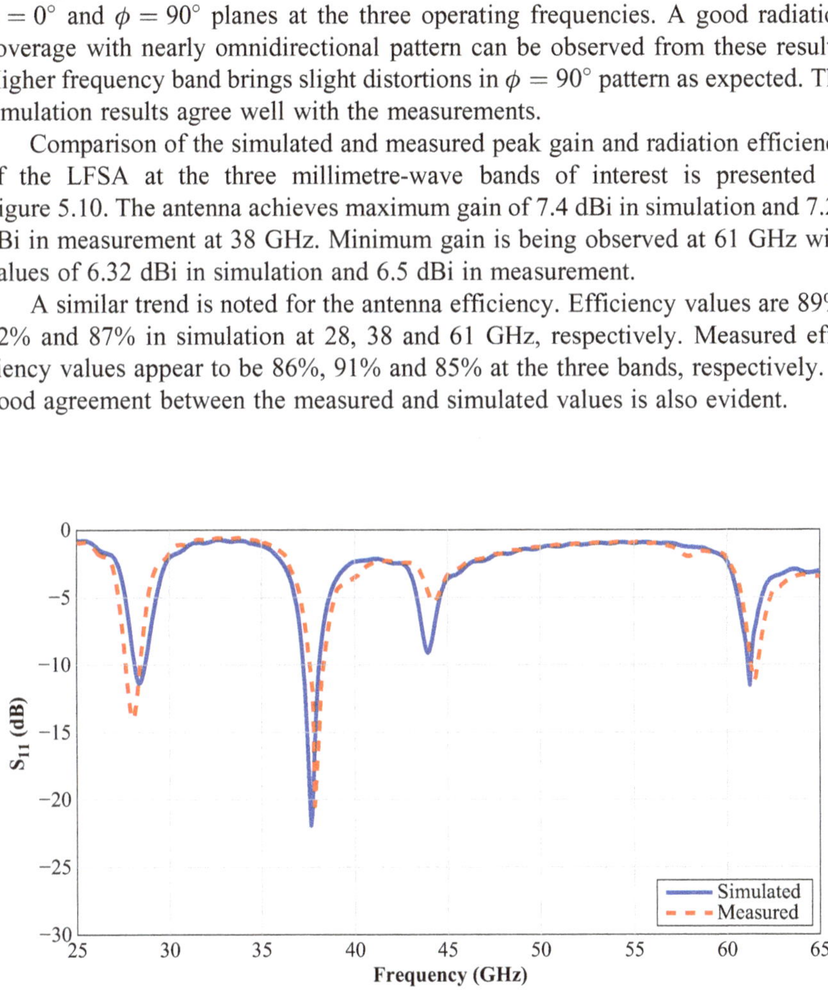

Figure 5.8 Simulated and measured reflection coefficient response of the LFSA [40]

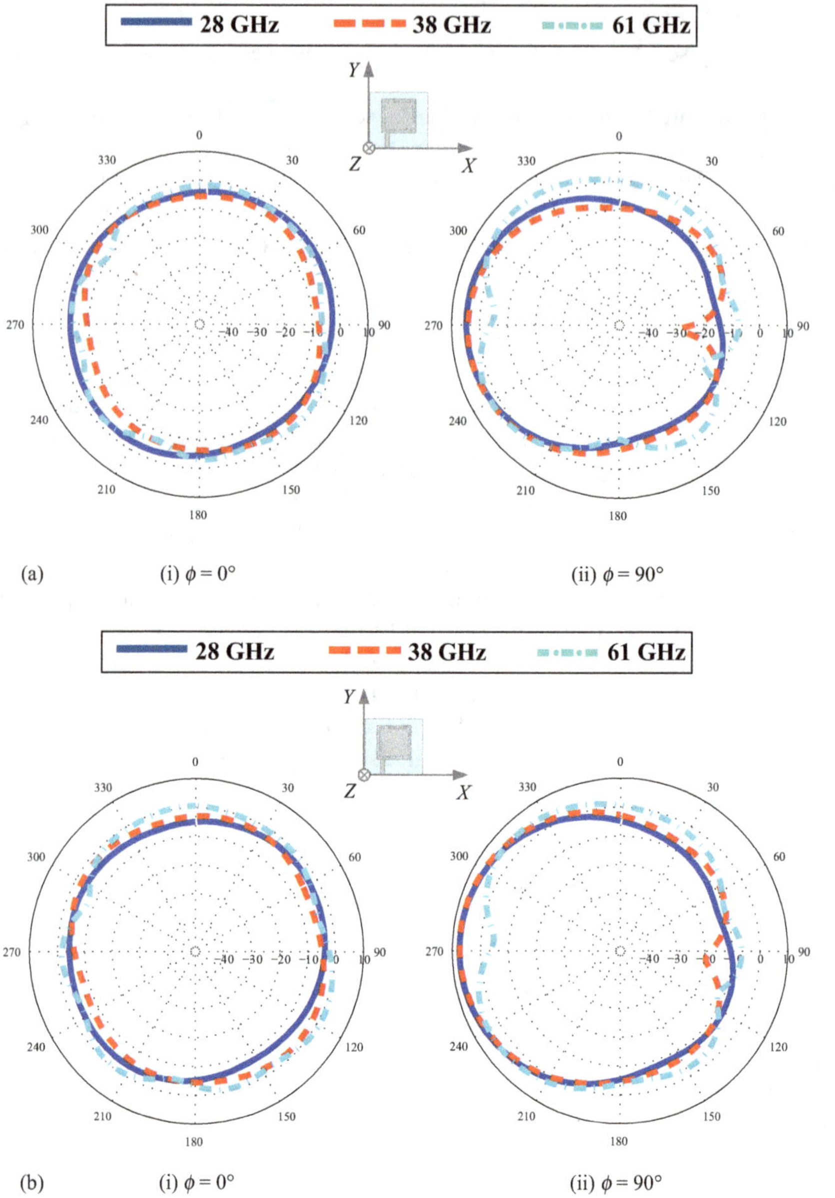

Figure 5.9 Simulated and measured radiation patterns of the tri-band LFSA at the three bands of interests of 28, 38 and 61 GHz: (a) simulated and (b) measured [40]

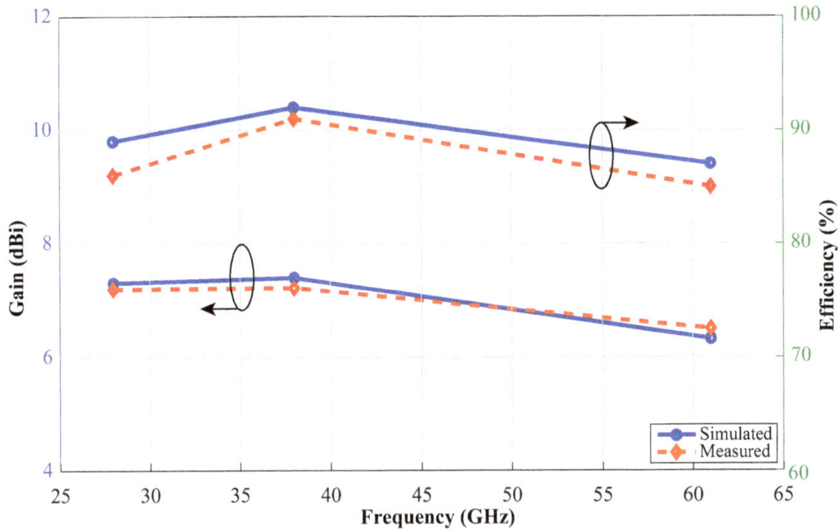

Figure 5.10 Simulated and measured gain and efficiency values for the LFSA [40]

5.3.3.2 Wearable scenarios

The human user is an integral part of the 5G and beyond BCNs, whose presence brings severe distortions in the antenna performance. Implanted sensors are also one of the fast growing BCN applications. It is, therefore, essential to analyse the BCN antennas in potential body-worn scenarios. This section presents a detailed investigation on the antenna impedance and radiation characteristics in on-body and in-body (implanted) configurations to establish its usability for BCN applications.

Maximum electromagnetic absorption by the human body at millimetre-wave frequencies takes place in the skin tissues due to a penetration depth of around 0.5 mm. A single-layer homogeneous torso phantom with electric properties of dry skin is, therefore, used in this investigation to study the effects of human body presence on the performance of the proposed tri-band antenna.

The skin-equivalent torso phantom has a realistic organ shape as it has been taken out of a high-resolution whole-body model as shown in Figure 5.11. Overall dimensions of the phantom are $288 \times 100 \times 40$ mm^3. An adaptive meshing scheme is used to reduce the overall number of cell volumes (voxels) in the computational domain. It helps to reduce the computation time for the model. The perfectly matched layer absorbing boundary conditions is used [39]. The model has a maximum resolution of 2 mm near the boundaries of the computational domain.

Dielectric properties of the dry skin tissue up to a frequency of 110 GHz have been well characterised through extrapolation of experimental data measured up to 20 GHz [41,42]. Debye model with a single relaxation time is used to model the permittivity data with good level of accuracy to the measured values in our

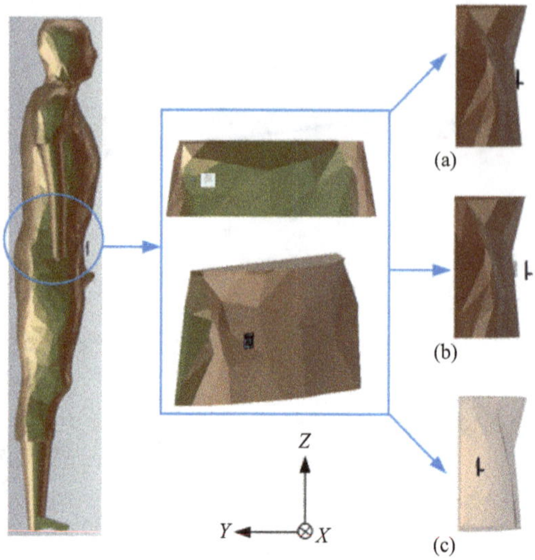

Figure 5.11 Complex permittivity of dry skin in 25–65-GHz band [40]

frequency range of interest [43]:

$$\varepsilon^* = \varepsilon_o(\varepsilon' - j\varepsilon'') = \varepsilon_o\left(\varepsilon_\infty + \frac{\varepsilon_s - \varepsilon_\infty}{1 + j\omega\tau}\right) \tag{5.1}$$

where ε_o is the free-space permittivity (8.85×10^{-12} F/m), ε_s represents the static permittivity, ε_∞ is the optical permittivity, ω depicts the angular frequency and τ is the relaxation time. Optimised values for the best fit in the frequencies of interest are $\varepsilon_s = 34.8$, $\varepsilon_\infty = 4.1$ and $\tau = 6.9 \times 10^{-12}$ [44]. Modelled permittivity values are plotted in Figure 5.12.

Two typical body-worn scenarios and one implanted configuration are studied. The considered body-worn scenarios include placement of the antenna at a gap of 1 mm ($d = 1$ mm) from the surface of the phantom to replicate antenna placed directly on-body and a gap of 5 mm ($d = 5$ mm) to incorporate clearance for the casing of the BCN device as shown in Figure 5.11. For the in-body (implanted) scenario, the LFSA is placed into the skin phantom at a depth of 20 mm.

Reflection coefficient responses and radiation patterns of the LFSA at the three operating frequencies in the off-body (free-space), on-body and implanted scenarios are compared numerically and illustrated in Figures 5.13 and 5.14. These results show that the antenna is quite resilient to the effects of human body presence. Though the presence of the phantom introduces impedance mismatch and detuning, the antenna successfully keeps its operation intact at 38 GHz for all on-body scenarios. The $d = 1$-configuration affects the matching and reduces the bandwidth at the three frequencies very significantly. The impact is more drastic at 28 and 61-GHz band where the matching is reduced to −6 dB due to change in electrical properties of the substrate. Resonance shift and mismatch are minimal at all three

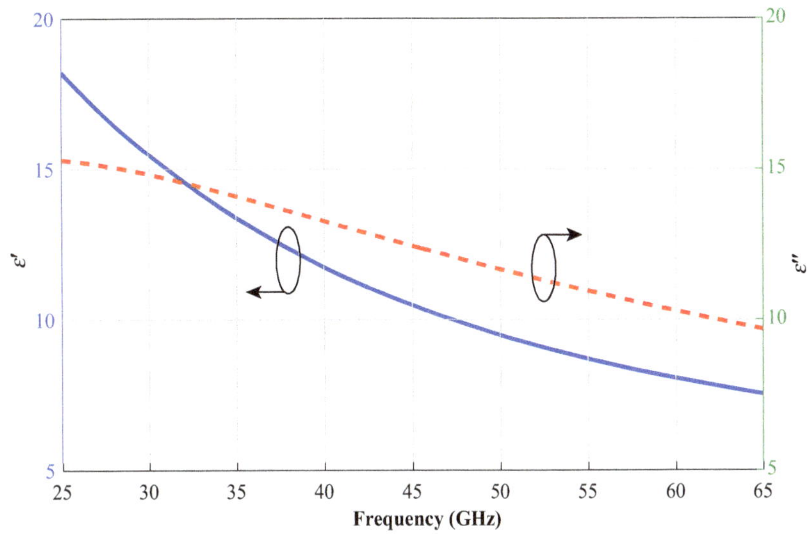

Figure 5.12 Structure of the realistic skin-equivalent human phantom taken out of a high-resolution human body model [40]: (a) d = 1 mm, (b) d = 5 mm and (c) implanted

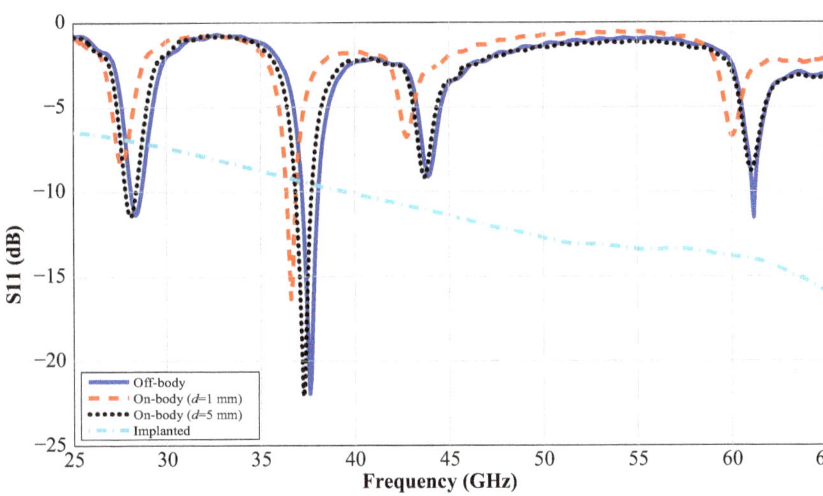

Figure 5.13 Reflection coefficient response of the LFSA in wearable and implantable scenarios in comparison with free-space performance [40]

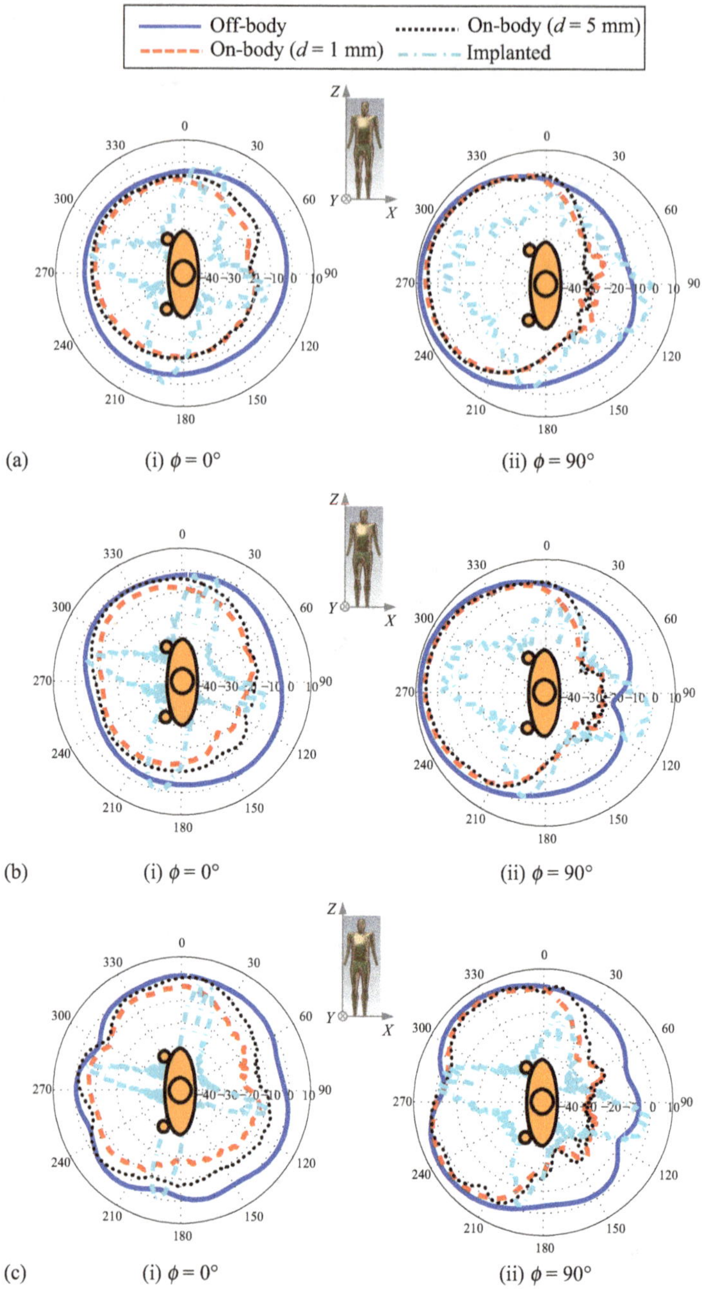

Figure 5.14 Radiation patterns of the tri-band LFSA in wearable and implantable scenarios at (a) 28, (b) 38 and (c) 61 GHz [40]

bands when antenna is placed at $d = 5$ mm. The LFSA performance is significantly affected when implanted in the body as matching is changed completely due to the presence of a high volume of body tissues. Operation at 28 GHz is the worst affected, but the antenna has retained −6-dB impedance matching which can be used in most of the commercial applications.

A similar trend is being observed for the antenna radiation pattern. The $d = 1$-mm configuration affects the radiation more out of the two on-body scenarios due to higher reflections from the body surface. Level of the pattern distortion reduces for $d = 6$ mm. In on-body placements, the human body backscatters energy changing induced current on the antenna structure, lowering back radiations due to absorptions and increasing maximum gain. In the implanted configuration, the LFSA loses its directivity completely due to the presence of the body mass in all directions. High absorptions are also visible. Overall, the radiation pattern is degraded more in $\phi = 0°$ plane due to the presence of higher body mass.

The peak gain values are observed to be 8.1, 8.3 and 7.5 dBi at 28, 38 and 61 GHz for $d = 1$ mm, while 7.2, 7.9 and 7.6 dBi for $d = 5$ mm placements, respectively. These values are 13.3, 14 and 14.3 dBi for the implanted configuration. The antenna radiation efficiency in the presence of the human body also decreases due to absorptions in the lossy human body tissues. Simulated efficiency varies from 54%, 60% and 58% for $d = 1$ mm to 55%, 62% and 60% for $d = 5$ mm at the three frequencies, respectively. In implanted situation, efficiency is observed to be 12%, 14% and 10%.

Specific absorption rate (SAR) averaged over 1 g of the body tissue volume for the proposed antenna having 0.5 W (peak) incident power has been evaluated to assess the level of exposure. For $d = 1$ mm, peak SAR values are noted to be 14.6, 13.9 and 18.1 W/kg, while for $d = 5$ mm, these values are observed to be 3.1, 2.5 and 4.4 W/kg at 28, 38 and 61 GHz, respectively. In the implanted configurations, SAR values are noted to be 310, 280 and 332 W/kg.

These results show that in typical wearable configurations, the LFSA can work very well. It is, therefore, suitable for body-worn applications where antenna is mounted on the circuit board or on top edge of the sensor/device. Implanted configurations need further exploration as to the best of our knowledge, no results at these frequencies are currently reported.

5.3.4 Comparative analysis

The LFSA is compared with related millimetre-wave antennas reported in open literature [27,28,30–32], in terms of structure, size and radiation characteristics to show its effectiveness. Particular emphasis is given on multiband solutions. Table 5.1 presents a summary of this comparative study.

It is evident that the LFSA exhibits a tri-band operation with gain and efficiency values higher than most of the available designs. Solutions that offer better gain employ complex array structures. Moreover, the LFSA makes the use of a simple geometry offering ease of fabrication via traditional low-cost techniques. It also has a size smaller than most of the reported designs except [31]. Furthermore, the LFSA works excellently in body-mounted configurations. This comparison

Table 5.1 Comparison between the LFSA antenna and state-of-the-art millimetre-wave designs

Parameter	Antenna designs					
	[27]	[28]	[30]	[31]	[32]	LFSA
Structure	LTCC array	Dual-feed square loop	L slot	Slotted patch	SIW array	L and F slots
Size (mm)	15.3 × 28.3 × 1.3	30 × 15 × 0.254	6.8 × 6.8 × 0.254	5 × 5 × 0.127	27.5 × 20 × 0.254	5.1 × 5 × 0.254
Operating frequency (GHz)	29.7–40	38.5	28	28	28	28
			38	38	38	38
						61
−10 dB BW (GHz)	10.3	4.6	0.85	5	0.32	0.84
			0.75	6	1.9	0.37
						0.9
Off-body gain (dBi)	9.3	2.9	4	3.6	11.9	7.2
			4.5	4.4	11.2	7.22
						6.5
On-body gain (dBi)	N/I*	N/I*	N/I*	N/I*	N/I*	8.1
						8.3
						7.5
Off-body efficiency (%)	50	90	N/R**	N/R**	N/R**	86
						91
						85
On-body efficiency (%)	N/I*	N/I*	N/I*	N/I*	N/I*	54
						60
						58

*The paper does not investigate human body effects on the antenna.
**The authors have not reported this value.

clearly shows the advantages of the LFSA for millimetre-wave BCN applications at 28, 38 and 61-GHz bands.

5.4 Summary

The multiband antennas offer a number of benefits to 5G and beyond systems. The antenna designers, however, face huge challenges due to growing demand of miniaturised devices as maintaining high-performance levels while offering small form factor is a complex task. This chapter has presented a detailed discussion on the need, requirements and operation of multiband antennas for 5G and beyond networks. The benefits of the multiband antennas are established as these antennas not only offer high performance and efficient operation at multiple frequencies but are also of small size. Performance of a slotted patch antenna consisting of L and F slots for the tri-band operation for wearable millimetre-wave devices operating at 28, 38 and 61 GHz is also studied. The antenna has been tested for off-body (free-space), on-body and implanted configurations and appeared to be good in containing the degradations caused by the presence of the human body for 5G and beyond networks.

References

[1] Q. H. Abbasi, M. Ur-Rehman, K. Qaraqe, and A. Alomainy, "Advances in body-centric wireless communication: Applications and state-of-the-art," *The IET*, London, 2016.
[2] X. Chen, C. G. Parini, B. Collins, Y. Yao, and M. Ur Rehman, "Antennas for global navigation satellite systems," *John Wiley and Sons, Inc.*, Chichester, 2012.
[3] P. S. Hall and Y. Hao, "Antennas and propagation for body-centric wireless networks (2nd Edition)," *Artech House*, London, 2012.
[4] Tactica, "Wearable device market forecasts," *Tactica*, 2015.
[5] L. Sorensen and K. Skouby, "User scenarios 2020—a worldwide wireless future," *WWRF*, 2009, http://www.wwrf.ch/files/publications.
[6] Z. Pi and F. Khan, "An introduction to millimeter-wave mobile broadband systems," *IEEE Communications Magazine*, vol. 49, no. 6, pp. 101–107, 2011.
[7] T. S. Rappaport, S. Sun, R. Mayzus, et al., "Millimeter wave mobile communications for 5G cellular: It will work!" *IEEE Access*, vol. 1, pp. 335–349, 2013.
[8] D. G. Fang, "Antenna theory and microstrip antennas," *CRC Press*, Boca Raton, FL, 2009.
[9] Z. N. Chen, "Antennas for portable devices," *John Wiley and Sons, Inc.*, New York, NY, 2007.
[10] T. Zhang, R. Li, G. Jin, G. Wei, and M. Tentzeris, "A novel multiband planar antenna for GSM/UMTS/LTE/Zigbee/RFID mobile devices," *IEEE Transactions on Antennas and Propagation*, vol. 59, no. 11, pp. 4209–4214, 2011.

[11] W. J. Liao, S. H. Chang, and L. K. Li, "A compact planar multiband antenna for integrated mobile devices," *Progress in Electromagnetics Research*, vol. 109, pp. 1–16, 2010.

[12] M. M. M. Ali, O. Haraz, S. Alshebeili, and A. R. Sebak, "Broadband printed slot antenna for the fifth generation (5G) mobile and wireless communications," in *2016 17th International Symposium on Antenna Technology and Applied Electromagnetics (ANTEM)*, July 2016, pp. 1–2.

[13] S.-B. Chen, Y.-C. Jiao, W. Wang, and F.-S. Zhang, "Modified T-shaped planar monopole antennas for multiband operation," *IEEE Transactions on Microwave Theory and Techniques*, vol. 54, no. 8, pp. 3267–3270, 2006.

[14] Q. Rao and W. Geyi, "Compact multiband antenna for handheld devices," *IEEE Transactions on Antennas and Propagation*, vol. 57, no. 10, pp. 3337–3339, 2009.

[15] O. Falade, M. Ur Rehman, Y. Gao, X. Chen, and C. Parini, "Single feed stacked patch circular polarized antenna for triple band GPS receivers," *IEEE Transactions on Antennas and Propagation*, vol. 60, no. 10, pp. 4479–4484, 2012.

[16] J.-W. Kim, T.-H. Jung, H.-K. Ryu, J.-M. Woo, C.-S. Eun, and D.-K. Lee, "Compact multiband microstrip antenna using inverted-L- and T-shaped parasitic elements," *IEEE Antennas and Wireless Propagation Letters*, vol. 12, pp. 1299–1302, 2013.

[17] A. Mehdipour, A. Sebak, C. Trueman, and T. Denidni, "Compact multiband planar antenna for 2.4/3.5/5.2/5.8-GHz wireless applications," *IEEE Antennas and Wireless Propagation Letters*, vol. 11, pp. 144–147, 2012.

[18] Y. Huang and K. Boyle, "Antennas: From theory to practice," *Wiley*, Chichester, 2008.

[19] M. A. Matin (Ed), "Wideband, multiband, and smart reconfigurable antennas for modern wireless communications," *IGI Global*, Hershey, PA, 2015.

[20] M. Ur Rehman, Y. Gao, Z. Wang, *et al.*, "Investigation of on-body bluetooth transmission," *IET Microwaves, Antennas and Propagation*, vol. 4, no. 7, pp. 871–880, 2010.

[21] G. A. Conway and W. G. Scanlon, "Antennas for over-body-surface communication at 2.45 GHz," *IEEE Transactions on Antennas and Propagation*, vol. 57, no. 4, pp. 844–855, 2009.

[22] Q. Abbasi, A. Alomainy, and Y. Hao, "Characterization of MB-OFDM-based ultrawideband systems for body-centric wireless communications," *IEEE Antennas and Wireless Propagation Letters*, vol. 10, pp. 1401–1404, 2011.

[23] H. Wong, W. Lin, L. Huitema, and E. Arnaud, "Multi-polarization reconfigurable antenna for wireless biomedical system," *IEEE Transactions on Biomedical Circuits and Systems*, vol. 11, no. 3, pp. 652–660, 2017.

[24] M. Ur-Rehman, Q. Abbasi, M. Akram, and C. Parini, "Design of band-notched ultra wideband antenna for indoor and wearable wireless communications," *IET Microwaves, Antennas Propagation*, vol. 9, no. 3, pp. 243–251, 2015.

[25] C. C. Y. Poon, B. P. L. Lo, M. R. Yuce, A. Alomainy, and Y. Hao, "Body sensor networks: In the era of big data and beyond," *IEEE Reviews in Biomedical Engineering*, vol. 8, pp. 4–16, 2015.

[26] A. Pellegrini, A. Brizzi, L. Zhang, et al., "Antennas and propagation for body-centric wireless communications at millimeter-wave frequencies: A review [wireless corner]," *IEEE Antennas and Propagation Magazine*, vol. 55, no. 4, pp. 262–287, 2013.

[27] M. Du, Y. Dong, J. Xu, and X. Ding, "35-GHz wideband circularly polarized patch array on LTCC," *IEEE Transactions on Antennas and Propagation*, vol. 65, no. 6, pp. 3235–3240, 2017.

[28] S. Li, T. Chi, Y. Wang, and H. Wang, "A millimeter-wave dual-feed square loop antenna for 5G communications," *IEEE Transactions on Antennas and Propagation*, vol. 65, no. 12, pp. 6317–6328, 2017.

[29] M. Khalily, R. Tafazolli, T. A. Rahman, and M. R. Kamarudin, "Design of phased arrays of series-fed patch antennas with reduced number of the controllers for 28-GHz mm-wave applications," *IEEE Antennas and Wireless Propagation Letters*, vol. 15, pp. 1305–1308, 2016.

[30] H. Aliakbari, A. Abdipour, R. Mirzavand, A. Costanzo, and P. Mousavi, "A single feed dual-band circularly polarized millimeter-wave antenna for 5G communication," in *2016 10th European Conference on Antennas and Propagation (EuCAP)*, April 2016, pp. 1–5.

[31] M. M. M. Ali, O. Haraz, and S. Alshebeili, "Design of a dual-band printed slot antenna with utilizing a band rejection element for the 5G wireless applications," in *2016 IEEE International Symposium on Antennas and Propagation (APSURSI)*, June 2016, pp. 1865–1866.

[32] N. Ashraf, O. Haraz, M. A. Ashraf, and S. Alshebeili, "28/38-GHz dual-band millimeter wave SIW array antenna with EBG structures for 5G applications," in *2015 International Conference on Information and Communication Technology Research (ICTRC)*, May 2015, pp. 5–8.

[33] R. Xie, J. Cao, R. Wang, X. Wang, Z. Xu, and S. Zhu, "A study of dielectric resonator antenna array applied to 5G communication system," in *2016 Progress in Electromagnetic Research Symposium (PIERS)*, August 2016, pp. 454–457.

[34] L. Lu, K. Ma, F. Meng, and K. S. Yeo, "Design of a 60-GHz quasi-Yagi antenna with novel ladder-like directors for gain and bandwidth enhancements," *IEEE Antennas and Wireless Propagation Letters*, vol. 15, pp. 682–685, 2016.

[35] J. Puskely, M. Pokorny, J. Lacik, and Z. Raida, "Wearable disc-like antenna for body-centric communications at 61 GHz," *IEEE Antennas and Wireless Propagation Letters*, vol. 14, pp. 1490–1493, 2015.

[36] M. Ur-Rehman, N. A. Malik, X. Yang, Q. H. Abbasi, Z. Zhang, and N. Zhao, "A low profile antenna for millimeter-wave body-centric applications," *IEEE Transactions on Antennas and Propagation*, vol. 65, no. 12, pp. 6329–6337, 2017.

[37] J. R. James, P. S. Hall, and C. Wood, "*Microstrip antenna theory and design*," IET, London, 1981.
[38] "RT/Duroid® 5880 laminates," 2017 http://ogerscorp.com/acs/products/32/rt-duroid-5880-laminates.aspx.
[39] *CST Microwave Studio® 2017 User Manual.*
[40] M. Ur-Rehman, M. Adekanye, and H. T. Chattha, "Tri-band millimetre-wave antenna for body-centric networks," *Nano Communication Networks*, vol. 18, pp. 72–81, 2018.
[41] S. Gabriel, R. W. Lau, and C. Gabriel, "The dielectric properties of biological tissues: II. Measurements in the frequency range 10 Hz to 20 GHz," *Physics in Medicine and Biology*, vol. 41, no. 11, pp. 2251–2269, 1996.
[42] "Dielectric properties," *ITIS website,* http://http://www.itis.ethz.ch/virtual-population/tissue-properties/database/dielectric-properties.
[43] O. G. Martinsen, S. Grimmes, and H. P. Schwan, "Interface phenomena and dielectric properties of biological tissue," in: "*Encyclopedia of Surface and Colloid Science*," *Marcel Dekker*, New York, NY, pp. 2643–2652, 2002.
[44] N. Chahat, M. Zhadobov, and R. Sauleau, "Broadband tissue-equivalent phantom for BAN applications at millimeter waves," *IEEE Transactions on Microwave Theory and Techniques*, vol. 60, no. 7, pp. 2259–2266, 2012.

Chapter 6
On-chip antenna: challenges and design considerations

Atif Shamim[1] *and Haoran Zhang*[1]

This chapter focuses on on-chip antennas (OCAs). The challenges associated with OCA design, implementation, and characterization are discussed in detail. In addition, this chapter highlights the research work that has been conducted in this field to date. This chapter concludes by proposing future research directions in the field of OCA.

6.1 Introduction

Presently, the most common approach for the realization of a wireless system is through the assembly of discrete components, such as digital integrated circuits (ICs), mixed-signal ICs, radio-frequency (RF) ICs (RFICs), and passive components, on printed circuit boards (PCB), as displayed in Figure 6.1. This arrangement is known as system-on-board (SoB) [1], and there are several problems associated with this conventional SoB approach. Since separate ICs are mounted on the PCB, each IC requires a bulky and costly IC package for physical protection and input–output connections from the chip set to the system board. This involves the use of on-chip bond pads and bond wires, which limits the chip's performance and reliability. The overall system in this approach occupies a large size, and the realization cost is high. In addition, since various types of ICs are used for one system, different design and fabrication processes are required for the realization of RFICs, digital ICs, and mixed-signal ICs, which unfortunately creates issues of compatibility and reliability. Other than the ICs, antennas are designed on a separate PCB, which must be integrated with the PCB-containing ICs and other discrete components. In some cases, the antenna can be on the same PCB as the ICs. This PCB antenna is connected to the RFIC through bond wires and bond pads, which unfortunately introduces additional RF insertion loss and transmission uncertainty because the RF properties of these bond wires and bond pads are not well characterized, especially for higher frequencies. To resolve the aforementioned

[1]Electrical Engineering Program, King Abdullah University of Science and Technology (KAUST), Thuwal, Saudi Arabia

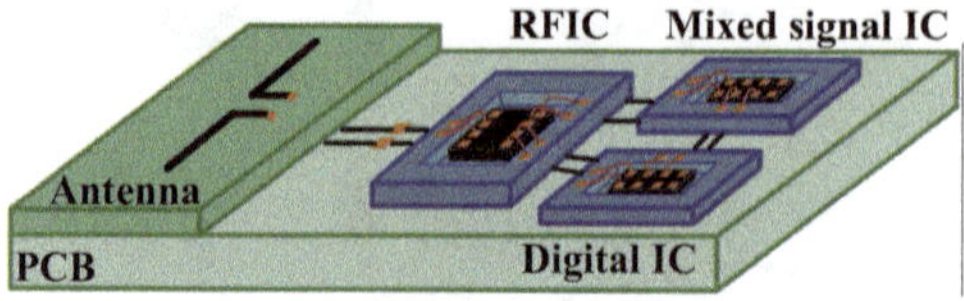

Figure 6.1 System PCB integrated with RFIC, mixed-signal IC, digital IC, and separate PCB antenna

problems associated with SoB, the best solution is to combine all discrete components on a single chip, an arrangement known as system-on-chip (SoC) [2].

Over the past few decades, the aggressive miniaturization of complementary metal oxide semiconductor (CMOS) transistors, followed by Moore's law, has promoted the development of ICs into a higher integration level, which makes it possible to integrate the complete system on a single chip, realizing an SoC. The highly integrated SoC provides the benefits of low cost, small physical size, light weight, and low power consumption, which facilitates the realization of portable and highly compact devices, such as Internet of Things (IoT) sensors and compact radar systems. Due to the rapid development of ICs, the digital module, analog module, and RF front-end can be integrated on a single chip. However, some issues remain in combining all different modules in one chip. One major issue is that each circuit module requires its own specialized process. If all the circuit modules are integrated in one chip, the process might not be suitable for all of them being in the best performance. Nonetheless, it is feasible to integrate all the circuit modules in a single chip; however, the bottleneck for a real wireless SoC is the realization of an on-chip antenna (OCA). The typical chip sizes are on the order of a few millimeter squares to around 100 mm^2. However, the antenna size is proportional to the wavelength, which is quite large at low frequencies. As a result, the typical antenna size is much larger than the size of the available chip space for low-frequency applications. For example, a typical WiFi antenna length is approximately 60 mm, so it cannot be efficiently realized on a chip. Moreover, after the integration of all the circuit modules, there is limited space for the realization of OCA. Over the past two decades, driven by the high demands of high bandwidth and system compactness, tremendous efforts have been made in the development of millimeter-wave (mm-wave) and sub-terahertz applications. The use of mm-wave and sub-terahertz frequency bands has significantly decreased the antenna sizes (due to shorter wavelengths), providing the possibility of implementing the antenna on the chip. Figure 6.2 illustrates the concept of OCA integration with ICs. Since the OCA can be fed by the circuit through an on-chip transmission line, the influence of bond wires and bond pads can be eliminated. Furthermore, integrating OCA and circuits in the same chip removes the restriction of 50 Ω impedance matching. Instead, more flexible methods of impedance matching can be applied, providing IC and antenna designers with greater design freedom. However, it is difficult to realize an OCA with high performance in a typical CMOS process. Many

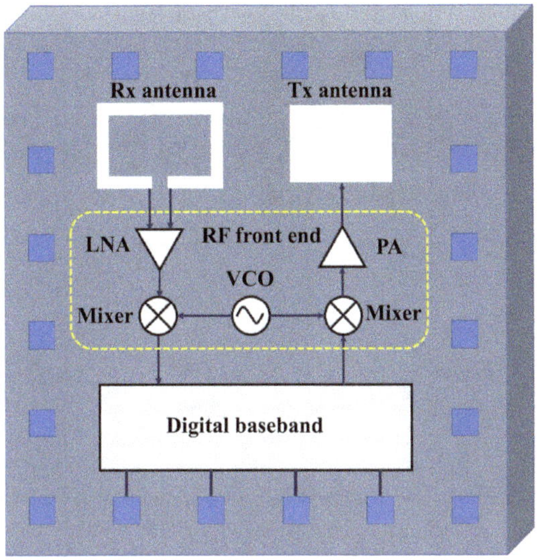

Figure 6.2 Concept of system-on-chip

challenges related to the OCA still must be tackled. These issues are discussed in detail in the following section.

6.2 On-chip antenna challenges

As mentioned in the previous section, OCA presents an excellent solution for the applications that require higher integration levels and lower cost. However, it is feasible to realize the OCA in the commercial CMOS processes, and there remain many challenges related to the OCA. In this section, challenges of OCA are studied from the perspectives of design, layout, and characterization.

6.2.1 Incompatible CMOS stack-up

Bulk silicon is the most commonly used substrate in standard CMOS processes. It has low resistivity and is highly suitable for the circuits. Its abundance ensures a lower cost for CMOS processes. However, the bulk material properties of silicon and the CMOS stack-up are not suitable for efficient realization of OCAs.

A typical CMOS stack-up is depicted in Figure 6.3. A 300–500 μm thick silicon substrate is located at the bottom. On top of the silicon substrate, thin layers of dielectrics (around 7–15 μm thick) are embedded with 6–9 metal layers, each of which has a thickness approximately 500 nm, except for the top metal layer, which has a thickness of 1–2 μm. The metal layers are typically used to realize the passive components, such as capacitors, inductors, and transmission lines. The interconnections are also implemented in these metal layers and use the interlayer vias. Among all the metal layers, the top layer is the best choice for the implementation

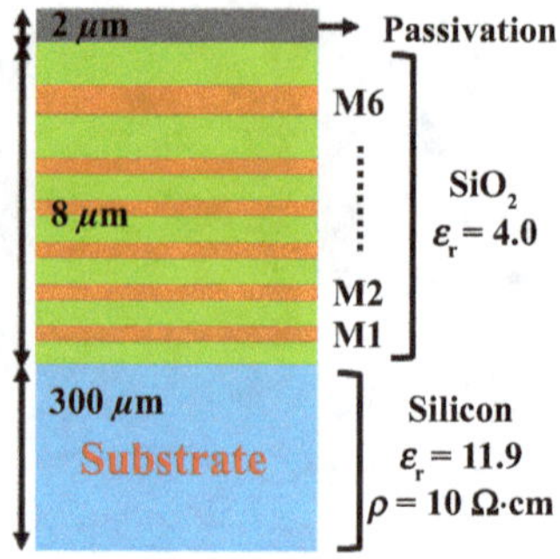

Figure 6.3 Silicon-based CMOS stack-up

of OCA because its relatively higher metal thickness can minimize the skin depth effect, especially for the antennas in mm-wave and sub-terahertz frequency bands. The dielectric layers (typically SiO_2) between the metal layers provide good insulation. On top, the chip is typically covered with a thin layer of silicon nitride or polyimide for protection, which is known as the passivation layer.

6.2.1.1 Low radiation efficiency

Low radiation efficiency is the most critical issue for the effective realization of an OCA. The primary reason for low radiation efficiency is the use of the low-resistivity silicon substrate in commercial CMOS processes.

As a semiconductor, the silicon substrate has a low resistivity of around 10–15 Ω cm, which is an advantage for IC design, since low resistivity is helpful for avoiding latch-up issues. Latch-up occurs due to current conduction by parasitic negative-positive-negative (NPN) or PNP devices formed in the substrate. By using low-resistivity substrates, the gain of these devices is reduced, which reduces the likelihood of latch-up [3]. However, the use of a low-resistivity substrate is catastrophic for OCA design. When the antenna is closely placed on top of this silicon substrate with low resistivity, the electromagnetic (EM) waves find a low-resistive path in the silicon substrate. Thus, the silicon substrate absorbs most of the RF power, which is dissipated in the silicon substrate in the form of heat [4] rather than being radiated in the air, as displayed in Figure 6.4. As a result, OCA offers very low radiation efficiency.

In addition to the property of high loss, the silicon substrate also has a high-permittivity value of 11.9. This results in more EM fields in the substrate compared with the free space radiation. In fact, the ratio of the power that is trapped in the substrate compared with the power that radiates into the air can be calculated approximately using the following equation [5]:

$$\frac{P_{air}}{P_{sub}} = \frac{1}{\varepsilon_r^{3/2}} \tag{6.1}$$

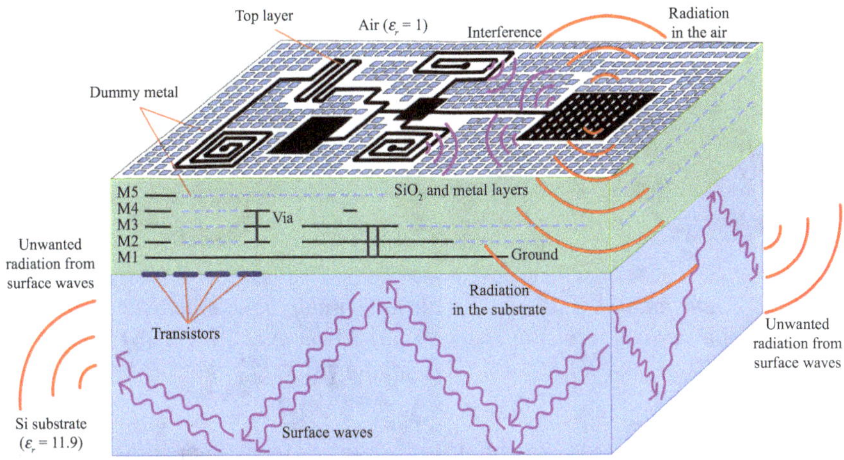

Figure 6.4 Overview of system-on-chip and on-chip antenna challenges

where P_{sub} is the power trapped/absorbed in the substrate, and P_{air} is the power that is radiated in the air. According to the equation, 97% of the total power is coupled to the typical bulk silicon substrate, while only 3% RF power is radiated in the air. The power coupled in the substrate is mostly dissipated as heat. Some power is radiated from the substrate in unintended directions; however, this is an issue that will be discussed in the following sub-section.

6.2.1.2 Surface waves

Other than the low radiation efficiency, the OCA suffers from the generation of surface waves due to the high permittivity of the silicon substrate and its large thickness. These high-order surface waves distort the desired antenna radiation pattern. The typical thickness of the silicon used in the CMOS foundries is in the range of 300–700 μm, which is considered electrically thick for mm-wave frequency applications. In addition, the high permittivity of the silicon substrate reduces the guided wavelength in the substrate. As a result, high-order surface waves can be generated in this silicon substrate. The antenna substrate always supports one surface wave mode at least. This surface wave mode is known as the TM_0 mode, which does not have a cutoff frequency, so they are always being excited. The following equation reveals the surface wave cutoff frequency for higher modes:

$$f_{cutoff} = \frac{nc}{4h\sqrt{\varepsilon_r - 1}} \tag{6.2}$$

where n is the surface wave mode number ($n = 0, 2, 4\ldots$ for TM_n modes; $n = 1, 3, 5\ldots$ for TE_n modes), c is the speed of light in free space, h is the thickness of the substrate, and ε_r is the substrate permittivity. According to the equation, the larger substrate thickness and higher permittivity lead to a lower cutoff frequency; as a

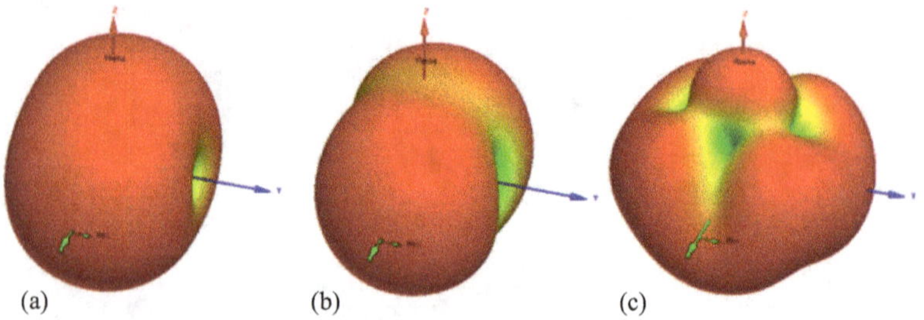

Figure 6.5 3D radiation patterns of a planar on-chip bowtie antenna at 77 GHz for various silicon thicknesses: (a) 20 μm thick silicon, (b) 50 μm thick silicon, and (c) 500 μm thick silicon [6]

result, more surface waves with higher modes can be excited. For example, a 77 GHz OCA can excite a higher order surface wave—TE_1 mode, which has a cutoff frequency of 56.8 GHz—if the antenna has a silicon substrate with a permittivity of 11.9 and thickness of 400 μm.

As displayed in Figure 6.4, surface waves can radiate at substrate side surfaces and surface discontinuities, such as truncations and cavities in the substrate, which cause changes in the radiation pattern and polarization characteristics. An investigation of how the surface waves distort the radiation pattern of OCA has been conducted in [6]. As presented in Figure 6.5, a satisfactory donut-shaped radiation pattern is obtained when the silicon substrate thickness is 20 μm. As the silicon substrate thickness increases, the radiation pattern begins to distort. Especially when thickness is increased to 500 μm, high-order surface waves are generated in the substrate, which completely distort the original radiation pattern.

6.2.2 Co-design of circuits and on-chip antenna

In an on-chip implementation, the antenna and circuits are placed very close to each other due to the limited space available on the chip. Thus, their influence on each other's performance can no longer be ignored. The coupling between the OCA and the other on-chip components, particularly on-chip inductor, capacitors, and RF interconnects, can be detrimental for system performance. Therefore, a codesign strategy for the circuits and antenna must be adapted to eliminate surprises after the chip has been fabricated; however, reliable and efficient simulation mechanisms, for the codesign approach, are still far from optimal. In this section, the issue of cross talk between the OCAs and circuits is discussed first. Afterward, the co-simulation approach and its challenges are presented.

6.2.2.1 Coupling effects between OCAs and circuits

As previously mentioned, the antenna is typically realized on a separate PCB; therefore, the cross talks or coupling effects between circuits and antennas are not significant due to the large distance between them. However, the SoC is a highly integrated environment in which the OCAs and circuits are integrated in a single chip with extremely small spacing between them. In this case, the cross talks or coupling effects cannot be ignored. Since the field of OCA is relatively young, few papers have investigated the coupling effects between the OCAs and on-chip circuits [7,8].

Unwanted coupling effects can distort the OCA's input impedance matching and radiation performance, such as radiation efficiency and radiation patterns. The close-by metals from the circuit part can alter the OCA's current distribution; thus, the OCA's input impedance and radiation performance can be affected. The circuit part primarily comprises active components, passive components, and some other elements, such as bond pads and dummy metal fills. The feature sizes of active components, such as transistors and diodes, are very small; therefore, they do not strongly impact the OCA's performance; however, the passive components, such as inductors, capacitors, and transmission lines, have relatively large sizes (proportional to the operation frequency), and their presence in the close vicinity of the OCA is detrimental. In addition, the side length of a bond pad is typically larger than 50 μm, which is considered large, particularly for an mm-wave OCA. On the other hand, dummy metal fills are used in all metal layers to fulfill the specific layout rules defined by the CMOS foundries. Though dummy metal fills have relatively small dimensions, they are placed in large numbers in close proximity and thus cause unwanted coupling effects. Figure 6.4 illustrates the coupling effects between an OCA and some close-by metals.

Negative effects can occur on the circuits' performance, as well, such as the frequency response of an amplifier's gain being distorted by unwanted coupling. Other than the coupling of EM energy through air, coupling can also occur through the substrate. The noise currents, generated by OCA, are injected into silicon substrate [9]. When the noise currents in the substrate reach the active region of the circuits, the circuit performance will be affected [10], as presented in Figure 6.6. Conversely, the noise currents generated by the circuits' active region can flow toward the critical area below the OCA and influence the antenna's performance.

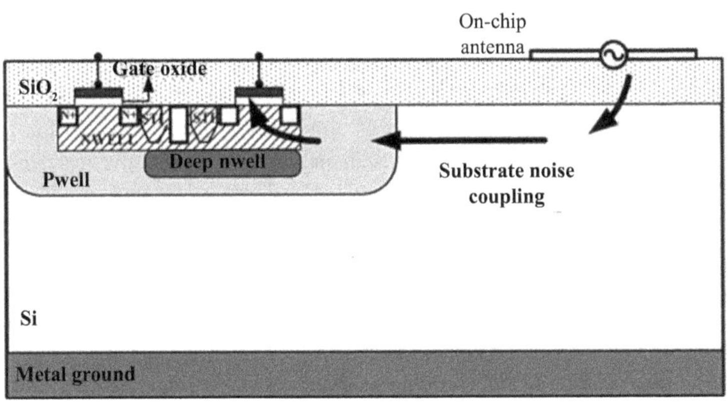

Figure 6.6 Substrate noise coupling [10]

One good solution to avoid the noise currents is through the use of guard rings that can isolate the OCA and circuits' active regions [11–13].

6.2.2.2 Co-simulations of OCAs and circuits

Typically, antenna design is performed in an EM simulator, such as ANSYS HFSS [14] or CST STUDIO SUITE [15], which relies on solving Maxwell's equations with appropriate boundary conditions. Circuit designers, on the other hand, do not require EM simulations and therefore use simulators such as Cadence [16] or Keysight ADS [17] for circuit simulations, which generally perform voltage-based or current-based nodal analysis. IC design in particular relies heavily on Cadence simulations that use the technology files for device models from the relevant foundry. Unfortunately, there are only a few mainstream simulation tools that can run both EM and circuit simulations simultaneously, but even in these simulators, capturing the coupling between OCA and the circuits is not completely feasible. For this reason, the OCA design is normally first optimized in an EM simulator; then, its layout is exported to an IC design tool, such as Cadence. However, the antenna structure cannot be simulated in the IC design tool, so the complete system (circuits and antenna) cannot be simulated together. This is because the EM simulators are based on solving Maxwell's equations in the meshed fields to obtain the exact EM performance of the 3D structures, which also accounts for all the coupling effects [18]. However, the coupling effects and antenna characteristics, such as radiation performance, cannot be modeled and simulated in circuit simulators because circuit simulators typically rely on nodal voltage-based and current-based calculations of lumped or distributed equivalent models. Similarly, circuit performances cannot be simulated in EM simulators.

The antenna must be simulated in the exact same manner as it is connected to the circuit, and vice versa. Although it is challenging to run a true co-simulation, several methods exist for partially investigating the effects of OCA and the circuits on each other. The most common method for including the OCA's effects on circuit performance is to extract the OCA's *S*-parameters from the EM simulator after optimization and import them into an IC design tool. Then, the circuit can be co-simulated with the OCA's input impedance characteristics. As presented in Figure 6.7(a), the OCA's *S*-parameters are imported into Cadence to run a co-simulation with an on-chip voltage-controlled oscillator (VCO). Normally, it requires several iterations between the EM simulator and IC design tool to achieve an optimized deign with good impedance matching between the circuit and the OCA. This, however, considerably increases the complexity and simulation time. In addition, the extracted *S*-parameters of the OCA are in the frequency domain and cannot be directly used in the time-domain analysis [19]. In addition, the OCA's *S*-parameters are influenced by the port impedance in the EM simulator, so it is important to ensure that the interface impedance is consistent with the port impedance in the EM simulator. To include the influence of circuit and other on-chip metal structures on the OCA's performance, the conventional method is to model the close-by metal structures, such as passive circuit components, bond pads, and dummy metal fills, in the EM simulator to ensure that the simulated OCA's

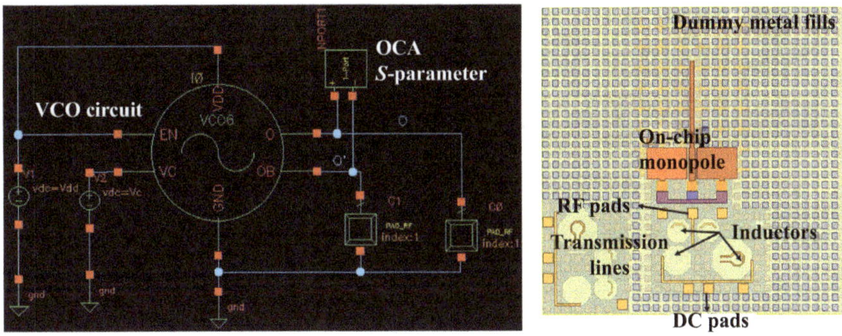

(a) (b)

Figure 6.7 (a) Co-simulation in Cadence and (b) co-simulation in HFSS

performance is reliable and accurate, as presented in Figure 6.7(b). However, this approach also encounters several challenges. First, layout is the last stage of the IC design, and since the antenna and circuit designs are done in parallel, the antenna designer must re-optimize the antenna after the IC layout is complete. Second, although the layout of the circuit part, including the passive components, bond pads, and dummy metal fills, can be imported in the EM simulator from the IC design tool, the circuit models are still not representative of the real case, as the circuit is not working in the EM simulator. For example, the coupling effects between an OCA and a close-by inductor without any currents differ from the case of the OCA being close to an inductor with currents. In addition, including the exact layout of the metal structures in EM simulation can dramatically increase the simulation time. Moreover, the EM simulators are unable to account for all the attributes in the semiconductor processes, which can be studied well only in the IC design tool [20].

This issue can be resolved only when the full-wave EM simulations and circuit simulations can be run together, in either a single simulator or two simulators running in conjunction.

6.2.3 On-chip antenna layout issue

In the IC design cycle, after the schematic design, there is another critical step that can significantly affect the OCA's performance. This step is the chip layout drawing, which is restricted by the design rule check (DRC). During the layout-drawing stage, DRC is an important process for ensuring that the chip layout fulfills the rules defined by the semiconductor foundries. Each semiconductor foundry has its own set of rules to ensure sufficient fabrication margins such that the chip fabrication has good repeatability and minimum variability in the fabrication process.

Among hundreds of rules in the DRC, three types of design rules are critical for OCA design. The first is the requirement of a certain metal density in each

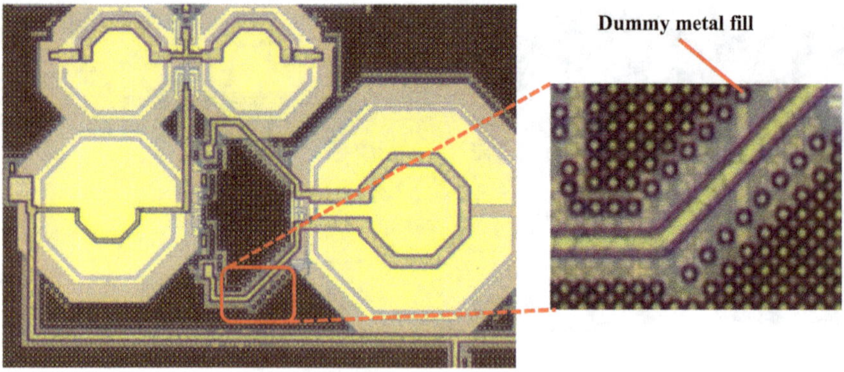

Figure 6.8 Microscope view of an RFIC with dummy metal fills

metal layer because the metal density impacts the etching rate in the fabrication stage. Global metal density rules indicate the metal density required for the entire chip, while local density rules define the metal density requirements for a certain area, normally an area of 100 μm×100 μm. The typical metal density range is 30%–70%. To meet the metal density requirement, a large number of dummy metal sheets should be placed in the empty area, while many slots should be etched on a large metal sheet surface. Figure 6.8 reveals that the blue dummy metal fills are placed in an empty area in all layers, while some slots are etched in the black large antenna patch metal. Once the slots and dummy metal sheets are introduced in the OCA area, the antenna's performance can deteriorate. In particular, the close-by dummy metal blocks can cause undesired interference with the OCA. The second type of critical design rules is the metal layer design rule, which indicates the maximum/minimum metal wire width, minimum spacing, and minimum size of the enclosure. Normally, the lower limits of the wire width, spacing, and enclosure size do not affect the OCA performance, as they are of the nanometer scale, while mm-wave and sub-terahertz antennas only require micron-scale fabrication resolution. However, the upper limit of the metal wire's width is approximately 20–30 μm, which can restrict the antenna design's flexibility. The third type is the via design rule, which includes the requirements for the size of a single via and the size of a via array. In the case of OCA design involving multilayers, a solid RF connection between the layers is highly desired, so the single via size and via array size play an important role.

Some other design rules could potentially affect the OCA. Thus, when designing the OCA, all these design rules should be considered and properly included in the simulations to avoid discrepancies between the simulations and the final measurements.

6.2.4 On-chip antenna characterization

Typically, the IC fabrication is outsourced to commercial CMOS foundries, such as Taiwan Semiconductor Manufacturing Company or GlobalFoundries, among others. After the chip is ready, the most important and challenging step is the accurate and reliable characterization of the OCA.

The conventional antenna characterization is performed in an anechoic chamber by connecting the antenna to a network analyzer through an SubMiniature version A (SMA) connector, as presented in Figure 6.9(a). This connector-based characterization method is convenient and accurate for measuring relatively large-sized antennas; however, this is not feasible for OCA characterization. The dimensions of the connector and its signal pin are on the scale of a few centimeters and millimeters, respectively, while the sizes of OCAs are typically on the orders of hundreds of micrometers, depending on their operation frequencies. Moreover, the RF pads for the measurement of OCAs are even smaller, typically in the range of 50×50 μm^2 to 100×100 μm^2. In such a case, the OCA is too small to be connected through the connector. Therefore, instead of the SMA connector, an RF probe is commonly used for getting access to the OCA, as displayed in Figure 6.9(b). The RF probe is specially manufactured for proper landing on the RF pads of RFICs and OCAs. The tiny tips of the RF probe can be landed on the RF pads accurately with the help of a microscope. The OCAs cannot be tested in standard anechoic chambers; thus, specialized chambers are required, which are very expensive. One such chamber is illustrated in Figure 6.10. It is a commercially available mm-wave anechoic chamber. Covered with an EM absorber, the chamber contains a theta/phi-direction rotatable disk attached with the reference antenna, a fixed arm for probe landing and placing the OCA at the chamber center, and a microscope as a part of the vision system to ensure accurate and reliable probe landing on chip pads. In addition, there are some customized chambers dedicated for OCA measurements [21–24]. Although such an anechoic chamber can be used for the characterization of an OCA's reflection coefficients and radiation patterns, there are still several challenges related to the measurement accuracy of probe-based characterization.

Among all the challenges related to probe-based OCA characterization, the interference between the RF probe and OCA is considered the most critical. As

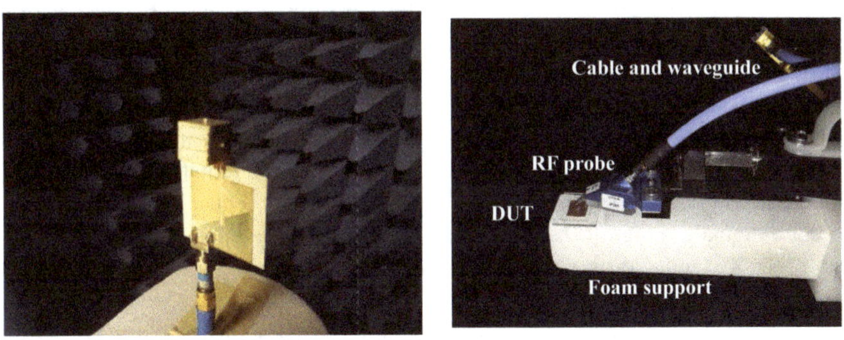

Figure 6.9 Antenna characterization methods: (a) connector-based and (b) probe-based

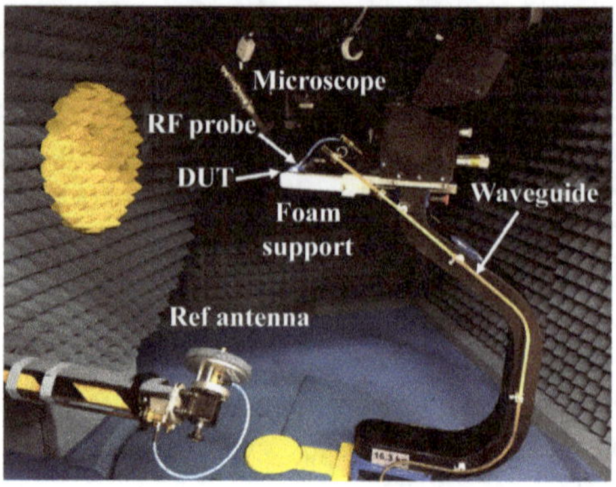

Figure 6.10 Probe-based OCA characterization setup in μ-Lab anechoic chamber

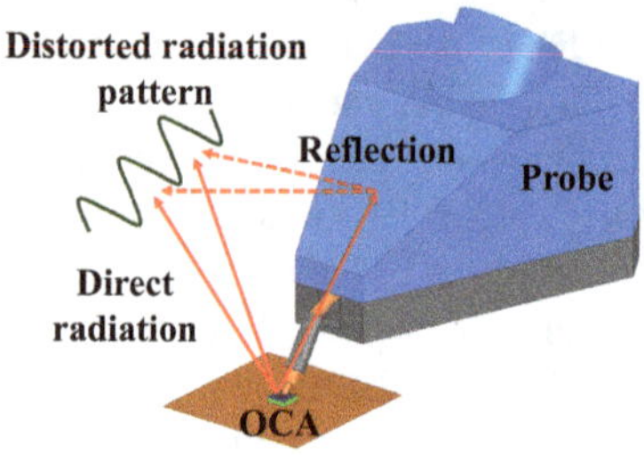

Figure 6.11 Interference between OCA's radiation and its reflection from probe

depicted in Figure 6.9(b), the RF probe consists of a large conductor body, and the probe tips are attached to a metallic arm, which is partially covered by the EM absorber. With the presence of the close-by large conductor body of the RF probe, EM waves get reflected from the probe body. The directly radiated EM waves from the antenna and the reflected waves from the probe conductor body interfere with each other. The final radiation pattern is the superposition of the radiated and reflected waves. The maximum and minimum radiations can be obtained at certain angles, depending on the phase difference between the radiated and reflected waves [25], as presented in Figure 6.11. As a result, the OCA's radiation pattern is distorted due to the presence of the probe's conductor body. Other than the influence on the antenna's radiation patterns, the unwanted interference between the OCA and probe's conductor body and uncovered probe tips can cause unexpected changes in the reflection coefficient, though the probe's inner impedance has been de-embedded through the impedance calibration [26]. Another challenge related to probe-based characterization is the self-radiation of the probe, itself. As previously mentioned, the probe tips are not covered by the EM absorber. Since the RF probe itself also radiates, the final radiation pattern is a superposition of the OCA's radiation pattern and the probe's radiation pattern [27]. To investigate the probe self-radiation, the probe's radiation pattern is characterized when the probe is connected to a calibration 50 Ω load, which represents a well-matched antenna. In such a case, only negligible radiation occurs from the 50 Ω load, so the measured radiation pattern can be approximately considered as the probe's self-radiation. The measured radiation pattern of a 60 GHz probe is presented in [27], which reveals that the maximum gain of the probe self-radiation is almost −21 dBi. This maximum defines the gain measurement sensitivity of the probe-based measurement system. This probe self-radiation is negligible for high-gain antenna measurement; however, it is considerable for the measurement of low-gain antennas, such as OCAs. Due to the low gain of the OCA, the probe self-radiation is comparable with the antenna's gain, thus causing the inaccuracy in the OCA's radiation pattern characterization. There is another drawback of probe-based OCA characterization: the radiation blockage in the probe area. Due to the presence of the probe, the OCA's radiation, in certain directions, is blocked by the probe body and its supporting arm. Thus, a complete radiation pattern cannot be obtained by the probe-based characterization method.

6.3 On-chip antenna overview

Despite the challenges mentioned in the previous section, much work has been conducted in the area of OCAs. Throughout the past two decades, tremendous efforts have been made to resolve the aforementioned challenges and issues related to OCAs. This section highlights the important work conducted in this area. It is organized with respect to the most prominent challenges and their solutions in this field. In this section, we begin by introducing several existing approaches for the gain and radiation efficiency enhancement of OCAs. These enhancement approaches primarily consist of the modification of lossy silicon substrate and the use of reflection surface, superstrate, focusing lens, and micro-electro-mechanical system (MEMS) structures, which are explained in detail in this section. On the other hand, regarding the antenna and circuit codesign issue, some important codesign strategies are discussed. These can be applied to obtain more accurate and reliable results. In addition to the codesign strategies, two codesign simulators, which can run both the circuit and EM simulation at the same time, are introduced. Toward the end of this section, the latest improvement methods for probe-based OCA characterization are presented, which can decrease the undesired probe interference and self-radiation.

6.3.1 Gain and radiation efficiency enhancement

As mentioned in Section 6.2, the low radiation efficiency of OCAs is one of the major challenges related to OCA design and is primarily caused by the incompatible silicon substrate. To minimize the silicon substrate's negative influence on OCA performance, many researchers have been working on enhancement methods for OCA gain and radiation efficiency.

6.3.1.1 Modification of incompatible silicon substrate

The lossy silicon substrate is the main cause of OCAs' low radiation efficiency. Essentially, there are three key negative-impact factors on OCAs' radiation efficiency due to the commercially used silicon substrate: low resistivity, high permittivity, and relative large thickness. A relatively straightforward approach is to alter the aforementioned properties of silicon substrate to decrease the losses in the substrate. Bulk micromachining and proton implantation are the two well-known methods for the modification of silicon substrate properties.

Bulk micromachining is a fabrication process used to form trenches or cavity structures in silicon or other substrates via selective etching [28]. It can be used to thin the silicon substrate under the OCA area. The overall relative permittivity of the mixture of air cavity and silicon substrate under the OCA area consequently decreases from 11.9 to a much lower value. However, more importantly, the overall loss due to the silicon substrate is also reduced. Bulk micromachining has been widely used for radiation efficiency enhancement [11,29,30]. For example, in [29], an air cavity was formed at the backside area of the on-chip patch antenna through selective etching, as presented in Figure 6.12. Comparison simulations between on-chip patch antennas with and without air cavity were performed, which presented around a 28% radiation efficiency enhancement with the help of the air cavity.

In addition to bulk micromachining, proton implantation is another nano-fabrication method for improving OCAs' radiation performance. Proton implantation is the process of implanting ions in the silicon substrate. Through proton implantation, the resistivity of the silicon substrate can be increased, resulting in improved radiation performance. For example, in [31], the substrate resistivity increases from 10 to 10^6 Ω cm by utilizing proton implantation. Eventually, the OCA demonstrates a 5 dB gain enhancement at 10 GHz.

From this point of view, bulk micromachining and proton implantation are capable of improving antenna radiation efficiency. However, both the methods require additional nano-fabrication processes, which lead to an increase in fabrication complexity and higher fabrication cost. Moreover, entirely changing the silicon substrate's resistivity is not good for the circuits' performance. This approach can be suitable if the change is made in a selective manner so that it changes the resistivity only in the area of the OCA.

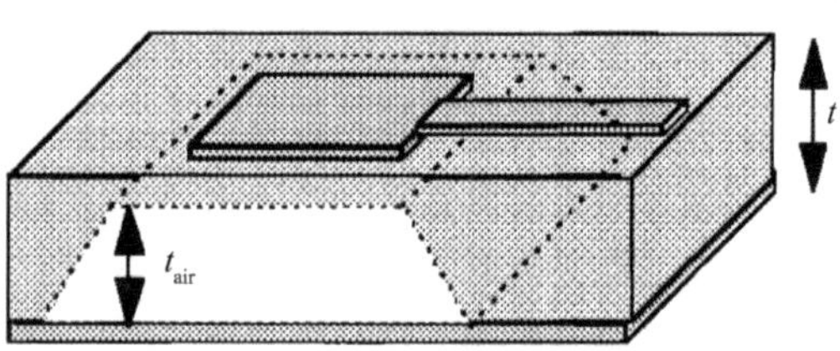

Figure 6.12 Geometry of micromachined patch antenna [29]

6.3.1.2 On-chip reflecting surface

Instead of altering the properties of silicon substrate, a proper on-chip reflecting surface can be utilized to isolate the OCA from lossy silicon substrate. One easy approach is to use a perfect electric conductor (PEC) surface in between the OCA and silicon substrate; however, the radiation from the image currents in the PEC surface negatively affects the radiation from the OCA due to the electrically small spacing between the OCA and the PEC surface, as mentioned in Section 6.2. Alternately, a perfect magnetic conductor (PMC) surface can be placed in the metal layers underneath the OCA to isolate it from the silicon substrate. This is beneficial, as in-phase reflection occurs on the PMC surface in contrast with a PEC surface and can boost the OCA's direct radiation. However, since PMC surfaces do not exist in nature, researchers have found an approximate equivalent in the form of artificial magnetic conductor (AMC) surfaces, which mimic PMC surfaces for a particular frequency range.

An AMC surface is realized through specially designed periodic structures over a PEC surface that acts a ground plane [32], as presented in Figure 6.13. The in-phase reflection performance of an AMC surface is primarily controlled by three important factors: the thickness of the substrate between the periodic structure and the ground plane, the size of the AMC surface, and the dimensions and shape of the AMC unit cell. Among the three factors, the substrate thickness is fixed by the CMOS stack-up. In theory, an infinite AMC surface behaves the same as a PMC surface for a certain frequency range; however, the AMC surface size is always restricted by the finite chip size. Therefore, the designed AMC surface size should be a reasonable compromise between AMC performance and limited chip size. Thus, for an on-chip AMC surface design, the dimensions and shape of the AMC unit cell are typically the only parameters that can be controlled by the designer to optimize the AMC performance. When the period of the periodic structure and the dimensions of its unit cell are sufficiently small compared with the free space wavelength in the operating frequency range, the structure can be assumed to be a homogeneous surface, which can be described by the parameter of surface impedance [33,34]. Theoretically, the surface impedance of this AMC structure can be tuned by making it resonant at the desired operating frequency through the dimensions and period of the AMC unit cell. In this way, a very high surface impedance can be achieved over the operating frequency range. In terms of an

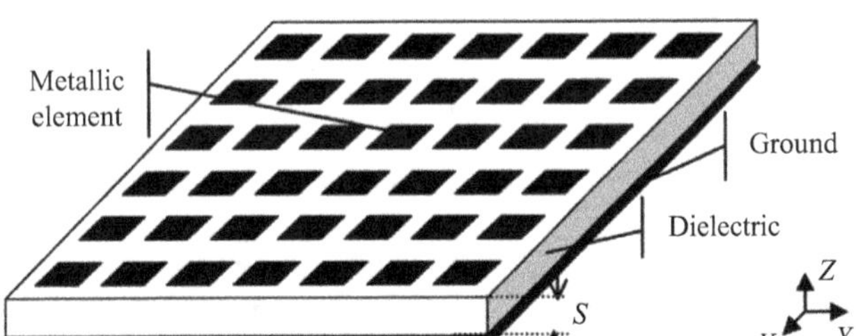

Figure 6.13 Geometry of a square patch-based AMC surface [32]

equivalent circuit, when an RF signal travels toward an open load, the reflection coefficient is positive, representing a zero-degree reflection phase shift. Similarly, in-phase reflection occurs from the AMC surface due to the high impedance it presents to the incident EM wave. Other than the period and dimensions of the AMC unit cell, its shape too can affect the overall performance. In [35], a comparison between the different shapes of an AMC unit cell has been made by studying their effect on bandwidth and return loss. Figure 6.14 displays five common structures for an AMC unit cell. It has been concluded that the square patch-based AMC surface has the widest bandwidth and lowest return loss among the five structures. Due to the capability of in-phase reflection, an AMC surface has been used to increase the radiation efficiency and boresight gain of OCAs [36–41]. For example, as presented in Figure 6.15, a 235 GHz double-layered dipole antenna is fed by the ground-signal-ground pads through a balun on the M6 layer, while the entire structure is backed by a specially designed AMC surface on the M1 layer. Eventually, the AMC-based on-chip dipole antenna presents a decent gain of 0 dBi and a high radiation efficiency of 63%.

6.3.1.3 Superstrate

As mentioned in Section 6.2, for a typical OCA, more EM energy goes to the high-permittivity silicon substrate instead of free space radiation. This is one of the major causes of OCAs' low radiation efficiency. One way to improve this situation is by adding a high-permittivity substrate on top of the OCA, also known as a superstrate.

Figure 6.16 illustrates the superstrate–substrate concept that can be beneficial for an OCA design. High boresight directivity can be achieved through this arrangement. With proper design, radiated power in air can be enhanced compared with the power coupled to the lossy silicon substrate, thus enhancing the radiation efficiency, as well. Although the use of a superstrate tends to somewhat increase the dielectric loss, it can be minimized by using low-loss material for the superstrate implementation. Nonetheless, through the appropriate use of a superstrate, the overall OCA gain can be enhanced at the boresight. To understand and analyze the high-gain condition, transmission line theory is considered convenient for theoretical investigation [42,43]. This high-gain condition is also known as the radiation resonance condition because the equivalent transmission lines at the boresight under this condition act as a resonance circuit [42]. In terms of transmission line theory, each dielectric layer can be treated as an equivalent transmission line with a certain characteristic impedance. The thickness of the superstrate layer greatly impacts the resonance condition. Theoretically, for the

Figure 6.14 Geometry of a square patch-based AMC surface [35]

Figure 6.15 3D view of the AMC-based on-chip dipole antenna [36]

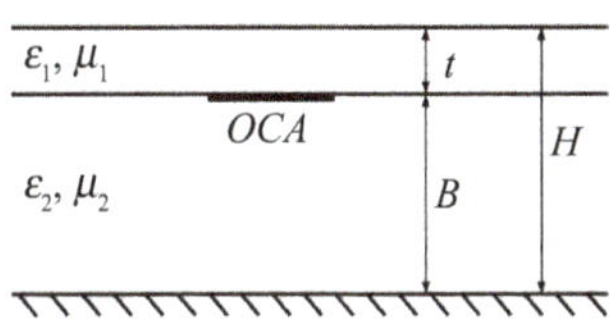

Figure 6.16 The superstrate–substrate concept

high-directivity resonance condition, the thickness of the high-permittivity superstrate must be equal to odd multiples of a quarter wavelength. In other words, the superstrate behaves similarly to a quarter-wavelength impedance transformer [42]. It is relatively easier to realize a superstrate, as it is simply a dielectric layer with no metallization patterns required.

Recently, on-chip implementations of superstrate layers have been reported for enhancing the OCA's radiation performance [37,44–46]. For example, in [44], an 89 GHz elliptical slot antenna has been realized in a CMOS 0.13 μm process with an off-chip quartz superstrate, as illustrated in Figure 6.17(a). Through the integration of the quarter-wavelength superstrate, the OCA achieves 6.4 dB of gain enhancement and approximately a 20% radiation efficiency enhancement. Since a relatively thicker superstrate is realized for OCA gain enhancement, it can also protect the chip from the environment. An example of the superstrate layer acting as the chip package, in addition to providing gain enhancement, has been demonstrated in [37]. However, one downside is that considerable reflection can occur at the interface between the high-permittivity superstrate and the air, if the design is not optimized appropriately. Furthermore, surface waves can be generated, which can distort the radiation pattern and degrade the radiation efficiency to some extent. To resolve the generation of surface waves in the superstrate, an artificial dielectric layer (ADL) has been proposed to be used as a superstrate layer for OCA [47,48]. A dielectric material's equivalent relative permittivity can be increased to a high value by embedding several layers of specially designed nonresonant periodic metallic structures, known as the ADL. The ADL can be used as a superstrate layer

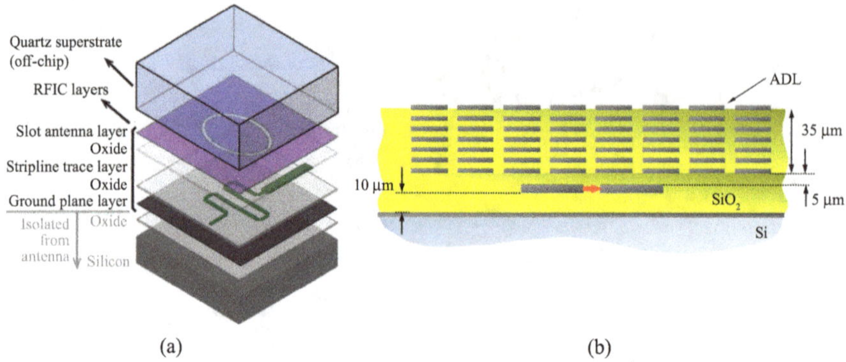

Figure 6.17 (a) Stack-up for on-chip elliptical slot antenna with off-chip quartz superstrate [44], (b) on-chip dipole antenna loaded with artificial dielectric superstrate [48]

for OCAs' gain enhancement without any surface waves due to its anisotropic properties [47]. In [48], an ADL superstrate with an equivalent relative permittivity of 32 has been designed and stacked on top of an on-chip dipole antenna, as presented in Figure 6.17(b). This resulted in a considerably high radiation efficiency of 87% and a decent gain of 6.8 dBi at 280 GHz.

6.3.1.4 Focusing lens

Lens structures have been widely used in optical systems; however, lens can also be used in the microwave or mm-wave range of frequencies for directivity enhancement. Similar to the superstrate case, lens can be beneficial for OCA radiation performance enhancement.

Primarily three different lens configurations have been investigated for OCAs' gain enhancement in the literature [5,49–56]. The first is the use of a silicon hemispherical lens at the backside of the chip [5,49–53], as displayed in Figure 6.18(a). As mentioned in Section 6.2, the generation of high-order surface waves is a critical issue for OCA design. In addition, most radiated waves are absorbed by the silicon substrate. The advantage of a backside silicon hemisphere lens is that it can help to transform the unwanted backside radiation and surfaces waves into useful radiation from the backside [5]. Moreover, the silicon lens behaves as a heat sink to dissipate the heat generated by the active on-chip circuits. The silicon lens must be specially designed with respect to the OCA's operational frequency. For example, in [49], a silicon lens has been designed and attached to the backside of the 120 GHz OCA's silicon substrate, as presented in Figure 6.18 (a). A matching layer has been coated on top of the silicon lens to match the characteristic impedance in silicon (110 Ω) to the characteristic impedance in air (377 Ω) [5]. This helps to achieve a high gain of 16.7 dBi and a radiation efficiency of 86.4% at 122 GHz. However, this approach restricts the OCA's radiation to the backside, and all the radiated EM waves must pass through the lossy silicon substrate. In addition, the unwanted small air gap between the chip and silicon lens can

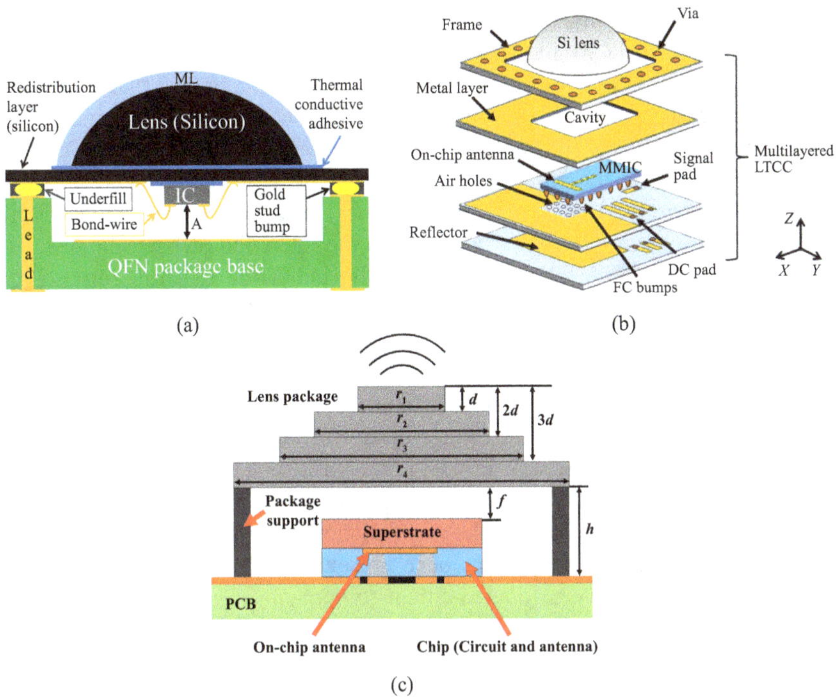

Figure 6.18 (a) Silicon hemispherical lens attached to the backside of IC [49], (b) structure of OCA with boresight silicon hemispherical lens in LTCC [54], (c) OCA with superstrate layer in a proposed Fresnel lens package [57]

cause impedance mismatching [55]. Alternately, a boresight hemispherical lens has also been proposed for OCAs' gain enhancement [54–56]. In this configuration, the dielectric lens is placed on top of the OCA at a certain distance, which is typically one focal length. Figure 6.18(b) displays a 300 GHz OCA placed in a multilayered low-temperature co-fired ceramic (LTCC), where the silicon hemispherical lens has been placed on top of the OCA with an air cavity between them [54]. In this design, the OCA achieved more than 90% radiation efficiency. One of the major issues for the conventional hemispherical lens is that it has a relatively large size compared with the OCA. To reduce the lens size, a Fresnel lens has been proposed to enhance the OCA's boresight gain [57]. For enhanced functionality, the chip package has been shaped into a Fresnel lens structure for gain enhancement. As presented in Figure 6.18(c), a Fresnel-lens-shaped package provides protection and gain enhancement.

6.3.2 Co-simulation of OCAs and circuits

As previously discussed, conducting accurate co-simulations of OCAs and circuits that account for all kinds of couplings between the two (through substrate and through air) is challenging. In this section, several state-of-the-art co-simulation methods and simulators are introduced.

CST STUDIO SUITE [15] provides a co-simulation approach that is named a transient EM/circuit co-simulation solution. In this solution, the 3D structures of the OCA can be symbolized by a block in a schematic, while the circuit components are connected to the ports of this 3D OCA's structure block. Then, a single simulation is performed with exactly the same excitation defined in the circuit simulation. The circuit's output is then fed to the OCA for EM simulations. The EM and circuit simulations run simultaneously, indicating that the currents and voltages are continuously exchanged at the interface of the circuit and OCA's block. This approach is beneficial because of its bidirectional simulation nature and faster simulation speed, as the OCA's full S-parameters are not required [18]. However, there does not yet exist an example that reveals that this simulation approach has been used for the co-simulation of the OCA and the circuits. Nonetheless, the main drawback is that this approach does not account for the OCA's radiation in addition to the coupling effects between the OCAs and the circuits, as this is only a schematic-level simulation. Similarly, in [19], the partial element equivalent circuit theory has been presented to model the antenna as an equivalent circuit with several lumped circuit components. This approach enables a co-simulation between the antenna and the circuits in both the frequency and time domain; however, this also does not include the coupling effects. To incorporate the coupling effects in the co-simulation, WiPL-D Microwave proposes a different co-simulation strategy [58]. The OCA's EM models can be pre-parameterized and imported into the circuit as any other circuit component. The inclusion of the EM models enables the investigation of an OCA's radiation and coupling effects between the OCA and the circuits. However, only limited options of antenna types are available in the 3D model library of the simulator, which restricts the design freedom for antenna designers.

In addition to including EM models of the antennas in the circuit simulations, the lumped circuit components can be imported in the EM simulator for co-simulations. In [59], the lumped circuit components and their characteristics are included in EM simulation through the finite difference time domain updating equations. The lumped circuit components contain active sources; nonlinear circuit elements, such as diodes and transistors; and linear circuit elements, such as resistors, capacitors, and inductors. When a circuit component is placed between the two nodes, its performance can be characterized by the I–V curve between the two nodes, where I and V represent the current flows through the circuit components and the voltage across the two nodes, respectively. According to Maxwell's curl equation (as presented next), the electric currents that flow through the circuit component, denoted as an impressed current density of J_i, are the sources that lead to the generation of electric and magnetic fields in the computational domain [60].

$$\nabla \times \vec{H} = \varepsilon \frac{\partial \vec{E}}{\partial t} + \sigma^e \vec{E} + \vec{J}_i \tag{6.3}$$

where the electric fields E and J_i can be expressed by the following two equations:

$$\begin{aligned} \vec{E} &= -\nabla V \\ I &= \int_S \vec{J} \cdot d\vec{s} \end{aligned} \tag{6.4}$$

where S is the cross section of a unit cell perpendicular to the flow direction of the current. With the help of this approach, the circuit simulation can be transformed into an EM simulation, making it possible to run a true co-simulation for the OCAs and the circuits in an EM simulator. This functionality has been integrated into an EM simulator: the computational EM simulator [59].

6.3.3 Advance on-chip antenna characterization methods

As mentioned in Section 6.2, the accurate and reliable characterization of the OCA is one of the major challenges. The OCA cannot be measured through connector-based methods in a conventional chamber. Although in recent years, a few dedicated chambers for OCA characterization have become commercially available, they are quite expensive and complicated to use. As OCAs' characterization requires RF probe feeding, which is much larger than the antenna itself, a number of issues arise, such as unwanted interference and self-radiation of the probe. To improve the accuracy and reliability of OCAs' characterization, several potential solutions are discussed in this section.

In some cases, OCAs are connected to the circuits on the same chip, making it impossible to directly measure the OCAs' reflection coefficient by landing the RF probe. In [26], a focus ion beam has been used to cut off the connection between the OCA and its driving circuits so that independent impedance characterization can be performed without the loading effects from the circuit part. However, cutting off the connection is an irreversible process. In [61], a de-embedding process for a balun circuit has been conducted in a simulator to eliminate its effects on the characterization of OCAs' reflection coefficient. In addition to the aforementioned issue, the unwanted coupling effects between the probe conductor body and the OCA can cause considerable discrepancies between the simulated and measured results, though the influence of the probe's inner impedance can be eliminated through proper calibration at the plane of the probe tips. In [62], an air coplanar probe has been modeled in HFSS, as illustrated in Figure 6.19, and its ABCD matrix has been obtained by exciting it from the coaxial side and landing the probe on short, open, and load calibration substrate in HFSS. Following this, a new simulation is performed after the calibration substrate is replaced with the OCA in HFSS. Based on the simulated results from the abovementioned two simulations, the de-embedding process is applied to remove the probe's inner impedance influence on the OCA reflection coefficient. The coupling effects between the probe and the OCA are also included in the simulation. Therefore, the simulated results with the probe model, after de-embedding, demonstrate satisfactory agreement with the measured results.

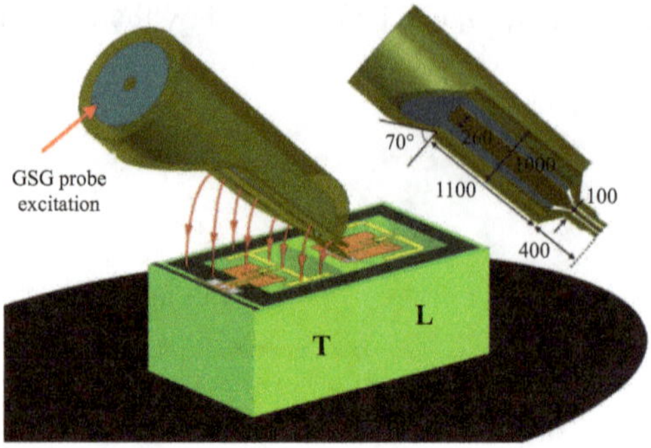

Figure 6.19 3D schematic of the chip and probe model feeding the L antenna [62]

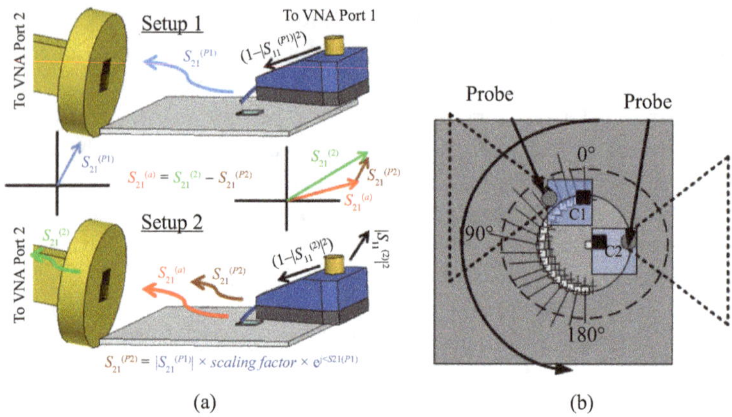

Figure 6.20 Probe radiation de-embedding methods: (a) one-probe-based, (b) two-probe-based [63]

Other than the characterization of the OCA's reflection coefficient, the radiation pattern's measurement is strongly affected by the presence of the RF probe, as discussed in Section 6.2. Several methods have been presented in the literature for minimizing the probe's negative influence on the measurement of the OCA's radiation pattern [23,25,63,64]. To eliminate the influence of probe self-radiation, two probe radiation de-embedding methods are proposed in [63]. Two measurements are required for the first method. As presented in Figure 6.20(a), the RF probe is placed in the air while remaining close to the OCA. In this setup, the probe's self-radiation can be measured, denoted as $S_{21}^{(P1)}$. In the second setup, the RF probe is landed on the OCA to excite it. Then, the overall radiation pattern $S_{21}^{(2)}$ can be obtained, which is a combination of probe self-radiation $S_{21}^{(P2)}$ and the OCA's radiation $S_{21}^{(a)}$. To obtain only the OCA's radiation, the probe self-radiation can be de-embedded from the measured radiation pattern. Although the probe self-radiation is not available in setup 2, it can be obtained by scaling the magnitude of the probe self-radiation in setup 1. Eventually, the antenna radiation can be calculated from the following equation [63]:

$$S_{21}^{(a)} = S_{21}^{(2)} - S_{21}^{(P2)} = S_{21}^{(2)} - scaling\,factor \times S_{21}^{(P1)} \tag{6.5}$$

Another method has been proposed in [63] for de-embedding the influence of the probe self-radiation. As displayed in Figure 6.20(b), this method requires two identical OCAs and RF probes. The S_{21} results are measured in four different measurement setups. For the first setup, both probes are landed on OCAs, indicated as $(P, A)(P, A)$. For the second and third measurement setups, one RF probe is landed, while the other probe is floating in the air, denoted $(P)(P, A)$. In the last measurement setup, the measured S_{21} parameter $(P)(P)$ is obtained when both probes are not landed and are in the air; then, the desired S_{21} parameter of the OCA to OCA, denoted as $(A)(A)$, can be calculated according to the phasor subtraction, as follows [63]:

$$(A)(A) = (P,A)(P,A) - (P)(P) - (P)(P,A) - (P,A)(P) \tag{6.6}$$

However, the de-embedding results of the S_{21} are obtained for only one direction. To obtain the radiation pattern based on this de-embedding method, the relative locations of the two OCAs can be adjusted to obtain the radiation pattern on the horizontal plane, as presented in Figure 6.20(b). However, this approach requires multiple probe landings, which is time consuming and has a high chance of damaging the probe and RF pads.

Although the probe self-radiation can be de-embedded by applying the aforementioned methods, the strong interference between the OCA and RF probe still exists. Several strategies have been employed to minimize the unwanted reflection from the probe conductor body [23,25,64]. A probe with less interference and smaller radiation blockage has been proposed in [64]. As illustrated in Figure 6.21(a), the probe tip feeder has been modified so that the large probe conductor body is moved away from the OCA. Therefore, the interference between the OCA and the RF probe is minimized. Moreover, the radiation blockage area is reduced to only 20 degrees.

A similar approach has been presented in [25], in which an extended waveguide is used to increase the distance between the probe tips and the probe conductor body, as displayed in Figure 6.21(b). To further reduce the interference and radiation blockage area, a bent RF probe has been proposed in [23]. Figure 6.21(c) reveals that the large probe conductor body is moved to the backside of the OCA by bending the waveguide between the probe tips and probe conductor body. In addition, an RF absorber is inserted between the OCA and the probe to further decrease the interference.

6.4 Emerging trends

It is well established that many advantages exist to integrating antennas on the chip, such as a lower overall cost, smaller physical size, lighter weight, and lower power

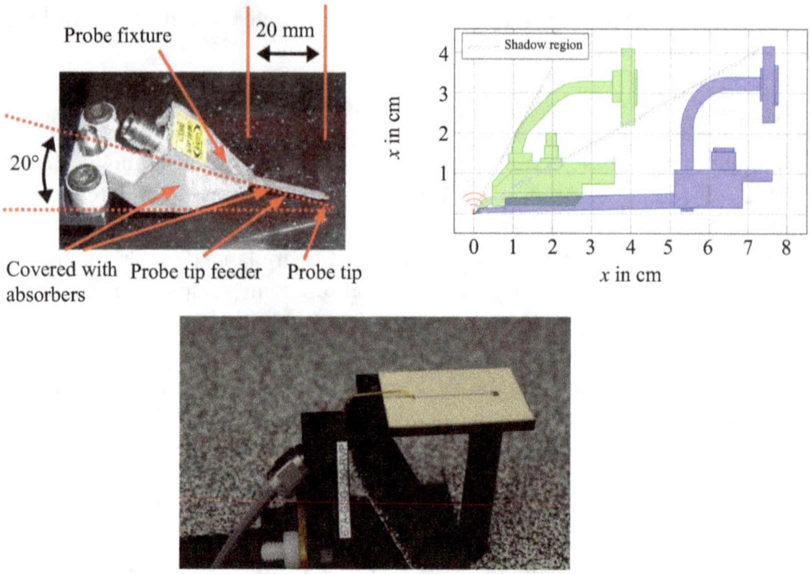

Figure 6.21 Novel probe configurations for less interference: (a) probe with less scattering [64], (b) extended probe model [25], (c) bent probe model [23]

consumption, among others. All these features are extremely important for portable and compact wireless devices, such as wireless sensors for IoT applications or compact radar systems, among others. With the aforementioned competitive advantages and due to the relevant emerging applications, it is expected that the field of OCA will surge further in the foreseeable future. This section describes some of the emerging trends that will dictate the future directions of research in the field of OCAs.

6.4.1 Drive toward higher frequencies reaching terahertz bands

With the maturity and refinement of CMOS processes, it is feasible to have operating frequencies that reach the sub-terahertz bands (100–300 GHz) or even early terahertz bands (300–3,000 GHz). The size of the antenna becomes extremely small, and thus, it becomes feasible to have antenna arrays on chip rather than only a single antenna. This is extremely attractive for phased arrays, particularly for imaging systems. Moreover, there is more available bandwidth for the OCA in the sub-terahertz and terahertz frequency bands, which is beneficial for extremely high-data-rate communication systems.

Recently, considerable research has focused on the sub-terahertz and terahertz OCA designs, which can be applied in multiple important applications, such as high-data-rate communication systems, short-range radar systems, medical detections, and imaging [53,65–71]. In [65], a 340 GHz patch OCA is realized and integrated with a receiver front-end in 40 nm CMOS technology, which is a good candidate for the applications of THz biomedical imaging. In another work, a 300 GHz dipole OCA is integrated with an amplitude-shift keying (ASK) receiver circuit for the application of high-data-rate communication [53]. This antenna achieves a high gain of 24 dBi by packaging the chip with a silicon lens. In addition, the antenna has a large bandwidth of more than 50 GHz that makes it possible for the overall system to achieve a high-data-rate of 24 Gbps for a communication range of 3 m. Other than the investigation of sub-terahertz OCAs, some examples demonstrate the CMOS OCA designs in terahertz frequencies [68–71]. For example, in [70], a graphene patch antenna is proposed for a wireless on-chip network at 2.5 THz. This graphene-based OCA is compatible with the CMOS process. In simulations, a gain of 5 dBi has been achieved.

Although there exist many attractive applications for sub-terahertz and terahertz OCAs, several challenges must be tackled in the near future. First, OCAs in the sub-terahertz and terahertz frequency range also suffer from low radiation efficiency. This is more critical for these high-frequency bands because the RF power generated on chip is low. The issue of surface waves is also more prominent at these frequencies. Second, as the frequency increases, the probe feeding also becomes complicated for the characterizations of the OCAs due to the size of the probe tips being comparable to the OCAs.

6.4.2 OCA becoming a key for biomedical wireless implants

Recently, with the rapid development of biomedical technology and the personalized health-care system, biomedical implantable devices have attracted much attention. The concept of implanting a medical monitoring or treatment device in the human body, which can send out information about health conditions in a wireless fashion, has opened many doors for health-care providers [72]. Naturally, the space is limited to biomedical-implantable devices; thus, compactness and low power consumption are two important features for such devices. The OCA, as an ultimate solution for compactness, is considered to be the best candidate for the realization of wireless biomedical implantable devices. It is important to design miniaturized OCAs while keeping an acceptable radiation performance to ensure efficient communication between the device and its reader. In addition to the challenge of OCA miniaturization, the human body effects must be considered during the design stage of the implantable OCA because the presence of human body tissue can strongly degrade the antenna's performance. Some examples of biomedical implantable OCA can be found in the literature. For example, in [73], a chip is proposed to be implanted in the human eye for monitoring the intraocular pressure, as presented in Figure 6.22. In this chip, a specially designed octagonal loop acts as both the inductor for the VCO and the antenna for the transmitter at 2.45 GHz. Designed in a 0.18 μm CMOS process, a 5.2 GHz receiving monopole OCA is also integrated in the same chip for wirelessly powering the chip. Overall, an acceptable read range of ~10 cm is achieved for this wireless sensing system. To

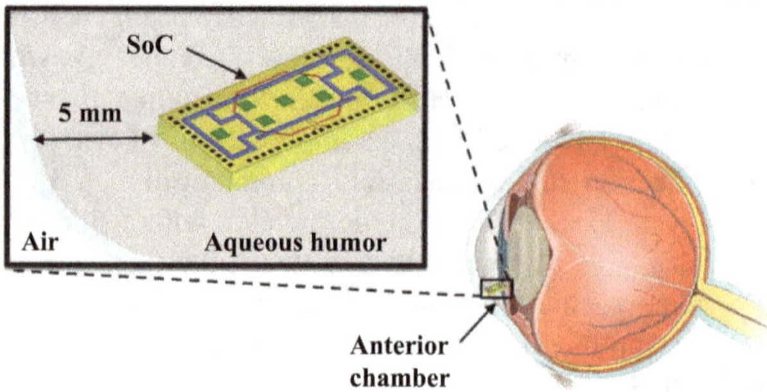

Figure 6.22 Implantation of the SoC in the anterior chamber of the human eye [73]

miniaturize the OCA size, a highly meandered differential wire OCA is presented in [74], which realizes a detection range of approximately 6 cm with a −42 dB link gain at 900 MHz. With improvements in the CMOS processes and maturity of OCA technology, improved performance can be achieved in future, which will open doors for the widespread use of OCA technology in biomedical applications.

6.4.3 Advanced simulation platforms for codesign of OCAs and circuits

A true codesign simulation platform is one that can include all aspects, such as coupling effects, the influence of OCA radiation on IC performance, and impedance matching, among others. The time-consuming iterative process between an OCA simulator and an IC simulator can be avoided through this approach. However, a simulator, which can provide the aforementioned simulation features simultaneously, is not available at present. Although several co-simulators have been developed recently [18,58], and a novel co-simulation strategy has been proposed in [59], there remain many unresolved issues. Issues include the port definition in the EM simulator not being compatible with the circuit simulation, or the requirement of extremely small meshing for the fine structures in the circuit layout, which increases the EM simulation time considerably, or truly capturing the coupling effects between the OCA and the circuits; these are some of the many deficiencies that the existing simulators have. This is an active area of research for computational tool developers on both the computational EM and circuits side.

6.4.4 Specialized CMOS process for OCA

As mentioned in Section 6.2, the incompatibility of the commercially used CMOS stack-up is the major reason for the OCA's low radiation efficiency and the excitation of unwanted surface waves. A silicon substrate with low resistivity and thin inter-metal dielectric layers is suitable for IC designs; however, the bulk material properties of the silicon substrate and the CMOS stack-up are not suitable for the efficient realization of OCAs. Nonetheless, with the growing interest in OCA, questions are being raised regarding specialized processes that offer the best of both worlds, referring to processes that can accommodate circuits and antennas efficiently. Some mainstream foundries are investigating this issue. However, there exist several examples that demonstrate considerable radiation efficiency enhancement of OCAs by applying specialized fabrication processes, such as bulk micromachining or proton implantation [11,29–31]. However, these modifications are for the entire substrate and thus are not useful for circuits. One possible solution is selective alteration of the silicon substrate (i.e., pockets of high resistivity are created in bulk Si, in which the OCA can be realized efficiently, while the resistivity remains unchanged in the circuit area). Other options, such as realization of OCA on the thick-field oxide rather than the metal layers on the thin oxides, may also be beneficial. Based on the previous discussion, it can be seen that it is required that antenna designers, IC designers, and nanofabrication engineers work in close collaboration to develop specialized CMOS processes that are suitable for the efficient integration of both OCAs and circuits.

References

[1] R. T. Rao, *System on Package: Miniaturization of the Entire System.* New York: McGraw-Hill Education (in Eng), 2008.

[2] Y. P. Zhang and D. Liu, "Antenna-on-Chip and Antenna-in-Package Solutions to Highly Integrated Millimeter-Wave Devices for Wireless Communications," *IEEE Transactions on Antennas and Propagation*, vol. 57, no. 10, pp. 2830–2841, 2009, doi: 10.1109/TAP.2009.2029295.

[3] F. Semiconductor, "Understanding Latch-Up in Advanced CMOS Logic," *Fairchild Semiconductor Corporation, Tech. Rep*, 1999.

[4] H. M. Cheema and A. Shamim, "The Last Barrier: On-Chip Antennas," *IEEE Microwave Magazine*, vol. 14, no. 1, pp. 79–91, 2013, doi: 10.1109/MMM.2012.2226542.

[5] A. Babakhani, X. Guan, A. Komijani, A. Natarajan, and A. Hajimiri, "A 77-GHz Phased-Array Transceiver With On-Chip Antennas in Silicon: Receiver and Antennas," *IEEE Journal of Solid-State Circuits*, vol. 41, no. 12, pp. 2795–2806, 2006, doi: 10.1109/JSSC.2006.884811.

[6] L. Marnat, A. A. A. Carreno, D. Conchouso, M. G. Martı´nez, I. G. Foulds, and A. Shamim, "New Movable Plate for Efficient Millimeter Wave Vertical on-Chip Antenna," *IEEE Transactions on Antennas and Propagation*, vol. 61, no. 4, pp. 1608–1615, 2013, doi: 10.1109/TAP.2013.2241720.

[7] K. Huang and D. D. Wentzloff, "60 GHz on-chip patch antenna integrated in a 0.13-μm CMOS technology," in *2010 IEEE International Conference on Ultra-Wideband*, 20–23 Sep. 2010 2010, vol. 1, pp. 1–4, doi: 10.1109/ICUWB.2010.5615699.

[8] K. Huang and D. D. Wentzloff, "A 60 GHz Antenna-Referenced Frequency-Locked Loop in 0.13 μm CMOS for Wireless Sensor Networks," *IEEE Journal of Solid-State Circuits*, vol. 46, no. 12, pp. 2956–2965, 2011, doi: 10.1109/JSSC.2011.2162768.

[9] J. Briaire and S. Krisch, "Principles of Substrate Crosstalk Generation in CMOS Circuits," *IEEE Transactions on Computer-Aided Design of Integrated Circuits and Systems*, vol. 19, no. 6, pp. 645–653, 2000, doi: 10.1109/43.848086.

[10] T. Deng and Y. P. Zhang, "On-Chip Antennas," in *Handbook of Antenna Technologies*, Z. N. Chen, D. Liu, H. Nakano, X. Qing, and T. Zwick Eds. Singapore: Springer, 2016, pp. 1565–1584.

[11] H. Chu, Y. Guo, T. Lim, Y. M. Khoo, and X. Shi, "135-GHz Micromachined On-Chip Antenna and Antenna Array," *IEEE Transactions on Antennas and Propagation*, vol. 60, no. 10, pp. 4582–4588, 2012, doi: 10.1109/TAP.2012.2209855.

[12] K. Kang, F. Lin, D. D. Pham, *et al.*, "A 60-GHz OOK Receiver With an On-Chip Antenna in 90 nm CMOS," *IEEE Journal of Solid-State Circuits*, vol. 45, no. 9, pp. 1720–1731, 2010, doi: 10.1109/JSSC.2010.2053095.

[13] S. Uemura, Y. Hiraoka, T. Kai, and S. Dosho, "Isolation Techniques Against Substrate Noise Coupling Utilizing Through Silicon Via (TSV) Process for RF/Mixed-Signal SoCs," *IEEE Journal of Solid-State Circuits*, vol. 47, no. 4, pp. 810–816, 2012, doi: 10.1109/JSSC.2012.2185169.

[14] Ansys.com. (n.d.). ANSYS HFSS: High Frequency Electromagnetic Field Simulation Software. [Online] Available at: https://www.ansys.com/products/electronics/ansys-hfss.

[15] CST Studio Suite 3D EM Simulation and Analysis Software. [Online] Available at: http://www.cst.com.

[16] Cadence: EDA Tools and IP for Intelligent System Design. [Online] Available at: https://www.cadence.com.

[17] Keysight: PathWave Advanced Design System (ADS). [Online] Available at: https://www.keysight.com/zz/en/products/software/pathwave-design-software/pathwave-advanced-design-system.html.

[18] A. Scott and V. Sokol, "True transient 3D EM/circuit co-simulation using CST STUDIO SUITE," in *Conference Proceedings*, 2008.

[19] Y. Zhang, N. H. W. Fong, D. C. W. Ng, and N. Wong, "Co-simulation of RFIC with bondwire antenna via retarded PEEC method," in *2012 IEEE International Symposium on Circuits and Systems (ISCAS)*, 20–23 May 2012 2012, pp. 229–232, doi: 10.1109/ISCAS.2012.6271751.

[20] M. R. Karim, X. Yang, and M. F. Shafique, "On Chip Antenna Measurement: A Survey of Challenges and Recent Trends," *IEEE Access*, vol. 6, pp. 20320–20333, 2018, doi: 10.1109/ACCESS.2018.2821196.

[21] R. N. Simons, "Novel On-Wafer Radiation Pattern Measurement Technique for MEMS Actuator Based Reconfigurable Patch Antennas," *NASA STI/Recon Technical Report N*, 2002, vol. 02. [Online].

[22] D. Titz, F. Ferrero, and C. Luxey, "Development of a Millimeter-Wave Measurement Setup and Dedicated Techniques to Characterize the Matching and Radiation Performance of Probe-Fed Antennas [Measurements Corner]," *IEEE Antennas and Propagation Magazine*, vol. 54, no. 4, pp. 188–203, 2012, doi: 10.1109/MAP.2012.6309179.

[23] A. C. F. Reniers, A. R. V. Dommele, A. B. Smolders, and M. H. A. J. Herben, "The Influence of the Probe Connection on mm-Wave Antenna Measurements," *IEEE Transactions on Antennas and Propagation*, vol. 63, no. 9, pp. 3819–3825, 2015, doi: 10.1109/TAP.2015.2452941.

[24] S. Beer and T. Zwick, "Probe based radiation pattern measurements for highly integrated millimeter-wave antennas," in *Proceedings of the Fourth European Conference on Antennas and Propagation*, 12–16 Apr. 2010 2010, pp. 1–5.

[25] L. Boehm, M. Hitzler, F. Roos, and C. Waldschmidt, "Probe influence on integrated antenna measurements at frequencies above 100 GHz," in *2016 46th European Microwave Conference (EuMC)*, 4–6 Oct. 2016 2016, pp. 552–555, doi: 10.1109/EuMC.2016.7824402.

[26] H. Zhang and A. Shamim, "Tackling the issues of millimeter-wave on-chip antenna measurements," in *2019 13th European Conference on Antennas and Propagation (EuCAP)*, 31 Mar.–5 Apr. 2019 2019, pp. 1–5.

[27] S. Ranvier, M. Kyrö, C. Icheln, C. Luxey, R. Staraj, and P. Vainikainen, "Compact 3-D On-Wafer Radiation Pattern Measurement System for 60 GHz Antennas," *COST 2100, Pervasive Mobile & Ambient Wireless Communications*, vol. 51, no. 2, pp. 319–324, 2009, doi: 10.1002/mop.24034.

[28] S. M. Spearing, "Micro Devices and Micro Systems, Materials for," in *Encyclopedia of Materials: Science and Technology*, K. H. J. Buschow, R. W. Cahn, M. C. Flemings, *et al.* Eds. New York: Elsevier, 2001, pp. 5580–5587.

[29] I. Papapolymerou, R. F. Drayton, and L. P. B. Katehi, "Micromachined Patch Antennas," *IEEE Transactions on Antennas and Propagation*, vol. 46, no. 2, pp. 275–283, 1998, doi: 10.1109/8.660973.

[30] W. T. Khan, A. Ç. Ulusoy, G. Dufour, *et al.*, "A D-Band Micromachined End-Fire Antenna in 130-nm SiGe BiCMOS Technology," *IEEE Transactions on Antennas and Propagation*, vol. 63, no. 6, pp. 2449–2459, 2015, doi: 10.1109/TAP.2015.2416751.

[31] K. T. Chan, A. Chin, Y. B. Chen, Y. Lin, T. S. Duh, and W. J. Lin, "Integrated antennas on Si, proton-implanted Si and Si-on-quartz," in *International Electron Devices Meeting. Technical Digest (Cat. No.01CH37224)*, 2–5 Dec. 2001 2001, pp. 40.6.1–40.6.4, doi: 10.1109/IEDM.2001.979659.

[32] G. Goussetis, A. P. Feresidis, and J. C. Vardaxoglou, "Tailoring the AMC and EBG Characteristics of Periodic Metallic Arrays Printed on Grounded Dielectric Substrate," *IEEE Transactions on Antennas and Propagation*, vol. 54, no. 1, pp. 82–89, 2006, doi: 10.1109/TAP.2005.861575.

[33] H. Mosallaei and K. Sarabandi, "Antenna Miniaturization and Bandwidth Enhancement Using a Reactive Impedance Substrate," *IEEE Transactions on Antennas and Propagation*, vol. 52, no. 9, pp. 2403–2414, 2004, doi: 10.1109/TAP.2004.834135.

[34] A. P. Feresidis, G. Goussetis, W. Shenhong, and J. C. Vardaxoglou, "Artificial Magnetic Conductor Surfaces and Their Application to Low-Profile High-Gain Planar Antennas," *IEEE Transactions on Antennas and Propagation*, vol. 53, no. 1, pp. 209–215, 2005, doi: 10.1109/TAP.2004.840528.

[35] B. S. Cook and A. Shamim, "Utilizing Wideband AMC Structures for High-Gain Inkjet-Printed Antennas on Lossy Paper Substrate," *IEEE Antennas and Wireless Propagation Letters*, vol. 12, pp. 76–79, 2013, doi: 10.1109/LAWP.2013.2240251.

[36] H. Zhu, X. Li, W. Feng, J. Xiao, and J. Zhang, "235 GHz on-Chip Antenna With Miniaturised AMC Loading in 65 nm CMOS," *IET Microwaves, Antennas & Propagation*, vol. 12, no. 5, pp. 727–733, 2018, doi: 10.1049/iet-map.2017.0710.

[37] H. Zhang and A. Shamim, "Gain and efficiency enhancement of a 77 GHz on-chip antenna through AMC and superstrate package," in *2018 IEEE International Symposium on Antennas and Propagation & USNC/URSI National Radio Science Meeting*, 8–13 Jul. 2018 2018, pp. 363–364, doi: 10.1109/APUSNCURSINRSM.2018.8608189.

[38] X. Bao, Y. Guo, and Y. Xiong, "60-GHz AMC-Based Circularly Polarized On-Chip Antenna Using Standard 0.18-μm CMOS Technology," *IEEE Transactions on Antennas and Propagation*, vol. 60, no. 5, pp. 2234–2241, 2012, doi: 10.1109/TAP.2012.2189725.

[39] H. Kuo, H. Yue, Y. Ou, C. Lin, and H. Chuang, "A 60-GHz CMOS Sub-Harmonic RF Receiver With Integrated On-Chip Artificial-Magnetic-Conductor Yagi Antenna and Balun Bandpass Filter for Very-Short-Range Gigabit Communications," *IEEE Transactions on Microwave Theory and Techniques*, vol. 61, no. 4, pp. 1681–1691, 2013, doi: 10.1109/TMTT.2013.2247622.

[40] H. Chu, Y. X. Guo, F. Lin, and X. Q. Shi, "Wideband 60GHz on-chip antenna with an artificial magnetic conductor," in *2009 IEEE International Symposium on Radio-Frequency Integration Technology (RFIT)*, 9 Jan.–11 Dec. 2009 2009, pp. 307–310, doi: 10.1109/RFIT.2009.5383667.

[41] S. Pan, D. Wang, C. Guclu, and F. Capolino, "High impedance layer for CMOS on-chip antenna at millimeter waves," in *2011 IEEE International Symposium on Antennas and Propagation (APSURSI)*, 3–8 Jul. 2011 2011, pp. 903–906, doi: 10.1109/APS.2011.5996422.

[42] D. Jackson and N. Alexopoulos, "Gain Enhancement Methods for Printed Circuit Antennas," *IEEE Transactions on Antennas and Propagation*, vol. 33, no. 9, pp. 976–987, 1985, doi: 10.1109/TAP.1985.1143709.

[43] N. Alexopoulos and D. Jackson, "Fundamental Superstrate (Cover) Effects on Printed Circuit Antennas," *IEEE Transactions on Antennas and Propagation*, vol. 32, no. 8, pp. 807–816, 1984, doi: 10.1109/TAP.1984.1143433.

[44] J. M. Edwards and G. M. Rebeiz, "High-Efficiency Elliptical Slot Antennas With Quartz Superstrates for Silicon RFICs," *IEEE Transactions on Antennas and Propagation*, vol. 60, no. 11, pp. 5010–5020, 2012, doi: 10.1109/TAP.2012.2207353.

[45] M. Uzunkol, O. D. Gurbuz, F. Golcuk, and G. M. Rebeiz, "A 0.32 THz SiGe 4x4 Imaging Array Using High-Efficiency On-Chip Antennas," *IEEE*

Journal of Solid-State Circuits, vol. 48, no. 9, pp. 2056–2066, 2013, doi: 10.1109/JSSC.2013.2262739.

[46] J. Xu and D. S. Ricketts, "Broadband W-band on-chip Yagi antenna using superstrate for high efficiency and endfire radiation," in *2014 Asia-Pacific Microwave Conference*, 4–7 Nov. 2014 2014, pp. 360–362.

[47] W. H. Syed, G. Fiorentino, D. Cavallo, M. Spirito, P. M. Sarro, and A. Neto, "Design, Fabrication, and Measurements of a 0.3 THz On-Chip Double Slot Antenna Enhanced by Artificial Dielectrics," *IEEE Transactions on Terahertz Science and Technology*, vol. 5, no. 2, pp. 288–298, 2015, doi: 10.1109/TTHZ.2015.2399276.

[48] D. Cavallo, W. H. Syed, and A. Neto, "Artificial dielectric enabled antennas for high frequency radiation from integrated circuits," in *2017 11th European Conference on Antennas and Propagation (EUCAP)*, 19–24 Mar. 2017 2017, pp. 1626–1628, doi: 10.23919/EuCAP.2017.7928187.

[49] B. Goettel, W. Winkler, A. Bhutani, F. Boes, M. Pauli, and T. Zwick, "Packaging Solution for a Millimeter-Wave System-on-Chip Radar," *IEEE Transactions on Components, Packaging and Manufacturing Technology*, vol. 8, no. 1, pp. 73–81, 2018, doi: 10.1109/TCPMT.2017.2758725.

[50] K. Statnikov, J. Grzyb, B. Heinemann, and U. R. Pfeiffer, "160-GHz to 1-THz Multi-Color Active Imaging With a Lens-Coupled SiGe HBT Chip-Set," *IEEE Transactions on Microwave Theory and Techniques*, vol. 63, no. 2, pp. 520–532, 2015, doi: 10.1109/TMTT.2014.2385777.

[51] M. Neshat, D. M. Hailu, M. Nezhad-Ahmadi, G. Z. Rafi, and S. Safavi-Naeini, "Gain Measurement of Embedded On-Chip Antennas in mmW/THz Range," *IEEE Transactions on Antennas and Propagation*, vol. 60, no. 5, pp. 2544–2549, 2012, doi: 10.1109/TAP.2012.2189772.

[52] J. Grzyb, K. Statnikov, N. Sarmah, B. Heinemann, and U. R. Pfeiffer, "A 210–270-GHz Circularly Polarized FMCW Radar With a Single-Lens-Coupled SiGe HBT Chip," *IEEE Transactions on Terahertz Science and Technology*, vol. 6, no. 6, pp. 771–783, 2016, doi: 10.1109/TTHZ.2016.2602539.

[53] H. Song, J. Kim, K. Ajito, M. Yaita, and N. Kukutsu, "Fully Integrated ASK Receiver MMIC for Terahertz Communications at 300 GHz," *IEEE Transactions on Terahertz Science and Technology*, vol. 3, no. 4, pp. 445–452, 2013, doi: 10.1109/TTHZ.2013.2252954.

[54] T. Tajima, H. Song, and M. Yaita, "Compact THz LTCC Receiver Module for 300 GHz Wireless Communications," *IEEE Microwave and Wireless Components Letters*, vol. 26, no. 4, pp. 291–293, 2016, doi: 10.1109/LMWC.2016.2537044.

[55] J. Shi, D. Cui, H. Hao, C. Deng, and X. Lv, "220 GHz High Spatial Resolution Detector With Above-Chip Mounted Dielectric Lens," *Electronics Letters*, vol. 53, no. 8, pp. 538–540, 2017, doi: 10.1049/el.2016.4755.

[56] J. Al-Eryani, H. Knapp, J. Kammerer, K. Aufinger, H. Li, and L. Maurer, "Fully Integrated Single-Chip 305–375-GHz Transceiver With On-Chip Antennas in SiGe BiCMOS," *IEEE Transactions on Terahertz Science and*

Technology, vol. 8, no. 3, pp. 329–339, 2018, doi: 10.1109/TTHZ.2018.2823202.

[57] H. Zhang and A. Shamim, "Gain Enhancement of Millimeter-Wave on-Chip Antenna through an Additively Manufactured Functional Package," *IEEE Transactions on Antennas and Propagation, to be published*, 2020.

[58] WiPL-D Microwave, *Effective Antenna Design Using EM and Circuit Co-Simulation in WIPL-D Microwave*. Available Online: www.wipl-d.com.

[59] A. Elsherbeni and V. Demir, *The Finite-Difference Time-Domain Method for Electromagnetics With MATLAB® Simulations: ACES series, 2nd edition*. Edison, NJ: SciTech Publishing, 2016, pp. 1–531.

[60] R. F. Harrington, *Time-Harmonic Electromagnetic Fields*. New York: Wiley, 2001.

[61] A. Shamim, L. Roy, N. Fong, G. Tarr, and V. Levenets, "On-chip antenna measurements: calibration and de-embedding considerations," in *2005 IEEE Instrumentation and Measurement Technology Conference Proceedings*, 16–19 May 2005 2005, vol. 1, pp. 463–466, doi: 10.1109/IMTC.2005.1604158.

[62] M. Seyyedesfahlan and I. Tekin, "ACP Probe Measurement of On-Chip Strip Dipole Antennas at W Band," *IEEE Transactions on Antennas and Propagation*, vol. 64, no. 4, pp. 1270–1278, 2016, doi: 10.1109/TAP.2016.2529642.

[63] J. Murdock, E. Ben-Dor, F. Gutierrez, and T. S. Rappaport, "Challenges and approaches to on-chip millimeter wave antenna pattern measurements," in *2011 IEEE MTT-S International Microwave Symposium*, Baltimore, MD, 5–10 Jun. 2011 2011, pp. 1–4, doi: 10.1109/MWSYM.2011.5972965.

[64] Z. Tsai, Y. Wu, S. Chen, T. Lee, and H. Wang, "A V-Band On-Wafer Near-Field Antenna Measurement System Using an IC Probe Station," *IEEE Transactions on Antennas and Propagation*, vol. 61, no. 4, pp. 2058–2067, 2013, doi: 10.1109/TAP.2012.2237091.

[65] C. Li, C. Ko, M. Kuo, and D. Chang, "A 340-GHz Heterodyne Receiver Front End in 40-nm CMOS for THz Biomedical Imaging Applications," *IEEE Transactions on Terahertz Science and Technology*, vol. 6, no. 4, pp. 625–636, 2016, doi: 10.1109/TTHZ.2016.2566580.

[66] P. H. Siegel, "Terahertz Technology," *IEEE Transactions on Microwave Theory and Techniques*, vol. 50, no. 3, pp. 910–928, 2002, doi: 10.1109/22.989974.

[67] D. L. Woolard, R. Brown, M. Pepper, and M. Kemp, "Terahertz Frequency Sensing and Imaging: A Time of Reckoning Future Applications?," *Proceedings of the IEEE*, vol. 93, no. 10, pp. 1722–1743, 2005, doi: 10.1109/JPROC.2005.853539.

[68] Z. Ahmad and K. O. Kenneth, "0.65–0.73THz quintupler with an on-chip antenna in 65-nm CMOS," in *2015 Symposium on VLSI Circuits (VLSI Circuits)*, 17–19 Jun. 2015 2015, pp. C310–C311, doi: 10.1109/VLSIC.2015.7231303.

[69] B. Khamaisi, S. Jameson, and E. Socher, "0.61THz radiating source with on-chip antenna on 65nm CMOS," in *2016 11th European Microwave Integrated Circuits Conference (EuMIC)*, 3–4 Oct. 2016 2016, pp. 389–392, doi: 10.1109/EuMIC.2016.7777573.

[70] S. Mrunalini and A. Manoharan, "Dual-Band Re-configurable Graphene-Based Patch Antenna in Terahertz Band for Wireless Network-on-Chip Applications," *IET Microwaves, Antennas & Propagation*, vol. 11, no. 14, pp. 2104–2108, 2017, doi: 10.1049/iet-map.2017.0415.

[71] Y. Kanazawa, E. Sano, S. Yokoyama, *et al.*, "CMOS terahertz imaging pixel with a wideband on-chip antenna," in *2018 43rd International Conference on Infrared, Millimeter, and Terahertz Waves (IRMMW-THz)*, 9–14 Sep. 2018 2018, pp. 1–2, doi: 10.1109/IRMMW-THz.2018.8509895.

[72] R. Karim, A. Iftikhar, B. Ijaz, and I. B. Mabrouk, "The Potentials, Challenges, and Future Directions of On-Chip-Antennas for Emerging Wireless Applications—A Comprehensive Survey," *IEEE Access*, vol. 7, pp. 173897–173934, 2019, doi: 10.1109/ACCESS.2019.2957073.

[73] L. Marnat, M. H. Ouda, M. Arsalan, K. Salama, and A. Shamim, "On-Chip Implantable Antennas for Wireless Power and Data Transfer in a Glaucoma-Monitoring SoC," *IEEE Antennas and Wireless Propagation Letters*, vol. 11, pp. 1671–1674, 2012, doi: 10.1109/LAWP.2013.2240253.

[74] C. Liu, Y. Guo, X. Liu, and S. Xiao, "An Integrated On-Chip Implantable Antenna in 0.18-μm CMOS Technology for Biomedical Applications," *IEEE Transactions on Antennas and Propagation*, vol. 64, no. 3, pp. 1167–1172, 2016, doi: 10.1109/TAP.2016.2518201.

Chapter 7

Reflectarray antennas: potentials for 5G and beyond

Muhammad I. Abbasi[1], *Muhammad H. Dahri*[2], *Mohd H. Jamaluddin*[3], *Muhammad R. Kamarudin*[2] *and Fauziahanim C. Seman*[2]

The array of elements combined together on a flat surface to reflect the incident signals coming from a properly distant feed defines the main architecture of a reflectarray antenna [1]. The reflection of the signals can be directed like a parabolic reflector with an additional advantage of a plane and light-weighted surface. Moreover, reflectarray can also perform beam scanning like a phased array antenna, but without the aid of any power divider or additional phase shifters [2]. The less complex designs of reflectarray make it more cost effective, especially for beam-scanning applications. The bulky and curvy design of parabolic antenna is not a good candidate for high-frequency applications [1]. Alternatively, a reflectarray antenna can easily be designed from as low as microwave [3] to as high as terahertz frequency range [4]. The adaptability of reflectarray to high frequencies makes it suitable for high-gain and high-bandwidth operation in a wide range of advanced applications, including 5G communications.

The performance of reflectarray antenna mainly depends on its design procedures [1]. Each expected output can behave differently for a same reflectarray antenna, for example a large reflecting element with a small element spacing is good for low loss performance but at the cost of narrow phase range [5]. Figure 7.1 shows the basic architecture of a reflectarray antenna with rectangular patch elements. The analysis of a reflectarray antenna can be performed by a full-wave technique [6] where a single element of reflectarray can be considered as unit cell, as depicted in Figure 7.1. The unit cell can be taken as an isolated element without mutual coupling effects or either as an array element placed in a periodic array environment with possible mutual coupling from surrounding elements [7]. The

[1]Faculty of Electrical and Electronic Engineering Technology (FTKEE), Universiti Teknikal Malaysia Melaka (UTeM), Melaka, Malaysia
[2]Faculty of Electrical and Electronic Engineering (FKEE), Universiti Tun Hussein Onn Malaysia (UTHM), Parit Raja, Malaysia
[3]Wireless Communication Centre (WCC), Universiti Teknologi Malaysia (UTM), Skudai, Malaysia

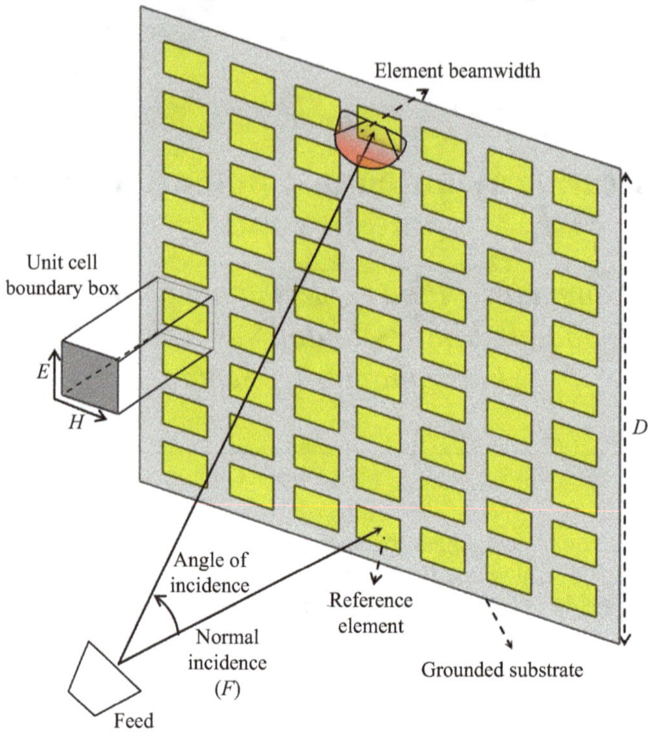

Figure 7.1 Basic architecture of a microstrip reflectarray with an offset feed

first approach is easy to perform, while the later one is more accurate for proper design and analysis. The performance of the unit cell reflectarray can be characterised in terms of its reflection loss, reflection phase and beamwidth [1]. Low reflection-loss and wide reflection phase range are essential for a wide bandwidth reflectarray. On the other hand, unit cell beamwidth is a crucial factor to determine the size of the reflectarray which can govern its gain performance.

As shown in Figure 7.1, a wide beamwidth is required by a corner element of reflectarray to properly accumulate the incidence signals from a distant feed. The reflection phase of a reflectarray element differs from the reference element with normal incidence, depending on the angle of incidence it holds from the feed. Therefore, each element on the reflectarray should be designed with proper progressive reflection phase in order to attain high-gain performance in a required direction. These reflection phase variations can be achieved by elements with variable size [8], variable rotation angle [9] or by the length of the stub attached to the unit cell element [10]. The incidence angle of feed from an element depends on the distance (*f*) it carries from the reflectarray. A large feed distance ensures the proper accumulation of corner elements, but at the cost of the large profile reflectarray antenna and high spillover losses. The feed distance is normally measured in *f*/*D* ratio, where *D* is the longest dimension of reflectarray. A reflectarray antenna can be designed with an offset feed [11] to avoid the feed shadow created by a centre feed. However, the offset feed may increase the angle of incidence for the corner elements. Conventionally reflectarray antennas are designed with a square or circular aperture based on the requirements.

7.1 Reflectarrays for 5G

The propagation issues related with mm-waves can be avoided by selecting a suitable type of antenna for 5G systems. Array antennas are considered as a decent candidate to compensate the issues regarding the path loss for short-range communications [12]. Two-dimensional arrays with large electrical apertures can provide narrow beamwidth, which is essential for 5G operations [13]. Large electrical aperture at mm-waves for 5G does not affect the physical profile of the antenna due to short wavelengths. Massive MIMO systems have also been suggested for 5G due to their possible integrity with small cells [13–16]. However, compared to array antennas, massive MIMO systems are not the potential candidate for 5G systems due to their design complexity and less adaptability with shorter wavelengths. There are many other types of antennas that can be found in the literature for proposed 5G operation [17–19]. Their main purpose is to achieve wide bandwidth to support high throughput of 5G systems [20]. The operation of antenna systems for 5G compatibility largely depends on the enhancement of its bandwidth performance. A massive bandwidth is required in mm-wave range to support high data requirements [14,21]. Bandwidth of the order of GHz is attainable at mm-wave frequency range, but some extra design efforts are still required to fully utilise it with other requirements.

However, only enhancing the bandwidth of proposed antenna does not solve all issues regarding 5G compatibility. Significant improvements in some other parameters like gain, efficiency, polarisation diversity and adaptive beam steering are also considered as a need of time [13,21,22]. It is because the antenna performance for 5G can directly depend on the mode of antenna operation. Antenna used for transmission or reception can significantly affect its required parameters for 5G operation. It is widely believed that the requirement of improvement in antenna parameters for transmission is higher than the same parameters for reception.

7.2 Reflectarray bandwidth enhancement

The upcoming future technologies will require a very wide frequency bandwidth to transfer high-definition data for various users at a time. Reflectarray antenna is generally considered as a narrow bandwidth antenna due to its high loss performance [23]. The conductor material used for the patch element along with the dielectric material are the main contributors for high loss performance [24]. Along with the high loss performance, differential spatial phase delays and resonant behaviour of reflectarray are also the main reasons behind its narrow-band performance [1]. Due to the planar structure of reflectarray, the path lengths between feed and its elements vary from its centre to its corner. This phenomenon produces undesired reflections of incident signals of the same frequency and limits the bandwidth performance of reflectarray. A reflectarray is required to be designed with progressive phase distribution with minimum phase errors to counter this problem. Additionally, the elements on the surface of reflectarray are commonly designed by considering a single resonant frequency, which also limits its bandwidth performance. The range of the operating frequencies of a reflectarray antenna mainly depends on its designing procedures. The bandwidth of reflectarrays can be divided into two sections; unit cell bandwidth and full reflectarray bandwidth [25]. Unit cell bandwidth depends on the slope of the reflection phase curve, while the full reflectarray bandwidth has usually been considered at 1-dB or 3-dB drop value on the gain vs frequency curve. The unit cell element is the basic building block of a full reflectarray and it is also responsible for the performance improvement of full reflectarray. Various design configurations based on unit cell element and full reflectarray have been thoroughly considered by many researchers for the purpose of bandwidth enhancement.

Some conventional techniques involving the bandwidth enhancement of reflectarrays include coupling of multiple resonances together [26,27], coupling of unit cells with true time delay lines or stubs [28] and multiband designs [29]. Various design configurations based on different patch elements, multilayered cascaded substrates and aperture-coupled elements were normally considered for the stated approaches. The variation in the thickness of the dielectric material has also been considered as an important parameter to improve the bandwidth performance of reflectarrays [23,30]. Some selected works based on the mentioned conventional approaches along with some latest techniques for reflectarray bandwidth improvement have been thoroughly discussed here.

As aforementioned, the design parameters of the reflectarray can be divided into two sections; the resonant element design and the full reflectarray design. The analysis of the techniques involving both approaches can further be utilised for the improvement of the bandwidth performance. Various resonant element designs with some advanced techniques have been recently noted from the literature, which include novel solo elements [31], combination of elements [32], parasitic elements [33], elements with open-ended stubs [34,35] and elements with aperture-coupled delay lines [36]. Figure 7.2 shows some of the selected resonant elements used for bandwidth-enhancement strategy.

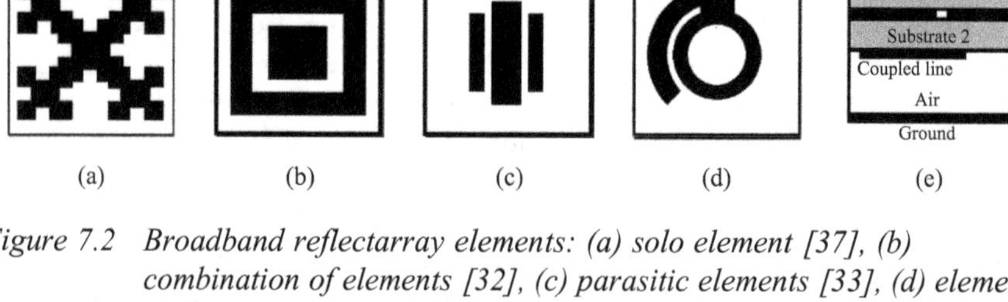

Figure 7.2 Broadband reflectarray elements: (a) solo element [37], (b) combination of elements [32], (c) parasitic elements [33], (d) element with open-ended stub [35], (e) element with aperture-coupled delay line [36]

The fractal element [37] shown in Figure 7.2(a) was used to get multi-resonance response due to the modification of current distributions on its surface, which leads to increase the reflection phase range and hence bandwidth. Square patch combined with square ring [32] shown in Figure 7.2(b) was set to get dual resonance response for bandwidth enhancement. Figure 7.2(c) depicts the dipole element surrounded by two parasitic dipoles [33]. The parasitic dipoles were not the part of resonant structure and only used to modify the surface capacitance and inductance of the dipole patch element. The modification in the reflection phase curve was then obtained by varying the lengths of the parasitic elements, which leads to enhance the bandwidth performance. In Figure 7.2(d), an open-ended phase tuning stub was attached with a circular ring [35]. The stub was used to introduce the time delay for the reflection of the signals, due to that delay the reflection phase curve smoothed and bandwidth increased. The aperture-coupled cascaded element [36] shown in Figure 7.2(e) also depicts a very common type of design for bandwidth enhancement. The coupled line was used to modify the capacitances between patch and ground plane. This modification leads to a change in the surface currents and electric field distributions, which are related to the frequency of operation and bandwidth performance.

The bandwidth of a full reflectarray antenna can also be improved by considering many factors. It can be evaluated based on its inter-element spacing [55], feeding mechanism [38] and its design approaches [39,40]. Two design approaches of a reflectarray antenna, other than the conventional approach, are shown in Figure 7.3. A full conductor reflectarray [39] is shown in Figure 7.3(a), where the height of the patches was varied to get the progressive phases for the maximum reflectivity in the desired direction. This non-resonant structure improves the bandwidth due to the extraction of dielectric losses. Figure 7.3(b) depicts a three-dimensional reflectarray element based on reflectarray DRA (dielectric resonator antenna) technology [40]. The dimensions of the strip printed over the DRA were varied to form a full reflectarray antenna with progressive reflection phase. The reflectarray DRA is also a good candidate for bandwidth enhancement due to its conductor-less design, which reduces mutual coupling between the elements.

The bandwidth enhancement of full reflectarray always requires some extraordinary efforts for the designing, which increases the possibility of higher design complexity. The design complexity could be the main limitation faced by these bandwidth-enhancement techniques in complying with the future 5G and beyond systems. The increment of the operating frequencies from microwave to millimetre waves can drastically affect the design sensitivity of the discussed techniques. The shorter wavelengths can be more prone to the performance degradation of reflectarrays with small fabrication errors. Table 7.1 summarizes the performance analyses of the discussed broadband approaches for reflectarrays. Although the performance of each tactic also depends on the selection of its patch element, a common hypothesis can be made based on the reported literature. As it is mentioned in Table 7.1 that the multilayer and aperture-coupled cascaded designs offer wider bandwidth performance compared to single layer due to their thick substrate. A thicker substrate as compared to single layer also allows them to draw less

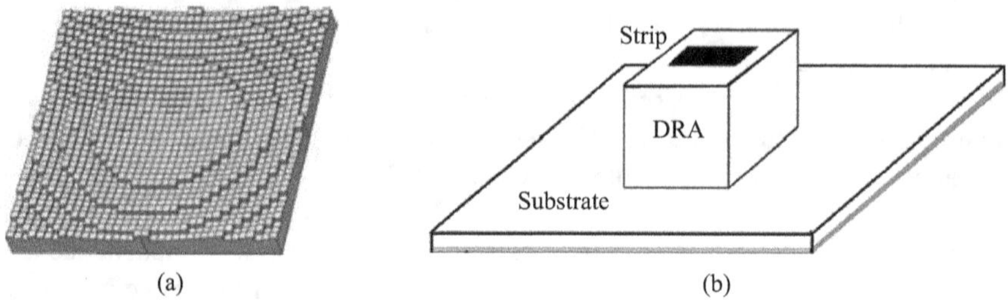

Figure 7.3 Some design configurations of reflectarrays: (a) full conductor reflectarray [39] and (b) reflectarray DRA element [40]

Table 7.1 Summary of the main bandwidth-enhancement techniques [41]

Parameter	Multi-resonance elements		Elements with coupled delay lines		Dual-band designs	
	Single element	Combination of elements	Open-ended stubs	Aperture coupled	Single layer	Multilayer
Bandwidth	N	H	N	H	N	H
Loss	N	H	N	L	H	L
Complexity	H	H	L	H	N	H
High-frequency compatibility	N	L	N	N	L	N

Symbols refer as H, high; N, neutral; L, low.

reflection losses for the reflectarray operation. Alternatively single-layer designs are more prone to get high reflection losses due to thin substrates. The loss performance of single-layer dual-band designs and combination of elements for multi-resonances is higher as compared to other single-layer techniques. It is mainly because of the less electrically conducting areas of their patch elements, which are commonly used to achieve wide reflection phase swing for bandwidth enhancement. The patch elements with less electrical areas such as rings and loops are more flexible in dimension to get wide, progressive phase distributions. The multilayer designs also face more design complexity due to the attachments of multiple layers together. The designs with combination of elements for multiple resonances and dual-bands are also complex due to the handling of two or more resonant structures for bandwidth enhancement. The elements with open-ended phase tuning stubs are less complex than other approaches because of the non-resonant behaviour of the tuning stubs.

The compatibility of these design approaches with high-frequency 5G and beyond reflectarrays depends on the consistency of their performance. The single-layer designs with combination of elements are less likely compatible with higher frequency, where two or more resonant structures could make it more difficult to perform progressive phase analysis with short dimensions. The other single-layer structure with multi-resonances and connected with open-ended stubs are possibly compatible with higher frequencies because of their simple designs. The multilayer designs could make it difficult to reach out the requirements for higher frequencies due to their fabrication sensitivity. On the other hand, the use of conventional patch elements with multilayer designs could possibly increase their chances of compatibility.

7.3 High-gain reflectarray design techniques

Along with bandwidth, the gain of a reflectarray antenna is also an important factor when a higher data rate with large throughput for wider coverage area is required. The reflectarray is a directional antenna; therefore, its gain in a particular direction is normally higher than an ideal isotropic radiator. Its gain depends on its aperture size; a large aperture size is essential for high-gain applications. A two-dimensional reflectarray with a pencil beam acquires higher gain than a linear reflectarray with fan-beam patterns. The spillover and ohmic losses are the main contributors for the degradation of the gain performance. Spillover losses depend on the electric aperture of reflectarray along with the position and the type of the feed used, while the ohmic losses are generated due to the dissipation of energy within the material used in its fabrication. Other factors like element type, element features and its position can also affect the gain of the reflectarray. Element gain and its beamwidth play an important role in gain enhancement. It is highly necessary to have high-gain and narrow-beam elements in the middle of the array, whereas wide-beam elements at the corners for high-gain reflectarray. It is because the middle elements can easily accumulate the incoming feed signals due to normal incidence, which is not possible for corner elements. However, this may increase the design complexity of the reflectarray antenna. A low side-lobe level (SLL) with negligible level of cross polarisation is also essential for high-gain value [1]. The gain and efficiency of reflectarray are correlated to each other; however, in this section, the main emphasis has been given to the gain-enhancement techniques, and approaches regarding efficiency enhancement will be discussed later in the next section.

The very common and simple approach to increase the gain of a reflectarray antenna is to increase its aperture size. A large two-dimensional aperture can achieve high-gain values due to a pointed pencil beam [42]. However, a large physical aperture can degrade the performance of the reflectarray antenna in a way that the signals coming from feed may not coincide with the edge elements. This effect has been explained in Figure 7.4(a), where feed signal is not reaching the edge elements and reducing the electrical aperture of the reflectarray by causing illumination losses. Increasing the f distance of the feed can somehow eliminate this issue, but a large f distance also has its own consequences. Another issue is depicted in Figure 7.4(b), where feed signals are exceeding the physical aperture of reflectarray and generating the spillover losses due to the diffracted waves from the

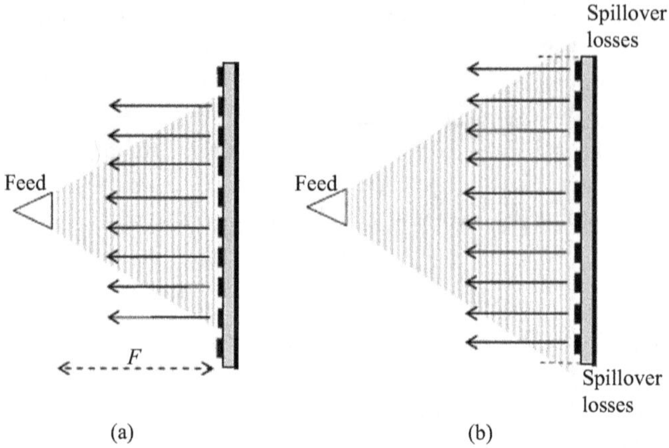

Figure 7.4 Reflectarray antenna with (a) electrical aperture less than its physical aperture and (b) spillover losses

edges [1]. The signals coming from the feed are not fully utilised due to the spillover losses.

It shows that the illumination and spillover losses are complementary to one another. The beamwidth of the feed antenna can be properly adjusted to control the illumination and spillover losses. Additionally, these losses can also be controlled, and gain can be improved by an additional sub-reflector with reflectarray [43]. The sub-reflector can be properly designed to reflect the signals from a feed to exactly pointing them to the aperture of the main reflectarray. This tactic eliminates the drawback of positioning the feed at larger distances from reflectarray to accumulate the physical aperture. However, the designing efforts can be doubled with increment in the complexity and cost of the system. Another way to increase the gain of the reflectarray antenna is to decrease its ohmic losses contributed by the conductor and dielectric materials. This can be done at the unit cell level where the reflection loss can be optimised by various design parameters [44]. However, low loss and wide bandwidth unit cells can contribute to large-phase errors [8,45], making it difficult to get low SLLs for high-gain performance.

The gain of a full reflectarray can be governed by giving the emphasis on its unit cell design parameters. The type of its patch elements, dielectric material and its scattering parameters can drastically affect the performance of a full reflectarray. Additionally, a high-gain full reflectarray can also be analysed by its profile, type (full metal, full dielectric or conventional) and feeding mechanism. Each of the ways of improving gain performance of reflectarray antenna that has been discussed in this section can also significantly affect some other reflectarray parameters. Table 7.2 summarizes the performance of all major techniques used for the gain enhancement in reflectarrays. The gain-enhancement performance of sub-reflector method is quite impressive due to the diminution of the phase errors. The reduced phase errors make it a potential way to achieve high gain with less SLL and

Table 7.2 Summary of the main gain-enhancement approaches [46]

Parameter	Gain enhancement by				
	Sub-reflector	Feeding tactic	Type		
			Microstrip	Dielectric	Metallic
Gain	H	N	N	H	H
Loss	N	N	N	L	L
Complexity	H	L	L	H	H
High-frequency compatibility	L	N	L	H	H

Symbols refer as H, high; N, neutral; L, low.

low cross polarisation. The reflection loss of this technique totally depends on the architecture of main and sub-reflectors. The suitable selection of the patches and substrate material is important for less loss performance. However, due to the involvement of dual reflectors, this technique is quite complex in terms of designing implementation. This also makes it less compatible for higher frequencies, which are required for 5G and beyond operation. The gain of reflectarray antenna can be well improved by the selection of its feeding mechanism with less design complexity. The position and type of the feed are important when expecting an improved gain performance. However, the dependency of resonant frequency on the feeding mechanism can entirely change the desirable output of the reflectarray.

The gain and loss performance of a conventional microstrip reflectarray antenna totally depends on its design parameters. On the other hand, due to the availability of various design approaches for its performance improvement, it offers less complexity for designing that is essential for high-frequency operation. However, at high frequencies the conventional methods are less reliable to attain best performances. Dielectric and metallic reflectarrays, on the other hand, are handy at higher frequencies, but at the cost of high design complexity. A special mechanism is required to fabricate the dielectric or metallic reflectarrays. The absence or reduction of conducting material in dielectric reflectarrays and dielectric material in metallic reflectarrays makes them suitable for optimised loss performance. The miniaturised designs of dielectric and metallic reflectarrays could become good candidates for future 5G and beyond communication if their design complexity is somehow reduced.

7.4 Techniques for high-efficiency reflectarrays

The efficiency of reflectarray antenna is directly related to the difference between its directivity and gain performance. This difference is actually called the overall loss of the reflectarray antenna. The total efficiency of reflectarray accounts all the losses of the system, including spillover loss, illumination loss, patch loss, delay line loss (if attached), feed loss and cross-pol loss [2]. The aperture efficiency accumulates illumination and spillover losses as it is shown earlier in Figure 7.4. These losses are related to the aperture size of the antenna and governed by a proper *f/D* ratio. The aperture efficiency is in direct proportion with the aperture size of the reflectarray, when a proper feed position is selected. But a large aperture size can limit the performance of the reflectarray antenna by introducing large-phase delays to the corner elements with slight change in frequency. The remaining sources of losses have been highlighted in Figure 7.5, where E_i is the direction of the incident electric field from the feed. The feed loss (F_d) is associated with the feed of the reflectarray antenna. The patch loss or reflection loss (RL) is the summation of conductor and dielectric losses of the patch element. The cross-pol loss is related with $E_{r\text{-}cross}$ component of the reflected field, which is normally lower in magnitude than the $E_{r\text{-}co}$ component.

The delay line and feed losses are mostly neglected; however, patch and cross-pol losses are related to the performance of reflectarray unit cell element. Patch loss is the reflection loss of the element, and cross-pol loss can occur due to dissimilarity or asymmetry in the design of the unit cell element. The efficiency of reflectarray antenna is largely affected by its phase error and cross-polarisation performance [1,11]. The reduction in the phase errors is linked with higher efficiency performance with low SLLs, while the unwanted reflections from reflectarray surface can be minimised by controlling its cross-polarisation level. These factors can significantly increase the efficiency of reflectarray. The high-efficiency reflectarrays can be analysed with various types of unit cell elements or by different techniques implemented to its full structure. The reflection phase range of unit cell

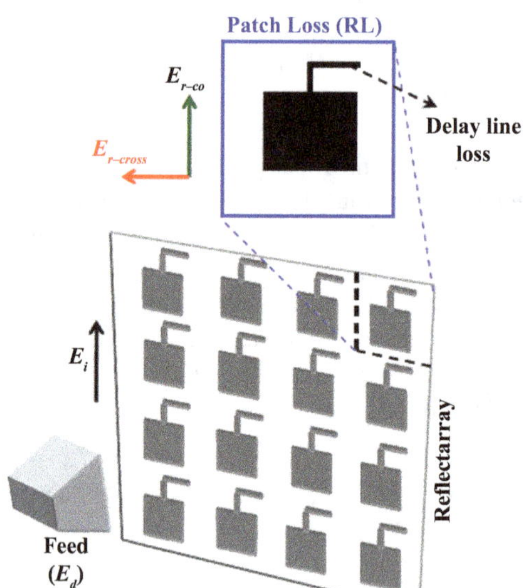

Figure 7.5 Representation of the sources of losses for a reflectarray antenna [46]

element can be increased to reduce the phase errors and hence increase the efficiency. On the other hand, a full reflectarray antenna can be set to reduce its cross-polarisation level by optimising its design mechanism in various ways.

The unwanted cross-polarisation reflections of reflectarray antenna can be controlled with various approaches used by researchers. The proper arrangement of the elements on the surface of the reflectarray and its feeding mechanism can be used to optimize its cross polarisation and hence efficiency performance. Likewise a potential work was reported in [47], where the cross polarisation of the reflectarray antenna was controlled by the proper arrangement of the elements on its surface. A circular element with two open-ended phase tuning stubs was used for this purpose. The elements were arranged in such a way that each element was set by mirroring the design of its adjoining element, as shown in Figure 7.6(a). This approach was proposed to reduce the dissimilar reflections from the reflectarray surface and hence enhance its performance in terms of gain and efficiency. A 21×31 reflectarray was tested with the proposed configuration, and significant improvement in cross-polarisation level was achieved compared to conventional arrangement of elements. The cross-polarisation level was reduced from 1 to 12 dB in E and H planes at different frequencies. This approach was also used to enhance the gain performance by 1.3 dB as compared to conventional design.

A dual-band design for X-band and Ka-band operation with two different layers of reflectarray and dual feed was proposed in [48]. The Ka-band reflectarray placed above the X-band reflectarray separated by an air gap has been shown in Figure 7.6(b). It was shown that the FSS-backed Ka-band reflectarray attained an efficiency of 42%, compared to a solid-grounded X-band reflectarray, which offered 60% efficiency. The frequency selective surface (FSS) ground was used in the Ka-band reflectarray in order to reduce the blockage of the signals reflected from X-band reflectarray. Additionally, the FSS ground plane could also introduce some back radiations, which were the main reasons behind the less efficiency of Ka-band reflectarray. Some other types of reflectarrays like full conductor reflectarray [39] and dielectric resonator reflectarray [49] were also tried to get high

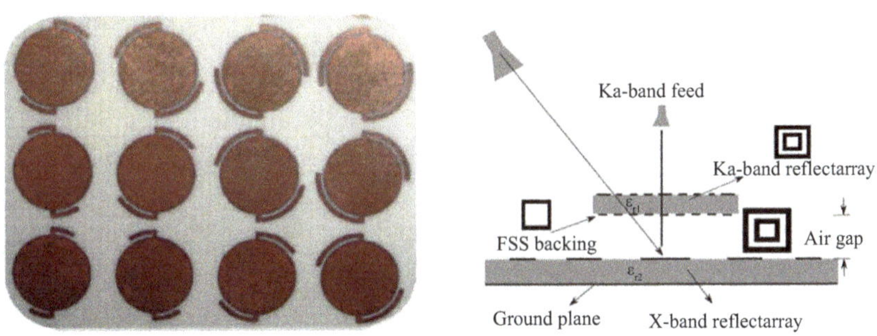

(a) (b)

Figure 7.6 (a) Mirroring of elements for cross-polarisation reduction [47] and (b) dual-layer reflectarray with Ka-band/X-band of operation [48] efficiency. Their main aim was to reduce the reflection losses for an efficient performance. The conductor cell reflectarray offered a higher efficiency of 50% as compared to a DRA reflectarray that offered 47%. Another advantage of conductor reflectarray was its millimetre wave operation at 95 GHz, while DRA reflectarray was designed at a lower frequency of 12 GHz.

As mentioned earlier, the feed mechanism can also affect the gain and hence the efficiency performance of reflectarray antenna. In a proposed design of a reflectarray antenna, feed distance from reflectarray was varied [50] along with some other amendments to get an optimised gain and efficiency performance. The resonant patch elements were also tested with two different sizes of $\lambda/2$ and $\lambda/3$ at 10 GHz frequency. The reflection phase value of the centre element with respect to a centre feed was taken as a phase reference value. That phase reference value was varied between 0°, 60° and 120° with a variable f/D ratio of 0.25–0.5 for the investigations. As it is depicted in Figure 7.7, the maximum gain performance with an efficiency of 60% was obtained at an element size of $\lambda/2$ with 60° phase reference and 0.33 f/D ratio. The stated work actually demonstrated that the gain and efficiency performance of a reflectarray antenna can be evaluated based on its element configuration, reflection phase and feeding mechanism.

Another useful technique regarding the feeding of a reflectarray antenna was proposed in [51], where a dipole antenna was used instead of a conventional feed horn antenna. Due to the omnidirectional characteristics of dipole antenna, the f/D ratio was drastically reduced to 0.3λ at a frequency of 1.87 GHz. But this reduction in the f/D ratio also reduced the efficiency performance of the reflectarray antenna. As described in Figure 7.7(b), the efficiency of the proposed reflectarray antenna was improved by combining the reflected waves of reflectarray and radiated waves of the dipole antenna together. Through this tactic, a gain of 11.2 dB was obtained with an efficiency of 52.6%.

To summarize the efficiency-improvement techniques of reflectarrays, it can be concluded that the unit cell element or the whole reflectarray antenna characteristics can be taken into account for the enhancement of its efficiency performance. A unit cell can contribute through a wide phase range element for the

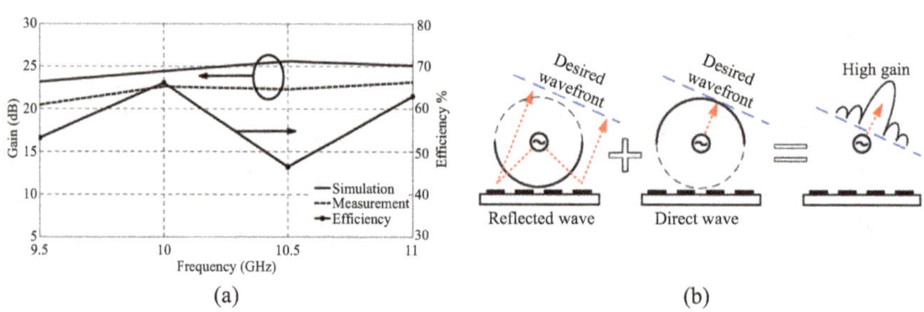

Figure 7.7 (a) Gain and efficiency with respect to frequency of the reflectarray with λ/2 element size [50] and (b) the combination of radiated and reflected waves for efficiency improvement of reflectarray [51]

performance improvement of the full reflectarray made of it. This task requires a lot more efforts than just a conventional design approach. Some new design approaches can actively perform the task. However, some complex derivations of the resonance behaviour along with its detailed analyses are required for each novel design of the unit cell element. Additionally, each design holds its own pros and cons for the performance improvement. It is an efficient but a time-consuming approach, which can contribute negatively to the designing of a 5G and beyond reflectarray antenna. On the other hand, controlling the characteristics of a full reflectarray antenna for the optimisation of its efficiency performance is much easier to perform. Considering the analysis of a reflectarray antenna as a whole reduces the designing efforts and time for the performance analysis. The main disadvantage of this tactic is that while improving one parameter of reflectarray antenna, it can also marginally affect its other parameters.

7.5 Polarisation diversity in reflectarray

The direction of the orientation of the reflected electric field from reflectarray represents its polarisation. It can be divided into horizontal polarisation (HP), vertical polarisation (VP), right-handed circular polarisation (CP) (RHCP) and left-handed CP (LHCP). HP and VP are collectively called linear polarisation (LP), whereas RHCP and LHCP can be represented as CP. Some conventional patch elements, like dipoles and rectangular patches, can easily be utilised for HP or VP, depending on their orientation to the incident field. On the other hand, CP can be achieved by using elements with angular rotation [52]. The CP is composed of two orthogonal components of the propagating electric field. The ratio of the larger component to the smaller one is called the axial ratio, which determines the performance of the CP [53]. A large value of axial ratio represents the elliptical polarisation, whereas a small value (ideally 1) is essential for good performance with CP. The CP is widely used for satellite communications due to its robustness towards environmental losses. Some common types of elements for CP used in the literature are split rings [54], rotated crossed-dipoles [55], aperture coupled patches [56] and circular patch [57]. In a recent work, mechanical rotation technique was utilised by motors attached to loaded split rings for wideband CP operation [58]. CP can also be achieved with a linearly polarised feed by performing polarisation transformation. A double-layer T-shaped element [59] and a circular patch with two elliptical cross slots [60] were used to convert a linearly polarised incident wave to circular, polarised, reflected wave.

The polarisation diversity is a state of operation when a single reflectarray can be used for more than a single polarisation. By doing so, it can be used for multiple applications at the same time. Reflectarrays with dual LP and dual CP can be designed using various techniques. A novel approach of getting four different polarisations from a single reflectarray by the rotation of feed horn was proposed in [61]. A single reflectarray consisting of rectangular elements with 0.3λ element spacing was used for HP, VP, RHCP and LHCP, respectively. VP and HP were obtained rotating the feed position by 90° as shown in Figure 7.8. The LP of the feed was then set parallel to the diagonal line of the patch element. In this way, the linearly polarised wave was decomposed into two orthogonal components parallel to the surface of reflectarray to generate a circularly polarised wave. As illustrated in Figure 7.8, RHCP was obtained by rotating the feed by 45° counter clockwise to HP and LHCP was attained by orienting the feed by 45° counter clockwise to VP. This method provided a simple way to achieve polarisation diversity with a single and conventional reflectarray antenna. However, only a single polarisation can be obtained at a time, and two or more polarisations cannot occur simultaneously through this tactic. Moreover, changing the orientation of the feed horn antenna can also degrade the performance of the reflectarray antenna in terms of its gain and efficiency.

The polarisation diversity in reflectarray can be achieved by double-layer designs with two different patch elements on each layer or with single-layer designs

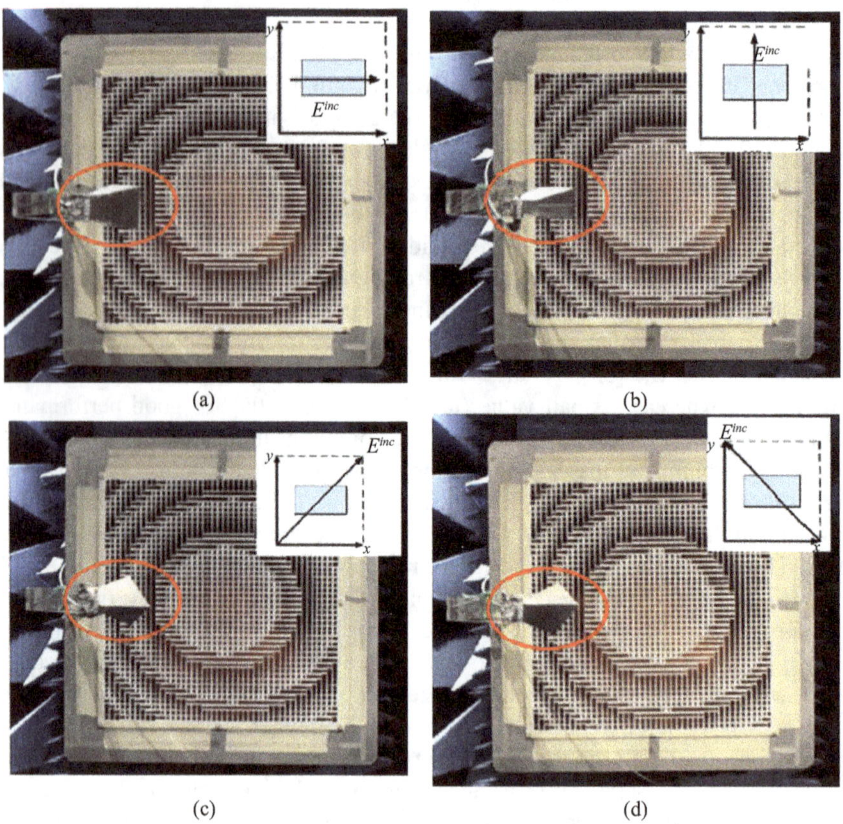

Figure 7.8 Different orientations of feed horn with same reflectarray for (a) horizontal polarisation, (b) vertical polarisation, (c) right-hand circular polarisation and (d) left-hand circular polarisation [61]

with the combination of elements. Similarly, it can be obtained at a single frequency or at two different bands of frequency. The main problem with single frequency dual polarised designs is their higher cross-polarisation level due to the double polarised reflections. The higher value of cross polarisation limits the reflectarray performance in terms of its efficiency, which is not good for a 5G and beyond reflectarray. This issue is more severe in double-layer designs, while this effect can be reduced by single-layer designs. However, the double-layer designs can operate at a single frequency for both desired polarisations, whereas single-layer designs support each polarisation with a different operating frequency. The reflectarray operation at two different frequencies eliminates the concept of frequency reuse for 5G and beyond communications. Producing polarisation diversity in reflectarrays requires some extraordinary efforts for the design and realisation of its unit cell element. It can possibly increase the design complexity along with the processing time for its completion. Moreover, the testing procedures of a reflectarray antenna with polarisation diversity are also more challenging than that of a single polarised reflectarray. A special kind of measurement setup is required for the scattering parameter measurements of a dual polarised unit cell reflectarray element. Additionally, the radiation-pattern measurements of full reflectarray also require special considerations for each individual polarisation [62].

7.6 Adaptive beam steering in reflectarrays

The fixed beam reflectarray with high gain and wide bandwidth is flexible to be used for many applications. However, its usefulness can increase to a new degree of freedom if its main beam can be steered in various directions. The coverage area can also be increased by pointing its beam in different directions. The conventional way of performing beam steering by an antenna is to physically rotate it, which is called mechanical beam steering. The mechanical beam steering can also be performed with reflectarray antenna, but basically reflectarray is not made for it. Alternatively, its counterpart parabolic antenna can easily attain beam steering linked with its mechanical movements. The beam steering in reflectarray antenna can be divided into passive and active beam steering. Passive beam steering is not very common and it involves in the physical orientation of its whole structure or a part of its structure. On the other hand, reflectarrays are well known for their capability of acquiring active or electronic beam steering.

The beam steering in reflectarrays totally depends on the reflection phase characteristics of its unit cell element. The variation in the reflection phase can make its main beam to move towards a certain direction. In active or electronic beam steering, the direction of the main beam can be controlled by an external biasing source. The amount of that biasing is usually negligible to affect the radiation parameters of the reflectarray. The active beam steering in reflectarrays can be performed either by controlling its material properties or by imposing capacitive or inductive loading to its radiating part [63]. The properties of dielectric or conductor part of the reflectarray can play important role in its resonant characteristics. The amplitude and reflection phase of the reflectarray depend on the behaviour of its material properties. Subsequently, its radiating characteristics can also be controlled by optimising the surface currents on its patch element.

The electronic beam steering in reflectarray antenna is a need of time, but usually it comes with drawbacks of high SLL and scan loss. As the main beam of the reflectarray antenna steered from its broadside direction, the SLL increases due to the undesired reflections from a portion of the reflectarray in the direction of generating side lobes. Moreover, the magnitude of the steered beam also deteriorates with beam steering due to the reduction of reflection energy in the direction of beam steering. These issues can be optimised by employing an amplifying circuit or by extending the range of progressive phase distribution of the elements. The first technique may increase the cost, profile and complexity of the system, while the second requires an extra amount of time for designing.

Beam-steering techniques have usually been employed in the conventional microstrip reflectarray antenna due to its ease of creating a current transition in the resonant structure. Other high-gain designs such as DRA reflectarrays and full metallic reflectarrays are still under development for beam-steering realisation. Moreover, the aperture-coupled reflectarray element designs are more liable to achieve electronic tunability with integrated lumped components as compared to a single-layer design. They provide wide bandwidth, low loss and high isolation between resonant and tuning structures, but at the cost of high design sensitivity. Alternatively, single-layer designs are easy to fabricate with surface-mounted lumped components, but with a high possibility of extra reflection-losses with limited bandwidth. Table 7.3 briefly compares the performance of major beam-steering techniques used in reflectarray antenna design. PIN and radio-frequency microelectromechanical system (RF-MEMS) are used for beam switching instead of a continuous beam steering due to their digitised nature. The number of PIN diodes or switching states in RF-MEMS increases in the unit cell reflectarray with an increasing demand of more switchable beam positions. On the other hand, the analogue nature of other beam-steering tactics allows them to be biased collectively in a combined manner. High loss is always an issue with RF-MEMS and liquid-crystal-based reflectarrays, which can significantly affect the gain performance of a

Table 7.3 Summary of adaptive beam-steering techniques [62]

Parameter	**Tunable materials**			**Lumped components**		
	Liquid crystal	**Ferroelectric**	**Graphene**	**PIN**	**Varactor**	**RF-MEMS**
Beam control	C	C	C	D	C	D
Biasing	A	A	A	D	A	D
Loss	H	N	L	L	L	H
Complexity	N	N	H	H	H	H
High-frequency compatibility	N	L	H	N	N	L

Symbols refer as C, continuous; A, analogue; D, discrete/digital; H, high; N, neutral; L, low.

5G and beyond reflectarray. The complex nature of RF-MEMS and the high value of dissipation factor of liquid crystal material are the main reasons behind that issue. Due to the fabrication and testing issues, the design complexity of lumped-components-embedded reflectarray unit cells is always high. It is also a challenging work to produce a graphene-based reflectarray antenna. The fabrication of a single graphene layer with high conductivity is a difficult task to achieve, but due to this reason it also holds a high compatibility with millimetre wave or even with THz frequencies. Due to this reason, graphene could be a potential candidate for future 5G and beyond communications.

7.7 Design of a mm-wave reflectarray antenna for 5G communication systems

Low losses, low cost, high efficiency and electronic beam steering are some of the significant features of reflectarray antennas that make them suitable for 5G communication systems. On the other hand, bandwidth of reflectarrays still has the margin of improvement and draws attention of various researchers. Numerous efforts for designing wideband reflectarrays are already being discussed in the earlier sections.

This section demonstrates the characterization of millimetre wave reflectarray antenna based on the requirements of modern 5G communication systems. Various performance parameters are characterized for different design configurations of unit cell reflectarray elements. Moreover, periodic reflectarray designs are also proposed in order to compare the performance of the selected design configurations. Furthermore, design and practical validation of a beam-steering reflectarray antenna based on mechanical movement of array designed at 26 GHz has been elaborated.

7.7.1 Design and fabrication of unit cells

In a reflectarray antenna, the total electric fields are the sum of incident fields, reflected fields from the ground plane and the fields scattered by the resonant patch elements. Mathematically, the total electric fields can be expressed as

$$\bar{E}_t = \bar{E}_i + \bar{E}_r + \bar{E}_s \tag{7.1}$$

where $\bar{E}_i = \bar{E}_0 e^{jk_0(xu_i + yv_i - z\cos\theta_i)}$ is the incident field from the reflectarray feed with $\bar{E}_0$ defining amplitude and polarization of the incident field, and u_i and v_i are the direction cosines of the wave. The reflected and scattered fields in [1] are also dependent on the incident field and can be given by

$$\begin{bmatrix} E_{r\theta} \\ E_{r\varphi} \end{bmatrix} = \begin{bmatrix} R_{\theta\theta} & 0 \\ 0 & R_{\varphi\varphi} \end{bmatrix} \begin{bmatrix} E_{0\theta} \\ E_{0\varphi} \end{bmatrix} e^{jk_0(xu_i + yv_i - z\cos\theta_i)} \tag{7.2}$$

$$\begin{bmatrix} E_{s\theta} \\ E_{s\varphi} \end{bmatrix} = \begin{bmatrix} S_{\theta\theta} & S_{\theta\varphi} \\ S_{\varphi\theta} & S_{\varphi\varphi} \end{bmatrix} \begin{bmatrix} E_{0\theta} \\ E_{0\varphi} \end{bmatrix} e^{jk_0(xu_i + yv_i - z\cos\theta_i)} \tag{7.3}$$

where $R_{\theta\theta}, R_{\varphi\varphi}, S_{\theta\theta}, S_{\theta\varphi}, S_{\varphi\theta}$ and $S_{\varphi\varphi}$ are the plane-wave coefficients of reflected and scattered electric field components, respectively. Hence, the total loss in a reflectarray has reflected and scattered components as

$$\bar{\bar{\Gamma}}(\theta_i, \varphi_i) = \bar{\bar{\Gamma}}_R(\theta_i, \varphi_i) + \bar{\bar{\Gamma}}_S(\theta_i, \varphi_i, D_i) \tag{7.4}$$

where D_i is one or more than one variable dimension of the ith patch element. This dimension is required to be varied in order to modify the phase of each unit cell and obtain a progressive phase distribution in a periodic reflectarray environment.

Therefore, in this work, different design configurations have been investigated in order to characterize the reflectarrays designed at 26 GHz. Reflection loss and reflection phase performance are compared for the proposed elements in accordance to the 5G communication system requirements. The studied design configurations are conventional rectangular patch elements (Figure 7.9(a)), rectangular patch with rectangular slot (Figure 7.9(b)), rectangular patch with circular slot (Figure 7.9(c)), rectangular ring (Figure 7.9(d)) and circular ring (Figure 7.9(e)).

Simulations of the unit cells have been carried out in CST MICROWAVE STUDIO® (CST MWS), using appropriate boundary conditions in order to represent an infinite array by a single element. Rogers RT/duroid® 5880 ($\varepsilon_r = 2.2$, tan $\delta = 0.0009$) has been used as a dielectric material with a copper cladding of

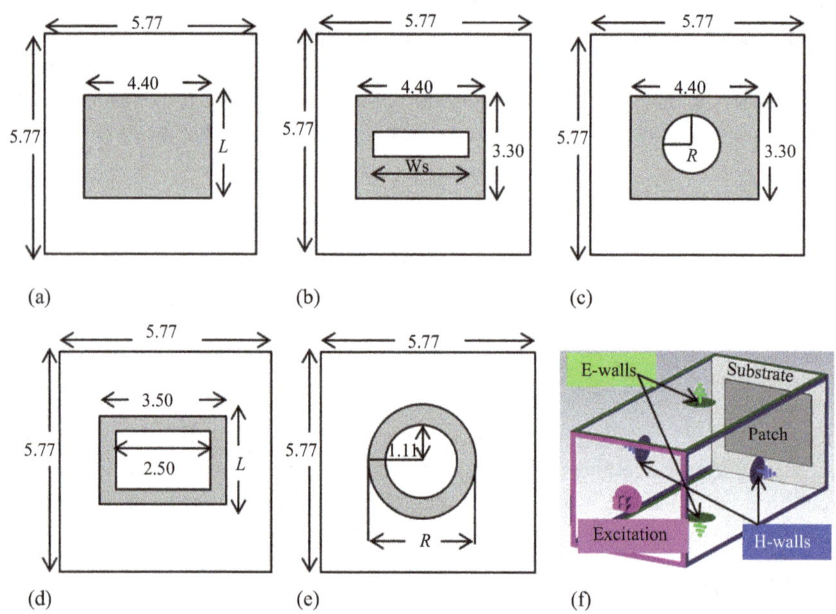

Figure 7.9 Investigated design configurations at 26 GHz: (a) rectangular patch, (b) rectangular patch with rectangular slot, (c) rectangular patch with circular slot, (d) rectangular ring, (e) circular ring and (f) unit cell with infinite boundaries

0.035 mm and substrate thickness of 0.254 mm. Progressive phase distribution was achieved by varying different dimensions of the unit cells designs. For conventional rectangular patch element, the element length was varied from 3.3 to 3.9 mm. In the case of rectangular and circular-slot-embedded patch elements, the slot width and radius were varied from 0.04 to $0.49w_p$ and 0.04 to $0.24w_p$, respectively (w_p represents the width of patch element). For rectangular ring, the outer length of the ring was varied from 2.0 to 2.5 mm, while in the case of circular ring the outer ring radius was varied from 1.51 to 1.71 mm. These variations in the dimensions provided a phase swing close to ideal 360° for each of the design configurations. The proposed unit cells reflectarray elements were fabricated using Rogers RT/duroid® 5880 substrate. Each unit cell was fabricated with two resonant elements in order to include the possible mutual coupling effects of the neighbouring elements. A waveguide simulator was also fabricated to operate between 24 and 28 GHz. Figure 7.10 shows the fabricated unit cell elements with different design configurations and the scattering parameter measurement setup. A number of unit cells for each design configuration with different dimensions were fabricated in order to achieve the progressive phase distribution required for reflectarray design. The fabricated unit cells were inserted in the aperture of waveguide which was connected to the network analyser from the other end through a waveguide to coaxial adapter. Reflection loss and reflection phase curves for different designed unit cells were obtained and compared with simulations in order to validate the results.

7.7.2 Scattering parameter measurements and analysis

As highlighted earlier, the main requirements for an antenna in order to achieve the standards of 5G communication are improvements in bandwidth (≈1 GHz), gain

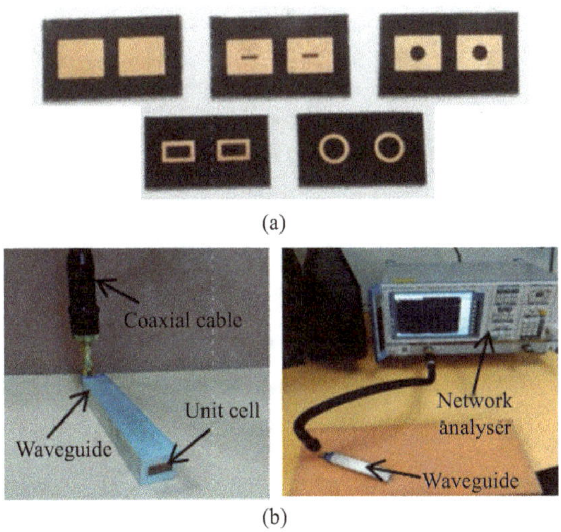

Figure 7.10 (a) Fabricated unit cell and (b) scattering parameter measurement setup

(>25 dB), efficiency (≈50%), polarization diversity and beam-steering capabilities. In the case of a reflectarray antenna, unit cell characterization is usually carried out in order to optimize the performance of the antenna without performing the design and analysis on a full reflectarray. The unit cell characterization saves time and resources and provides useful information required for the optimum design of complete periodic reflectarray antenna. Figure 7.11 provides a comparison between measured and simulated reflection parameter results, plotted against the variable dimension of the resonant element, for different design configurations. It can be observed from Figure 7.11 that, in all the cases, measured reflection-losses as reflection phases are in close agreement with the simulated results. The maximum discrepancy between the simulated and measured reflection-losses is less than 1 dB which validates the reliability of the simulated results. The measured data are tabulated in Table 7.4 in order to compare the performance of different design configurations and characterize them according to 5G antenna requirements.

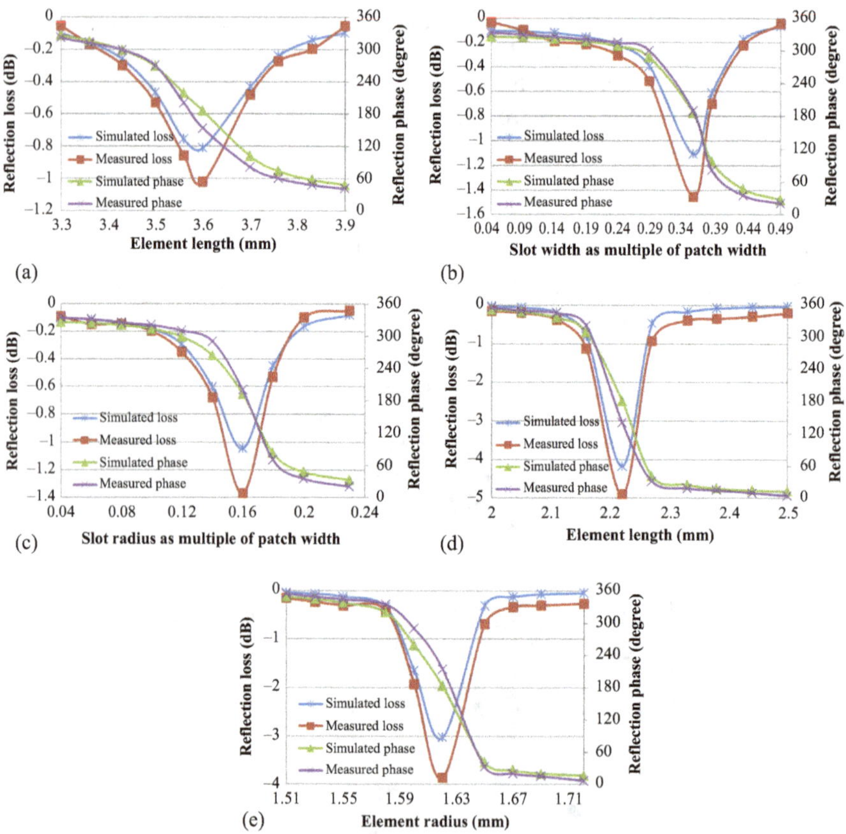

Figure 7.11 Comparison between simulated and measured reflection parameters at 26 GHz: (a) rectangular patch, (b) rectangular patch with rectangular slot, (c) rectangular patch with circular slot, (d) rectangular ring and (e) circular ring

Table 7.4 Measured figure of merit for different reflectarray design configurations

Unit cell design configuration	Reflection loss (RL) (dB)	20% Bandwidth (BW) (MHz)	Phase sensitivity (σ) (°/GHz)	Phase error (φe) (°)
Rect. Patch	1.02	560	122	80
Rect. Slot	1.45	420	160	47
Circ. Slot	1.36	455	157	48
Rect. Ring	4.89	200	263	10
Circ. Ring	3.86	225	238	13

Refection loss and reflection phase errors govern the antenna gain and efficiency, 20% bandwidth of the unit cells can provide an idea about the reflectarray antenna bandwidth, while phase sensitivity and phase error can also be linked to the polarization diversity and beam-steering capabilities of the antenna. It can be observed from Table 7.4 that rectangular patch element provides minimum reflection-loss (1.02 dB) with maximum bandwidth (560 MHz) and low phase sensitivity (122°/GHz). However, this configuration also exhibits the highest phase error (80°) among the investigated designs. On the other hand, because of higher slope of reflection phase curves, ring element configurations (both rectangular and circular) demonstrated higher losses, less bandwidth and low phase errors. This trade-off between reflection loss and phase error is crucial in designing the reflectarray antenna according to the requirement of the application. The higher amount of phase error will limit the available reflection phase of the element in the array. Hence, SLLs will be increased due to limitation in progressive phase distribution.

Additionally, high loss and narrow bandwidth can adversely affect the gain and efficiency performance of the antenna. Phase sensitivity is also important in terms of controlling the SLLs as well as beam steering in reflectarrays. Elements with lesser phase sensitivity provides larger number of dissimilar reflection phases which can be used to control mutual coupling as well as SLLs in the reflectarray antenna design. Hence, the selection of suitable resonant element design configuration for reflectarray antenna depends on the priority of certain performance parameter for a particular application.

7.7.3 Periodic reflectarray design

After characterization of reflectarray unit cell design configurations, periodic reflectarrays have been designed at 26 GHz. Two reflectarrays have been designed using 'rectangular patch element with variable length' (Rect. Patch) and 'circular ring elements with variable radius' (Circ. Ring) in order to compare the performance of these arrays. Rect. Patch was chosen because of its lower reflection loss and high unit cell bandwidth, while Circ. Ring was selected based on the lower phase errors and higher phase sensitivity. Figure 7.12 shows the schematic of the designed reflectarray where 20 × 20 resonant element arrays were designed on

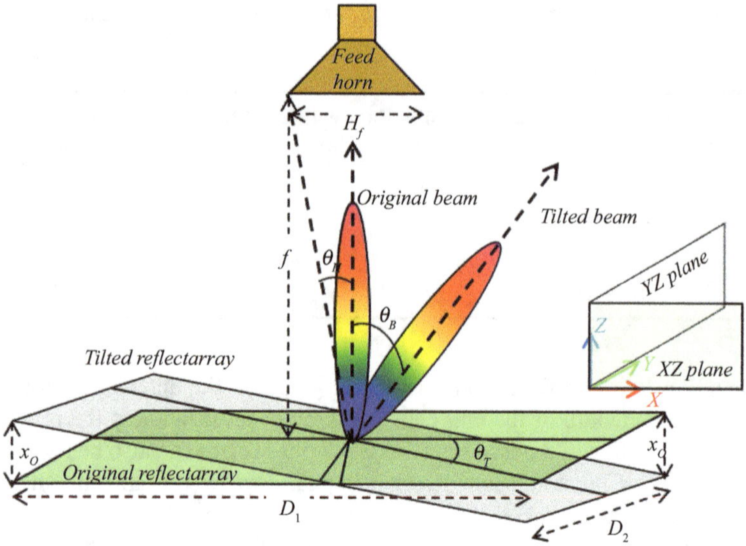

Figure 7.12 Schematic of the proposed reflectarray design

0.254-mm-thick Rogers RT/duroid® 5880 dielectric substrate. The inter-element was optimized to be kept constant at 0.5λ, which made the total size of the array to be 10λ or 115.4 mm. The required phase shift for different elements in the array was determined by very well-established ray tracing method. A centre feed horn was placed above the reflectarray according to $f/D = 1$. In order to avoid the complexities involved with offset feed horn, the designed reflectarray was tilted to a certain angle as shown in Figure 7.12. This helps to avoid the horn blockage and create a reflected beam at an angle θ_B, greater than the horn-blockage angle θ_H expressed as

$$\theta_H = \frac{1}{2}\text{HPBW} + \tan^{-1}\frac{H_f}{2f} \tag{7.5}$$

where HPBW is the 3-dB beamwidth of the desired reflected beam, H_f is the maximum aperture dimension of the feed horn and f is the distance between the array and the feed. The complete assembly of reflectarray antenna designed in CST MWS is presented in Figure 7.13 with simulated direction of the phase of surface-current distribution. It can be observed from Figure 7.13 that the maximum current was concentrated in the centre of the array in the case of 'No tilt'. In this case, the reflectarray will reflect a centre beam shown as 'original beam' in Figure 7.12. However, this beam will have maximum blocking in practical case because of horn-blockage factor. Therefore, the array was tilted to 20° from the *YZ* plane (around *Y*-axis) to obtain a clear reflected beam. It is demonstrated in Figure 7.13 that in the case of tilted array, the phase of the surface-current distribution travels linearly towards one side of the array producing an offset reflection beam. It has to

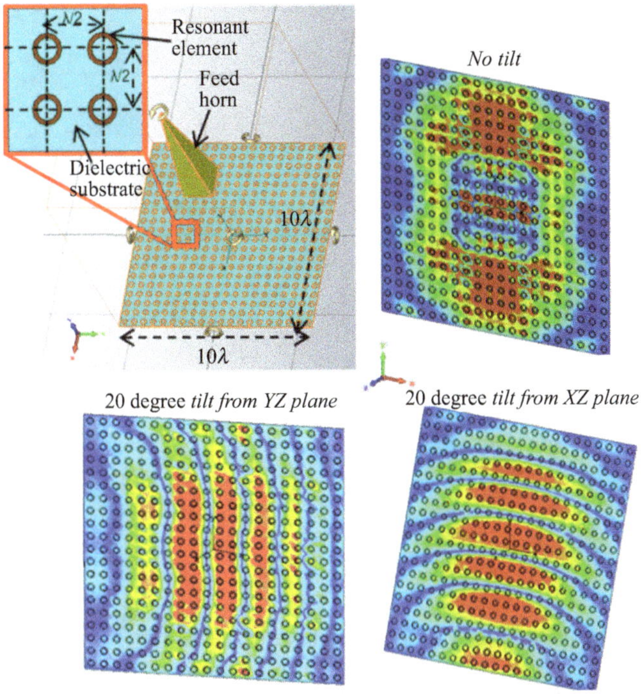

Figure 7.13 Reflectarray antenna with 20 × 20 circular ring elements and the phase of surface current with different orientations

be noticed here that while tilting the array, the progressive phase distribution and placement of individual resonant elements have to be taken into account for accurate beam forming.

The simulated radiation pattern results obtained for different tilt orientations are shown in Figure 7.14. It has been demonstrated in Figure 7.14 that the maximum gain achieved by 20 × 20 element reflectarray with circular ring elements was 26.7 dB at 26 GHz. The simulated gain is almost same in the case of different tilt orientations; however, the radiation pattern shows that in the case of No tilt, the SLLs are significantly higher at −12.4 dB as compared to −19.4 dB in the case of 20° tilt from *YZ* plane. This is because, the reflectarray was designed with elements placed in such a way that it can produce optimum beam with 20° tilt. Hence, changing the tilt angle disturbs the progressive phase distribution and causes radiations in the undesired directions as well. Figure 7.14 also demonstrates that even though the reflectarray is designed symmetrically, there are still higher side lobes (−12.6 dB) in the case of 20° tilt from the *XZ* plane (around *X*-axis). In this case, the reason for higher side lobes is the orientation of the feed horn where the E-fields are excited in the *Y*-direction, making the reflectarray more sensitive in the *YZ* plane.

In order to compare the performance of different investigated reflectarray element design configurations, periodic reflectarray was also designed, using

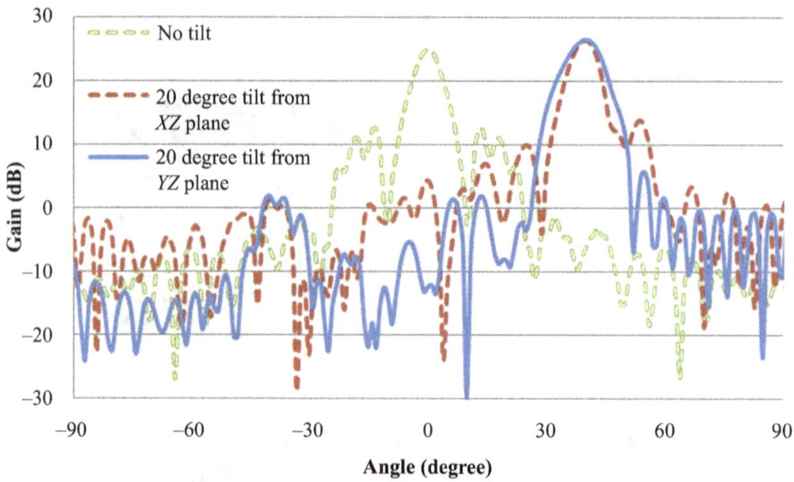

Figure 7.14 Simulated radiation patterns of reflectarray antenna with different tilt orientations at 26 GHz

rectangular patch elements with variable lengths, following the same principle as described previously. In both the cases, the reflectarrays were tilted 20° from *YZ* plane, creating a reflected beam directed 40° in the *XZ* plane. Figure 7.15 shows the comparison between the simulated radiation-pattern results of both the designed reflectarrays in azimuth and elevation planes at 26 GHz. As predicted by the unit cell scattering parameter results, the reflectarray designed with rectangular patches provided a lower gain of 25.7 dB as compared to 26.7 dB provided in the case of circular ring elements. It has to be noted here that although the reflection loss for rectangular patch unit cell was lower (1.02 dB) as compared to circular ring elements (3.86 dB), still the gain is higher for circular ring element reflectarray. This is due to the larger amount of phase error, as shown in Table 7.4 which caused rectangular patch element reflectarray to demonstrate higher SLLs of −18.4 dB as compared to circular ring reflectarray which provided SLLs of −19.4 dB in azimuth plane. The SLLs in the elevation plane are below −20 dB level for both the design configurations. Higher side lobes deteriorate the gain of the amount of energy in the main lobe decreasing the gain of reflectarray antenna.

7.7.4 Reflectarray fabrication and radiation-pattern measurements

Periodic reflectarrays with 20 × 20 elements were fabricated using rectangular patch and circular ring configurations on a 0.254-mm-thick Rogers RT/duroid 5880 substrate. Figure 7.16(a) shows the fabricated reflectarray breadboards, while complete far-field radiation-pattern-measurement setup in anechoic chamber is shown in Figure 7.16(b). A standard horn with WR34 waveguide adapter (22–33 GHz operation frequency range) was used as a feed with *f*/*D* ratio of 1. The

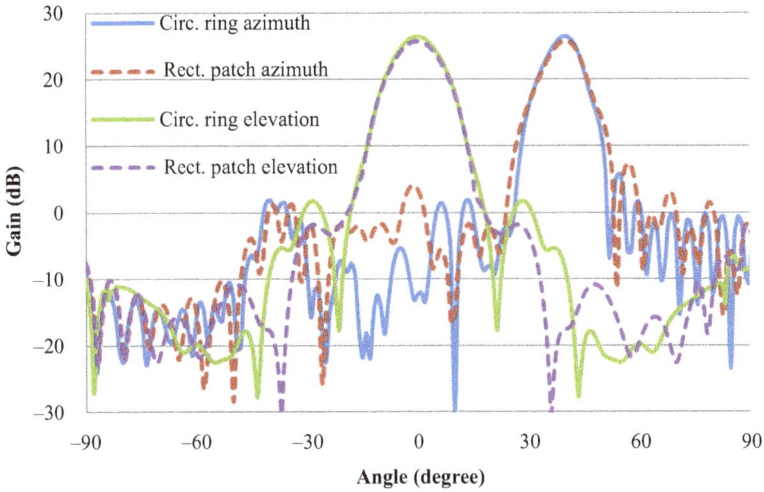

Figure 7.15 Radiation pattern in azimuth and elevation planes for reflectarray with different design configurations of unit cells

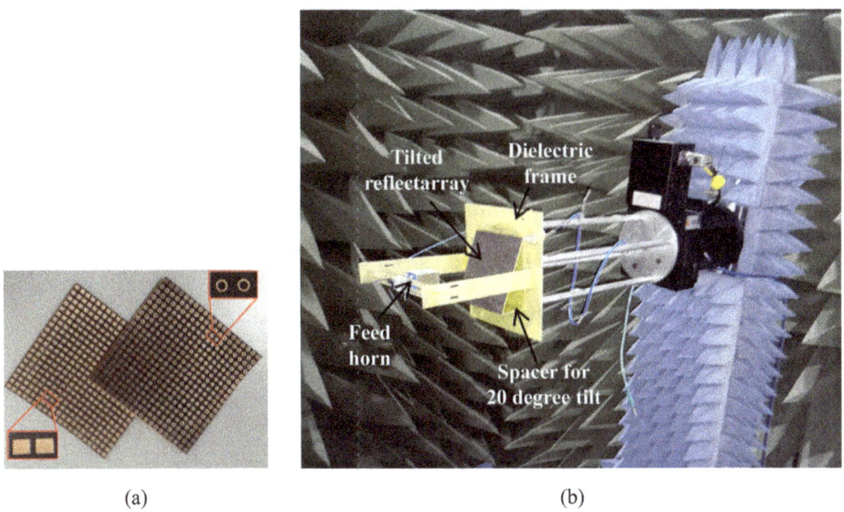

Figure 7.16 Developed hardware: (a) fabricated 20 × 20 reflectarrays with rectangular patch and circular ring configurations and (b) radiation-pattern-measurement setup

fabricated reflectarrays and the feed horn were mounted on a frame made of dielectric material to avoid any extra reflections. The reflectarray was tilted using a 21-mm spacer on one side which provided a 20° tilt as required by the design.

A comparison between the simulated and measured radiation-pattern results of the two designed reflectarray antennas is provided in Figure 7.17. It can be

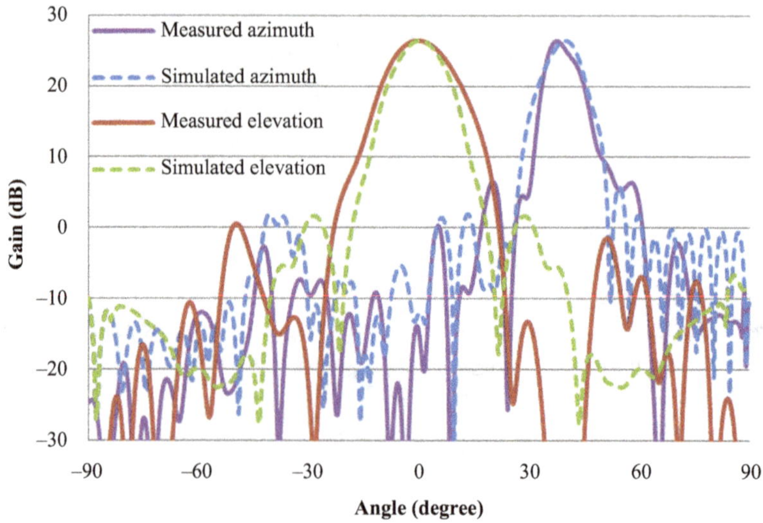

(a)

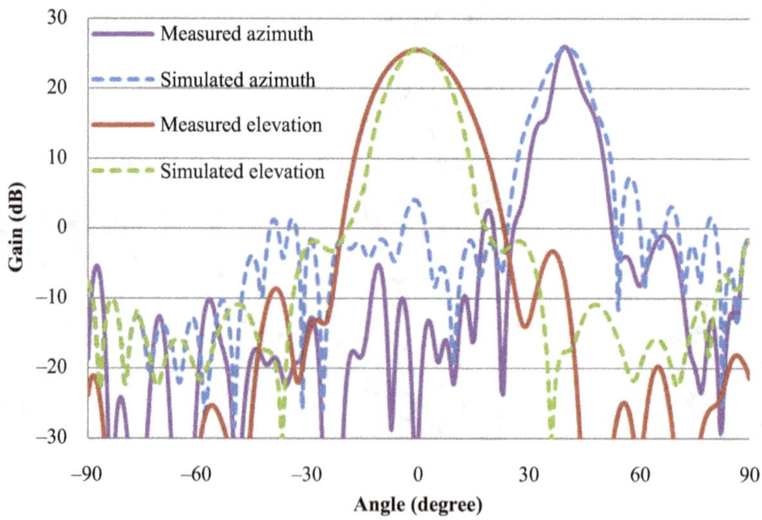

(b)

Figure 7.17 Comparison between measured and simulated radiation-pattern results: (a) circular ring element reflectarray and (b) rectangular patch element reflectarray observed that the measured and simulated results provided a very close agreement. In the case of circular ring elements reflectarray, the maximum measured gain was observed to be 26.45 dB at 37° in the azimuth plane as compared to 26.7 dB at 40° provided by simulations, while in the case of rectangular patch element configuration, the reflectarray provided a measured gain of 25.2 dB at 39.5° as compared to a simulated gain of 25.7 dB at 40° in the azimuth plane. The measured 3 dB beamwidths in the case of circular ring configuration are 9° and 12° as compared to 8° and 15° simulated 3 dB beamwidths for azimuth and elevation planes, respectively. Similarly, for rectangular patch elements reflectarray, azimuth and elevation plane measured 3 dB beamwidth are observed to be 6° and 17° as compared to simulated 3 dB beamwidth of 9.5° in azimuth and 13° in the elevation plane. The difference between simulated and measured gain can be attributed to the extra losses caused by the measurement setup as well as specular reflections and edge diffractions.

The edge diffraction occurs because of the approximation that the tangential fields are zero outside of the reflectarray. Although the diffraction at the edges of reflectarray should not be significant because of the low illumination level, however, more accurate radiation patterns can be obtained by considering both tangential electric and magnetic fields on the reflectarray surface. On the other hand, the slight discrepancy in the direction of main beam can be due to the minute difference in the alignment of antenna or placement of the spacer.

Figure 7.18 shows the measured three-dimensional radiation patterns of co-polar and cross-polar components of both the designed reflectarray antennas. Clear main beams around 40° in the azimuth plane and 0° in the elevation plane can be observed from Figure 7.18.

Moreover, as depicted in Figure 7.19, the measured cross-polar levels are demonstrated to be more than 25 dB below the maximum level of the main beam.

Finally, the bandwidth of a reflectarray antenna can be calculated by plotting the gain at a particular frequency range. Figure 7.20 shows the gains of both the designed reflectarrays obtained by gain comparison method at a frequency range of 24–28 GHz. As highlighted in Table 7.4, in the case of unit cells, rectangular patch elements provided broader bandwidth as compared to circular ring elements. Similarly in the periodic reflectarray design, the reflectarray based on rectangular patch elements provided a 1-dB gain drop bandwidth of 13.6% as compared to 13.1% in the case of circular ring elements reflectarray antenna.

7.7.5 Beam-steering reflectarray

As discussed previously, one of the techniques to steer the main beam of the reflectarray antenna is by mechanical movement of the antenna. In this work, the reflectarray with an attached horn antenna has been tilted to steer the main resultant antenna beam in order to demonstrate the capabilities of reflectarrays to be used as an adaptive beam-steering 5G antenna. Symmetricity of the array and the resonant elements has been exploited to steer the main beam by tilting the reflectarray at different angles. The design parameters and other factors were kept same as in

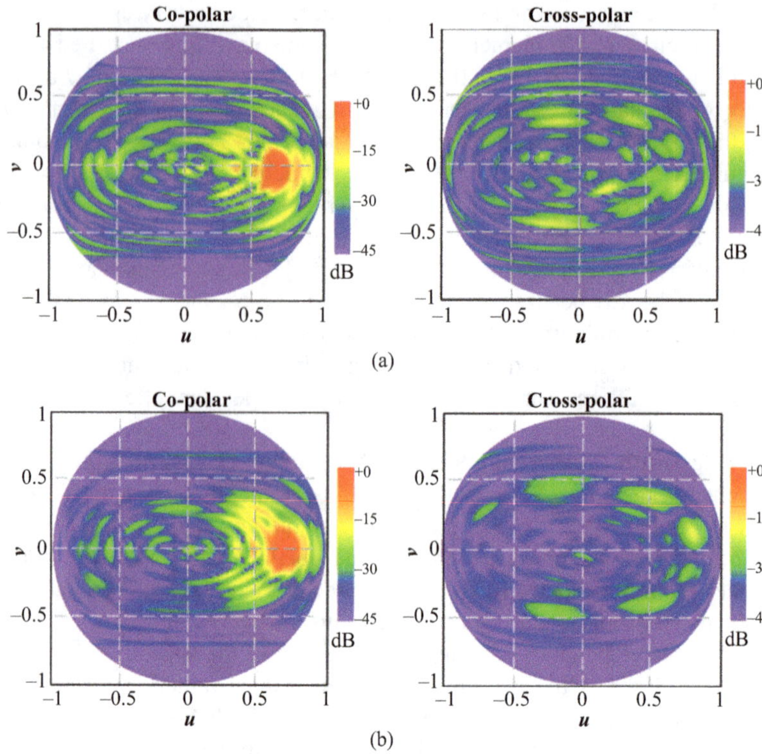

Figure 7.18 Three-dimensional measured radiation patterns for co-polar and cross-polar components: (a) circular ring elements reflectarray and (b) rectangular patch element reflectarray

Section 7.7.4. The progressive phase distribution for a reflectarray designed at 26 GHz in this work was calculated using the following equation:

$$\Delta\varphi = \varphi - \varphi_i = \frac{2\pi}{\lambda}(f - f_i) \tag{7.6}$$

where $\Delta\varphi$ is the change in reflection phase, φ is the progressive phase centre, φ_i is the phase of ith element, f is the distance between feed and array and f_i is the path delay of the ith element with respect to the array centre. According to the previous equation, the change in the reflection phase remains constant for the elements placed on a circle with a particular radius in the reflectarray. Keeping in view this theory, the planar reflectarray was tilted in this work from $-30°$ to $30°$ in the *YZ* plane, in order to steer the reflected beam. It should be noted that the original tilt angle of 20° in the *XZ* plane was still maintained. The tilting of the reflectarray modified the direction of the phase of surface-current distribution as shown in Figure 7.21. The red, solid arrow shows the direction of surface-current phase for the fixed beam reflectarray, while the black, dotted arrow shows the direction with

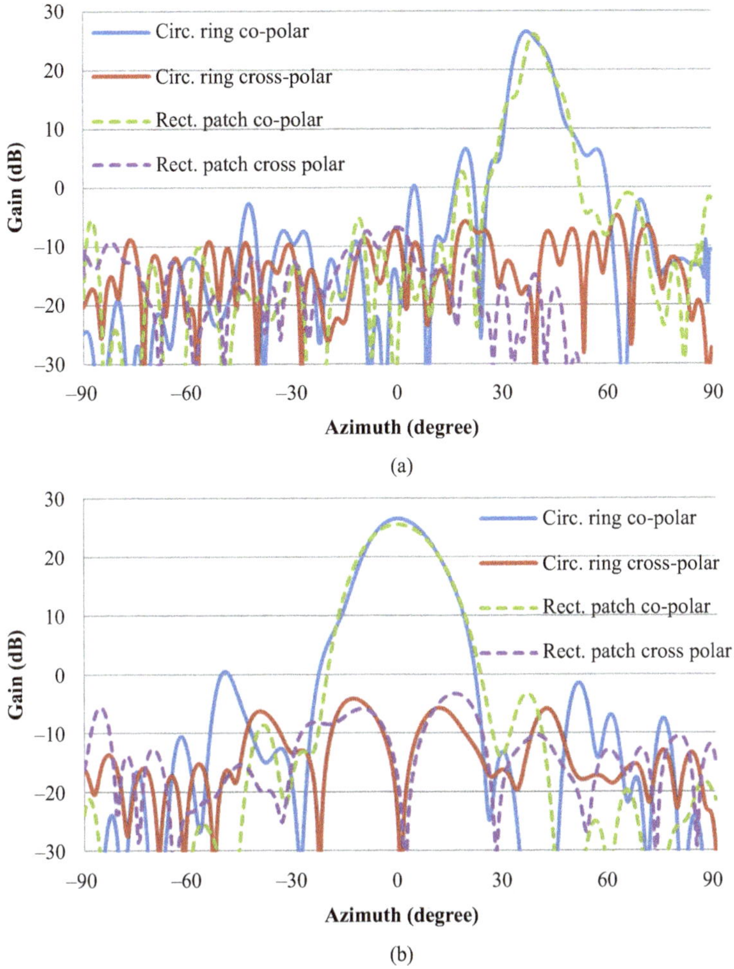

Figure 7.19 Comparison between circ. ring and rect.-patch-based reflectarray radiation-pattern results: (a) azimuth and (b) elevation

particular tilt angles. The phase currents travel in a direction (θs) which is almost double of the tilt angle. Moreover, according to (7.6), the change in the reflection phase is similar for the red, solid arrow as well as black, dotted arrows in all the cases because the feed centre is kept constant. Therefore, the reflected beam is steered in a direction governed by θs.

For the practical validation of this theory, radiation-pattern measurements were carried out for reflectarrays tilted at different angles in *YZ* plane using the same measurement setup as shown in Figure 7.16. The tilting of the array can easily be managed by using a single stepper motor. However, in this work, for demonstration purpose, the array was tilted manually at different angles from $-30°$ to $30°$. *AZ* and

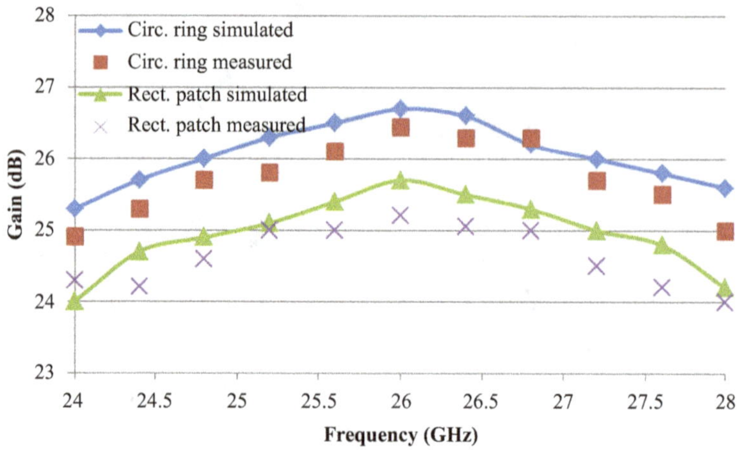

Figure 7.20 Gain variation with frequency for reflectarray with different design configurations

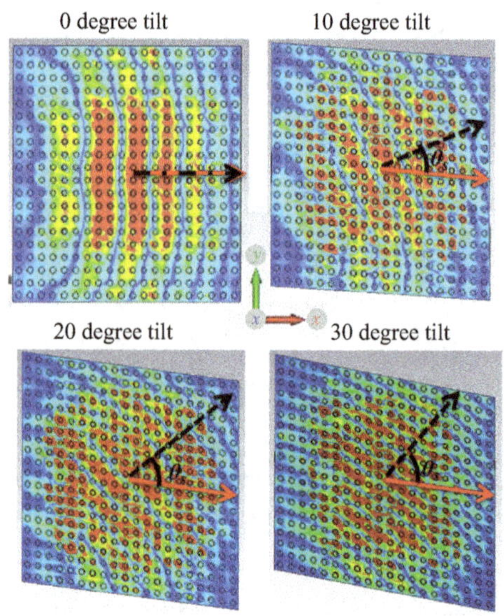

Figure 7.21 Phase of the surface-current distribution of different tilt angles

EL cut radiation patterns were obtained to demonstrate the tilting effect in *XZ* and *YZ* planes, respectively. For the clear visibility, results of only 0°–30° tilt in *YZ* plane are shown in Figure 7.22. The tilt of 0° to −30° demonstrated almost identical results because of symmetricity. Ideally, all the beams were expected to be at 40° in the *AZ* plane. However, due to tilting in the *YZ* plane, the beam direction was slightly disturbed as shown in Figure 7.22(a). The focus of the beam in the *AZ* plane

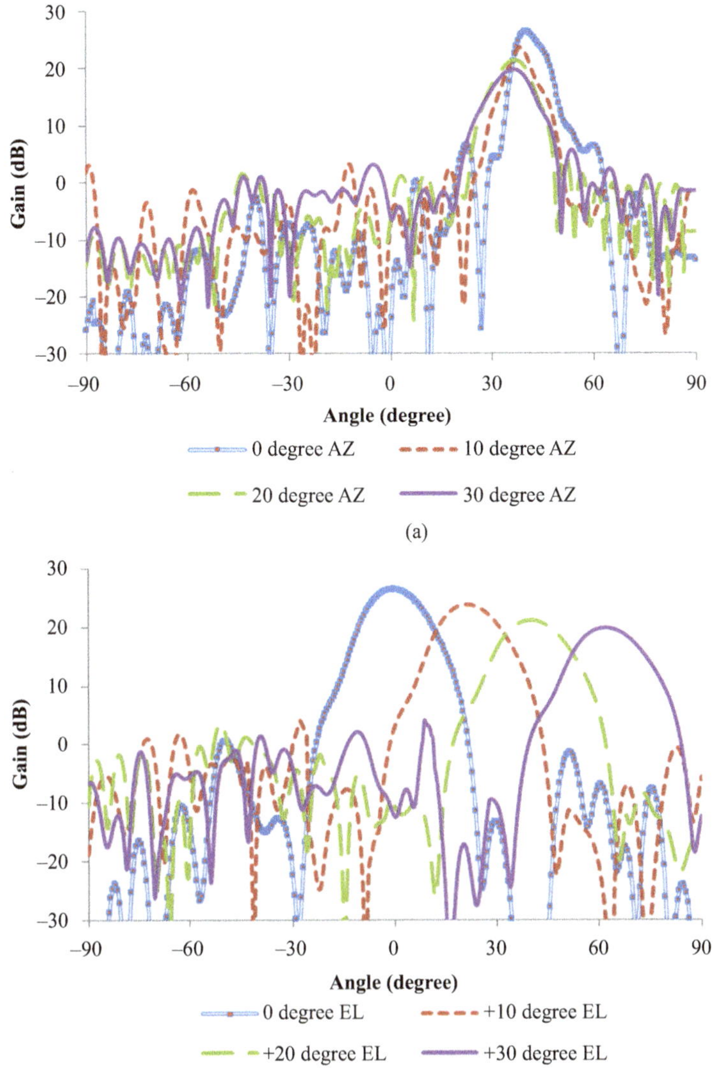

Figure 7.22 Measured radiation patterns for scanned beam reflectarray at different tilt angles: (a) azimuth plane and (b) elevation plane

varied from 40° to 36.3° when the reflectarray was tilted from 0° to 30° in the *YZ* plane. On the other hand, in the elevation plane, the beam steered from 0° to ±61.9° at different tilt angles of the reflectarray as depicted in Figure 7.22(b).

In order to highlight the focus of the beam, the SLL and gains obtained for each tilt angle from 0° to ±30° are shown in Figure 7.23. As per expectations, the maximum gain and minimum SLLs were obtained for 0° tilt in the *YZ* plane

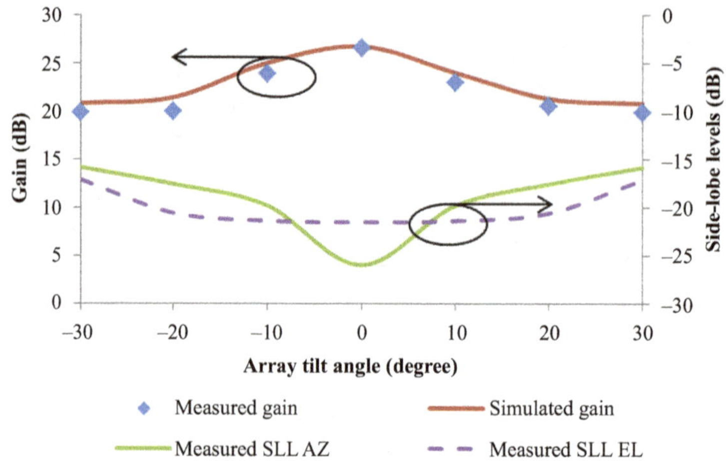

Figure 7.23 Gain and side-lobe levels for different reflectarray tilt angles

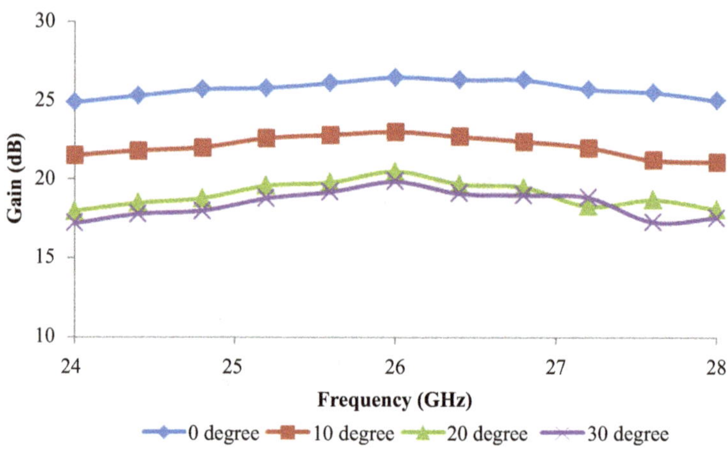

Figure 7.24 Measured gain vs frequency plots for different tilt angles

because the initial progressive phase distribution was optimized for this particular case. The gain reduced from 26.47 to 19.8 5dB, while the SLL increased from −25.9 to −15.85 dB, as the tilt angle was increased from 0° to 30°, respectively.

A 1-dB gain drop bandwidth has also been calculated for all the cases by doing radiation-pattern measurements in the frequency range of 24–28 GHz. Figure 7.24 shows the frequency vs gain plots for different tilt angles. A maximum 1-dB gain drop bandwidth of 13.1% was obtained in the case of 0° tilt, which reduced to 7.0% when the tilt angle was increased up to 30°. The summary of all the measured results is provided in Table 7.5. One of the advantages of this type of beam-steering

Table 7.5 Summary of the measured results of beam steerable reflectarray antenna

Tilt angle (°)	Scan angle (°)		Gain (dB)	SLL (dB)		Bandwidth (%)
	AZ	*EL*		*AZ*	*EL*	
0	40.0	0	26.47	−21.5	−25.9	13.1
10	38.3	21.3	23.83	−21.4	−19.8	9.2
20	36.8	39.8	21.17	−20.6	−17.5	7.6
30	36.3	61.9	19.85	−17.1	−15.8	7.0

technique is that the required tilt angles can be obtained by using a single stepper motor with the whole array instead of using motors with each individual element of the array as presented in previous research works. This helps to reduce the complexity as well as minimizes the power dissipation required for the reconfigurable antennas.

References

[1] Huang J, and Encinar J. *Reflectarray Antennas*. Piscataway, NJ, USA: IEEE-Wiley Inter Science; 2007.

[2] Huang J. Analysis of microstrip reflectarray antenna for microspacecraft applications. Spacecraft Telecommunications Equipment Section; 1995.

[3] Yu A, Yang F, Elsherbeni AZ, Huang J, and Kim Y. An offset-fed X-band reflectarray antenna using a modified element rotation technique. IEEE Trans Antennas Propag. 2012;60(3):1619–24.

[4] Carrasco E, and Perruisseau-Carrier J. Reflectarray antenna at terahertz using graphene. IEEE Antennas Wirel Propag Lett. 2013;12:253–6.

[5] Ismail MY, Dahri MH, and Zaihasra WN. Characterization of material properties for tunable reflectarray antenna design. In: American Institute of Physics (AIP) conference proceedings 1528 [Internet]. National conference on physics (PERFIK 2012); 2013. pp. 237–42. Available from: http://proceedings.aip.org/resource/2/apcpcs/1528/1/237_1?isAuthorized=no

[6] Pozar DM, and Metzler TA. Analysis of a reflectarray antenna using microstrip patches of variable size. Electron Lett [Internet]. 1993;29(8):657–8. Available from: http://link.aip.org/link/ELLEAK/v29/i8/p657/s1&Agg=doi

[7] Jamaluddin MH, Gillard R, Sauleau R, and Milon M-A. Perturbation technique to analyze mutual coupling in reflectarrays. IEEE Antennas Wirel Propag Lett [Internet]. 2009;8:697–700. Available from: http://ieeexplore.ieee.org/document/5061733/

[8] Pozar DM, Targoski D, Syrigos HD, Targonski SD, and Syrigos HD. Design of millimeter wave microstrip reflectarrays. IEEE Trans Antennas Propag. 1997;45(2):287–96.

[9] Huang J, and Pogorzelski R. A Ka-band microstrip reflectarray with elements having variable rotation angles. IEEE Trans Antennas Propag. 1998;46(5):650–6.

[10] Chang D-C, and Huang M-C. Multiple-polarization microstrip reflectarray antenna with high efficiency and low cross-polarization. IEEE Trans Antennas Propag. 1995;43(8):829–34.

[11] Chang D-C, and Huang M-C. Microstrip reflectarray antenna with offset feed. Electron Lett [Internet]. 1992;28(16):1489–91. Available from: http://digital-library.theiet.org/content/journals/10.1049/el_19920946

[12] Malkowsky S, Vieira J, Liu L, *et al.* The world's first real-time testbed for massive MIMO: design, implementation, and validation. IEEE Access. 2017;5:9073–88.

[13] Boccardi F, Heath R, Lozano A, Marzetta TL, and Popovski P. Five disruptive technology directions for 5G. IEEE Commun Mag. 2014;52(2):74–80.

[14] Andrews JG, Buzzi S, Choi W, *et al.* What will 5G be? IEEE J Sel Areas Commun [Internet]. 2014;32(6):1065–82. Available from: http://ieeexplore.ieee.org/ielx7/49/6862080/06824752.pdf?tp=&arnumber=6824752&isnumber=6862080

[15] Vook FW, Ghosh A, and Thomas TA. MIMO and beamforming solutions for 5G technology. In: 2014 IEEE MTT-S Int Microwave Symposium [Internet].; 2014. pp. 1–4. Available from: http://ieeexplore.ieee.org/lpdocs/epic03/wrapper.htm?arnumber=6848613

[16] Zhao X, Li S, Wang Q, Wang M, Sun S, and Hong W. Channel measurements, modeling, simulation and validation at 32 GHz in outdoor microcells for 5G radio systems. IEEE Access. 2017;5:1062–72.

[17] Haraz OM, Elboushi A, Alshebeili SA, and Sebak AR. Dense dielectric patch array antenna with improved radiation characteristics using EBG ground structure and dielectric superstrate for future 5G cellular networks. IEEE Access. 2014;2:909–13.

[18] Haraz OM, Ali MM, Alshebeili S, and Sebak AR. Design of a 28/38 GHz dual-band printed slot antenna for the future 5G mobile communication networks. In: IEEE international symposium on antennas and propagation & USNC/URSI National Radio Science meeting; 2015. pp. 1532–3.

[19] Ban YL, Li C, Sim CYD, Wu G, and Wong KL. 4G/5G multiple antennas for future multi-mode smartphone applications. IEEE Access. 2016;4:2981–8.

[20] Ka Ming M, Hau Wah L, Kwai Man L, and Chi Hou C. Circularly polarized patch antenna for future 5G mobile phones. IEEE Access. 2014;2:1521–9.

[21] Roh W, Seol JY, Park J, *et al.* Millimeter-wave beamforming as an enabling technology for 5G cellular communications: Theoretical feasibility and prototype results. IEEE Commun Mag. 2014;52(2):106–13.

[22] Dinh Thuy PH, Sternad M, and Svensson T. Making 5G adaptive antennas work for very fast moving vehicles. IEEE Intell Transp Syst Mag. 2015;7(2):71–84.

[23] Rajagopalan H, and Rahmat-Samii Y. On the reflection characteristics of a reflectarray element with low-loss and high-loss substrates. IEEE Antennas Propag Mag [Internet]. 2010;52(4):73–89. Available from: http://ieeexplore.ieee.org/xpl/freeabs_all.jsp?arnumber=5638237

[24] Rajagopalan H, and Rahmat-Samii Y. Dielectric and conductor loss quantification for microstrip reflectarray: simulations and measurements. IEEE Trans Antennas Propag [Internet]. 2008;56(4):1192–6. Available from: http://ieeexplore.ieee.org/lpdocs/epic03/wrapper.htm?arnumber=4483612

[25] Pozar DM. Bandwidth of reflectarrays. Electron Lett. 2003;39(21):1490–1.

[26] Chaharmir MR, Shaker J, Gagnon N, and Lee D. Design of broadband, single layer dual-band large reflectarray using multi open loop elements. IEEE Trans Antennas Propag. 2010;58(9):2875–83.

[27] Encinar JA, and Zornoza JA. Broadband design of three-layer printed reflectarrays. Antennas Propag IEEE Trans. 2003;51(7):1662–4.

[28] Carrasco E, Encinar JA, and Barba M. Bandwidth improvement in large reflectarrays by using true-time delay. IEEE Trans Antennas Propag [Internet]. 2008;56(8):2496–503. Available from: http://ieeexplore.ieee.org/lpdocs/epic03/wrapper.htm?arnumber=4589079

[29] Yu A, Yang F, Elsherbeni AZ, and Huang J. Experimental demonstration of a single layer tri-band circularly polarized reflectarray. In: IEEE antennas and propagation society international symposium (APSURSI); 2010. pp. 1–4.

[30] Inam M, and Ismail MY. Reflection loss and bandwidth performance of X-band infinite reflectarrays: simulations and measurements. Microw Opt Technol Lett. 2011;53(1):77–80.

[31] Moustafa L, Gillard R, Peris F, Loison R, Legay H, and Girard E. The phoenix cell: a new reflectarray cell with large bandwidth and rebirth capabilities. IEEE Antennas Wirel Propag Lett [Internet]. 2011;10:71–4. Available from: http://ieeexplore.ieee.org/lpdocs/epic03/wrapper.htm?arnumber=5702267

[32] Li QY, Jiao YC, and Zhao G. A novel microstrip rectangular-patch/ring-combination reflectarray element and its application. IEEE Antennas Wirel Propag Lett. 2009;8:1119–22.

[33] Li L, Chen Q, Yuan Q, *et al.* Novel broadband planar reflectarray with parasitic dipoles for wireless communication applications. IEEE Antennas Wirel Propag Lett. 2009;8:881–5.

[34] Malfajani RS, and Atlasbaf Z. Design and implementation of a dual-band single layer reflectarray in X and K bands. IEEE Trans Antennas Propag [Internet]. 2014;62(8):4425–31. Available from: http://ieeexplore.ieee.org/lpdocs/epic03/wrapper.htm?arnumber=6823164

[35] Li Y, Bialkowski ME, and Abbosh AM. Single layer reflectarray with circular rings and open-circuited stubs for wideband operation. IEEE Trans Antennas Propag. 2012;60(9):4183–9.

[36] Venneri F, Costanzo S, and Di Massa G Bandwidth behavior of closely spaced aperture-coupled reflectarrays. Int J Antennas Propag. 2012;2012:1–11.

[37] Sayidmarie KH, and Bialkowski ME. Fractal unit cells of increased phasing range and low slopes for single-layer microstrip reflectarrays. IET

Microwaves Antennas Propag [Internet]. 2011;5(11):1371. Available from: http://ieeexplore.ieee.org/xpl/freeabs_all.jsp?arnumber=6012985

[38] Mohammadirad M, Komjani N, Chaharmir MR, Shaker J, and Sebak AR. Impact of feed position on the operating band of broadband reflectarray antenna. IEEE Antennas Wirel Propag Lett [Internet]. 2012;11:1104–7. Available from: http://ieeexplore.ieee.org/lpdocs/epic03/wrapper.htm?arnumber=6301676

[39] Yi M, Lee W, Yoon YJ, and So J. Non-resonant conductor reflectarray element for linear reflection phase. Electron Lett. 2015;51(9):669–71.

[40] Jamaluddin MH, Sauleau R, Castel X, *et al.* Design, fabrication and characterization of a dielectric resonator antenna reflectarray in Ka-band. Prog Electromagn Res B [Internet]. 2010;25:261–75. Available from: http://www.jpier.org/PIERB/pier.php?paper=10071306

[41] Dahri MH, Jamaluddin MH, Inam M, and Kamarudin MR. A review of wideband reflectarray antennas for 5G communication systems. IEEE Access. 2017;5:17803–15.

[42] Liu D, Pfeiffer U, Grzyb J, and Gaucher B. *Advanced Millimeter-wave Technologies: Antennas, Packaging and Circuits*. Chichester, UK: John Wiley & Sons, Ltd; 2009.

[43] Tienda C, Encinar JA, Arrebola M, Barba M, and Carrasco E. Design, manufacturing and test of a dual-reflectarray antenna with improved bandwidth and reduced cross-polarization. IEEE Trans Antennas Propag. 2013;61(3):1180–90.

[44] Rajagopalan H, and Rahmat-Samii Y. Loss quantification for microstrip reflectarray: issue of high fields and currents. In: 2008 IEEE antennas and propagation society international symposium [Internet].; 2008. pp. 1–4. Available from: http://ieeexplore.ieee.org/stamp/stamp.jsp?tp=&arnumber=4619755&isnumber=4618896

[45] Pozar DM, Targonski SD, and Syrigos HD. Analysis and design of millimeter wave microstrip reflectarrays. IEEE Antennas Propag Soc Int Symp 1995 Dig [Internet]. 1995;1(2):287–96. Available from: http://ieeexplore.ieee.org/lpdocs/epic03/wrapper.htm?arnumber=560348

[46] Dahri MH, Abbasi MI, Jamaluddin MH, and Kamarudin MR. A review of high gain and high efficiency reflectarrays for 5G communications. IEEE Access. 2018;6:5973–85.

[47] Hasani H, Kamyab M, and Mirkamali A. Low cross-polarization reflectarray antenna. IEEE Trans Antennas Propag. 2011;59(5):1752–6.

[48] Chaharmir MR, Shaker J, and Legay H. Dual-band Ka/X reflectarray with broadband loop elements. IET Microwaves Antennas Propag. 2010;4(2):225.

[49] Zainud-Deen SH, Gaber SM, Abd-Elhady AM, Awadalla KH, and Kishk AA. Wideband perforated rectangular dielectric resonator antenna reflectarray. In: 2011 IEEE international symposium on antennas and propagation (APSURSI) [Internet]. IEEE; 2011. pp. 113–16. Available from: http://ieeexplore.ieee.org/lpdocs/epic03/wrapper.htm?arnumber=5996653

[50] Maddahali M, and Forooraghi K. High efficiency reflectarray using smooth tapering in phase pattern on antenna surface. Microw Opt Technol Lett. 2013;55(4):747–53.

[51] Kim D, and Park IY. A miniaturized reflectarray antenna for scanned beam applications. IEEE Trans Antennas Propag [Internet]. 2016;64(3):960–7. Available from: http://ieeexplore.ieee.org/lpdocs/epic03/wrapper.htm?arnumber=7381648

[52] Zhong XJ, Chen L, Shi Y, and Shi X-W. A dual-frequency single layer circularly polarized reflectarray with frequency selective surface backing. Prog Electromagn Res C. 2014;51:87–93.

[53] Lim EH, and Leung KW. *Compact Multifunctional Antennas for Wireless Systems*. Hoboken, NJ, USA: John Wiley & Sons, Inc; 2012. 241 p.

[54] Zhong X-J, Chen L, Shi Y, and Shi X-W. Single-layer broadband circularly polarized reflectarray with subwavelength double-ring elements. Electromagnetics [Internet]. 2015;35(4):217–26. Available from: http://www.tandfonline.com/doi/full/10.1080/02726343.2015.1027632

[55] Zhong X-J, Chen L, Shi Y, and Shi X-W. A dual-band single-layer circularly polarized reflectarray composed by elements having variable rotation angles. Int J RF Microw Comput Eng [Internet]. 2015;25(1):39–47. Available from: http://doi.wiley.com/10.1002/mmce.20821

[56] Albooyeh M, Komjani N, and Mahani MS. A circularly polarized element for reflectarray antennas. IEEE Antennas Wirel Propag Lett. 2009;8:319–22.

[57] Malfajani RS, and Atlasbaf Z. Design and implementation of a broadband single layer circularly polarized reflectarray antenna. IEEE Antennas Wirel Propag Lett. 2012;11:973–6.

[58] Martinez-Lopez L, Rodriguez-Cuevas J, Martynyuk AE, and Martinez-Lopez JI. Wideband-reconfigurable reflectarrays based on rotating loaded split rings. J Electromagn Waves Appl [Internet]. 2016;29(2):218–32. Available from: http://www.tandfonline.com/action/journalInformation?journalCode=tewa20

[59] Ren LS, Jiao YC, Li F, Zhao JJ, and Zhao G. A dual-layer T-shaped element for broadband circularly polarized reflectarray with linearly polarized feed. IEEE Antennas Wirel Propag Lett. 2011;10:407–10.

[60] Ahmadi F, Namiranian A, and Virdee B. Design and implementation of a single layer circularly polarized reflectarray antenna with linearly polarized feed. Electromagnetics [Internet]. 2015;35(2):93–100. Available from: http://www.tandfonline.com/doi/abs/10.1080/02726343.2014.987637

[61] Guo L, Tan PK, and Chio TH. A simple approach to achieve polarization diversity in broadband reflectarrays using single-layered rectangular patch elements. Microw Opt Technol Lett [Internet]. 2015;57(2):305–10. Available from: http://doi.wiley.com/10.1002/mop.28833

[62] Dahri MH, Jamaluddin MH, Khalily M, Abbasi MI, Selvaraju R, and Kamarudin MR. Polarization diversity and adaptive beamsteering for 5G reflectarrays: a review. IEEE Access [Internet]. 2018;6:19451–64. Available from: http://ieeexplore.ieee.org/document/8328804/

[63] Hum SV, and Perruisseau-Carrier J. Reconfigurable reflectarrays and array lenses for dynamic antenna beam control: a review. IEEE Trans Antennas Propag. 2014;62(1):183–98.

Chapter 8

Performance modelling of wireless Xhaul and associate impact on network provisioning for 5G and beyond

Mona Jaber[1], *Francisco Javier Lopez Martinez*[2] *and Akram Alomainy*[1]

5G is no longer a futuristic vision but has become today's reality whilst the research community has commenced the design of the next mobile wireless network generation, dubbed 6G. A successful 5G deployment will pave the way to a better 6G requirements definition but demands an imminent and efficient approach for tackling the related deployment challenges, not the least the backhaul (BH). The performance evolution of the current realistic BH infrastructure towards meeting the expectations of next-generation mobile networks is a lengthy and costly process. Various efforts towards bridging the performance gap between available and stipulated solutions are ongoing as reported in [1]. Some attempt to evolve the existing BH technologies (e.g. [2] for copper, [3] for fibre, and [4] for wireless). Others explore the integration of information technology techniques to alleviate the BH performance requirements (e.g. [5] for caching and [6] for software-defined network architectures). Theoretically, fibre is believed to be the ultimate BH solution; however, it is also known to be scarce in many 5G leading countries. In these cases, network-wide deployment is also highly impractical due to the cumbersome and very slow process of laying new fibre in urban areas. At the fibre to the home (FTTH) Council Europe Conference in 2019, 5G was identified as a leading factor for the promotion of fibre deployment and a catalyst for boosting investments from public and private stakeholders. At the same time, the advances in evolving legacy BH technologies, particularly copper-based G-fast,* are expected to slow down the deployment of FTTH or fibre to the building (FTTB). Based on 2019 data, there are less than ten countries worldwide that have achieved FTTH/FTTB higher than 50% coverage, with Asia-Pacific countries (e.g. China and

[1]The School of Electronic Engineering and Computer Science, Queen Mary University of London, London, United Kingdom
[2]Departamento Ingeniería de Comunicaciones, Universidad de Malaga, Málaga, Spain

*Fast Access to Subscriber Terminal; the letter G stands for ITU-T G series of recommendations

Japan) and modern-built countries (e.g. UAE and Qatar) leading the charts and Latvia being the only European country just above 50% penetration.

As such, wireless links are seen as the optimum solution to fill in the gap where fibre optic links are not available, by virtue of the minimal required installation work and relative fast deployment. Academic and industry researchers see great potential in the traditional microwave and novel extremely high-frequency (EHF) bands and are working on evolving the technologies to meet 5G requirements. On the other hand, self-backhauling, whereby small cells reuse the radio access spectrum as means of wireless backhauling to the closest hub, is gaining more credibility with the emergence of beamforming antennas. Wireless technology is evolving and expanding to offer fibre-like throughput in the order of few Gbps. Evolution is based on improved spectrum efficiency, e.g. high adaptive modulation, improved system gain such as multiple-input–multiple-output (MIMO) technologies and antenna array beamforming, and increased throughput efficiency with advances such as multilayer compression schemes. The capacity of mobile BH wireless links is expected to reach beyond 10 Gbps, partly as a result of the adopted new EHF spectrum [7]. The main impediment of wireless BH networks is their vulnerability to propagation conditions, not the least fading and loss of line of sight (LOS). On the other hand, the radio spectrum is scarce, and wireless BH solutions increase the contention for radio resources, which is already aggravated due to the explosive increase in connected devices resulting from the Internet of Things (IoT). A multi-hop heterogeneous BH network is, therefore, unavoidable and perhaps favourable for the launch of 5G services and for readying the BH infrastructure for the emergence of 6G, as shown in Figure 8.1. Moreover, dynamic solutions for optimising the radio resource usage are imperative for enabling efficient and effective wireless BH deployment. The main trends for addressing these concerns explore dynamic mechanisms for sharing radio resources between radio access and wireless self-BHs such as [8–13], where some exploit the advances in MIMO technologies (e.g. [14,15]), caching (e.g. [16,17]) or full-duplex transmission (e.g. [16–21]).

NGMNs are deployed using multiple radio access technologies and are composed of various coverage layers, often referred to as ultra-dense heterogeneous networks. Thus, the radio access of NGMS is, similarly to the BH, heterogeneous and differing performance characteristics. On the other hand, an unseen spectrum of types of devices is anticipated to connect to NGMNs and an even larger range of applications running on each of these devices. The abundance of these devices and applications is an unprecedented diversity in stipulated service requirements. For instance, an e-health device and a smart meter are both IoT devices that require low-rate transmission; however, the first is latency-critical whereas the second is not. Another e-health IoT device used in a remote surgery solution or a video download application running on a mobile device both stipulate high data rates, but only the first is latency and resilience critical. To this end, the novel user-centric allocation scheme exploits the diversity of users' performance exigences to mitigate the limitations in some hauls in the radio and BH access network in order to maximise quality of experience (QoE) of most users whilst maximising the network

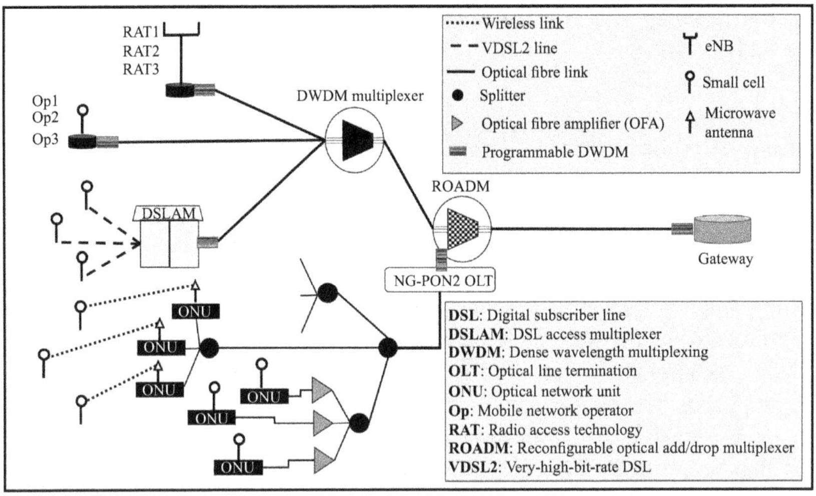

Figure 8.1 Example of a multi-hop hybrid BH network that employs a fibre-based architecture with optional xDSL and wireless links used to cover the last mile

operator's return on investment [22]. Innovation in machine learning approaches in the implementation of such user-centric methods is paving the way towards the provisioning of elastic NGMNs that adjust according to the users' demands and network limitations such as [23–31].

Nonetheless, such elastic or self-optimised solutions are often met with some level of scepticism as for their ability to yield stable and optimised networks. To this end, it is of paramount importance to enable extensive simulations that could validate the performance of novel proposed schemes before implementing these in test beds or real networks. In this chapter, we present a comprehensive and modular analytical framework that captures the performance characteristics of any multi-hop hybrid BH network, based on the network topology, relevant distances, and mix of transport technologies used. We focus on modelling the wireless BH, as part of a multi-hop hybrid solution and derive a tractable analytical model that represents the nature of wireless BH propagation with fidelity.

This chapter is organised as in four sections. Section 8.1 describes the modular approach for modelling the performance of a multi-hop hybrid BH network. Section 8.2 surveys the wireless BH solutions and offers a tractable model for simulating the distinct characteristics of each type. Section 8.3 offers a case study that leverages the described performance modelling approach to identify bottlenecks in the realistic BH network and design pertinent solutions. Section 8.4 surveys the state-of-the-art research on intelligent wireless BH provisioning and offers a user-centric approach for in-band wireless BH resource allocation using reinforcement learning techniques.

8.1 Modelling the performance of a multi-hop hybrid BH

Designing the next-generation BH network is a major challenge. Apart from the pervasive extensions required to cater for the anticipated network densification, the performance requirements of these links are also highly exigent. Optical fibre networks are believed to be the only technology currently developed for migrating users fully to the NGMN era. Nonetheless, fibre is not widely available and cannot be laid at the same pace as network growth due to the complicated process of trenching and the associated permit acquisition and cost. Wireless technologies in the millimetre-wave bands and innovation in copper-based solutions are being developed as alternatives to fibre cables, for certain use cases, with promising potential [32]. A heterogeneous BH network is, therefore, inevitable and favourable when used in conjunction with innovative elastic features. There are many recent works that propose possible ways for addressing this need, often based on stochastic geometry which models the network using elements of point process (PP) theory. Authors in [33] propose analytical models that capture the delay and cost of four types of BH links: copper-based, fibre-based, millimetre-wave, and sub-6-GHz technologies. However, their work assumes that a multi-hop BH connection consists of one type of technologies, while different connections may employ different technologies. In practice, though, BH connections are often hybrid in addition to being heterogeneous. References [34,35] focus on fibre-based BH performance modelling. In these works, analytical models for the delay incurred and availability in various topologies are proposed. The models are then used to validate novel schemes for BH-aware user association and WiFi offloading. Coverage and throughput derivations in a wireless cellular network are elaborated in [36] and used to develop a general and tractable framework for modelling and analysing joint radio access resource partitioning and offloading in a two-tier cellular network. Outage probability of multi-link wireless BH connections is studied in [37], and an analytical model is proposed that accounts for retransmissions and selection combining.

Each of these works partially sheds light on some aspects of the problem at hand but none offers a scalable model that represents the hybrid nature of existing and anticipated multi-hop BH connections. To this end, we present the first generic and modular analytical framework that is able to capture the performance characteristics of any multi-hop hybrid BH, based on the network topology, relevant distances, and mix of transport technologies used [32].

8.1.1 System model

The expectations of NGMN users with regards to QoE differ greatly from what is offered by incumbent cellular generations. First, NGMN users cover a broad spectrum of things which include sensors, driverless cars, machines, and humans, each with very different capabilities and moreover, the applications running on devices have a wider set of QoE requirements, as opposed to the conventional

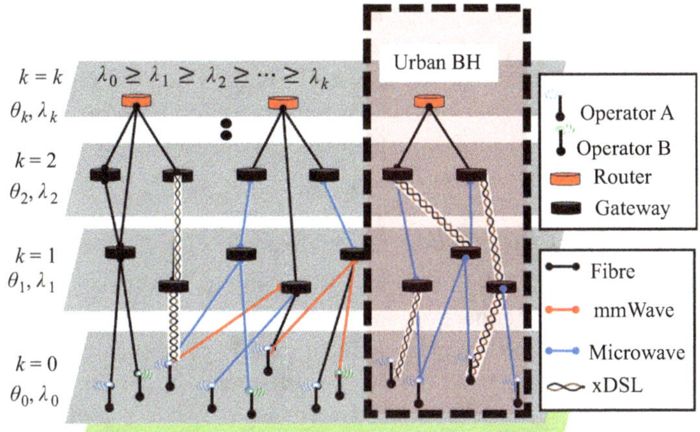

Figure 8.2 System model representing a multi-hop and hybrid 5G BH. Shown nodes may be pre-aggregation ($k \neq K$) or aggregation ($k = K$) routers. This architecture allows for the representation of transport loops in the BH at any stage of the network. The highlighted section represents the urban BH example adopted [32]

quality measures, such as the user throughput or mean opinion score (MOS). For instance, the focal quality aspects in 5G are throughput, delay, cost, resilience, security, and energy efficiency. These attributes are in principle all dynamic but are likely to vary at different rhythms, and their fluctuation is triggered by different phenomena. In order to allow for the user-centric service provisioning, it is thus vital to capture the BH network's performance with regards to each of these attributes.

8.1.2 BH constraints and characteristics

The end-to-end performance of cells is determined by both the radio and the BH networks; in this work, we propose an analytical modelling of the hybrid BH network characteristics. The current BH network, which transports incumbent cellular traffic, consists mostly of optical fibre, copper, and microwave links. Some of these links are owned by the cellular operator, others are leased, and some are shared among multiple cellular operators. The NGMN BH will build on the existing infrastructure and will expand its capillaries and reach to connect the numerous fast-spreading small cells, by relying on new technologies to fill in the gap where needed. The end result is a hybrid BH network composed of a mix of various types of wired and wireless links. The approach we propose for modelling the performance characteristics of such a network is modular. In other words, the BH network consists of layers of nodes connected through various links, as shown in Figure 8.2.

According to Figure 8.2, each small cell s connects to a gateway g in layer $k = K$ through a number of links K_s which differs for every small cell s. Let

$\mathscr{C}_l = \{\Theta_l, D_l, C_l, O_l\}$ and $\mathscr{C}_r$ be the sets of characteristics that describe the performance of link, l, and node (or router), r, respectively, where Θ_l (or Θ_r) is the throughput, D_l (or D_r) is the delay, C_l (or C_r) is the cost, and O_l (or O_r) is the outage probability. A hop, h_k, is defined as the connection between a node in layer $k-1$ and a node in layer k, and its performance is determined by both, that of the link and that of the router in layer k.

8.1.3 Topology of hybrid BH

Until such times when globally ubiquitous BH solutions become available, cell-to-core mobile network BH connections are likely to be composed of multi-hops with multiple technologies. Several factors define the topology of such a BH and need to be analysed and modelled. First, the number of layers K_s, corresponding to each node layer s, needs to be determined. Urban deployments are often connected through less than three or four layers; whereas, rural cells may require daisy chains, hence up to six or seven layers [38]. The number of layers is thus a variable with a probability density function that can be approximated by a Gaussian distribution and which differs according to the operator, country, and deployment scenario. Let λ_k be the density of nodes in layer k which follow a Poisson PP (PPP) such that $\lambda_k \leq \lambda_{k-1}$. Let each layer k be characterised by a set of probabilities $\{p_{k,1}, p_{k,2}, p_{k,3}, \ldots, p_{k,T}\}$ such that $\sum_{t=1}^{T} p_{k,t} = 1$, where each $p_{k,t}$ represents the fraction of links using technology t between layer $k-1$ and layer k, for $k > 0$. Let $t \in [1, T]$ be the index corresponding to the technology used, e.g. $t = 1$ corresponds to VDSL,[†] $t = 2$ to G-fast,[‡] $t = 3$ to EPON,[§] $t = 4$ to XG-PON,[¶] and $t = 5$ to E-band. For instance, for layer $k = 1$ (i.e. the last mile), the probability $p_{1,5}$ of wireless links would be higher than that of fibre links, whereas for layer $k = K$, the probability $p_{K,4}$ of fibre XG-PON links is highest. Let X be the variable distance between two layers $k-1$ and k; it may be assumed to follow the following distribution, considering that nodes in lower layers associate with the closest upper nodes [36]:

$$f_X(x) = 2\pi\lambda_k \cdot x \cdot e^{-\pi\lambda_k \cdot x^2} \tag{8.1}$$

Thus, the mean distance between layers $k-1$ and k is $\mathrm{E}(X) = 1/(2\sqrt{\lambda_k})$ which increases when the density of nodes, λ_k, in the upper layer decreases. According to (8.1), the expected length of last-mile links connecting to nodes at 1 node/km^2 (typical density of urban small cells' gateway) is 500 m, as shown in the zoomed window of Figure 8.3. This matches actual data from a typical Western European mobile network operator. The data represents the distribution of the length of all the hops in the network and shows that BH links connected to the core network are the fewest and could reach more than 25 km in length.

[†]Very-high-bit-rate digital subscriber line 2
[‡]Very-high-bit-rate digital subscriber line (over copper); Fast access to subscriber terminal (over copper).
[§]Ethernet Passive Optical Network
[¶]Ethernet Passive Optical Network (PON); 10 Gbps capable PON.

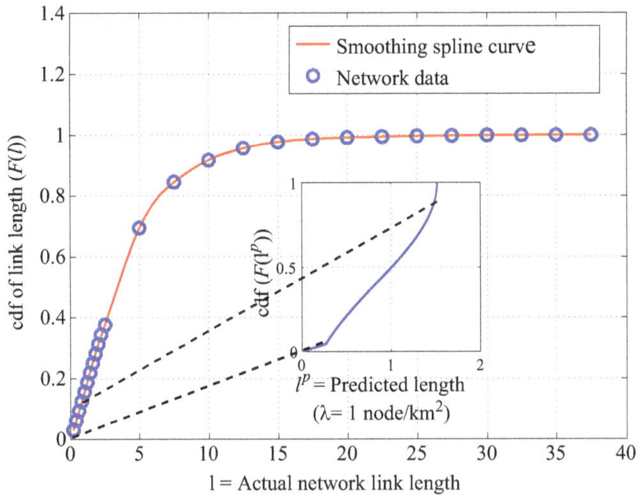

Figure 8.3 Cumulative distribution function (cdf) of actual network-wide backhaul link length taken from a typical Western European operator. The zoom-in figure is generated assuming a 1 node/km² density and matches the actual data [32]

8.1.4 Hybrid BH performance models

The performance metrics of the hybrid multi-hop BH are derived in this section and capture the expected end-to-end throughput, latency, cost, and outage probability. For demonstration, a four-layer urban network is assumed, in which layer zero contains the small cells and layer three the gateways. In this case, all links connected to the gateways are fibre-based, whereas links in other layers are a mix of VDSL and wireless (microwave) links, as shown in Figure 8.2.

8.1.4.1 Throughput

The effective throughput of a hybrid multi-hop BH connection is constrained by the most limiting hop. Thus, the expected hybrid downlink throughput from gateway g at level $k = K$ to small cell s at level $k = 0$ can be expressed as follows:

$$E[\Theta_{s,g}] = E[\min_k\{\Theta_{k,t}\}] = \min_k\{E[\Theta_{k,t}]\}$$

$$\forall t \in [1, T] | p_{k,t} > 0; \forall k \in [1, K]$$

where $E[\Theta_{k,t}]$ is the expected throughput of the hop h_k that uses technology t. This reflects any adopted topology between the layers, e.g. redundant or paralleled hops. Figure 8.4 shows $E[\Theta_{s,g}]$ of the urban BH in Figure 8.2 for different penetrations of microwave/VDSL links and Figure 8.5 for different node densities λ_k. Higher penetration of VDSL or lower node density in Layer 2 have higher impact on $E[\Theta_{s,g}]$ due to longer incurred distances; an impediment in both copper and wireless technologies, but more so in copper.

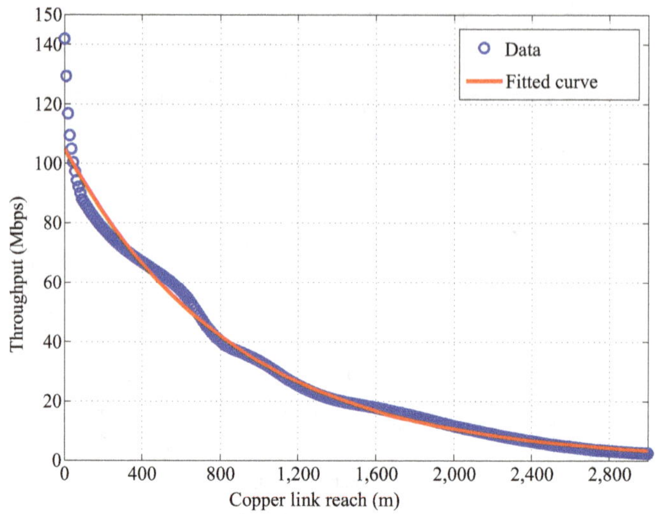

Figure 8.4 Impact of varying the share of wireless/VDSL links on the expected throughput based on urban backhaul in Figure 8.2 [32]

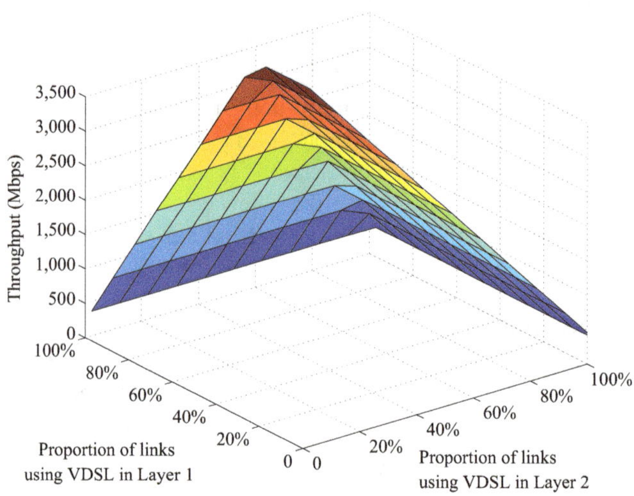

Figure 8.5 Impact of node density on throughput for 50% wireless links in Layer 1 and 70% in Layer 2 based on urban backhaul in Figure 8.2 [32]

In these figures, the expected throughput of wireless hops is modelled based on (8.15) in [36], assuming a propagation exponent equal to four (worst case scenario). Fibre hops' throughput is not modelled as it is not limiting. We propose to model the throughput over a single-hop VDSL link as an exponential decay: $\Theta_c(x) = \Delta \cdot e^{-\delta \cdot x}$, where x is the length of the copper line, Δ is the maximum achievable throughput, and δ is the decay rate, both dependent on the technology and bandwidth used. As seen in Figure 8.6, the proposed model fits with the actual measurements, once the parameters have been tuned. In the case of VDSL, these are found to be $\Delta = 105$ Mbps and $\delta = 1.14 \cdot 10^{-3}$. The expected unshared throughput over a VDSL link is thus derived as follows:

$$\begin{aligned} \mathrm{E}[\Theta_c] &= \int_0^{\infty} (\Delta \cdot e^{-\delta \cdot x}) \cdot (2\pi\lambda_k \cdot x \cdot e^{-\pi\lambda_k \cdot x^2}) d(x) \\ &= \Delta\left(e^{-\delta^2/4\pi\lambda_k} - \frac{\delta}{\sqrt{\lambda_k}} \cdot \left(1 - erf\left(\frac{\delta}{2\sqrt{\pi\lambda_k}}\right)\right)\right) \end{aligned} \tag{8.2}$$

8.1.4.2 Latency

The effective latency of a hybrid multi-hop BH connection is the cumulative effect of all delays incurred along the path, as shown in the following:

$$\mathrm{E}[D_{s,g}] = E\left[\sum_{k=1}^{K}\sum_{t=1}^{T} p_{k,t} D_{k,t}\right] = \sum_{k=1}^{K}\sum_{t=1}^{T} p_{k,t} \mathrm{E}[D_{k,t}]$$

where $\mathrm{E}[D_{k,t}]$ is the expected delay of the hop h_k that uses technology t. The expected latency of the BH in Figure 8.2 is shown in Figure 8.7 for different shares of wireless/wired links.

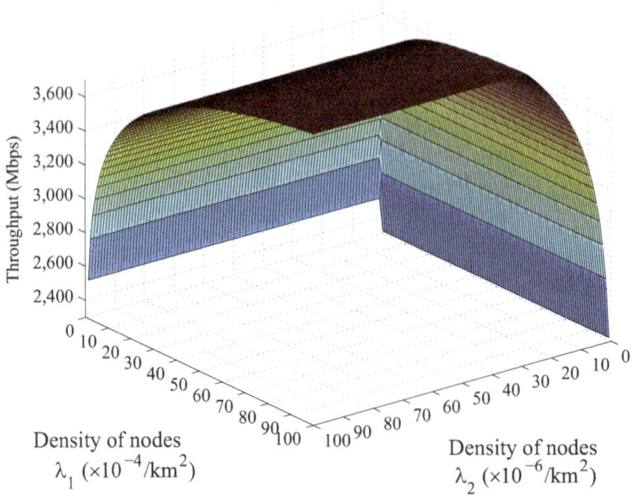

Figure 8.6 Typical VDSL throughput variation over a range of copper line length. Blue dots represent measurements from a European internet service provider, and the red curve is an approximation model assuming exponential decay [32]

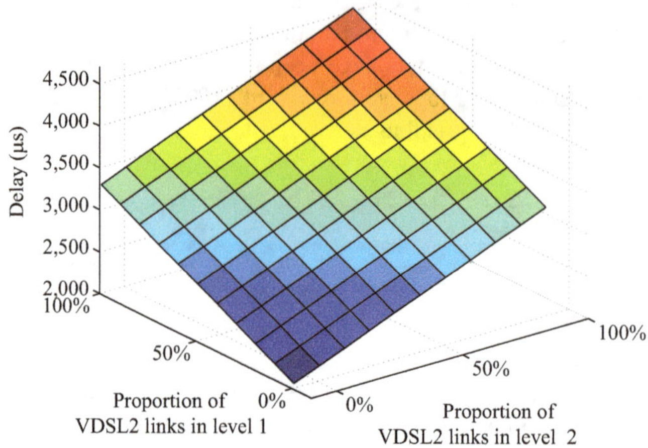

Figure 8.7 Impact of varying the share of wireless/VDSL links on the expected delay based on urban backhaul in Figure 8.2 [32]

Calculating the average packet delay over any link entails looking at four phenomena: the propagation delay (D^p), the transmission delay (D^τ), the processing delay (D^r), and the queuing delay (D^q). The propagation delay is the time difference between the moment a packet is placed on one end of the line and the time it is received at the other end. Propagation delay is thus a function of the distance travelled and the refractive index $\widetilde{n}$ of the propagation medium as $\mathrm{E}[D^p] = (1/2\sqrt{\lambda_k})/(\mathfrak{c}/\widetilde{n})$, where $\mathfrak{c}$ is the speed of light. The transmission delay is the time taken to push all the packet bits on to the link, hence, is the ratio of the average packet size over the nominal capacity of the channel such as $\mathrm{E}[D^\tau] = \mathrm{E}[\varepsilon]/\mathrm{E}[\Theta]$, where $\mathrm{E}[\varepsilon]$ is the average length of a packet. The processing and queuing delays occur at the router and are often approximated jointly as a gamma distribution with parameters that depend on the router, the given load, and technology as in [33]: $\mathrm{E}[D^{r,q}] = \phi \cdot (1 + 1.28 \cdot \lambda_{k-1}/\lambda_k) \cdot (\alpha + \mathrm{E}[\varepsilon])\beta)$, where ϕ and α are representative of the processing power of the router, and β is a factor representing the processing capability of the router with respect to the packet size $\mathrm{E}[\varepsilon]$. The processing and queuing delays are often the dominant causes for latency in all technologies, even for wireless links that employ decode-and-forward mechanisms; thus, $\mathrm{E}[D_{k,t}] \approx \mathrm{E}[D^{r,q}_{k,t}]$.

8.1.4.3 Cost

The cost of a hybrid BH connection is also the aggregate effect of all expenditures incurred in all comprised links and routers, as expressed in the following:

$$\mathrm{E}[C_{s,g}] = \mathrm{E}\left[\sum_{k=1}^{K}\sum_{t=1}^{T} p_{k,t} C_{k,t}\right] = \sum_{k=1}^{K}\sum_{t=1}^{T} p_{k,t}\mathrm{E}[C_{k,t}]$$

where $\mathrm{E}[C_{k,t}]$ is the expected cost of the hop h_k connecting layers $k-1$ and k using technology t.

8.1.4.4 Resilience

A multi-hop hybrid BH connection is only available if none of its hops is out of service. Thus, the collective outage probability of a multi-hop hybrid BH connection is expressed as follows:

$$\begin{aligned}\mathrm{E}[O_{s,g}] &= 1 - E\left[\prod_{k=1}^{K}\prod_{t=1}^{T}(1 - p_{k,t}O_{k,t})\right] \\ &= 1 - \prod_{k=1}^{K}\prod_{t=1}^{T}(1 - p_{k,t}E[O_{k,t}])\end{aligned} \tag{8.3}$$

where $\mathrm{E}[O_{k,t}]$ is the expected outage probability of the hop h_k that uses technology t and reflects the redundancy scheme adopted. As the outage probabilities of different hops are uncorrelated, the expected value of the product can be assumed equal to the product of expected values.

8.2 Modelling the performance of the wireless BH

Wireless technology is evolving and expanding to offer fibre-like throughput in the order of few Gbps. Currently, the average capacity of mobile BH wireless links is less than 200 Mbps but is expected to reach beyond 10 Gbps, partly as a result of the adopted new EHF spectrum [7]. The V-band (71–76 and 81–86 GHz), E-band (57–66 GHz), and D-band (141–174.8 GHz), all in the EHF category, have different spectrum characteristics and nature. The V-band benefits from a continuous block of nine GHz and immunity to inter-link interference due to high oxygen absorption; its primary usage is envisaged to cater for the small cell last-mile BH/fronthaul. The E-band has been regulated since the year 2000 with a large amount of spectrum (10 GHz). On the contrary, it does not suffer from oxygen absorption, hence, allows longer links in the order of few kilometres. The E-band is often seen as the ultimate wireless solution where fibre is unavailable and is believed to allow throughput and latency on par with fibre links. It is expected that the E-band would constitute 20% of the wireless family of mobile BH and the V-band 10% [7]. Simulation results in [4] demonstrate the usage of D-band over 1.2 km LOS assuming a transmit power of 10 dBm and an antenna gain of 45 dBi. While the technology is still under development, it is expected to deliver a nominal throughput in the range of 30–50 Gbps as the propagation in this band is not affected by oxygen absorption. It is thus essential to derive tractable analytical models that represent the nature of wireless BH propagation with fidelity. There are emerging efforts towards addressing this need, albeit, these remain sporadic and incomplete.

A tractable approach was proposed in [36] (and used in [32]) to model the radio coverage and throughput in a cellular network; however, it does not take into account BH-specific characteristics (e.g. dominant LOS condition). The authors in [39,40] propose analytical models to capture the cumulative radio access and BH delay as perceived by users in the network. However, the authors consider two extreme cases: the first models the noise-limited wireless BH delay as a constant and the second models the in-band interference-limited scenario where interference is received from macrocells and small cells. None of these assumptions fits the wireless BH characteristics as these are likely to be noise limited and are vulnerable to the fluctuations of the desired signal. The same authors propose an approximation to the noise-limited wireless BH, assuming log-normal shadowing in [33,41]. However, the mean packet delay is approximated by using the mean value of the link distance, instead of capturing the actual distribution of the variable distance. The availability and survivability of wireless connections are studied and modelled in [37] but do not capture the presence of dominant LOS.

In this section, we present the first framework for analysing the performance of a wireless BH hop in LOS conditions with pertinent assumptions and tractable expressions where possible [42]. As wireless BH hops are the first contenders to spread the BH in breadth and depth in a timely manner, we next look at the impact of such hops on the multi-hop hybrid BH performance. Accordingly, in Section 8.2.3, we model the performance of a multi-hop hybrid BH that comprises wireless segment(s) by employing the expressions derived in Section 8.2.2.

8.2.1 System model

Typically, a wireless BH link that is properly planned should allow for a LOS between the transmitting and receiving antennae. In addition to the dominant LOS path, a diffuse component is also present as a result of signal scattering. Such propagation is referred to as LOS fading and is described and modelled in Section 8.2.1.1.

In this section, we use $\mathrm{E}\{\cdot\}$ to represent the expectation operator. The probability distribution function (pdf), the cumulative distribution function (cdf), and the complementary cdf of a random variable X are represented as $f_X(x)$, $F_X(x)$, and $\overline{F}_X(x)$, respectively. Given a function $g(x,y)$ of two random variables X and Y, $g_x(y) \triangleq g(y,x)|x$ denotes the function conditioned to a particular value x. The probability of an event z is denoted as $Pr(z)$.

8.2.1.1 LOS-fading representation

In a LOS-fading scenario, the diffuse component may be modelled as zero-mean Gaussian variate with variance equal to $2\sigma^2$. The sum of the dominant LOS component (with mean equal to $|\mu|$) and the diffuse component is the random variable G. The power $Z = |G|^2$ of such a signal produces Rice fading that is characterised by the Rician factor $K = |\mu|^2/2\sigma^2$ which represents the ratio between the power of the LOS component and that of the sum of the multipath components. The cdf of the power of fading is a key expression in the derivations of performance metrics of wireless channels; in the case of Rice fading, it is represented as follows:

$$F_Z(z) = 1 - Q\left(\sqrt{2K}, \sqrt{2\frac{1+K}{\Omega}z}\right), \tag{8.4}$$

where $\Omega \triangleq \mathrm{E}\{z\} = |\mu|^2 + 2\sigma^2$ and $Q(\cdot,\cdot)$ is the Marcum-Q function.¶¶

Clearly, analytical modelling that involves Rician fading, hence the Marcum-Q function, is mathematically very challenging. A workaround was recently proposed in [43], in which the LOS component is assumed to randomly fluctuate, i.e. $K_u = u \cdot K$, where u is a unit-mean gamma-distributed random variable with pdf given by

$$f_U(u) = \frac{m^m \cdot u^{m-1}}{(m-1)!} e^{-m \cdot u}, \tag{8.5}$$

and $m \in \mathbb{N}$. This distribution was originally formulated in the context of modelling the joint effect of small-scale fading and shadowing, often referred to as *Rician shadowed distribution*. We propose to employ it in this chapter as a more tractable approximation to the Rician distribution owing to the additional degree of freedom, the parameter m, and we refer to as *fluctuating Rician distribution*. In fact, it was recently demonstrated that this approximation converges to the Rician distribution as $m \to \infty$ in [44,45]. Given that Z follows the fluctuating Rician distribution, the authors derive the following simple forms of the pdf $f_Z(z)$ and cdf $F_Z(z)$, as a finite sum of powers and exponentials by restricting the parameter m to take positive integer values:

$$f_Z(z) = \sum_{i=0}^{m-1} C_i \cdot \frac{1}{B^{m-i}} \frac{z^{m-1-i}}{(m-i)!} e^{-\frac{z}{B}}, \tag{8.6}$$

$$F_Z(z) = 1 - \sum_{i=0}^{m-1} C_i e^{-\frac{z}{B}} \sum_{r=0}^{m-1} \frac{1}{r!} \left(\frac{z}{B}\right)^r, \tag{8.7}$$

with $B \triangleq \Omega(m+K)/m(1+K)$ and

$$C_i \triangleq \binom{m-1}{i} \left[\frac{m}{m+K}\right]^i \left[\frac{K}{m+K}\right]^{m-1-i}. \tag{8.8}$$

In this chapter, we use the traditional distribution of Rician fading (8.32) where mathematically possible, and the fluctuating Rician fading distributions (8.6) and (8.7), otherwise.

8.2.1.2 BH network topology

We assume that the BH network is composed of Λ layers of nodes connected with $\Lambda - 1$ hops as in [32]. In Figure 8.8 (left), a BH topology that comprises two hops is

¶¶J. Marcum, Table of Q Functions, Memorandum (Rand Corporation), Rand Corporation, 1950.

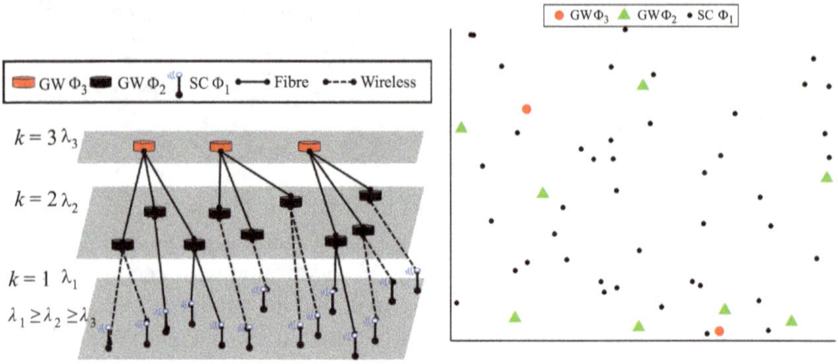

Figure 8.8 Example of multi-hop hybrid BH network deploying fibre-based or wireless last mile [42]

shown, and (right) a simulated example of this topology that is generated using PPP to represent three layers: $\mathbf{\Phi}_3$ for the gateways in the core network, $\mathbf{\Phi}_2$ for the BH aggregation gateways, and $\mathbf{\Phi}_1$ for the small cells' layer.

The performance of a single wireless hop between layers $\kappa - 1$ and κ is then modelled assuming LOS fading. The distribution of the distance between two layers $\kappa - 1$ and κ of the BH network is given by [32]:

$$f_X(x) = 2\pi\lambda_\kappa \cdot x \cdot e^{-\pi\lambda_\kappa x^2}, \tag{8.9}$$

where λ_κ is the intensity of the PPP that represents the node distribution in layer κ in a two-dimensional plane. For the purpose of simplifying the derived expressions, in this article, we use λ to refer to λ_κ, the density of gateways in layer κ that aggregates wireless BH links from layer $\kappa - 1$. Although the models derived account for parallel links, these are considered as alternative hot-backup connections and, thus, do not allow simultaneous transmission. As such, the throughput derivations assume one active link, hence the scheduling delay is not considered. Nonetheless, the topology and redundancy schemes are captured in the resilience modelling derivations.

Wireless point-to-point BH deployments are often noise limited, i.e. the interference from neighbouring links within the designated spectrum is minimal, thus, the signal-to-interference-and-noise ratio (SINR) can be approximated with a signal-to-noise-ratio (SNR). Indeed, the traditional wireless spectrum is licensed and careful frequency planning limits inter-link interference. On the other hand, in-band backhauling is enabled through the usage of beamforming antennae which, by virtue, limit the inter-beam interference [46]. Moreover, the propagation characteristics of the EHF waves suppress unwanted signals from neighbouring links, naturally. For these reasons, the received SINR of wireless links in any of the mentioned bands can be approximated with the SNR. The instantaneous SNR at the receiver side, denoted as γ, can be expressed as

$$\gamma(h,x) = \frac{P \cdot A \cdot h \cdot x^{-\alpha}}{L_0 \cdot N}, \tag{8.10}$$

where L_0 is the propagation loss constant, α is the propagation exponent, x is the propagation distance of the signal, h is the random component of small-scale fading, P is the equivalent isotropic radiated power, A is the receive antenna gain, and N is the thermal noise power. Note that we explicitly indicate the dependence of γ on h and x, although for the sake of notational simplicity, we will simply refer to γ unless necessary to avoid any confusion.

Given that a wireless BH is an inherently LOS-based transmission, we assume that the distribution of h is that of a Rician-distributed fading channel. Thus, assuming a normalised channel $\mathrm{E}\{h\} = 1$, the distribution of the instantaneous SNR conditioned to a particular value of x, denoted as $\gamma_x \triangleq \gamma(h,x)|x$, follows a power (or squared) Rician distribution, as in Section 8.2.1.1, with mean $\Omega_x = P \cdot A \cdot x^{-\alpha}/L_0 \cdot N$ (replacing h by $\mathrm{E}\{h\} = 1$ in (8.10)).

8.2.2 Wireless BH performance

8.2.2.1 Throughput

The upper bound of a wireless channel capacity (in symbols per second) depends first on the channel bandwidth (W) and the SNR. The actual bit-per-second throughput is derived from the selected modulation and coding scheme which relates to the number of bits per symbol. Higher modulation schemes require higher SNR in order to avoid bit confusion which would lead to unwanted high bit error rates. Defining $T_x \triangleq W \cdot \log_2(1+\gamma_x)$, then the average rate for a given value of x can be derived as follows, by using the definition of the expectation operator:

$$\mathrm{E}\{T_x\} = \int_0^\infty W \cdot \log_2(1+\gamma_x) \cdot f_{\gamma_x}(z)dz. \tag{8.11}$$

The average capacity of a wireless link under Rician fading at a given distance x can be expressed as follows:

$$\mathrm{E}\{T_x\} = \frac{W}{\ln(2)} e^{-K} \sum_{n=0}^{\infty} \frac{K^n}{n! \cdot n!} \times G_{2,3}^{3,1}\left[\frac{(1+K)}{\Omega_x} \middle| \begin{matrix} 0,1 \\ n+1,0,0 \end{matrix}\right], \tag{8.12}$$

where $G_{p,q}^{m,n}[\cdot]$ is the Meijer-G function [49, eq. (9.301)].

Even though the expression in (8.12) can be computed with state-of-the-art mathematical packages such as Mathematica,** it necessitates, nonetheless, the evaluation of an infinite series expression involving the unwieldy Meijer-G function. To this end, we propose the usage of the fluctuating Rician approximation,

**https://www.wolfram.com/mathematica/

presented in Section 8.2.1.1, in order to reduce the complexity of computation, and as will be later shown, to facilitate further analytical derivations.

The average capacity of a wireless link under Rician fading at a given distance x can be approximated as follows, using the fluctuating Rician fading pdf (8.6):

$$\mathrm{E}\{T_x\} \approx \mathrm{E}\{T'_x\} = \frac{W}{\ln(2)} e^{1/B_x} \sum_{i=0}^{m-1} C_i \sum_{l=0}^{m-i-1} \frac{\Gamma(-l, 1/B_x)}{B_x^l}, \tag{8.13}$$

where $B_x \triangleq \Omega_x(m+K)/m(1+K)$ and $\Gamma(\cdot,\cdot)$ are the upper incomplete gamma functions, which can be computed in this specific case as [49, eq. (8.352.3)]:

$$\Gamma(-n, x) = \frac{(-1)^n}{n!}\left[E_1(x) - e^{-x}\sum_{r=0}^{n-1}(-1)^r \frac{r!}{x^{r+1}}\right], \tag{8.14}$$

where $E_1(\cdot)$ is the exponential integral function.

For sufficiently large m, the fluctuating Rician distribution collapses to the Rician distribution as shown in [45] (please refer to figure 1 in [45]). This is illustrated in Table 8.1(a) and (b), where we first find the value of m that would yield approximated results (8.13) within $\varepsilon = 0.1\%$ of those obtained with (8.12) for different values of K and α ($\varepsilon = (\mathrm{E}\{T'_x\} - \mathrm{E}\{T_x\})/\mathrm{E}\{T_x\}$). With the selected value of m, the approximation using (8.13) is validated for a given wireless BH with different values of K, as shown in Figure 8.9.

Table 8.1 (a) Default parameters used in all figures, unless otherwise stated

Parameter	**Value**
Rice factor (K)	10
Channel bandwidth (W) (MHz)	112
Propagation exponent (α)	2.5
Operating frequency (f) (GHz)	38
Propagation loss constant	dB
($L_0 = 10\alpha\log_{10}(4\pi/\mathbf{c}^{\mathrm{m/s}}) + 10\alpha\log_{10}(f^{\mathrm{GHz}})$)	80.0468
Equivalent isotropic radiated power (EiRP P) (dBm)	20 dBm + 43 dBi
Receive antenna gain (A) (dBi)	43
Thermal noise power (N) (mW/MHz)	$4 \cdot 10^{-12}$

Table 8.1 (b) Minimum value of m required to reduce the error of approximation

K	**$\alpha = 2$**	**$\alpha = 2.5$**	**$\alpha = 3$**
0	1	1	1
1	4	5	8
5	15	19	28
10	15	20	30
20	15	20	30

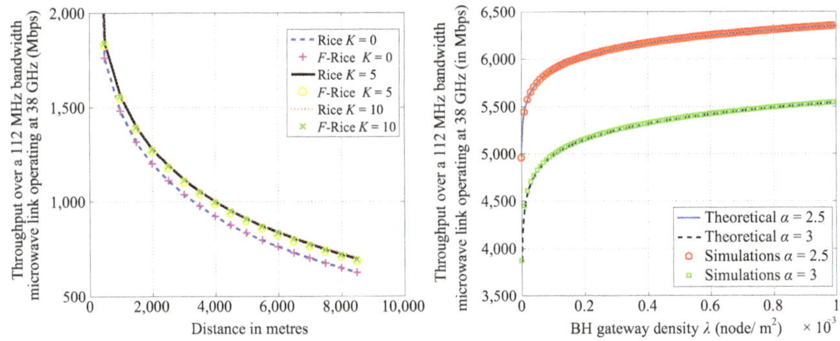

Figure 8.9 *(Left) Optimum setting of m is used to generate approximated throughput within 0.1% error of the actual throughput using Rice fading for three values of K and α = 2.5 (refer to Table 8.1(a) and (b)). (Right) Average throughput over any distance using the proposed approximation (8.13) with m values as defined in Table 8.1(a) and (b) [42]*

The expected throughput over all possible values x is $\mathrm{E}\{T\}$, which can be derived as follows and approximated using (8.13):

$$\begin{aligned} \mathrm{E}\{T\} &= \int_0^\infty T_x \cdot f_x(x)dx \\ &\approx \frac{W}{\ln(2)} \sum_{i=0}^{m-1} C_i \sum_{l=0}^{m-i-1} \int_0^\infty e^{x^\alpha/B} x^{\alpha \cdot l} \frac{\Gamma(-l, x^\alpha/B)}{B^l} 2\pi\lambda \cdot x \cdot e^{-\pi\lambda x^2} dx, \end{aligned} \tag{8.15}$$

where $B \triangleq ((m+K)/m(K+1))((P \cdot A)/(L0 \cdot N))$. The expression still requires numerical integration but can be solved with, relatively, manageable complexity. However, for the special case of $\alpha = 2$, we can further simplify the expression to a closed form.

The average capacity of a wireless link under Rician fading at any distance and for a special case of $\alpha = 2$ can be approximated as in (8.16) next.

$$\begin{aligned} \mathrm{E}\{T\} \approx \frac{W}{\ln(2)} \sum_{i=0}^{m-1} C_i \sum_{l=0}^{m-i-1} \pi\lambda B \frac{(-1)^l}{l!} \\ \times \left\{ \frac{\Gamma(l+1)}{l+1} {}_2F_1(l+1, l+1; l+1; 1-\pi\lambda B) - \sum_{r=0}^{l-1} (-1)^r r! (\pi\lambda B)^{r-l} (l-r-1)! \right\}. \end{aligned} \tag{8.16}$$

$$\Pr\{\gamma > \tau\} = \int_0^\infty Q\left(\sqrt{2K}, \sqrt{2(1+K)\frac{\tau N L_0 x^\alpha}{PA}}\right) 2\pi\lambda \cdot x \cdot e^{-\pi\lambda x^2} dx \tag{8.17}$$

$$\begin{aligned}\Pr\{\gamma_{\alpha=2} > \tau\} &= \int_0^\infty Q\left(\sqrt{2K}, x \cdot \sqrt{2(1+K)\frac{\tau N L_0}{PA}}\right) 2\pi\lambda \cdot x \cdot e^{-\pi\lambda x^2} dx \\ &= 1 - \frac{(1+K)\tau N L_0}{(1+K)\tau N L_0 + \pi\lambda PA} \times exp\left(-\frac{\pi\lambda KPA}{(1+K)\tau N L_0 + \pi\lambda PA}\right)\end{aligned} \tag{8.18}$$

The expression (8.15) is used in Figure 8.9 (right) to derive the average throughput over a range of gateway densities for two settings of the propagation exponent: $\alpha = 2.5$ and $\alpha = 3$. As can be expected, higher gateway densities indicate shorter wireless BH ranges, consequently, higher γ and higher throughput. Similarly, a higher propagation exponent induces higher loss per decade, hence, lower achievable throughput, as seen in Figure 8.9 (left).

8.2.2.2 Latency

Latency in a communication system is a combined result of four distinct processes: propagation, transmission, processing, and queuing, as detailed in Section 8.1.4.2. The propagation delay is the time it takes a packet to travel between transmitting and receiving antennae; hence, it can be ignored in wireless communication as the propagation speed is equal to speed of light. The transmission delay is the time taken to push all the packet bits onto the link. The dominant factor causing transmission delays in wireless links is the decode-and-forward mechanism, typically applied at each hop, which results in retransmissions when the received SNR value is below the required threshold (τ). Accordingly, the success transmission probability of one hop in a wireless point-to-point link is equivalent to the probability $\Pr(\gamma > \tau)$ which can be directly obtained from (8.32) as $1 - F_Z(z)$ for Rician fading, at a given distance. Accordingly, the average success probability over all possible values of x can be obtained as in (8.17). For a special case in which $\alpha = 2$, a closed-form solution for (8.17) can be obtained using [50, eq. (B.27)], as shown in (8.18). Considering a time-slotted transmission over wireless BH links, let θ be the timeslot duration corresponding to given wireless technology. If the first attempt of transmission is successfully received, the transmission delay would be equal to θ, which is the ideal case. Otherwise, retransmissions occur until a successful reception is recorded. Each retransmission would incur an additional delay equal to the timeslot duration, if we assume an ideal errorless reception of acknowledgement in the opposite direction. For instance, if the probability of successful reception is 0.5, the expected latency over the wireless hop would be $2 \times \theta$. In general, the mean transmission delay is inversely proportional to the transmission success probability $\Pr(\gamma > \tau)$, following the approach in [41], as shown here:

$$\mathrm{E}\{D^{t}\} = \frac{\theta}{\Pr\{\gamma > \tau\}}. \tag{8.19}$$

Figure 8.10 shows the average success probability of transmission over a wireless link for a range of SNR target values, 4 node densities, and two propagation exponents $\alpha = 2.5$ and $\alpha = 3$. For $\alpha = 2.5$, the success probability is quasi-guaranteed (to the tenth significant digits). For the latter setting ($\alpha = 3$), the success probability is predominantly constrained by the propagation but is also limited by node density and setting of τ. For a high setting of $\theta = 100$ µs, the transmission delay is associated with a high SNR target $\tau = 15$ dB, and for $\alpha = 3$ and $\lambda = 1$ gateway per km^2, the effective transmission delay would be ~103.1 µs.

In addition to the transmission delay, communication over wireless links requires node processing which incurs additional delays. These are the processing and queuing delays which are often approximated jointly as a gamma distribution with parameters that depend on the router, the given load, and technology, as discussed in Section 8.1.4.2:

$$\mathrm{E}\{D^{r,q}\} = \phi \cdot (1 + 1.28 \cdot \lambda_{k-1}/\lambda_k) \cdot (a + \mathrm{E}\{\varepsilon\}b), \tag{8.20}$$

where ϕ is a delay-scaling factor that reflects the processing power of the node per connected line; higher values of ϕ incur higher delays. The expression $(1 + 1.28 \cdot \lambda_{k-1}/\lambda_k)$ denotes the mean number of nodes in layer $k - 1$ served by the designated node (router) in layer k to which the wireless BH link in connected. The parameter a (in µs) represents the traffic-independent processing delay of the

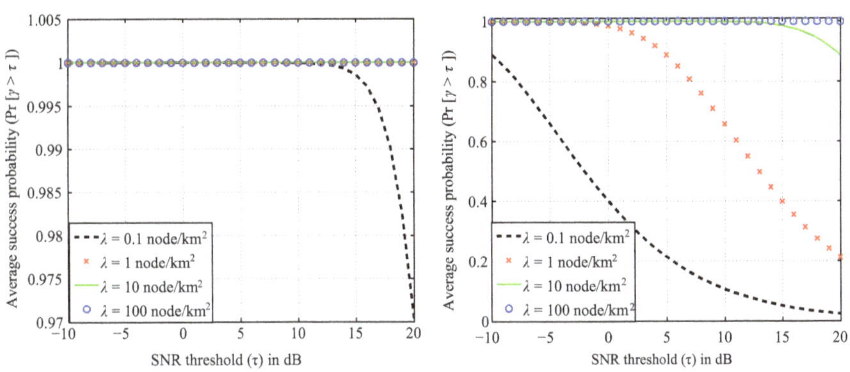

Figure 8.10 The average transmission success rate over a wireless link with Rice fading and variable distance is shown for two propagation exponents $\alpha = 2.5$ (left) and $\alpha = 3$ (right) for $\lambda = \{0.1, 1, 10, 100\}$ gateway(s) per km^2. Parameters settings are as listed in Table 8.1(a) and (b) except $P = 10 + 43$ dBm as for higher values the success probability is $> 99.999\%$ [42]

router, and b (in µs /bit) represents the processing delay with respect to the packet size $\mathrm{E}\{\varepsilon\}$. Accordingly, a point-to-point wireless link using decode-and-forward scheme incurs a total delay D, as shown in the following equation:

$$\begin{aligned} \mathrm{E}\{D\} &= \mathrm{E}\{D^t + D^{r,q}\} = \mathrm{E}\{D^t\} + \mathrm{E}\{D^{r,q}\}, \\ &= \frac{\theta}{\mathrm{E}\{\Pr(\mathrm{SNR} > \tau)\}} + \phi \cdot \left(1 + 1.28 \cdot \frac{\lambda_{k-1}}{\lambda_k}\right) \cdot (a + \mathrm{E}\{\varepsilon\}b). \end{aligned} \tag{8.21}$$

Figure 8.11 shows that the number of connected lines (function of $\lambda_{\kappa-1}$) has a major impact on increasing the processing delay (hence, the total delay). The total delay also seems to increase linearly with increasing delay-scaling factor (ϕ), as shown in Figure 8.11 (left). The delay also increases with increase in packet length, as seen in Figure 8.11 (right), but the end value is dominated by the effect of the delay-scaling factor and node density. Results shown in both figures advocate that the joint processing and queuing delay (tens of ms) is order of magnitude higher than the transmission delay in wireless BH links (~0.1 ms).

8.2.2.3 Resilience

There are several performance metrics that may be used to characterise the resilience of a link [37]. (i) Reliability is the probability of a link being operational and is measured as the mean time before failure (MTBF). (ii) Survivability measures the performance of recovering from failure in terms of the mean time to repair (MTTR). (iii) Availability is the proportion of time the link is operational but,

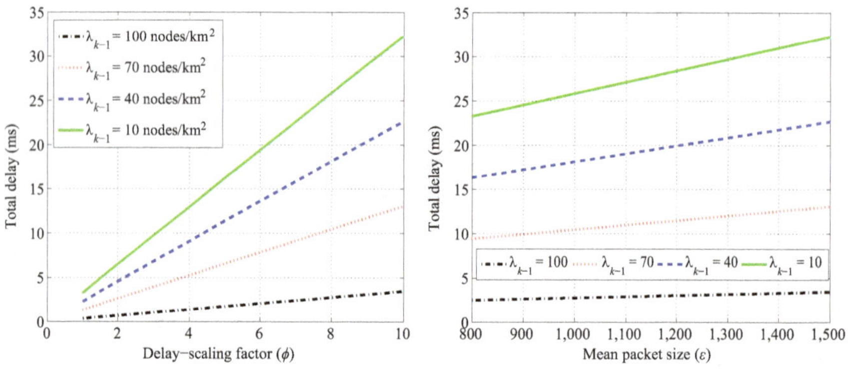

Figure 8.11 The expected total delay over a wireless link is represented in these figures: (left) for variable node delay-scaling factor and node density. Parameters based on Table 8.1(a) and (b) with P = 100 mW, {ε} = 1,500 bits, a = 10 µs, b = 0.01 µs/bit, τ = 10 dB, λk = 1 node/km² and (right) for variable packet length (in bits) and node density. Parameters based on Table 8.1(a) and (b) with P = 100 mW, ϕ = 10, a = 10 µs, b = 0.01 µs/bit, τ = 10 dB, λk = 1 node/km² [42]

differently from (i), it takes into account failures and repair; it is quantified as uptime/(uptime+downtime), e.g. MTBF/(MTBF+MTTR). In this paper, we model the outage probability, i.e. downtime/(uptime+downtime) which is affected by the radio link failure and the equipment failure. Features such as automatic coding and modulation (ACM) are commercially employed to increase the robustness of the radio link at the cost of spectrum efficiency reduction. Figure 8.12 shows experimental results that highlight the vulnerability of EHF signals to fades and weather conditions [4,49].

In this case, mechanisms such as decode-and-forward dampen the effect of loss of radio link through repeated retransmissions. These improve the tolerance of a wireless link which may be recovered if the fade lasts less than the time allowed for maximum retransmission.

Consequently, modelling the resilience of a wireless link entails an analysis of the fade duration distribution and determining the probability of having a fade duration that is less than the critical limit. The average receive power is characterised by $\mathrm{E}\{J\}$, and the minimum power level threshold is $J_{\min}$ (e.g. receiver sensitivity or minimum received signal strength to guarantee a target bit error rate). The fading margin is defined as $\rho \triangleq \sqrt{J_{\min}/\mathrm{E}\{J\}}$ which is equal to $\sqrt{J_{\min}}$ for $\mathrm{E}\{J\} = 1$. The time during which the signal power is below $J_{\min}$ is the fade duration. The first step is to determine the level crossing rate R_L and the average fade duration $\mathrm{E}\{F_d\}$, equivalent to mean-time-to-repair or MTTR. These are

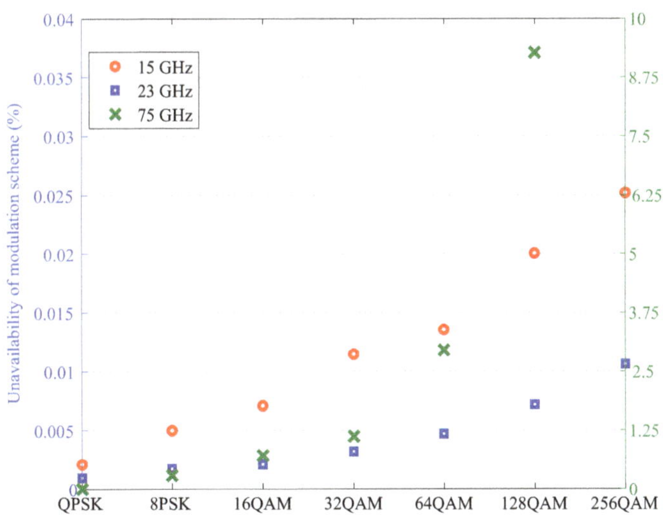

Figure 8.12 Percentage of time the indicated ACM is not available for three scenarios based on data from different trials using automatic coding and modulation in different frequency bands [42]

derived in [50] for Rician fading, as shown in the following equations:

$$R_L = \sqrt{2\pi(K+1)} \cdot f_m \cdot \rho e^{-K-(K+1)\rho^2} \cdot I_0(2\rho\sqrt{K(K+1)}) \tag{8.22}$$

$$\mathrm{E}\{F_d\} = \frac{1 - Q(\sqrt{2K}, \sqrt{2(K+1)\rho^2})}{\sqrt{2\pi(K+1)} \cdot f_m \cdot \rho e^{-K-(K+1)\rho^2} \cdot I_0(2\rho\sqrt{K(K+1)})} \tag{8.23}$$

where $f_m = vf/c$ is the maximum Doppler shift, v denotes the relative motion of scattering objects between the transmit and receive antennae, f is the carrier frequency of the signal, c is the speed of light, and K is the Rice factor.

The none-fade duration $\mathrm{E}\{\overline{F_d}\} = 1/R_L - \mathrm{E}\{F_d\}$ (equivalent to the MTBF) is then defined and used in a Gilbert-Elliot two-state-model to find the probability of fades being larger than a threshold $J_{\min}$ as in [37]. Owing to the retransmissions mechanism, such fades do not necessarily cause a link outage unless they persist for a duration $\Delta_f > \Delta_{\max}$. The outage probability of a wireless link is thus $\Pr(\Delta_f > \Delta_{\max})$ which can only be determined if the fade duration distribution is known. Albeit, this distribution is largely unknown and remains an open problem in the literature ever since the original formulation by Rice [51]. In this work, we assume that the fade duration distribution exhibits an exponential-like behaviour for high values of ρ, similar to [52]. The radio outage probability of a wireless BH link ($\mathrm{E}\{O^r\}$) can thus be expressed as follows:

$$\mathrm{E}\{O^r\} = \Pr(\Delta_f > \Delta_{\max}) = e^{-\Delta_{\max}/\mathrm{E}\{F_d\}}. \tag{8.24}$$

In Figure 8.13, the radio outage probability (i.e. $\mathrm{E}\{O^r\}$) of a wireless link that follows Rician fading is simulated as a function of the critical time $\Delta_{\max}$ and the fade margin ρ. Higher values of $\Delta_{\max}$ improve the radio link availability and vice versa. On the other hand, higher fade margins render the probability of fades higher; consequently, the link is more vulnerable which increases the outage probability.

Often, key radio links that affect a large service area are designed with redundancy schemes that employ at least two independent radio links, such as [1+1] or [2+0] schemes. In the former example, one link is active at a given time with seamless switching to the backup link in case the first is in outage. In the latter example, both links are used simultaneously to enhance the effective throughput; if one radio link fails, the system switches to a single link automatically without causing service interruption. In both examples, the wireless connection is only in outage if both radio links fail simultaneously. In mesh wireless networks, there are often more than one path between two points; these points are considered disconnected if all possible paths fail simultaneously. The outage expression, with any redundancy scheme, can be extended as follows, according to (8.24):

$$\mathrm{E}\{O^{r'}\} = Pr'(\Delta_f > \Delta_{\max}) \stackrel{(a)}{=} \prod_{i}^{L} e^{-\left(\Delta_{\max}/\mathrm{E}\{F_{d,i}\}\right)} \stackrel{(b)}{=} e^{-L(\Delta_{\max}/\mathrm{E}\{F_d\})} \tag{8.25}$$

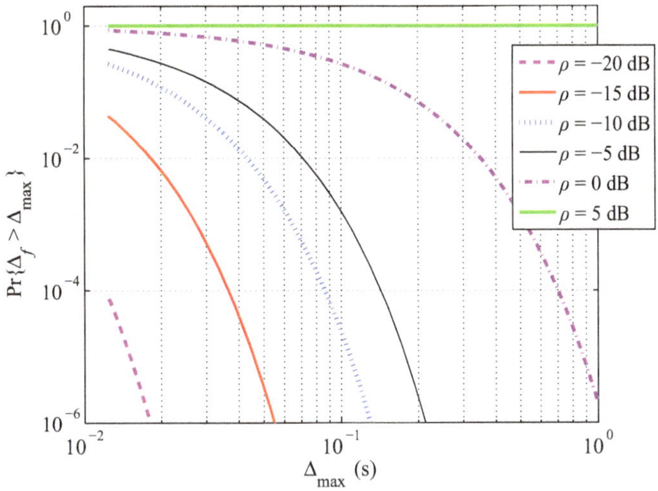

Figure 8.13 Outage probability of a Rice fading channel link based on (8.24), assuming $f_m = 10$ Hz for different values of ρ. Other parameters used in the simulations are listed in Table 8.1(a) and (b) [42]

where L denotes the number of links and $\mathrm{E}\{F_{d,i}\}$ is the mean fading duration over link i. The equality (a) indicates that, in case parallel links are used as backup between layers $\kappa - 1$ and κ, the outage occurs only when all the links fail simultaneously. The notation $\Pr'$ refers to the link outage probability with redundancy deployments. If the same value can be assumed for the mean fading duration of all L links, then the expression is further simplified as in ((b) of (8.25)) [37].

Equipment failure is another reason for wireless link outages and concerns both indoor units (IDUs) and outdoor units (ODUs). The MTBF of such units is normally very high (>12 years), and redundancy schemes are often used to improve the reliability. As such, a unique wireless link could be equipped with an active ODU, and a hot-backup passive ODU that is activated whenever the prime ODU fails. The MTTR for wireless equipment varies significantly based on the location and accessibility; in an urban environment, it is often within 3 h but may take up to 3 days in remote areas with difficult access. The outage probability of wireless links, including radio equipment and radio link failures, and any redundancy plan can be expressed as follows:

$$
\begin{aligned}
\mathrm{E}\{O\} &\triangleq 1 - \mathrm{E}\{(1 - O^{r'}) \times (1 - O^{o'}) \times (1 - O^{i'})\} \\
&=^{(a)} 1 - (1 - \mathrm{E}\{O^{r'}\}) \times (1 - \mathrm{E}\{O^{o'}\}) \times (1 - \mathrm{E}\{O^{i'}\})
\end{aligned}
\tag{8.26}
$$

where $O^{o'}$ and $O^{i'}$ are the outage probabilities of ODUs and IDUs, respectively, such that $O^{o'} \triangleq \prod_i O_i^o$, where i indicates the redundancy index of the hardware, and $O_i^o(\mathrm{MTTR}/(\mathrm{MTBF} + \mathrm{MTTR}))_i^o$ is the outage probability of the ODU with

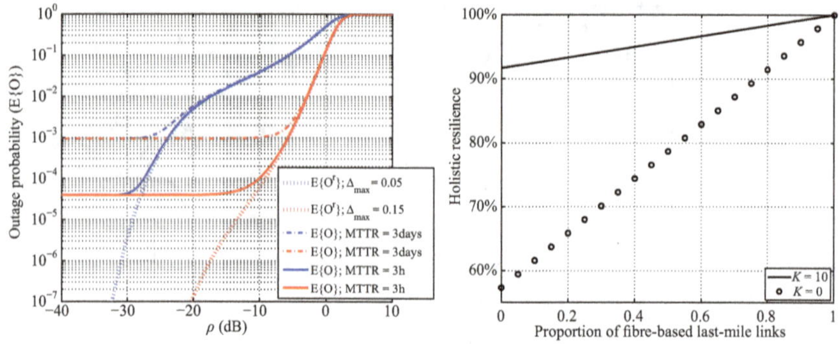

Figure 8.14 *(Left) The outage probability of the wireless link is shown as a function of ρ for two values of* $\Delta_{\max} = \{0.05, 0.15\}$ s. *The impact of equipment failure is dominant for lower values of ρ, and has minimal effect otherwise. In these regions, the impact of the MTTR value is also importance. (Right) The holistic resilience of the BH network with* $\rho = 3$ dB *and* $\Delta_{\max} = 0.1$ *s; comparison between Rice fading* $(K = 10)$ *and Rayleigh fading* $(K = 0)$ *[42]*

redundancy index i based on the MTTR and MTBF values. The equality (a) in (8.26) is possible because the concerned probabilities are independent of each other.

The cause of a wireless link outage is often the radio failure, as shown in Figure 8.14, unless the fade margin is very low. For these low values, the wireless link outage probability is dominated by equipment failure. For higher fade margin values, the ODU and/or IDU failures have limited impact on the total wireless link availability.

8.2.3 Wireless BH in a multi-hop hybrid network

In this section, we integrate the derived performance models of a wireless BH hop in the estimation of the multi-hop hybrid BH performance. Referring to the topology in Figure 8.8, with parameters as defined in Table 8.2, we derive the expectations of the holistic BH resilience, capacity, and latency following the approach in [32]. Based on [32], the resilience of a multi-hop BH is affected by outages on any hop. Differently, the capacity of a multi-hop hybrid BH is determined by the single hop which is the most limiting, i.e. the last mile. The latency, on the other hand, is the aggregate effect of delay occurring on all hops. We consider that the BH network between the gateways in Layer 2 and the core network has a cumulative mean delay of 10 ms.[††] The parameters employed to model the wireless links' capacity, latency, and resilience are listed in Table 8.3, in which we consider

[††]Assuming the BH gateway to the core network has similar processing delay parameters to the gateway in layer 2 and 99 connected lines, the delay can be computed as: $5 \cdot (1 + 99) \times (5 + 0.01 \cdot 1500) = 10$ ms

Table 8.2 Simulation parameters

Parameter	Value	Remark
λ_0	400 users/km^2	Density of Φ_0 PPP representing users
λ_1	40 cells/km^2	Density of Φ_1 PPP representing small cells
λ_2	4 gateways/km^2	Density of Φ_2 PPP representing BH aggregation points
λ_3	1 gateway/km^2	Density of Φ_3 PPP representing BH gateways
p_m	50%	50% of last-mile links are randomly allocated microwave (38 GHz) technology
p_f	50%	50% of last-mile links are randomly allocated fibre-based technology

Table 8.3 Performance modelling parameters in addition to those listed in Table 8.1(a) and (b)

Parameter	Value	Remark
ϕ_3^f	5	Scalar factor representing the processing power of BH gateway (lower value indicates better processing power)
ϕ_2^f	10	Scalar factor representing the processing power of fibre aggregation point
ϕ_2^m	10	Scalar factor representing the processing power of microwave aggregation point
d_3^f	5 μs	Traffic-independent processing delay of BH gateway
d_2^f	10 μs	Traffic-independent processing delay of fibre aggregation point
a_2^m	10 μs	Traffic-independent processing delay of microwave aggregation point
b_3^f, b_2^f, b_2^m	0.01 μs/bit	Processing delay of all routers relative to the packet size
ε	1,500 bits	Ethernet packet size

a microwave band operating in the 38 GHz spectrum. The expected capacity of a last-mile wireless microwave hop that exhibits LOS fading is derived using (8.15). The expectations of delay and resilience in microwave links are based on (8.21) and (8.26), respectively.

We adopt the network topology that was presented in Figure 8.8 and consists of two hops. The first is the last-mile connecting the small cells to an aggregation point and uses either microwave links or fibre-based links (with mean capacity of 10 Gbps). The second hop connects the aggregation point to the gateway in the core network and is always provisioned with fibre-based links (10 Gbps). We study the impact of increasing the density of aggregation points and/or the proportion of fibre-based last-mile links on the holistic BH performance. In parallel, we compare the results with both Rice and Rayleigh fading considerations.

We vary the proportion of last-mile fibre links and the density of the aggregation gateways in Layer 2 and examine the resulting impact on the expectations of the multi-hop BH throughput and latency in Figure 8.15 (left) and (right), respectively. The node density λ_2 has limited effect on the resulting holistic throughput, whereas the increase of fibre links steps up the throughput by a factor of two. In contrast, the increase of node density λ_2 (i.e. proportion of fibre last-mile links) has equal share in reducing the holistic latency of the multi-hop BH. As for resilience, the impact of increasing the share of fibre in the last mile doubles the resilience of the BH network, as seen in Figure 8.14 (right).

Had Rayleigh fading ($K = 0$) been considered instead of LOS fading (e.g. $K = 10$), the resulting capacity estimation would be pessimistic as clearly seen in the zoom-in window of Figure 8.15 (left). On the other hand, the type of fading does not have a significant impact on the latency, as expected from the results obtained in Section 8.2.2.2. Essentially, the delay in the holistic BH link is dominated by the processing and queuing delays, as such, the difference in delay caused by the error in modelling the fading is negligible. In contrast, resilience of wireless links is the most vulnerable to the type of fading, as seen in Figure 8.14 (right). Consequently, the error in adopting Rayleigh fading models instead of Rice fading when estimating the multi-hop BH resilience is significant and would result in invalid conclusions.

8.3 Case study on using modular approach to unlock the realistic BH

The user-centric BH (UCB) is the state-of-the-art solution for intelligent user-cell-BH association that is context aware, radio network aware, and BH aware and, as such, affords the network the chameleon-like characteristics [22]. It was demonstrated in [22] that the UCB reveals the hard limits of the network that cannot be circumvented by intelligent user-cell-BH matching. It is vital for operators deploying 5G networks to determine the weak aspects of the incumbent BH and

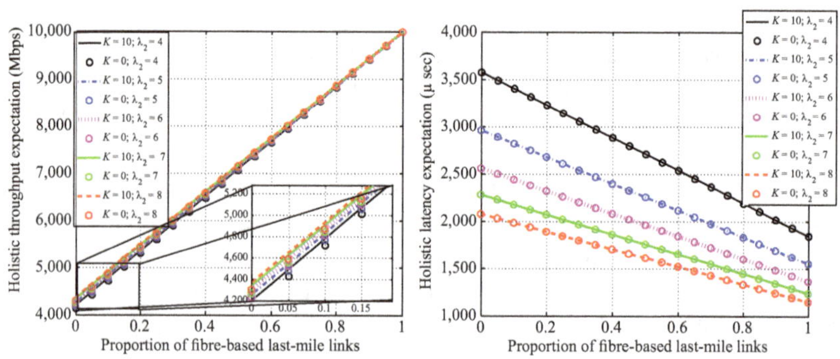

Figure 8.15 Holistic BH network throughput (left) and latency (right) for LOS fading with $K = 10$ and Rayleigh fading with $K = 0$ [42]

identify pertinent solutions that yield the highest benefits at the minimum cost. In this section, we present a case study in which we implement the multi-hop BH performance model described in this chapter to identify a cost-effective and quality-aware BH upgrade using the UCB [27].

A realistic dense network of small cells is created based on elements of stochastic geometry, as in [32]. The performance of the multi-hop BH links is generated based on technology-specific analytical models and according to the topology. The UCB is first used to gauge the gap between the available and required network performance by looking at two attributes: throughput and latency. The performance modelling approach in [32] is then used to obtain alternative solutions for reducing this gap. These alternatives are then ranked according to their incurred gains and total cost of ownership (TCO).

We consider a dense network of small cells which have identical radio characteristic. Each cell connects back to a BH aggregation point using one of three possible transport technologies: VDSL2 (copper-based), microwave (28 GHz), and 10G-PON (10 Gbps). Aggregation points are linked to the core network through a fibre-based BH. The last mile is thus the network throughput bottleneck. The described system is simulated in MATLAB® as a multilayered network in which users belong to Layer 0, small cells to Layer 1, and aggregation points to Layer 2. Each of these layers is described by a PPP Φ_l with density λ_l for $l \in [0, 1, 2]$. The link between Layers 0 and 1 is an LTE-based radio interface. The link between Layers 1 and 2 is the last-mile BH network which consists of p_v share of VDSL2, p_m of microwave, and p_f of fibre links, such that $p_v + p_m + p_f = 1$. An example network is shown in Figure 8.8.

8.3.1 Monte Carlo simulations

In each Monte Carlo run, an instance of the described network model is generated with the following three parameters: (i) number, location, and quality requirements of users; (ii) number, location, and type of last mile of small cells; and (iii) number and location of aggregation points. The distance between any cell in Φ_1 and any gateway in Φ_2 is then computed and the shortest is selected. Consequently, the capacity and latency expectations of each cell's last mile are derived. In each iteration, the users' parameters are updated and the corresponding metrics stored. The UCB consists of dynamically setting two bias factors for each cell that corresponds to the capacity and latency attributes, respectively. A higher bias value indicates better performance and vice versa. Let u denote a user in Layer 0, each user u associates two sets of parameters that are randomly generated: $\{Q_{u,c}, Q_{u,l}\}$ and $\{\omega_{u,c}, \omega_{u,l}\}$. The first set indicates the target values of throughput and latency, respectively. The second represents the weights associates with each attribute. The UCB endeavours to find the optimum setting of each bias factor in each cell is such a way that maximise the users' satisfaction and the network throughput. The users' and network' key performance indicators (KPIs) that are used to gear this process are described next.

The case study is built based on the system model described in the previous section with parameters as defined in Table 8.4, which constitute the baseline

Table 8.4 Simulation parameters

Parameter	Value	Remark
λ_0	400 users/km^2	Density of Φ_0 PPP representing users
λ_1	40 cells/km^2	Density of Φ_1 PPP representing small cells
λ_2	4 gateways/km^2	Density of Φ_2 PPP representing BH aggregation points
Q_c	[0,5] dB	50% of users are randomly allocated a low target SINR and 50% a high target
Q_l	[30,100] ms	50% of users are randomly allocated a low target latency and 50% a high target
ω_c	[0.1,0.9] dB	50% of users are randomly allocated a low weight to SINR target and 50% a high weight
ω_l	[0.1,0.9] dB	50% of users are randomly allocated a low weight to latency target and 50% a high weight
p_v	40%	40% of last-mile links are randomly allocated VDSL2 technology
p_m	30%	30% of last-mile links are randomly allocated microwave (28 GHz) technology
p_f	30%	30% of last-mile links are randomly allocated fibre-based technology
N_i	10	Number of iterations per run
N_r	100	Number of Monte Carlo runs

scenario. The capacity and latency modelling of the BH network are described in Section 8.3.3.1 and the cost model in Section 8.3.3.2. The corresponding modelling results are shown in Section 8.3.3.3 together with the best solution that maximises the performance improvement whilst minimising the associated cost.

8.3.2 Users' and network's KPIs

Assessing the effectiveness of a user-cell-BH scheme may only be performed after identifying adequate and representative metrics. Traditionally, service-based metrics have been used to assess the network's performance, such as peak data rates, coverage, and spectral efficiency, termed *quality of service* or QoS. However, 5G networks are about the always sufficient rate to give users the perception of infinite capacity,[‡‡] and thus, the key metric should gauge subjective users' perception of the quality and not objective QoS. ITU-T G.114 recommends a maximum one-way delay of 150 ms for voice-over-IP beyond which the degradation may be discerned. On the other hand, delay requirements for online gaming differ greatly, depending on the game played (e.g. *Battlefield 2* and *MMORPGs*[§§]). Elite performance of a maximum delay of 50 ms is recommended for intense first-person shooters games such as Battlefield 2. Other real-time strategy games, e.g. MMORPGs, can run with a delay of 150 ms [53]. From a user-centric point of

[‡‡]http://www.surrey.ac.uk/5gic/about/5gic-vision

[§§]http://www.mmorpg.com/games-list and http://www2.ea.com/battlefield-2

view, a delay of 100 ms (QoS value) leads to excellent QoE for this game or voice-over-IP. The same QoS value would result in an unacceptable QoE for a Battlefield 2 player.

To this end, 5G optimisation demands novel metrics that are centred on users' QoE and are context aware [54]. The MOS is the most popular indicator for measuring such perceived media quality. Nonetheless, gathering sufficient scores from users, referred to as *subjective* MOS, is often a delicate task that is time-consuming and costly. *Objective* MOS methods are automated alternatives that may be calibrated to reflect the quality reported by the subjective MOS and can readily be employed in optimisation schemes of cellular systems [55]. One 5G challenge is to define a context-agnostic mapping between QoE and easy-to-use QoS, which is beyond the scope of this work. In this case study, we assume that the QoE-to-QoS mapping is a given for all 5G use cases. For each use-case, we associate a set of required QoS parameters with corresponding weights. It is assumed that these QoS and weights have been tuned to reflect the target QoE of the use cases. These are compared to the achieved QoS, i.e. delivered by the network. The metric we propose to measure the users' satisfaction gauges the gaps between each of the target and delivered QoS parameters and is affected by the corresponding weight. Let $\{Q'_{u,c}, Q'_{u,l}\}$ represent the measured QoS values delivered by the network to user u. We define the level of dissatisfaction $(\widehat{Q}_{u,q})$ of a user u with respect to quality attribute $q = \{c(\text{capacilty}), l(\text{latency})\}$ as the ratio of the gap $(Q'_{u,q} - Q_{u,q})$ to the target QoS $(Q_{u,q})$. The cumulative dissatisfaction of users who associate a high weight to the given quality is $\widehat{Q}_q$, shown next. We also estimate the cumulative weighted users' dissatisfaction $\overline{Q_q}$ for both types of users. It should be noted that the measured $Q'_{u,q}$ is clamped to the target value, i.e. it is not beneficial to exceed it nor does it compensate for quality shortage of other users.

$$\widehat{Q}_q \triangleq \frac{1}{\lambda_0} \sum_u \left| \frac{Q'_{u,q} - Q_{u,q}}{Q_{u,q}} \right| |\omega_{u,q} = High \tag{8.27}$$

$$\overline{Q_q} \triangleq \frac{1}{\lambda_0} \sum_u \omega_{u,q} \times \widehat{Q}_{u,q}, \forall q \in \{c, l\} \tag{8.28}$$

On the other hand, maximising the infrastructure usage efficiency remains an operator's prime target. To this end, it is desirable for a novel BH scheme to maximise the BH load without overloading. By doing so, the total system throughput is also maximised, and the best utilisation of the infrastructure is achieved. Consequently, a key network-centric QoS metric is the total system throughput T_{tot} and the rate of unsatisfied users O_c with respect to capacity and O_l to latency.

8.3.3 Adopted models and results

8.3.3.1 Performance models

The performance modelling is focused on the BH network as the radio access is purposely over-dimensioned. Thus, the capacity of a multi-hop hybrid BH is

Table 8.5 Performance modelling parameters

Parameter	**Value**	**Remark**
Δ	105 Mbps	Highest possible throughput over the shortest copper-line reach
δ	$1.14 \cdot 10^{-3}$	Throughput decay per unit length of copper
P	10 dBm	Transmitting power of last-mile point-to-point microwave link
A	43 dBi	Antenna gain on either ends of the microwave link
L_0	120 dB	Minimum path loss incurred by the shortest microwave reach
α	2.5	Propagation exponent of the microwave link
f	38 GHz	Operational frequency of the microwave link
W	112 MHz	Bandwidth of the microwave link channel
σ^2	$4 \cdot 10^{-15}$ mW/KHz	Thermal noise power intensity
K	10	Rice factor of the LOS fading in the microwave link channel
κ_v, κ_m	10	Parameter that reflects the processing power of BH gateway [57]
a_v, a_m	10 μs	Parameter that reflects the processing power of BH gateway [57]
b_v, b_m	0.01 μs/bit	Parameter that reflects the processing power of BH gateway re packet size [57]
ε	1,500 bits	Ethernet packet size

determined by the hop that is the most limiting: the last mile. The latency, on the other hand, is the aggregate effect of delay occurring on all hops. We consider that the BH network between the small cell BH gateway and the core network has a cumulative mean delay of 10 ms. The parameters employed to model the VDSL2 and microwave links' capacity and latency are listed in Table 8.5. The expression characterising the expected capacity of a VDSL2 hop for r metres of copper is $\mathrm{E}[T_v(r)] = \Delta \cdot e^{-\delta \cdot r}$ [32]. The expected capacity of a wireless microwave hop that exhibits LOS fading (h) conditioned on r is shown in the following [60, eq. (5)]:

$$\mathrm{E}[T_m(\gamma(r))|r] = \frac{\mathrm{W}}{\ln(2)} e^{-K} \sum_{n=0}^{\infty} \frac{K^n}{n! \cdot n!} \times G_{2,3}^{3,1}\left[\frac{(1+K)}{\Omega_r}\middle|\begin{matrix}0,1\\ n+1,0,0\end{matrix}\right] \tag{8.29}$$

where K is the Rice factor, W is the channel bandwidth in KHz, $G_{p,q}^{m,n}[\cdot]$ is the Meijer-G function, and $\Omega_r \triangleq \mathrm{E}[\gamma|r] = P \cdot A \cdot r^{-\alpha}/L_0 \cdot \sigma^2$ is the expectation of the SNR (γ) conditioned to a particular value of r.

The expected delay over a VDSL2 link is mainly determined by the multiplexing/demultiplexing box and is modelled according to the following expression which does not depend on the reach of the copper line [32]:

$$\mathrm{E}[L_v] = \kappa_v \cdot (1 + 1.28 \cdot \lambda_1/\lambda_2) \cdot (a_v + \varepsilon \cdot b_v) \tag{8.30}$$

where $(1 + 1.28 \cdot \lambda_1/\lambda_2)$ denotes the mean number of small cells served by the designated aggregation point. Moreover, the parameters κ_v and a_v are representative of the generic processing power of the router, and b_v represents the router's processing capability with respect to the packet size ε. The delay in microwave links is due to two factors: processing and queuing delay $L_{m,1}$ (8.30) and the delay caused by the forward-and-retransmit mechanism $L_{m,2}$. When the data is not correctly detected at the receiving end of the hop (i.e. $\Pr([\gamma|r] < \tau_t)$), during the technology-specific timeslot θ, it is retransmitted for a number of times before it is declared lost. The waiting period and the retransmission(s) incur an additional delay $L_{m,2}$; for a wireless link assuming Rician LOS fading, it is expressed as follows, where $\mathcal{Q}(\cdot)$ is the Marcum-Q function [56]:

$$\mathrm{E}[L_{m,2}] = \frac{\theta}{\Pr([\gamma|r] > \tau_t)} \tag{8.31}$$

$$\Pr([\gamma|r] > \tau_t) = \mathcal{Q}\left(\sqrt{2K}, \sqrt{2\frac{1+K}{\Omega_r}\gamma}\right) \tag{8.32}$$

The fibre-based last mile is considered to have higher capacity than the cells' maximum throughput and lower latency than the lowest Q_l. Consequently, the measured QoS of users associated with a fibre-based cell are determined by the radio conditions and are never limited by the BH.

8.3.3.2 Cost model

We propose the following model that captures the cost of BH gateways and last-mile hops to the small cells, as shown in the following equation:

$$C = \frac{\lambda_1}{\lambda_2}\sum_i p_i(E_i + \mathrm{E}[H_i(r)]) = \frac{\lambda_1}{\lambda_2}\sum_i p_i\left(E_i + \frac{\Gamma(\rho_i/2+1)}{(\pi\lambda_2)^{\rho_i/2}}\right)$$

where $i = \{m,f\}$, E_i is the equipment cost and $H_i(r) = r^{\rho_i}$ is the rate of cost increase as a function of distance r. The cost calculation is limited to microwave and fibre technologies as no upgrade would require increase in copper-based hops. We assume $E_m = \$2,000$ and $E_f = \$50,000$ and $\rho_m = \rho_f = 2$ [58].

8.3.3.3 Results

Based on the analytical multi-hop hybrid BH performance model, the baseline scenario offers a system capacity expectation of 1,137 Mbps and corresponding delay of 12.04 ms (see Figure 8.16). The corresponding simulation results (100 runs) are shown in Figure 8.17 and indicate higher throughput (mean 1,442.2 Mbps), albeit, with high users' dissatisfaction of $\mathrm{E}[\widehat{Q}_c] = 171.2\%$ and $\mathrm{E}[\widehat{Q}_l] = 173\%$, on average, with respect to throughput and latency, respectively. Effectively, the BH network is overloaded, and hence the surge is throughput and

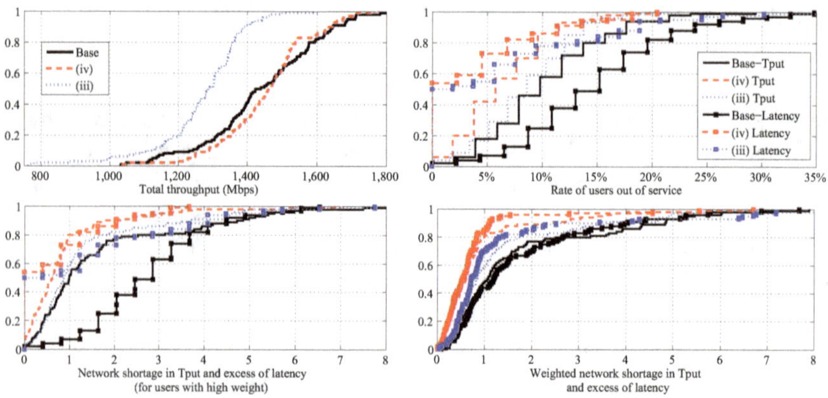

Figure 8.16 Effect of key factors on BH network capacity and latency [27]

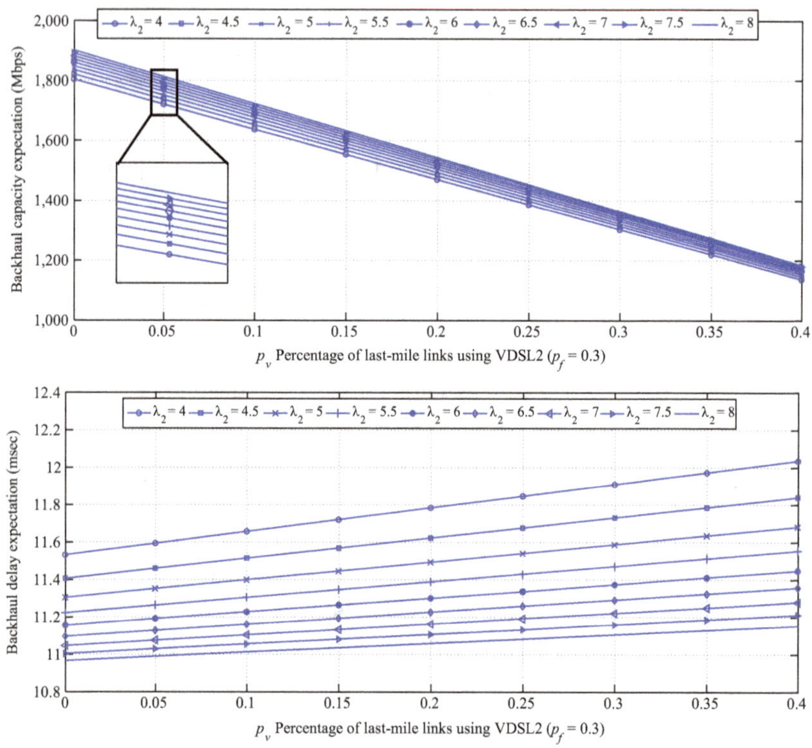

Figure 8.17 Simulation results of baseline scenario and potential upgrade solutions. The cumulative distribution function probabilities of each metric are shown over 100 runs on the y-axis [42]

Table 8.6 Solution shortlist

Changes	Tput (Mbps)	Delay (ms)	Cost (10^6 \$)
(i) $p_v = 0.05$, $p_m = 0.65$	1,721	11.55	0.919
(ii) $p_v = 0.1$, $p_m = 0.4$, $p_f = 0.5$	1,693	11.22	0.974
(iii) $p_v = 0.15$, $p_m = 0.05$, $p_f = 0.8$	1,697	10.63	1.077
(iv) $p_v = 0.1$, $p_m = 0.6$, $\lambda = 6.5$	1,697	11.16	0.371

Table 8.7 Simulation results

Mean	BL	(iii)	(iv)	Mean	BL	(iii)	(iv)
T_{tot}	1,442	1,455	1,270	$\overline{Q_c}(\%)$	170	89	136
$O_c(\%)$	10.22	6.4	8.57	$\overline{Q_l}(\%)$	276	68	131
$O_l(\%)$	14.72	3.80	6.04	$\widehat{Q}_c(\%)$	171.2	90	140
				$\widehat{Q}_l(\%)$	173	73	142

deterioration of quality. Priority users are those that associate high weight to the given attribute. There are 10.22 and 14.72 such users with under par performance re-throughput and latency, respectively.

8.3.4 Upgrade solution

In this section, we use our analytical modelling to find the optimum solution for upgrading the capacity of the BH network by jointly analysing the offered performance improvement and associated cost of each potential solution. Starting with the baseline scenario described in Table 8.4, we study the effect of replacing VDSL2 links with microwave, increasing the share of fibre links in the last mile, and increasing the density of BH gateways in Figure 8.16 (left) and (right) for BH network capacity and latency, respectively.

8.3.4.1 Shortlisting of potential upgrades

We set our upgrade target as *the reduction of user dissatisfaction with respect to throughput by* 50%. To this end, we aim to increase the BH capacity to $1{,}137 \times 1.5 = 1{,}705.5$ Mbps. There are four possible upgrades that offer this capacity, as shown in Table 8.6. Based on the performance and cost analytical modelling, two solutions are short-listed for implementation in the simulator: solution (iii) because it gives the highest improvement and solution (iv) because it is the cheapest but still delivers second best delay expectation.

8.3.4.2 Simulation results and final selection

Each scenario is simulated over 100 Monte Carlo runs, as described in Section 8.3.2. The cdfs of each metric are shown in Figure 8.17 and the average values of the results obtained in Table 8.7. Both scenarios (iii) and (iv) succeed in significantly improving the user-centric metrics. In addition, solution (iii) maintains a similar network throughput compared to the baseline, whereas solution (iv) results in 12% throughput reduction. Nonetheless, solution (iv) targets users with high quality expectations and reduces the number of unsatisfied users (compared to the baseline) by 37% and 74% for throughput and latency, respectively. More importantly, the cumulative QoE gap of those users is reduced by 21% and 52.5%, respectively, at the cost of 0.371 million dollars. On the other hand, solution (iii) reduces the QoE gap even further by 46% and 75%, respectively, in comparison with the baseline, but at a TCO that is almost three times higher than that of solution (iv). Accordingly, money conscious operators would benefit from adopting solution (iv) as a first upgrade to offer better QoE to their existing users and reduce the churn. In contrast, operators that have the means should upgrade using solution (iii) as it offers better quality to users and higher capacity for attracting new users.

8.4 Intelligent wireless backhauling

We have demonstrated the key role of wireless BH in the success of NGMNs and presented state-of-the art approaches to modelling various performance aspects of individual wireless links and their respective impact on the overall performance of multi-hop heterogeneous BH networks. We have also highlighted that the scarcity of the radio resources and the importance of intelligent and dynamic sharing of the bandwidth to accommodate the explosive number of devices whilst meeting their respective stipulated QoE. The main trends for addressing these concerns explore dynamic mechanisms for sharing radio resources between radio access and wireless self-BHs such as [8–13], where some exploit the advances in MIMO technologies (e.g. [14,15]). The authors in [59] propose a novel wireless backhauling strategy for small-cell networks based on dynamic base station cooperation that they refer to as fog radio access network backhauling strategy. This strategy enables base stations to combine and process signals received from diverse paths and demonstrate the gain in the transmission efficiency of the BH network when compared to, direct transmission, decode and forward, and cloud-RAN. The work in [60] proposes a novel-distributed solution for the problem of joint frame design, resource allocation, and user association to maximise the sum rate of a massive MIMO HetNet with wireless BHs. The authors prove that the game-based solution converges to a Nash equilibrium and the simulation results show that the proposed schemes outperform several benchmark schemes. The third generation partnership project is envisioning an integrated access and BH (IAB) architecture for 5G and beyond in which the same infrastructure and spectral resources are used for both access and BH. To this end, authors in [61] present an analytical framework for IAB-enabled network and study the performance of three BH partition strategies: (i) equal partition: when all cells have the same BH bandwidth allocation, (ii) instantaneous load-based partition: when the BH allocation is proportional to the instantaneous radio access load, and

(iii) average load-based partition: when the BH allocation is proportional to the average radio access load. The study shows that the optimum strategy depends on the traffic scenario among other conditions. Unmanned aerial vehicle mounted base stations (UBS) are gaining popularity owing to the flexibility of deployment and agility in optimisation. The authors in [62] present a placement algorithm that jointly optimises the resource allocation and user association for multiple UBSs with in-band wireless BH with promising results. On the other hand, authors in [63] exploit the emerging deep reinforcement learning technologies to predict blockage patterns and enable an efficient allocation of wireless BH resource. Full-duplex transmission is another promising line of research that offers significant advantages in its application to wireless BH. The authors in [64] explore the effectiveness of full-duplex scheduling mechanism in the context of mm-wave wireless BH network with the limited timeslot resources based on coalition game for maximising the quality of service on most flows.

As highlighted earlier in this chapter, perhaps the most disruptive aspect of NGMNs is the broad spectrum of connected devices with significantly differing service requirements. Thus, any intelligent wireless BH solution should take into consideration the user-centric performance indicators in addition to the network centric indicators in order to bridge the gap in BH provisioning until such times when fibre-based solutions are ubiquitous. To this end, we present a feasibility study of intelligent resource allocation between radio access and in-band wireless BH in the context of a heterogeneous network serving smartphones and IoT devices with various applications.

8.4.1 System model and simulations settings

For the purpose of demonstrating the potential gain of intelligent user-centric in-band wireless backhauling, we adopt a heterogeneous network that consists of two layers: macrocells contiguous layer overlapping with a small-cell patchy layer. Macrocells are deployed following a grid structure where each macrocell consists of three directional sectors. The small cells are omnidirectional and are located in hotspot areas. We assume that each macrocell sector covers five hotspots, hence five small cells. Both macrocells and small cells share the same radio access spectrum of 10 MHz. Moreover, the small cells are equipped with the UCB scheme to maximise the cumulative throughput and the users' satisfaction according to the cells' capabilities or limitations. We assume fibre-based BH connections to all macrocells with quasi-ideal characteristics. On the other hand, the small cells are provisioned with in-band wireless BH links to the closest macrocell sector that reuses the same spectrum as the shared radio access. However, resource blocks that are used from in-band backhauling are not reused for radio access. Similarly, all small cells within a given macrocell sector are allocated the same number of resource blocks at any time. The capacity and latency of the wireless BH links of each small cell are calculated using the expressions derived in Section 8.2 based on the distance between the small cell and the macrocell and assuming a 16×16 MIMO with high gain point-to-point antennas.

Two broad categories of users are randomly generated within the coverage area: smartphones and IoT devices. Each application, running on any device, is described based on the stipulated performance targets with respect to two attributes: throughput and latency. Each performance attribute is described by the target value and its importance in the QoE perception with regards to the application. Thus, each instance of smartphone users is randomly generated by selecting an application as follows:

- one target throughput from a pool of two values,
- one importance level with regards to throughput from a pool of two values,
- one target latency from a pool of two values, and
- one importance level with regards to latency from a pool of two values.

The same is done for any instance of IoT device with different pools of values from the smartphone applications.

In this feasibility study, we test seven different splits of resource blocks between radio access and wireless BH as listed in Table 8.8. For each scenario, we run the UCB scheme on all small cells to optimise two CREO values per cell. Each CREO value reflects the cell's end-to-end capabilities or limitation with regards to one of the two attributes: throughput and latency.

8.4.2 Results and analysis

We have implemented the above-mentioned system in MATLAB with the parameters as listed in Table 8.9. Two sets of simulations were conducted. In the first, the small cells are randomly generated in every run during which the penetration of IoT devices is increased from 10% to 90% and for each penetration, all seven resource block (RB) split scenarios are tested. In the second set of simulations, the penetration of IoT devices is fixed to 50%, and the location of small cells is changed in every iteration.

8.4.2.1 Varying the penetration of IoT devices

In this case study, we investigate the impact of IoT devices on the optimum split of resources between radio access and wireless BH. We generate a random

Table 8.8 Tested scenarios for splitting the 50 resource blocks between radio access and wireless BH

Scenario	Number of RA RBs	Number of WBH RBs
1	42	8
2	37	13
3	33	17
4	25	25
5	17	33
6	12	38
7	10	40

Table 8.9 Main parameters used in simulations [65]

Number of small cells	5	
Transmit power of macrocell	46 dBm	EiRP = 57 dBm 61 degrees half-power beamwidth
Transmit power of small cell	20 dBm	EiRP = 24 dBm
Conducted transmission power of each transmission path for wireless BH	30 dBm	EiRP = 59.75 dBm [66]
Path loss between macrocell and users	$128.1 + 37.6 \cdot \log_{10}(d)$	d (km)
Path loss between small cell and users	$140.7 + 36.7 \cdot \log_{10}(d)$	d (km)
Path loss for wireless BH	$100.7 + 25.5 \cdot \log_{10}(d)$	d (km)
Log-normal standard deviation for macrocells	8 dB	For simulating shadowing
Log-normal standard deviation for small cells	10 dB	For simulating shadowing
UE noise figure	7 dB	IoT and smartphones
Wireless BH noise figure	10 dB	[66]
Penetration loss	18 dB	All users are assumed indoor
Hot spot radius	100 m	20 users per hotspot
Total number of active users	150	20 users per hotspot and 50 elsewhere

distribution of five small cells and compute the wireless BH capacity and latency accordingly. The small cells remain in the same location throughout this case study. We perform five sets of simulations, in which the IoT penetration increases as 10%, 30%, 50%, 70%, and 90%. For each simulation set, we perform ten runs, and for each run, we conduct five learning iterations where each consists of each small cell learning the optimum setting of its CREO factors. The network performance results are shown in Figure 8.18, and the user-centric performance results are shown in Figures 8.19 and 8.20. In view of the network-centric results in Figure 8.18, Scenarios 5–7 are not acceptable due to the number of users in outage. Studying the results of user-centric performance, the best scenario for RB splitting differs for different IoT penetration levels. For instance, for an IoT penetration of 10%, Scenario 4 generates the best network performance, followed by Scenario 3, then Scenario 2 (see Figure 8.18). However, from a user perspective, the best performance is achieved with Scenario 2 for both attributes if we disregard Scenarios 5–7 (see Figures 8.19 and 8.20). The network loss by selecting Scenario 2 instead of Scenario 4 is 4.3%, whereas the drop in satisfaction level incurred by selecting Scenario 4 instead of Scenario 2 is 13.85% and 13.33%, respectively, for throughput and latency. Accordingly, an intelligent wireless BH solution should be able to gauge both network and user performance indicators and dynamically select

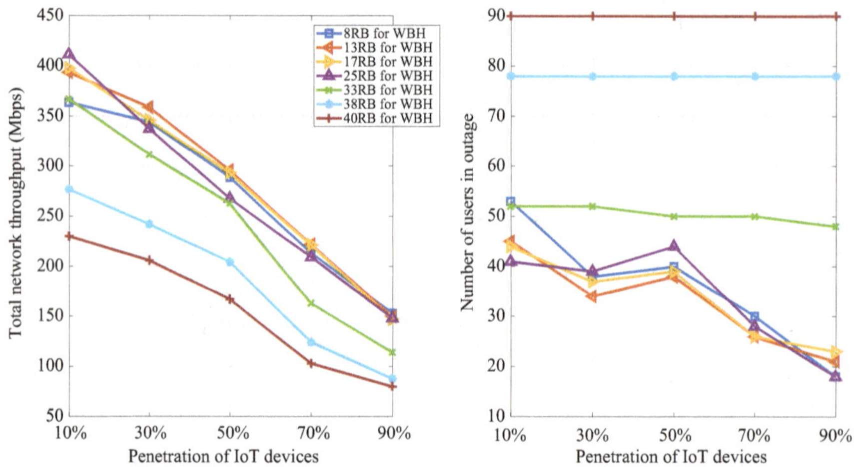

Figure 8.18 This figure summarises the results of each RB split scenario after the completion of the UCB scheme for each of the IoT penetration levels. We notice that the total system throughput decreases as the number of IoT devices increases as the IoT-generated throughput is limited. On the other hand, we notice that for Scenarios 6 and 7, the design is radio access limited with a maximum of 78 and 60 available RB for users' allocation, respectively

the best settings; in the case of 10% IoT penetration, the best settings are those of Scenario 2. However, for other IoT penetration levels, different best scenarios can be found by examining these results. For instance, for 70% IoT penetration, the best is Scenario 1, whereas for 50%, it is Scenario 3, and for both 70% and 90% it is Scenario 1.

8.4.3 Concluding remarks

We have presented a simple study to highlight the importance of intelligent wireless BH solutions, particularly in the context of in-band BH. The results show that the joint dynamics between the users' diversity and network diversity has a significant impact on the suitability of any intelligent wireless BH solution. To this end, it is essential to capture the user-centric perceived performance in addition to the network centric results to be able to optimise the wireless BH. In this chapter, we have presented an exhaustive search approach and have subscribed the reasoning for reaching the optimum solution. The next challenge is to design a machine learning algorithm that can reproduce this thinking and realise an optimum wireless BH solution that automatically adjusts to the changing circumstances in the network.

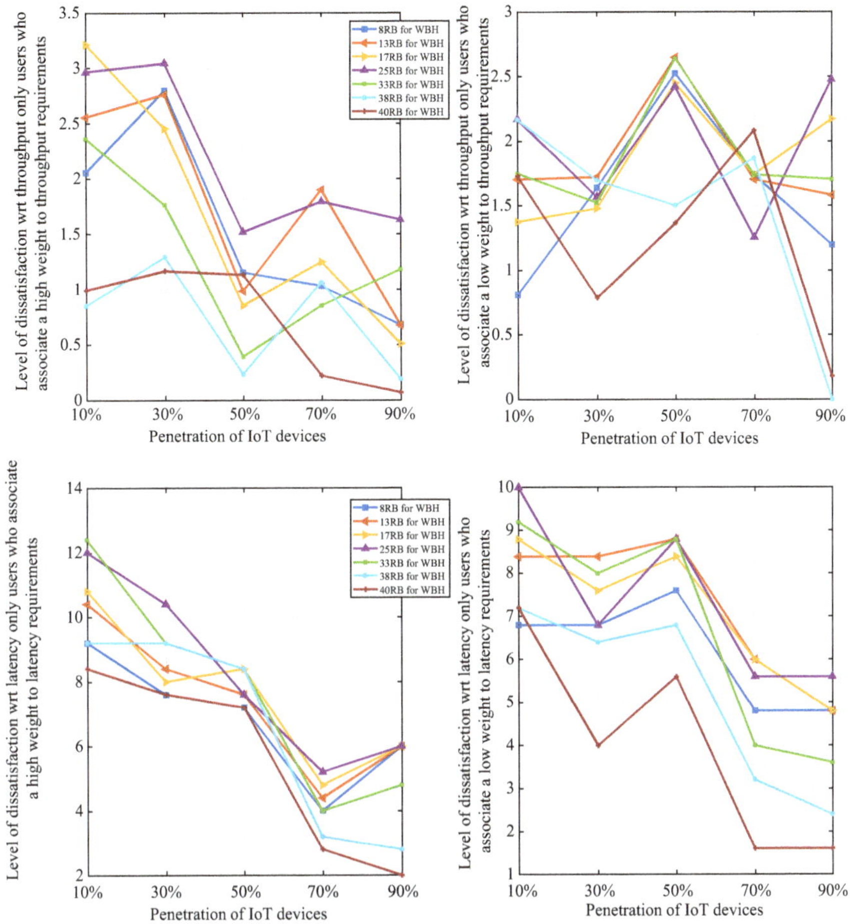

Figure 8.19 This figure summarises the user-centric perception of the throughput-related attribute. The figure on the left shows the level of dissatisfaction of users who associate high importance to this attribute. The UCB scheme attempts to improve the level of satisfaction of these users. IoT devices are easier to satisfy with respect to throughput compared to smartphones, hence the decreasing trend in dissatisfaction with more IoT devices

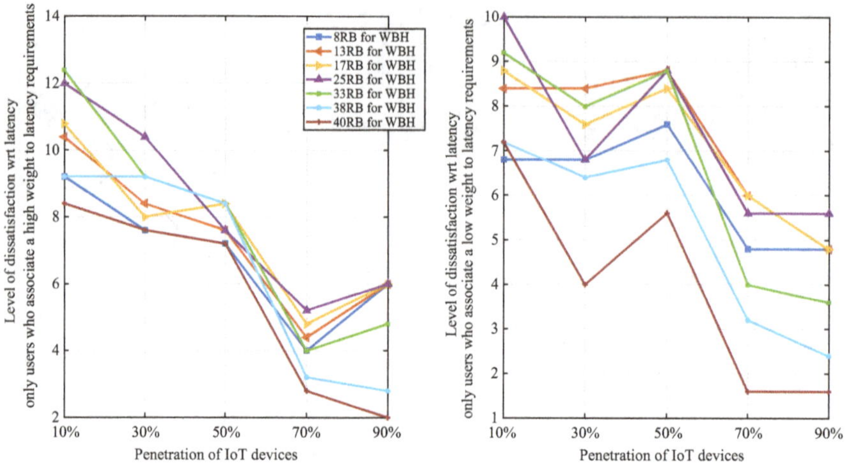

Figure 8.20 This figure summarises the user-centric perception of the latency-related attribute. The figure on the left shows the level of dissatisfaction of users who associate high importance to this attribute. The UCB scheme attempts to improve the level of satisfaction of these users. Fewer IoT devices are latency-critical compared to smartphones, hence the decreasing trend in dissatisfaction with more IoT devices

References

[1] Jaber M, Imran M, Tafazolli R, *et al.* 5G Backhaul Challenges and Emerging Research – A survey. IEEE Access. 2016;4(2016):1743–1766.

[2] Coomans W, Moraes RB, Hooghe K, *et al.* XG-FAST: Towards 10 Gb/s copper access. In: IEEE Globecom Workshops (GC Wkshps); 2014. p. 630–635.

[3] Van Uden R, Correa RA, Lopez EA, *et al.* Ultra-High-Density Spatial Division Multiplexing With a Few-Mode Multicore Fibre. Nature Photonics. 2014;8(11):865–870.

[4] Coldrey M. Maturity and field proven experience of millimetre wave transmission. ETSI—European Telecommunications Standards Institute; 2015.

[5] Paschos G, Bastug E, Land I, *et al.* Wireless Caching: Technical Misconceptions and Business Barriers. IEEE Communications Magazine. 2016;54(8):16–22.

[6] Costa-Perez X, Garcia-Saavedra A, Li X, *et al.* 5G-Crosshaul: An SDN/NFV Integrated Fronthaul/Backhaul Transport Network Architecture. IEEE Wireless Communications. 2017;24(1):38–45.

[7] Ericsson. Microwave Towards 2020, White Paper; 2015. [Online], Available: https://www.ericsson.com/res/docs/2015/, Accessed on 30/06/2016.

[8] Vu TK, Bennis M, Debbah M, *et al.* Joint Path Selection and Rate Allocation Framework for 5G Self-Backhauled mm-Wave Networks. IEEE Transactions on Wireless Communications. 2019;18(4):2431–2445.

[9] Liu Y, Fang X, and Xiao M. Discrete Power Control and Transmission Duration Allocation for Self-Backhauling Dense mmWave Cellular Networks. IEEE Transactions on Communications. 2018;66(1):432–447.

[10] Kulkarni MN, Andrews JG, and Ghosh A. Performance of Dynamic and Static TDD in Self-Backhauled Millimeter Wave Cellular Networks. IEEE Transactions on Wireless Communications. 2017;16(10):6460–6478.

[11] Andrews M, Ghosh A, Pupala RN, *et al.* Performance Evaluation of Self Backhauled Small Cell Heterogeneous Networks. IEEE Transactions on Wireless Communications. 2017;16(5):3102–3110.

[12] Tang L, Wei Y, He L, *et al.* Queue-Aware Dynamic Resource Reuse and Joint Allocation Algorithm in Self-Backhaul Small Cell Networks. IEEE Access. 2018;6:61077–61090.

[13] Ortiz A, Asadi A, Sim GH, *et al.* SCAROS: A Scalable and Robust Self-Backhauling Solution for Highly Dynamic Millimeter-Wave Networks. IEEE Journal on Selected Areas in Communications. 2019;37(12):2685–2698.

[14] Vu TK, Bennis M, Samarakoon S, *et al.* Joint Load Balancing and Interference Mitigation in 5G Heterogeneous Networks. IEEE Transactions on Wireless Communications. 2017;16(9):6032–6046.

[15] Bonfante A, Galati Giordano L, Lpez-Prez D, *et al.* 5G Massive MIMO Architectures: Self-Backhauled Small Cells Versus Direct Access. IEEE Transactions on Vehicular Technology. 2019;68(10):10003–10017.

[16] Wang L, Wong K, Lambotharan S, *et al.* Edge Caching in Dense Heterogeneous Cellular Networks with Massive MIMO-Aided Self-Backhaul. IEEE Transactions on Wireless Communications. 2018;17(9): 6360–6372.

[17] Jia X, Xu W, Chen Y, *et al.* Hybrid Self-Backhaul and Cache Assisted Millimeter Wave Two-Tier Heterogeneous Networks with MIMO Equipped Backhaul Access Points. IEEE Access. 2019;7:59963–59983.

[18] Sharma A, Ganti RK, and Milleth JK. Joint Backhaul-Access Analysis of Full Duplex Self-Backhauling Heterogeneous Networks. IEEE Transactions on Wireless Communications. 2017;16(3):1727–1740.

[19] Korpi D, Riihonen T, Sabharwal A, *et al.* Transmit Power Optimization and Feasibility Analysis of Self-Backhauling Full-Duplex Radio Access Systems. IEEE Transactions on Wireless Communications. 2018;17(6):4219–4236.

[20] Lei L, Lagunas E, Chatzinotas S, *et al.* NOMA Aided Interference Management for Full-Duplex Self-Backhauling HetNets. IEEE Communications Letters. 2018;22(8):1696–1699.

[21] Chen E, Tao M, and Zhang N. User-Centric Joint Access-Backhaul Design for Full-Duplex Self-Backhauled Wireless Networks. IEEE Transactions on Communications. 2019;67(11):7980–7993.

[22] Jaber M, Imran M, Tafazolli R, *et al.* A Distributed SON-Based User-Centric Backhaul Provisioning Scheme. IEEE Access. 2016;4:2314–2330.

[23] Jaber M, Onireti O, and Imran MA. Backhaul-aware and context-aware user-cell association approach. In: ICC 2019 – 2019 IEEE International Conference on Communications (ICC); 2019. p. 1–7.

[24] Farooq H, Imran A, and Jaber M. AI Empowered Smart User Association in LTE Relays HetNets. In: 2019 IEEE International Conference on Communications Workshops (ICC Workshops); 2019. p. 1–6.
[25] Valente Klaine P, Jaber M, Souza RD, *et al.* Backhaul Aware User-Specific Cell Association Using Q-Learning. IEEE Transactions on Wireless Communications. 2019;18(7):3528–3541.
[26] Pervez F, Jaber M, Qadir J, *et al.* Memory-Based User-Centric Backhaul-Aware User Cell Association Scheme. IEEE Access. 2018;6:39595–39605.
[27] Jaber M, Imran MA, Tukmanov A, *et al.* Case study on using the user-centric-backhaul scheme to unlock the realistic backhaul. In: 2017 IEEE 86th Vehicular Technology Conference (VTC-Fall); 2017. p. 1–5.
[28] Pervez F, Jaber M, Qadir J, *et al.* Fuzzy Q-learning-based user-centric backhaul-aware user cell association scheme. In: 2017 13th International Wireless Communications and Mobile Computing Conference (IWCMC); 2017. p. 1840–1845.
[29] Jaber M, Imran MA, Tafazolli R, *et al.* Energy-efficient SON-based user-centric backhaul scheme. In: 2017 IEEE Wireless Communications and Networking Conference Workshops (WCNCW); 2017. p. 1–6.
[30] Jaber M, Imran MA, Tafazolli R, *et al.* A multiple attribute user-centric backhaul provisioning scheme using distributed SON. In: 2016 IEEE Global Communications Conference (GLOBECOM); 2016. p. 1–6.
[31] Jaber M, Owens D, Imran MA, *et al.* A joint backhaul and RAN perspective on the benefits of centralised RAN functions. In: 2016 IEEE International Conference on Communications Workshops (ICC); 2016. p. 226–231.
[32] Jaber M, Imran M, Sutton A, *et al.* Modular Approach for Modelling the Hybrid Multi-Hop Backhaul Performance. IEEE Wireless Communications Letters. 2017;6(2):262–265.
[33] Zhang G, Quek T, Kountouris M, *et al.* Fundamentals of Heterogeneous Backhaul Design – Analysis and Optimization. IEEE Transactions on Communications. 2016;64(2):876–889.
[34] Aurzada F, Levesque M, Maier M, *et al.* FiWi Access Networks Based on Next-Generation PON and Gigabit-Class WLAN Technologies: A Capacity and Delay Analysis. IEEE/ACM Transactions on Networking. 2014;22(4): 1176–1189.
[35] Beyranvand H, Lim W, Maier M, *et al.* Backhaul-Aware User Association in FiWi Enhanced LTE-A Heterogeneous Networks. IEEE Transactions on Wireless Communications. 2015;14(6):2992–3003.
[36] Andrews JG, Baccelli F, and Ganti RK. A Tractable Approach to Coverage and Rate in Cellular Networks. IEEE Transactions on Communications. 2011;59(11):3122–3134.
[37] Ohmann D, and Fettweis GP. Minimum duration outage of wireless Rayleigh-fading links using selection combining. In: IEEE Wireless Communications and Networking Conference (WCNC); 2015. p. 681–686.

[38] Allen J, Chevalier F. and Mason A, editor. Report for Vodafone – Mobile Backhaul Market: Phase 1 Report; 2014.
[39] Chen DC, Quek TQS, and Kountouris M. Wireless backhaul in small cell networks: modelling and analysis. In: IEEE 79th Vehicular Technology Conference (VTC Spring); 2014. p. 1–6.
[40] Chen DC, Quek TQS, and Kountouris M. Backhauling in Heterogeneous Cellular Networks: Modeling and Tradeoffs. IEEE Transactions on Wireless Communications. 2015;14(6):3194–3206.
[41] Zhang G, Quek TQS, Huang A, *et al.* Delay modeling for heterogeneous backhaul technologies. In: IEEE 82nd Vehicular Technology Conference (VTC Fall); 2015. p. 1–6.
[42] Jaber M, Lopez-Martinez FJ, Imran MA, *et al.* Wireless Backhaul: Performance Modeling and Impact on User Association for 5G. IEEE Transactions on Wireless Communications. 2018;17(5):3095–3110.
[43] Romero-Jerez JM, Lopez-Martinez FJ, Paris JF, *et al.* The Fluctuating Two-Ray Fading Model: Statistical Characterization and Performance Analysis. IEEE Transactions on Communications. 2017;16(7):4420–4432.
[44] Paris JF. Closed-Form Expressions for Rician Shadowed Cumulative Distribution Function. Electronics Letters. 2010;46(13):952–953.
[45] Lopez-Martinez FJ, Paris JF, and Romero-Jerez JM. The k-μ Shadowed Fading Model with Integer Fading Parameters. IEEE Transactions on Vehicular Technology. 2017;66(9):7653–7662.
[46] Singh S, Kulkarni MN, Ghosh A, *et al.* Tractable Model for Rate in Self-Backhauled Millimeter Wave Cellular Networks. IEEE Journal on Selected Areas in Communications. 2015;33(10):2196–2211.
[47] Gradshteyn IS, and Ryzhik IM. Table of Integrals, Series and Products. 7th ed. Boston, MA: INCAP; 2007.
[48] Simon MK. Probability Distributions Involving Gaussian Random Variables: A Handbook for Engineers and Scientists. Berlin: Springer Science & Business Media; 2007.
[49] Nadiv R, and Rosenhouse T. Flex Your Backhaul Network with Adaptive Coding & Modulation; 2010. [Online], Available: http://www.winncom.com/, Accessed 07/11/2016.
[50] Stuber G. Principles of Mobile Communication. 2nd ed. New York: Kluwer Academic Publishers; 2001.
[51] Rice SO. Distribution of the Duration of Fades in Radio Transmission: Gaussian Noise Model. The Bell System Technical Journal. 1958;37(3):581–635.
[52] Ramos-Alarcon F, Kontorovich V, and Lara M. On the Level Crossing Duration Distributions of Nakagami Processes. IEEE Transactions on Communications. 2009;57(2):542–552.
[53] Berliner B, Clark B, and Hartono A. QoS Requirements of Multimedia Applications; 2011. [Online], Available: https://cse.osu.edu/, Accessed 26/04/2017.

[54] Bangerter B, Talwar S, Arefi R, *et al.* Networks and Devices for the 5G Era. IEEE Communications Magazine. 2014;52(2):90–96.

[55] Streijl RC, Winkler S, and Hands DS. Mean Opinion Score (MOS) Revisited: Methods and Applications, Limitations and Alternatives. Multimedia Systems. 2016;22(2):213–227.

[56] Sagias NC, Tombras GS, and Karagiannidis GK. New Results for the Shannon Channel Capacity in Generalized Fading Channels. IEEE Communications Letter. 2005;9(2):97–99.

[57] Zhang G, Quek TQS, Huang A, *et al.* Delay modeling for heterogeneous backhaul technologies. In: Vehicular Technology Conference (VTC Fall), 2015 IEEE 82nd; 2015. p. 1–6.

[58] Suryaprakash V, and Fettweis GP. Modeling backhaul deployment costs in heterogeneous radio access networks using spatial point processes. In: 12th International Symposium on Modeling and Optimization in Mobile, Ad Hoc, and Wireless Networks (WiOpt); 2014. p. 725–732.

[59] Li Z, Sichitiu ML, and Qiu X. Fog Radio Access Network: A New Wireless Backhaul Architecture for Small Cell Networks. IEEE Access. 2019;7:14150–14161.

[60] Feng M, Mao S, and Jiang T. Joint Frame Design, Resource Allocation and User Association for Massive MIMO Heterogeneous Networks with Wireless Backhaul. IEEE Transactions on Wireless Communications. 2018;17(3): 1937–1950.

[61] Saha C, Afshang M, and Dhillon HS. Bandwidth Partitioning and Downlink Analysis in Millimeter Wave Integrated Access and Backhaul for 5G. IEEE Transactions on Wireless Communications. 2018;17(12):8195–8210.

[62] Qiu C, Wei Z, Feng Z, *et al.* Joint Resource Allocation, Placement and User Association of Multiple UAV-Mounted Base Stations with In-Band Wireless Backhaul. IEEE Wireless Communications Letters. 2019;8(6): 1575–1578.

[63] Feng M, and Mao S. Dealing with Limited Backhaul Capacity in Millimeter-Wave Systems: A Deep Reinforcement Learning Approach. IEEE Communications Magazine. 2019;57(3):50–55.

[64] Jiang H, Niu Y, Zhang J, *et al.* Coalition Game Based Full-Duplex Concurrent Scheduling in Millimeter Wave Wireless Backhaul Network. China Communications. 2019;16(2):59–75.

[65] Jaber M, Imran M, Tafazolli R, *et al.* An adaptive backhaul-aware cell range extension approach. In: IEEE International Conference on Communications (ICC), Workshops. London, UK; 2015.

[66] Leinonen ME, Destino G, Kursu O, *et al.* 28 GHz Wireless Backhaul Transceiver Characterization and Radio Link Budget. ETRI Journal. 2018;40(1):89–100. Available from: https://onlinelibrary.wiley.com/doi/abs/10.4218/etrij.2017-0231.

Chapter 9

OTA test methods and candidates for 5G and beyond

Tian Hong Loh[1]

9.1 Introduction

The digital economy is essential for wealth creation. Over the past few decades, the market for modern wireless communication systems has grown rapidly in response to consumer demand. As 'connectivity' becomes more and more ubiquitous and reliable and offers higher capacity, it will become the new 'lifeblood' of the digital economy and connected society [1] with over 8.9 billion mobile subscriptions by 2022 [2]. Focusing on user experience, the fifth generation (5G) network is envisaged to offer a dramatic improvement over fourth generation (4G)'s network capabilities. 5G network promises more reliable, faster, and near-instant connections. With strong global momentum, multiple worldwide 5G research projects are defining, developing and standardizing the technology for 5G and beyond. To develop the necessary infrastructure to meet the aforementioned promises, significant efforts have been devoted in international liaison on seeking for a global consensus over visions, applications, standards and the identification of suitable spectrums [3]. Following the approval of 3GPP (Third-Generation Partnership Project) Release 15 5G new radio (NR) standard in June 2018, 5G commercial networks were quickly launched in the United States (Verizon and AT&T) and South Korea (KT, LG UPlus and SK Telecom) by the end of the year [4]. While the standardization process for 5G NR is ongoing, industry had already begun efforts to implement infrastructure compliant with the draft 5G NR standard [5].

Operating at a higher radio frequency (RF) is a route to achieve the 5G promise on higher data bandwidth. 5G is the first mobile system to use millimetre-wave (mm-wave) frequencies in which more contiguous spectrum is available to send more data through the channel. However, the use of mm-wave frequency bands poses new measurement challenges in signal quality due to high path loss, and blockage under non-line-of-sight (LOS) (NLOS) scenarios. To improve the bit

[1]5G & Future Communications Technology Group, Electromagnetic & Electrochemical Technologies Department, National Physical Laboratory, Teddington, United Kingdom

efficiency and solve the problems caused by the RF channel, a raft of emerging wireless technologies for 5G and beyond are becoming available and increasingly implemented into today's wireless environment [6]. Among these technologies, the multiple-input–multiple-output (MIMO) technology plays an important role, due to their rich scattering nature, ability to perform beamforming and to reuse the same frequency within each cell, which provides improved spectral efficiencies and increased network capacity. As a natural progression for the next-generation systems, it was found that greatly increasing the number of antennas gives a commensurate capacity increase at the cost of computational overhead. Beamforming massive MIMO (mMIMO) and mm-wave MIMO technologies have been identified as important elements for addressing high data demands required for 5G and future communications.

To understand the end-to-end reception performance of a wireless communication system or device, over-the-air (OTA) testing, which include both antenna and propagation characteristics at the same time, is needed [7]. Throughput is often being considered as the key system-level link performance measurement metric. Typically, OTA test of single-input-single-output (SISO) wireless communication system only investigates the demodulation performance under the LOS scenario within a fully anechoic laboratory environment. The addition of MIMO capability has driven towards the need for OTA testing which is becoming a critical test methodology to accurately evaluate wireless communication system performance in realistic scenarios in which emulation of the spatial environment is the key for the test. Furthermore, with the adoption of key technologies such as beamforming mMIMO and mm-wave in 5G, several worldwide standards bodies (e.g. 3GPP [8], ETSI (European Telecommunications Standards Institute) [9], IEEE (The Institute of Electrical and Electronics Engineers) [10], and CTIA (International Association for the Wireless Telecommunication Industry) [11]), industries and research communities are facing new measurement challenges on OTA test for verifying products that meet desired performance parameters for fulfilling the diverse 5G technological requirements. This chapter presents an overview on the current MIMO OTA test methods for characterizing wireless communication system (further referred to in the rest of this chapter as the system under test (SUT)) and the candidate methodologies for 5G and beyond as well as their associated challenges. The main SUT considered are mobile user equipment (UE) and base station (BS).

9.2 OTA test methods

The link performance of wireless communication systems may degrade in the real-world operation due to various reasons, for example the effects induced from proximity lossy objects (such as hands and human head, which are not part of the final product), multipath fading, unwanted interferences, etc. With the use of adaptive beamforming antennas in emerging wireless system and the need to characterize propagation channels at mm-wave bands, OTA measurement methodologies have become very important research topics for 5G and beyond [12–14]. Many different OTA test methods have been proposed, which vary widely with respect to their implementation, equipment used, chamber size, running cost and measurement time.

9.2.1 Definition of OTA test

OTA is a synonym for wireless [15]. It generally is referred to any type of wireless transmission. By characterizing the inherent antenna performance together with emulating the entire signal propagation path, OTA test is a system-level measurement method to evaluate the reception performance of wireless products, which is an important test to ensure compliance with the necessary standards. To assess how wireless communication SUT responds in various real-world scenarios, the OTA test method could be implemented in a screened chamber by establishing in single or multiple simultaneous OTA links to a communication tester (also known as BS emulator (BSE)) via fading-emulator (also known as channel emulators (CEs)) whereby reception performance of SUTs could be measured either by conducted (using cables) or radiated approaches (placing SUT within an appropriate free-space test environment). This enables assessment of the reliability and realistic performance characteristics of wireless products, both for mobile and fixed location scenarios in the real-world operation. Tables 9.1 and 9.2 show the required compliance testing methods for different mobile generations, respectively, for BS and UE.

9.2.2 Definition of figures of merits

In general, figures of merit (FoM) can be defined as the numerical value representing a measure of effectiveness, efficiency, performance or other important factor, and ascertained or approximated from analysis, appraisal or estimation techniques. The following shows some typical test FoMs that could be assessed when the SUT operates as transmitter:

1. EIRP (equivalent (or effective) isotropically radiated power)
2. TRP (total radiated power)

Table 9.1 BS compliance testing methods required for different mobile generations

Generation	Antenna technology	Compliant testing methods
2G	SISO	Cable conducted
3G	SISO/smart antenna	Cable conducted
4G	MIMO	Cable conducted
5G	Massive MIMO/mm-wave	OTA radiated: • RF/measurement • System-level link performance test

Table 9.2 UE compliance testing methods required for different mobile generations

Generation	Antenna technology	Compliant testing methods
2G	SISO	Cable conducted
3G	SISO	SISO OTA
4G	MIMO	SISO/MIMO OTA
5G	MIMO/mm-wave	OTA radiated: • RF/antenna measurement • Demodulation test • SISO/MIMO system-level link performance test • Radio resource management (RRM) test

3. Antenna efficiency
4. EVM (error vector magnitude)
5. Adjacent channel leakage ratio

On the other hand, when the SUT operates as receiver some typical test parameters are as follows:

1. TIS (total isotropic sensitivity)
2. System-level link performance test (e.g. throughput under specific scenario)

9.2.3 SISO OTA test methods

With a focus on mobile UE, the SISO OTA method for characterizing single-antenna SUTs using a far-field fully anechoic chamber (AC) was first defined by CTIA in 2001 [16] and then adopted by 3GPP in 2006 [17], with additional specification on device performance requirements [18]. Through these wireless industry groups, the wireless industry has chosen the TRP and TIS (also known as total reference sensitivity (TRS)) as the FoMs at physical layer to represent the reception performances of a wireless product that operates with SISO link. Both TRP and TIS over 4π space could be calculated from the EIRP and EIS (effective (or equivalent) isotropic sensitivity), using the following equations:

$$TRP = \frac{1}{4\pi}\iint_{4\pi}(EIRP_{\theta}(\theta,\varnothing) + EIRP_{\varnothing}(\theta,\varnothing))d\Omega \tag{9.1}$$

$$TIS = \frac{4\pi}{\iint_{4\pi}(1/(EIS_{\theta}(\theta,\varnothing)) + 1/(EIS_{\varnothing}(\theta,\varnothing)))d\Omega} \tag{9.2}$$

where θ, $\varnothing$ and Ω denote, respectively, the polar angle, the azimuth angle and the spherical domain of the spherical coordinate centred around the SUT. Their subscripts associated with a parameter refer to the relevant θ-polarized and $\varnothing$-polarized components for that parameter.

EIRP is the total power that would have to be radiated by a hypothetical isotropic antenna to give the same signal strength as the actual source in the direction of the antenna's strongest beam [19], and EIS is sensitivity of a hypothetical isotropic antenna when illuminated with a plane-wave transmitted from the test equipment at the power level recorded for the SUT. Both EIRP and EIS are typically expressed in watts and could be measured on the full 3D (three-dimensional) sphere in an AC using a reference antenna in two orthogonal polarizations. Figure 9.1 shows the example set-up for SISO OTA test where further details are provided in CTIA test plan in [20].

The TRP is a measure of how much power a SUT radiates over a sphere centred around it, whereas the TIS is the average over the same sphere of the minimum received power to achieve a specific bit error rate (BER). An alternative test method using a reverberation chamber (RC) was also specified in [17]. This was possible since it was shown empirically that the results obtained by averaging many individual measurements using a point source within an AC were very similar to results achieved by averaging in a mode-stirred RC [18,21].

9.2.4 MIMO OTA test methods

MIMO antenna technology has been used in modern wireless standards, including 3GPP 4G LTE (long-term evolution), IEEE 802.11n and WiMAX (Worldwide Interoperability for Microwave Access). It overcomes the detrimental effects of signal multipath and fading and can achieve high data throughput in limited-bandwidth channels. When the first mobile MIMO UE devices were commercially deployed in 2010, there were no standards for testing their MIMO OTA performance [18]. The SISO OTA measurements set-up (see Figure 9.1) is not suitable for MIMO OTA testing as when testing MIMO systems, it is necessary to include both antennas and propagation characteristics in the system set-up [18,22]. Nevertheless, the work to standardize MIMO OTA measurement methods and performance requirements has evolved from the SISO OTA standards by CTIA [20] and 3GPP [23] for testing of third-generation (3G) and 4G multiple antenna SUTs. To date, several MIMO OTA test methods have been adopted within standard for 3G and 4G SUT, which consider both the antenna performance and radio propagation channel's spatial characteristics [23,24]. One of the commonly used key FoM for MIMO OTA tests is the system-level absolute data throughput.

Figure 9.2 shows an example of basic set-up for measuring antenna gain, TIS and TRP for SISO and MIMO OTA test configuration [25]. As depicted in Figure 9.2, for antenna gain measurement, a vector network analyser could be used where a known reference antenna would be required for the antenna calibration procedure [26]. For TRP measurement, the signal analyser could be used to measure the SUT-transmitted power at the reference antenna, whereas the BSE can be connected to a link antenna for capturing *I*/*Q* data or a channel estimate associated with the received signal. Further details for TRS measurement are provided in [25]. There are mainly two categories of screened laboratory environments being adopted for 3G and 4G LTE MIMO OTA, namely RC and far-field AC [23,24]. For 5G

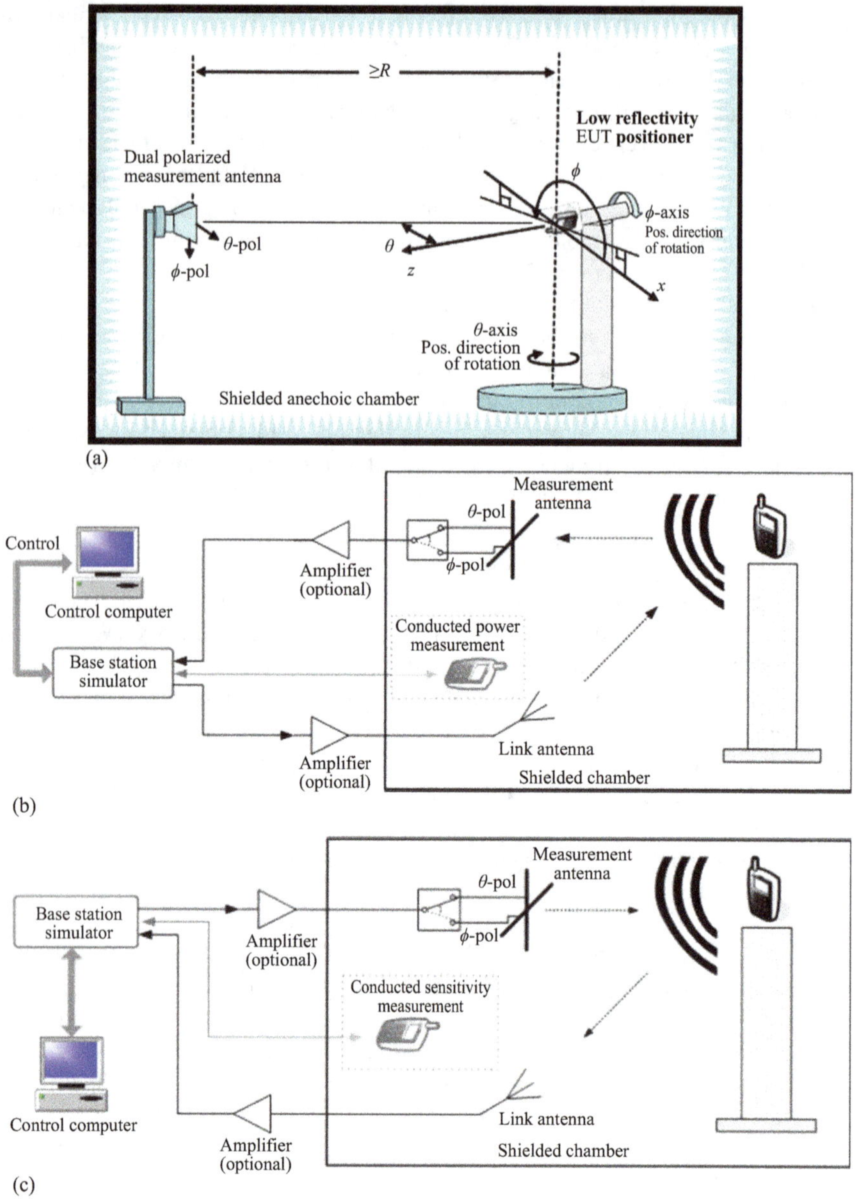

Figure 9.1 Example of typical set-up for SISO OTA test (figures A-26, A-27, & A-28 in CTIA OTA Test Plan v3.8.2 [20]. © 2001-2019. CTIA-The Wireless Association. All Rights Reserved): (a) set-up using roll-over-azimuth positioning system; (b) configuration using BSE for TRP measurements; (c) configuration for using BSE for TIS measurements

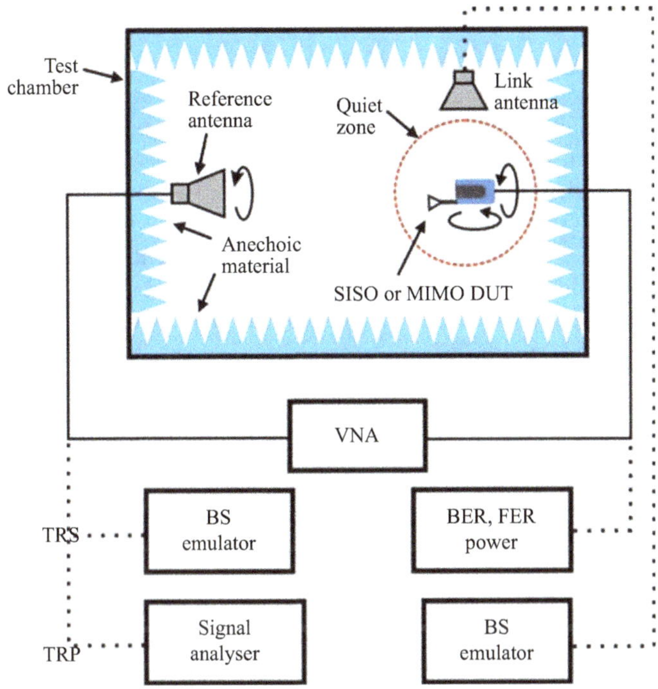

Figure 9.2 Basic OTA test configuration for measuring antenna gain, TRS and TRP of SISO and MIMO devices (Figure 9 in [25]. © Keysight Technologies, Inc.)

NR MIMO OTA testing, including RC and far-field AC, several other categories of candidate-screened laboratory environments are currently being considered, namely compact antenna test range (CATR) chamber, near-field chamber and sectored multi-probe AC (MPAC) [27–29]. For more details see Section 9.3.3.

9.3 Test methodologies

SISO OTA testing is a relatively straightforward method for assessing reception performance of single-antenna SUTs as it only considers the functionality of the SUT. As both the antenna performance and radio propagation channel's spatial characteristics play the key roles in the multiple-antenna system link performance, the accurate evaluation of antenna radiation performance and the correct implementation of radio CE within these testing environments is a crucial part of the MIMO OTA test system. Thus, making accurate MIMO OTA testing is much more challenging and complex as compared with SISO OTA testing. Therefore, the rest of this subsection will focus on discussion of MIMO OTA test methodologies.

9.3.1 Key figure of merits

A MIMO SUT can operate in diversity according to the signal-to-noise ratio (SNR) level under a multipath rich environment. If the SNR is low, the diversity can be applied to combat the fading. The characterization of such multiple antennas SUT would require both the propagation characteristics of the radio channel and the radiation characteristics of the SUT receive antenna. The two main FoMs, namely the TRP and the TIS of a mobile device tested with SISO OTA methods, cannot fully characterize the spatial behaviour of MIMO antenna system. Therefore, introduction of additional MIMO OTA FoMs is needed. For the standardized MIMO OTA testing, the selection of FoMs is critical to be able to distinguish a good device from a bad device as the aim is to evaluate the performance of the whole SUT, including the full receiver and/or transmitter chain. For example table 5-1 in [30] presents categories of FoM for both HSPA (high-speed packet access) and LTE systems. It is noted that for field-domain testing, FoMs would include accuracy of the reproduced electromagnetic field, and the spatial correlation, whereas for the channel characterization and wireless link performance metrics such as diversity gain, spatial correlation, MIMO capacity, throughput, BER, EVM and channel quality indicator would be useful. Generally, absolute throughput performance is considered as the key FoM for MIMO OTA evaluation obtained through 3GPP/CTIA round robin campaigns in order to be able to compare the different proposed methodologies in their ability to distinguish good and bad MIMO SUT [23].

9.3.2 Standardization and ongoing work

Significant research effort in support of the development of appropriate MIMO OTA compliance test method for standardization purposes has been ongoing in the wireless industries and research communities through COST (European Cooperation in Science and Technology) actions [31] (e.g. COST actions IC1004, CA15104), CTIA [11] and 3GPP [8]. These groups work very closely together in which the proposed methodologies are similar. The wireless industry, through 3GPP and CTIA standardization bodies, has then developed the MIMO OTA standards for testing of 3G HSPA and 4G LTE multiple antenna systems/terminals [23,24]. The development of MIMO OTA test for 5G NR is ongoing, and it is envisaged that it may take some time to reach recognized test plans widely. Depending on applications, relevant propagation channel characteristics, implementing laboratory design and cost, several candidate MIMO test methods have now been proposed. For example mobile UE [32,33], mMIMO BS [34–37], automotive vehicle [38–41], IoT systems [42–48], etc.

9.3.3 Candidate methodologies

As the MIMO OTA methodologies proposed by CTIA, 3GPP and COST groups are similar, the concentration here is given on the description based on definition specific in 3GPP specifications in which several candidate solutions have been proposed for creating the necessary environment for conformance testing of MIMO

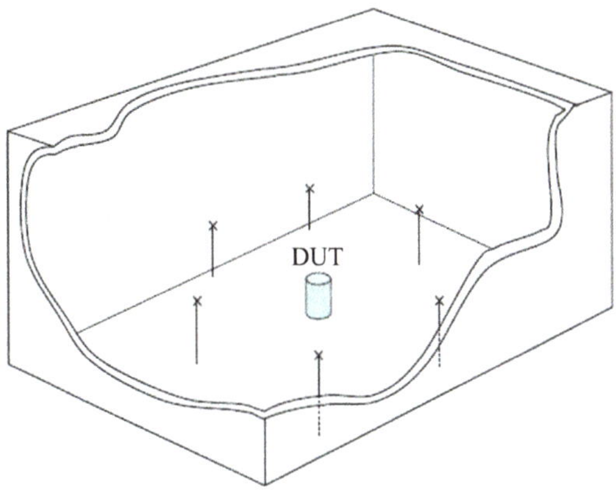

Figure 9.3 N-element AC approach (absorbing tiles and cabling not shown) (figure 6.3.1.1-1 in TR 37.977 [23], reprinted with permission from 3GPP)

OTA performance [30]. This subsection provides an overview and brief description of the different existing 3G and 4G LTE MIMO OTA candidate methodologies [23,24] that could potentially be considered as for 5G NR MIMO OTA testing as well as other new candidate methodologies that are currently being considered [27–29]. These are grouped based on the categories of screened laboratory environment used.

9.3.3.1 Methodologies based on the far-field fully anechoic chamber

Far-field fully AC is an ideal RF free-space (i.e. reflectionless) environment which typically provides direct far-field (DFF) condition for which the SUT is assessed. A MIMO OTA method based on the use of a far-field AC is described consisting of several test antenna probes located in the chamber-transmitting signals with temporal and spatial characteristics for testing multiple antenna devices. This subsection describes the MIMO OTA methodologies based on far-field AC, where several test antennas are in different positions of the chamber, and the SUT is located at centre position and is tested OTA without RF cables.

Arbitrary number multi-probe method

This candidate methodology is defined, consisting of several source elements located in different positions of the chamber surrounding the SUT, which is located at the centre position. Using geometry-based stochastic channel models (GSCM), this method creates a realistic geometry-based spatio-temporal radio channel for testing MIMO OTA performance. The latter implementation is illustrated in Figure 9.3, in which the SUT is tested OTA without RF cables.

By utilizing specific geometries of the test probes in the chamber, a range of possible channels are emulated. The exact number and positioning of the source antenna probes will be fixed in a final design; however, they may be optimized for the best performance when the OTA channel models are defined. In other words, depending on the types of channel models required and the range of parameters needed for testing, the number of probes may be optimized to produce the best performance with the fewest number of probes. For example based on the range of channels defined by the SCM (spatial channel model), SCME (SCM extended), WINNER (wireless world initiative new radio) I and II channel models [49], the optimized number of probes may vary from 6 to 8 for a given polarization. In general, the most flexible configurations require the higher number of probes. Azimuth spread is created by energizing sets of probes separated in azimuth with signals that will combine OTA at a specific delay to emulate a path or cluster (an illustrative diagram is shown in figure 6.3.1.1.2-1 in [23]). Elevation spread may be created by installing probes at different elevations; however, doing this tends to constrain their flexibility.

A system-level block diagram is shown in Figure 9.4, which includes the system simulator to generate the M branch MIMO signal, and an RF CE with an OTA channel generation feature to properly correlate, fade, scale, delay and distribute the signal to each test probe in the chamber. The system enables a range of possible channel models and parameterizations to be used to specify the most generic and versatile antenna test probe configuration. For performance and conformance testing, the channel model is expected to be limited in scope or simplified, which may allow an optimized design to reduce the number of test probes required. Thus, the number of test probes will be selected to best meet the requirements of the test so that the most efficient and economical design can be achieved.

Symmetrical ring of probes method

This candidate set-up is composed of several OTA antennas, a multidimensional fading emulator, an AC, communication tester/BSE and a SUT (shown as device under test (DUT) in Figure 9.5). Figure 9.5 depicts an example of the OTA concept where the SUT is located at centre of the AC. The purpose of the figure is not to restrict the implementation, but rather to clarify the general idea of the MIMO OTA concept. For simplicity, uplink cabling is not drawn here.

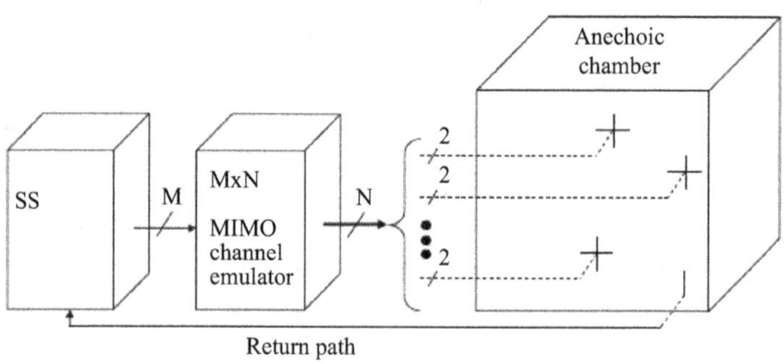

Figure 9.4 System-level block diagram for multi-probe OTA approach (figure 6.3.1.1-2 in TR 37.977 [23], reprinted with permission from 3GPP)

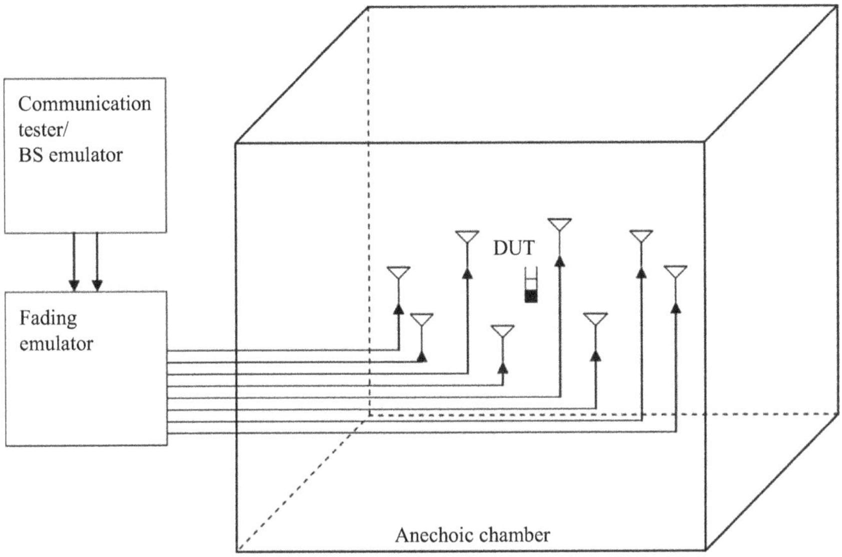

Figure 9.5 Example of MIMO/multiantenna OTA test set-up (uplink signal path omitted in the figure) (figure 6.3.1.2-1 in TR 37.976 [30], reprinted with permission from 3GPP)

Having the SUT locating at the centre enables the creation of a radio channel environment where the signal can arrive simultaneously from various directions to the SUT, which is the key aspect of the wideband MIMO radio channel models implemented today. The crucial challenge is to generate realistic and accurate angular, and polarization behaviour creates appropriate correlation at the SUT antennas within the AC. The correlation for per-path would need to be defined implicitly via the measured results of angle of arrival, Doppler, angle of departure, delay, polarization and real antennas characteristics. Geometry-based channel modelling methodology models BS and UE antenna arrays and the propagation between them as well as the angular power spectra. The family of GSCM, which includes 3GPP SCM, SCME, WINNER and International Mobile Telecommunications-Advanced channel models, is well suitable for MIMO OTA testing [30].

Two-channel decomposition method

This candidate method shown in Figure 9.6 is a special case of the multi-probe method and used just two probes without CE. In this candidate method, both conducted and radiated test will be performed. The conducted test is considered as the conformance test that assesses many properties of the SUT, whereas the radiated OTA test is considered as test to provide information that could not have been achievable via conducted cable set-up. On this basis, it is enough to define some abstract channel environment during the test which does not need to be very close to reality. Abstraction on the environment makes it easier to interpret the obtained

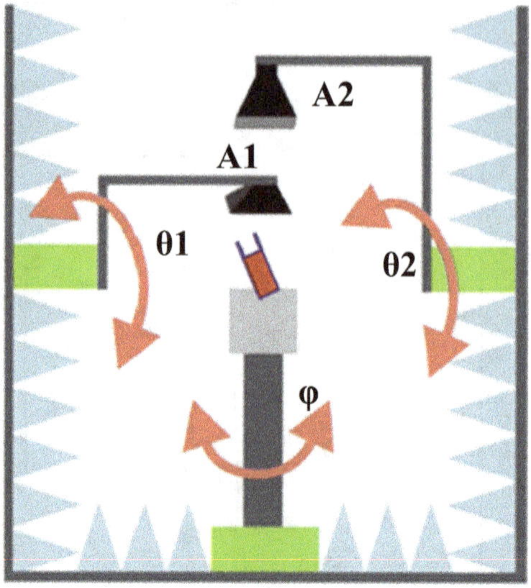

Figure 9.6 Two-channel method, antenna arrangement in anechoic chamber (figure 6.3.1.4.1-1 in TR 37.976 [30], reprinted with permission from 3GPP)

results. The channel information available in the UE can be used to deliver a quick answer to the test system about the current receive quality.

During the OTA test, an assessment of the antenna's performance in MIMO or diversity operation could be performed. Several simplifications could be considered in order to optimize the testing. This method proposes a test of the UE in an anechoic environment with the help of a BSE, with a limited number of faded channels and transmitting antennas, and in a simple geometrical set-up. The underlying principle is to decompose the task for evaluating MIMO performance by evaluating the impact of the direction and angular separation of the two signals on the SUT performance. Such a measurement is much faster than the evaluation of a throughput figure, but it is nevertheless closely related to it. If necessary, an explicit scaling from one quantity onto the other one can be made. By carrying out many tests using different combinations of angles, statistical analysis can be used to derive FoM for the SUT. Direct comparison with the results achieved using more complex spatial signals with temporal variations is not possible, but this method provides similar SUT ranking [30].

Conducted and radiated two-stage method

This candidate method takes a fundamentally different approach to creating the necessary conditions to test MIMO performance. Both conducted and radiated two-stage methods are illustrated, respectively, in Figures 9.7 and 9.8. The principle of two-stage MIMO OTA method is based on the assumption that the far-field antenna radiation pattern will contain all the necessary information for the evaluation of the antenna's performance like radiation power, efficiency and correlation and that

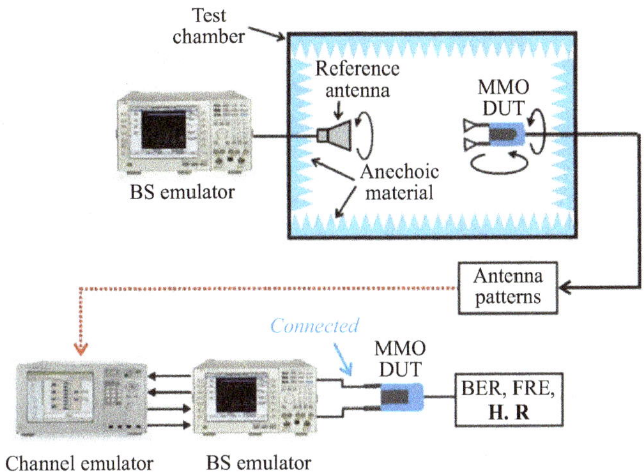

Figure 9.7 Proposed conducted two-stage test methodology for MIMO OTA test (figure 6.3.1.3.1-2 in TR 37.977 [23], reprinted with permission from 3GPP)

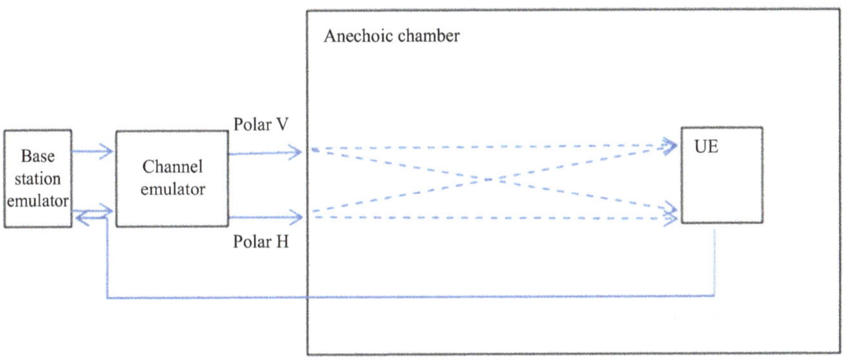

Figure 9.8 Proposed radiated two-stage (RTS) test methodology for MIMO OTA test (figure 6.3.1.3.1-3 in TR 37.977 [23], reprinted with permission from 3GPP)

with the use of CE, the influence of antenna radiation pattern can be incorporated into the channel model [23].

The first stage involves the measurement of the MIMO antenna patterns, and the second state incorporates the measured antenna patterns with chosen MIMO OTA channel models for real-time emulation. In order to accurately measure the antenna pattern of the intact device, the chipset needs to support amplitude and relative phase measurements of the antennas. If the SUT has dynamic antenna-tuning elements, the detailed information on the implementation would be required in order to understand the consequences for the pattern measurement. The BSE and

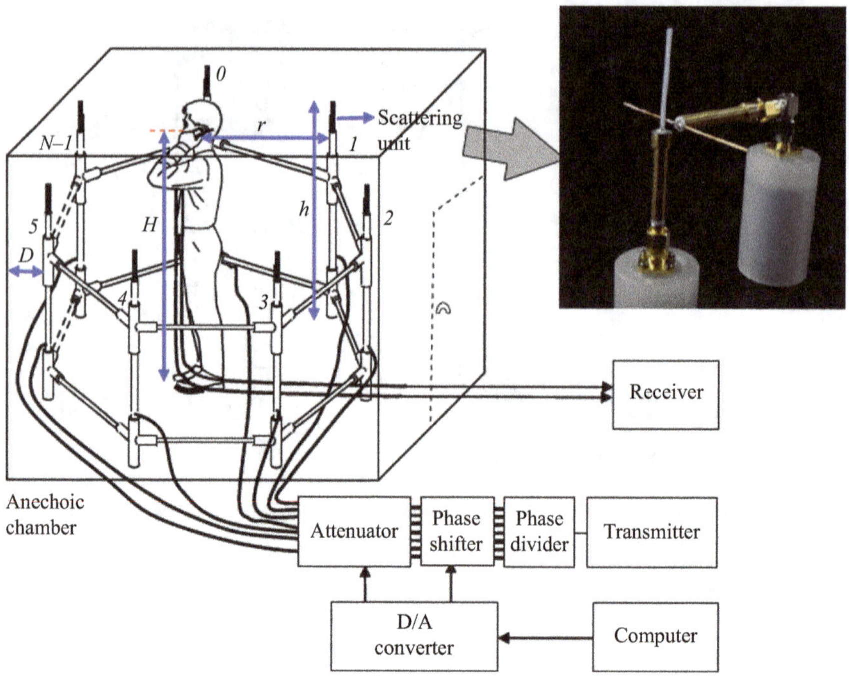

Figure 9.9 Experimental set-up of the spatial fading emulator (figure 6.3.1.5.1-1 (a) in TR 37.977 [23], reprinted with permission from 3GPP)

SUT can then be connected to the real-time CE through the standard temporary antenna connectors to do the test on throughput, etc., either using conducted (Figure 9.7) or radiated (Figure 9.8) approach to test how the MIMO antennas will influence the performance. Further details are provided in 3GPP TR 37.977 [23]. It should be noted that should this methodology be chosen for conformance testing, the methods and procedures for antenna gain and phase measurement would need to be considered and standardized.

Spatial channel emulation method

This candidate method could be considered as a variation of the ring of probes method, where the channel emulation function is provided by a much simpler programmable attenuator and phase shifter per antenna. This is depicted in Figure 9.9. The RF-controlled spatial fading emulator can directly reproduce a multipath radio propagation environment by radio waves emitted from antenna-probe units arranged around a SUT. Both LOS and NLOS scenarios could be implemented by the emulator. Moreover, the emulator has an advantage of measuring radiation characteristics of a SUT as well as the multipath testing because of its RF operation. By controlling the amplitude and phase in real time, a Rayleigh distribution or other relevant multipath distribution can be obtained. This method can generate an equal angular distribution for all propagation delays. Further set-up and measurement procedure details are provided in 3GPP TR 37.977 [23].

9.3.3.2 Methodologies based on the reverberation chamber

The RC is an overmoded reflecting environment providing statistically uniform and statistically isotropic field within its working volume, which resembles the multipath fading scenarios of wireless communication systems. The Rayleigh environment in an RC is well known as a good reference for urban and indoor environments but does not well represent rural and suburban environments.

Reverberation chamber method

Figure 9.10 shows a standard RC, shown with open door and the set-up for measuring radiated power of a mobile phone close to the head phantom. The test set-up for testing UE receiver diversity performance is composed of a BSE, a RC equipped with fixed BS wall-mounted antennas, a switch to direct the BS signal to/from one of the BS wall-mounted antennas, mechanical metallic stirrers and a rotating platform to hold the SUT. Alternatively, the chamber may contain one or more cavities coupled through waveguides or slotted plates (see Figure 9.11).

Mechanical stirring (using the stirrers) and polarization stirring (switching among different fixed BS wall-mounted monopoles antennas) allow simulating the Rayleigh fading at each antenna of the terminal inside the chamber. Accuracy can be increased by rotating the platform holding the device. Each position of the mechanical stirrers for each position of the platform and each fixed BS antenna represents a point of the Rayleigh distribution in terms of receive power on the device antennas. In that way, a Rayleigh fading is artificially created. For each point of the Rayleigh distribution created by the different configurations of the

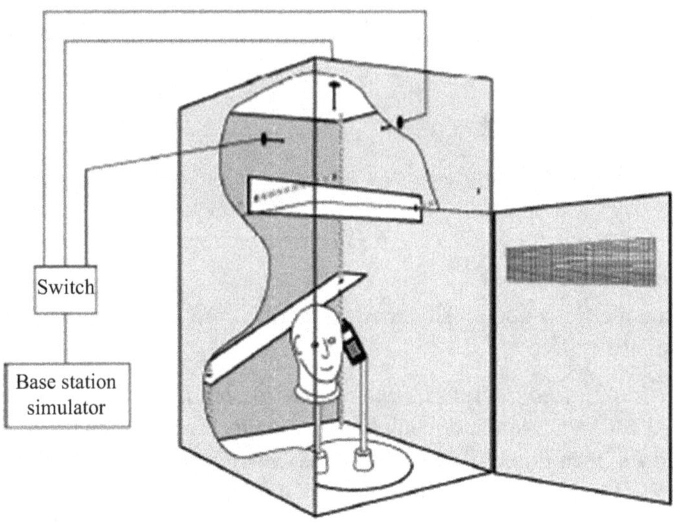

Figure 9.10 Reverberation chamber set-up for SUT testing with single cavity (figure 6.3.2.1-1 in TR 37.977 [23], reprinted with permission from 3GPP)

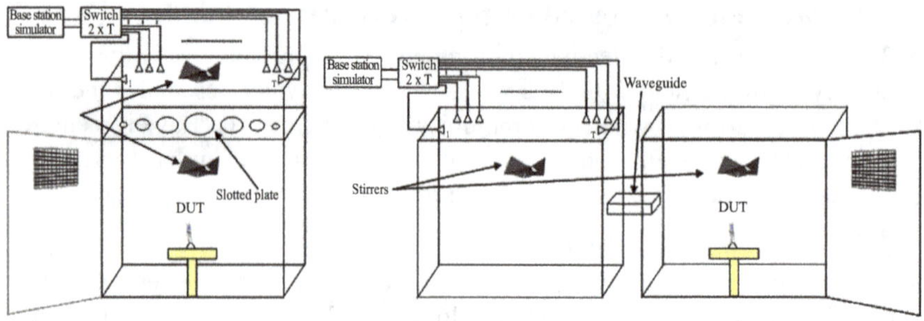

Figure 9.11 Reverberation chambers with multiple cavities (figure 6.3.2.1-2 in TR 37.977 [23], reprinted with permission from 3GPP)

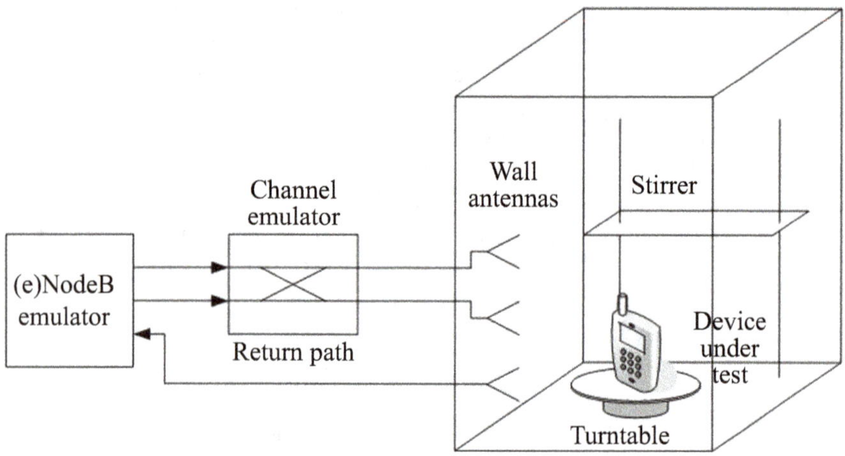

Figure 9.12 Test bench configuration for test using channel emulator and reverberation chamber for a 2 × 2 MIMO configuration (figure 6.3.2.2.1-1 in TR 37.976 [30], reprinted with permission from 3GPP)

chamber, the metric is noted. This method can be used to measure UE sensitivity and UE radiated power.

Reverberation chamber and channel emulator method

This method adds a CE to the downlink prior to launching the signals into the basic or cascaded RC (see Figure 9.12). This allows the temporal aspects of the desired channel model to be fully controlled, although the underlying natural and very short decay time of the chamber will unavoidably spread the power-delay profile. With the use of a CE capable of negative time delay (inverse injection), it is possible to accurately emulate the power-delay profiles of 3GPP SCME channel models [18,50].

9.3.3.3 Methodologies based on the compact antenna test ranges (CATR) chamber

By means of a reflector, a CATR is a collimator system (see Figure 9.13) in which the spherical wave is transformed in plane-wave within the desired quite zone. This method has recently been considered by 3GPP TR 38.810 as an indirect far-field method [27]. Inherently, CATR supports reciprocal operation, i.e. it could be used to perform OTA test in both transmit and receive modes. This makes it very attractive for the direct testing of far-field RF parameters and, especially it has been applied in 3GPP TR 37.842 for testing active antenna systems that describe BS conformance testing for EIRP, TRP as well as for EIS test [51].

The quality of the pattern measurement using this method is primarily determined by the purity and uniformity of the phase and amplitude of the transformed plane-wave. Despite this method can create a plane-wave field in much less space than a DFF method, the cost of implementing a CATR facility is currently much higher than DFF method. The method seems ideal for 5G mm-wave OTA testing, but it has difficulties providing different frequency ranges.

9.3.3.4 Methodologies based on the near-field chamber

This method also known as near-field to far-field transform (NFTF) method, which employs a near-field measurement system that could measure the near-field amplitude and phase on a surface around the SUT. Based on Huygen's principle,

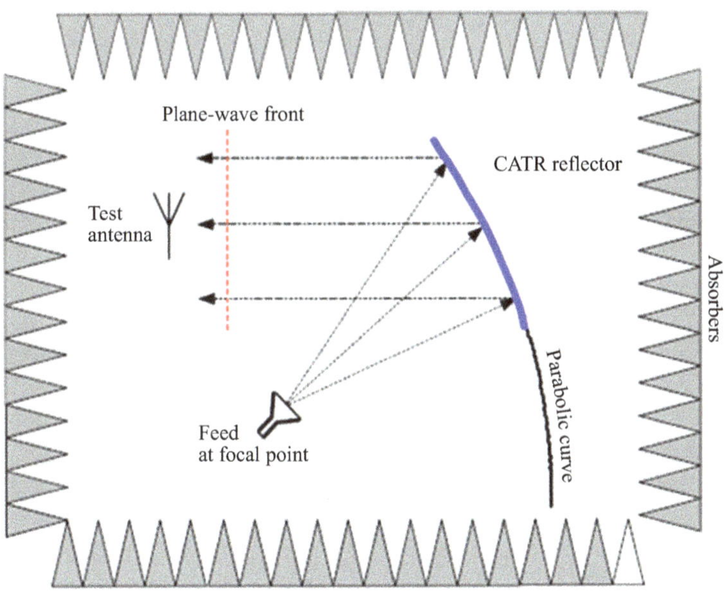

Figure 9.13 Compact range – working principle (figure E.1-1 in TR 38.810 [27], reprinted with permission from 3GPP)

the NFTF method relies on using mathematical transformation to determine the key performance indicators in the far field from the acquired near-field pattern scan results.

Figure 9.14 shows a circular probe array near-field measurement set-up considered by 3GPP TR 38.810 for EIRP and TRP measurements. The correspondent 3D far-field pattern could be obtained by using a modal spherical wave expansion. The NFTF method shows deficits with respect to OTA testing during real-time operation of the SUT.

9.3.3.5 Methodologies based on the sectored MPAC chamber

A simple-sectored MPAC (SS-MPAC) method [28,52] has been proposed for OTA testing of mMIMO systems, which inherently has beamforming capabilities (see Figures 9.15 and 9.16). This method requires good understanding on the optimized probe configuration and allocation on the sectored MPAC [29]. With a limited number of active probes employed, the SS-MPAC method could reduce the total set-up costs, including the hardware and the size of the chamber as compared with multi-probe methods mentioned in Section 9.3.3.1. The application of SS-MPAC method for mm-wave adaptive beamforming OTA evaluations has been discussed in [28,52]. The requirements for the test system design were analysed in [53], which considers factors such as the measurement range *R*, number of OTA probe antennas, number of active OTA probe antennas and amount of CE resource. Other research works have been carried out on design methodologies for the probe set-up [29,33,36,52].

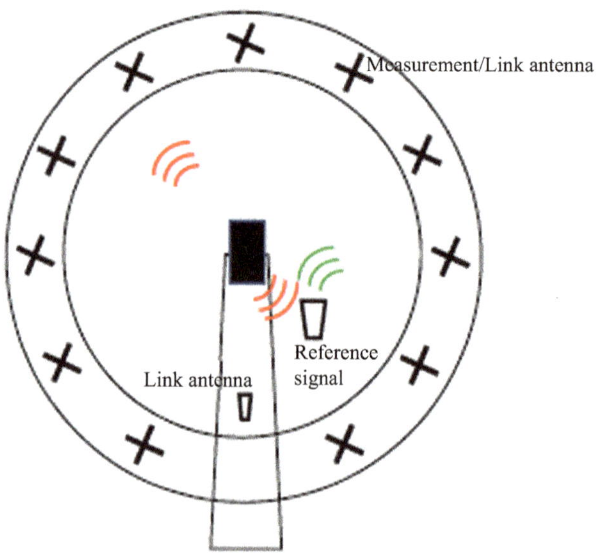

Figure 9.14 A circular probe array near-field measurement set-up for EIRP/TRP measurements (figure 5.2.4.1-1 in TR 38.810 [27], reprinted with permission from 3GPP)

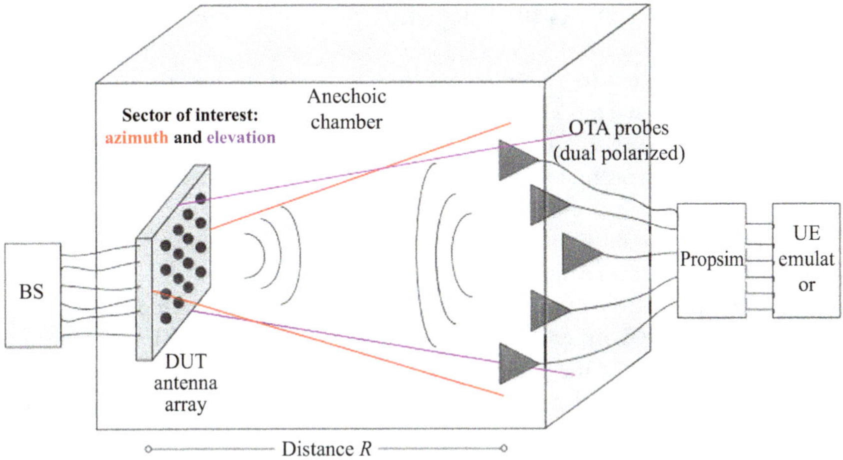

Figure 9.15 An illustration of the MPAC set-up for massive MIMO BS performance evaluation (figure 2 in [28])

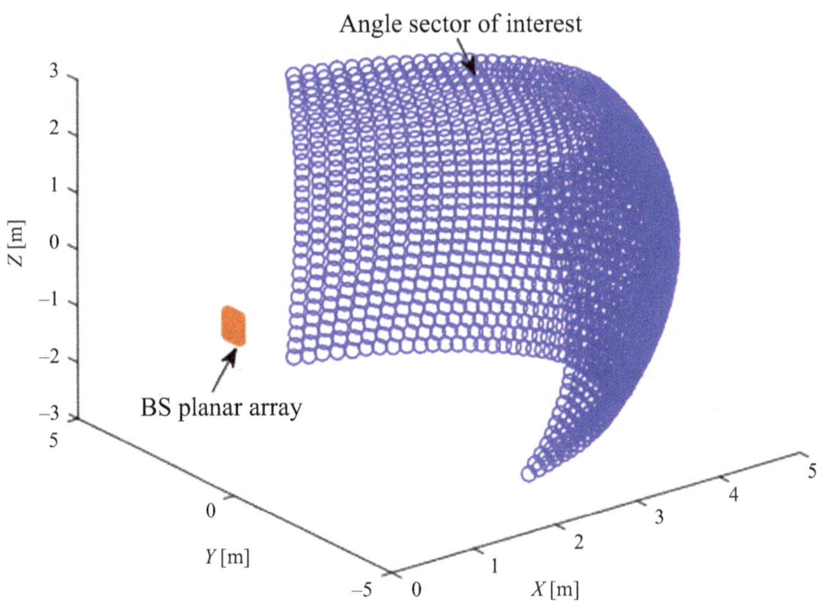

Figure 9.16 Typical chamber set-ups (figure 4 in [28])

9.4 Challenges for 5G and beyond

OTA testing is critical to accurately evaluate device performance in realistic scenarios. MIMO test-equipment is available but intense activity within several EU programmes such as COST CA1504, COST IC1004 and COST 2100 illustrates that the measurement issues have not been fully resolved. Operating emerging wireless technologies at a higher RF such as mm-wave and quasi-THz is a route to higher data bandwidth. Testing at these frequency bands, their traceability, propagation channel, interference characterization (especially for sparse multipath) and system-level testing present new measurement challenges for assessing their performance in a real-world environment as RF beamforming is essential to overcome the higher losses at these frequencies, and the antenna and RF circuitry are envisaged to be highly integrated, i.e. without RF connection port for conducted test resulting that OTA is the only measurement method possible [54]. Towards the development of emerging wireless technologies, both BS and UE are expected to perform beamforming by using different types of antenna elements. Especially measuring the performance of the wireless systems with adaptive antennas and characterizing propagation channels at mm-wave bands create challenges where innovative test and measurement methods are required. Depending on the correlation between the antennas in the test environment, this emulated channel matrix may be well-conditioned or ill-conditioned [55].

With the industrial adoption of multi-user MIMO (MU-MIMO), mMIMO beamforming, beam reconfigurable intelligent antennas and mm-wave in the emerging wireless systems, several worldwide industries, research communities, standards bodies (e.g. 3GPP, CTIA, ETSI and IEEE) are now facing new measurement challenges on NR OTA test for verifying products that meet desired performance parameters for fulfilling the diverse technological requirements. Note that, for 5G NR FR1 (410–7,125 MHz), the traditional conducted testing will still be used for most compliance test requirements except TRP and the system-level link performance throughput test. For 5G NR FR2 (24,250–52,600 MHz), all tests are required to be carried out by OTA-radiated methods. Furthermore, general public's risk concerns have led strengthen compliance regulation in some countries. The time-burden issue on employing current RF exposure assessment methods applicable to specific absorption rate (SAR) and power density (PD) defined in The International Electrotechnical Commission (IEC) and IEEE International standards will need to be addressed, especially, for high-volume beam reconfigurable product verifications. The key challenges remain on lack of accurate, fast, low-cost, and traceable methods for NR high-volume product verifications. Despite some key concepts for their validation have been discussed for production verification in standardization process but test methods and equipment remained undefined in the industry and there is immediate need to provide underpinning metrology to support the development of these technologies covering all aspects from the signals to systems.

9.5 Conclusions

5G OTA testing will evolve quickly following release of defined 5G NR standards so to enable 5G products go into production. Several methods are currently being assessed for the component and system verifications and final product compliance testing. The primary trade-offs for cost, and accuracy will need to be determined and the test methods standardized so to support acceleration of 5G deployments. 3GPP, COST and CTIA have published several technical reports and updated time by time. This chapter has provided an overview of the current and potential candidates methodologies for 5G and beyond to date for the OTA testing of SUT (mainly UE and BS) as well as the associated challenges for 5G and beyond. A brief description has also been provided over the different environments that are used to mimic the actual conditions in which these SUTs are operated. These proposed OTA test methods vary widely with respect to their implementation, equipment used, chamber size, running cost and measurement time.

Acknowledgement

The work of T. H. Loh was supported by the 2014–16 and the 2017–20 National Measurement System Programme of the UK government's Department for Business, Energy and Industrial Strategy (BEIS), under Science Theme Reference EMT of that programme.

References

[1] European Commission *Why the EU is Betting Big on 5G – Research*eu Focus Magazine* [online]. February 2015. Available from: https://ec.europa.eu/digital-single-market/en/news/why-eu-betting-big-5g-researcheu-focus-magazine [Accessed January 2020].

[2] Ericsson *Mobility Report, the Future of Mobile Subscriptions* [online]. August 2018. Available from: https://www.ericsson.com/en/mobility-report/future-of-mobile-subscriptions [Accessed January 2020].

[3] 5GPPP *Global 5G Actions* [online]. Available from: https://5g-ppp.eu/global-5g-actions/ [Accessed January 2020].

[4] Hindle P. *Test & Measurement Industry Tackles 5G Over-the-Air Testing* [online]. March 2019. Available from: https://www.microwavejournal.com/articles/31906-test-measurement-industry-tackles-5g-over-the-air-testing [Accessed January 2020].

[5] Wikipedia *5G NR* [online]. Available from: https://en.wikipedia.org/wiki/5G_NR [Accessed January 2020].

[6] Wikipedia *5G* [online]. Available from: https://en.wikipedia.org/wiki/5G [Accessed January 2020].

[7] Loh T. H., Li C., Wang H., and Qin F. 'A software-defined-radio platform for multiple-input-multiple-output over-the-air measurement'. *The 10th European Conference on Antennas and Propagation (EuCAP 2016)*, Davos, Switzerland, 11th–15th April 2016.

[8] Third Generation Partnership Project (3GPP) *Release-15 and Release-16*, [online]. Available, respectively, from: https://www.3gpp.org/release-15 and https://www.3gpp.org/release-16 [Accessed May 2020].

[9] European Telecommunications Standards Institute (ETSI) *5G* [online]. Available from: https://www.etsi.org/technologies/5g [Accessed May 2020].

[10] The Institute of Electrical and Electronics Engineers (IEEE) Standards Association *IEEE Standards Activities in 5G* [online]. Available from: https://standards.ieee.org/content/dam/ieee-standards/standards/web/documents/other/5G.pdf [Access May 2020].

[11] International Association for the Wireless Telecommunication Industry (CTIA) *Wireless Standards and Certification: A Brief Explainer* [online]. Available from: https://www.ctia.org/news/blog-wireless-standards-and-certification-a-brief-explainer [Accessed May 2020].

[12] Gustafsson M., Jämsä T., and Högberg M. 'OTA methods for 5G BTS testing – survey of potential approaches'. *The 32nd URSI GASS (URSI 2017)*, Montreal, Canada, 19–26 August 2017.

[13] Jing Y., Rumney M., Kong H., and Wen Z. 'Overview of 5G UE OTA performance test challenges and methods'. *The 2018 IEEE MTT-S International Wireless Symposium (IWS)*, Chengdu, China, 2018.

[14] EDN *5G OTA Testing: Key Concepts and Definitions* [online]. August 2019. Available from: https://www.edn.com/5g-ota-testing-key-concepts-and-definitions/ [Accessed January 2020].

[15] Wikipedia *Over-the-Air* [online]. Available from: https://en.wikipedia.org/wiki/Over-the-air [Accessed January 2020].

[16] CTIA *Test Plan for Mobile Station Over the Air Performance*, v1.0, October 2001.

[17] 3GPP *Measurements of Radio Performances for UMTS Terminals in Speech Mode*, Tech. Rep. 25.914, v.7.0.0., 2006.

[18] Rumney M., Pirkl R., Landmann M. H., and Sanches-Hernandez D. A. 'MIMO Over-the-Air Research, Development, and Testing'. *International Journal of Antennas and Propagation*, 2012, 2012, pp. 1–8.

[19] Wikipedia *EIRP* [online]. Available from: https://en.wikipedia.org/wiki/Effective_radiated_power [Accessed January 2020].

[20] CTIA *Test Plan for Wireless Device Over-the-Air Performance*, v3.8.2, April 2019.

[21] Kildal P., Chen X., Orlenius C., Franzen M., and Patane C. S. L. 'Characterization of Reverberation Chambers for OTA Measurements of Wireless Devices: Physical Formulations of Channel Matrix and New Uncertainty Formula'. *IEEE Transactions on Antennas and Propagation*, 2012, 60(8), pp. 3875–3891.

[22] Scannavini A., Foged L. J., Gross N., and Noren P. 'OTA measurement of wireless devices with single and multiple antennas in anechoic chamber'. *The Eighth European Conference on Antennas and Propagation (EuCAP 2014)*, The Hague, the Netherlands, 2014.

[23] 3GPP TR 37.977 V15.0.0 (2018-09) *Verification of Radiated Multiantenna Reception Performance of User Equipment (UE)*, September 2018. © 2018. 3GPP™ TSs and TRs are the property of ARIB, ATIS, CCSA, ETSI, TSDSI, TTA and TTC who jointly own the copyright in them. They are subject to further modifications and are therefore provided to you "as is" for information purposes only. Further use is strictly prohibited.

[24] CTIA *Test Plan for 2x2 Downlink MIMO and Transmit Diversity Over-the-Air Performance, Version 1.2*, April 2018.

[25] Agilent Application Note *Theory, Techniques and Validation of Over-the-Air Test Methods for Evaluating the Performance of MIMO User Equipment*, December 2010.

[26] Balanis C. A. *Antenna Theory: Analysis and Design*. Hoboken, NJ: Wiley-Blackwell, May 2005.

[27] 3GPP TR 38.810 V16.5.0 (2019-12) *NR; Study on Test Methods*, December 2019. © 2020. 3GPP™ TSs and TRs are the property of ARIB, ATIS, CCSA, ETSI, TSDSI, TTA and TTC who jointly own the copyright in them. They are subject to further modifications and are therefore provided to you "as is" for information purposes only. Further use is strictly prohibited.

[28] Kyösti P., Fan W., Pedersen G. F., and Latva-Aho M. 'On Dimensions of OTA Setups for Massive MIMO Base Stations Radiated Testing'. *IEEE Access*, 2016, 4, pp. 5971–5981.

[29] Gao H., Wang W., Wu Y., Liu Y., and Pedersen G. F. 'A Virtual Over-the-Air Method for 5G Massive MIMO Base Station Testing With Flexible Virtual Probes'. *IEEE Access*, 2019, 7, pp. 108474–108485.

[30] 3GPP TR 37.976 V15.0.0 (2018-09) *Measurement of Radiated Performance for Multiple Input Multiple Output (MIMO) and Multi-Antenna Reception for High Speed Packet Access (HSPA) and LTE Terminals*, September 2018. © 2018. 3GPP™ TSs and TRs are the property of ARIB, ATIS, CCSA, ETSI, TSDSI, TTA and TTC who jointly own the copyright in them. They are subject to further modifications and are therefore provided to you "as is" for information purposes only. Further use is strictly prohibited.

[31] European Cooperation in Science and Technology (COST) *Actions IC1004 and Actions CA15104* [online]. Available, respectively, from: http://www.ic1004.org/ and http://www.iracon.org/ [Accessed May 2020].

[32] 3GPP TS 38.521-2 V16.2.0 (2019-12) *NR; User Equipment (UE) Conformance Specification; Radio Transmission and Reception; Part 2: Range 2 Standalone*, December 2019.

[33] Fan W., Kyosti P., Rumney M., Chen X., and Pedersen G. F. 'Over-the-Air Radiated Testing of Millimeter-Wave Beam-Steerable Devices in a Cost-Effective Measurement Setup'. *IEEE Communications Magazine*, 2018, 56(7), pp. 64–71.

[34] Qiao Z., Xie Y., Wang Z. and Miao J. 'Exploring OTA testing for massive MIMO base stations in small region'. *The 2017 Sixth Asia-Pacific Conference on Antennas and Propagation (APCAP)*, Xi'an, China, 2017, pp. 1–3.

[35] 3GPP TS 38.141-2 V16.2.0 (2019-12) *NR; Base Station (BS) Conformance Testing Part 2: Radiated Conformance Testing*, December 2019.

[36] Fan W., Carton I., Kyosti P., *et al.* 'A Step Toward 5G in 2020: Low-Cost OTA Performance Evaluation of Massive MIMO Base Stations'. *IEEE Antennas and Propagation Magazine*, 2017, 59(1), pp. 38–47.

[37] Kong H., Wen Z., Jing Y., and Yau M. 'Midfield Over-the-Air Test: A New OTA RF Performance Test Method for 5G Massive MIMO Devices'. *IEEE Transactions on Microwave Theory and Techniques*, 2019, 67(7), pp. 2873–2883.

[38] Kildal P., Glazunov A. A., Carlsson J., and Majidzadeh A. 'Cost-effective measurement setups for testing wireless communication to vehicles in reverberation chambers and anechoic chambers'. *The 2014 IEEE Conference on Antenna Measurements & Applications (CAMA)*, Antibes Juan-les-Pins, 2014.

[39] Artner G., Kotterman W., Del Galdo G., and Hein M. A. 'Automotive Antenna Roof for Cooperative Connected Driving'. *IEEE Access*, 2019, 7, pp. 20083–20090.

[40] Glazunov A. A., Kildal P., and Kildal M. S. 'Devising a horizontal chamber array for automotive OTA tests in random line-of-sight'. *The 2015 International Symposium on Antennas and Propagation (ISAP)*, Hobart, Australia, November 2015.

[41] Razavi A., Glazunov A. A., Moghaddam S. M., Maaskant R., and Yang J. 'Characterization Method of an Automotive Random-LOS OTA Measurement Setup'. *Progress In Electromagnetics Research C*, 2018, 84, pp. 47–60.

[42] Wu X., Liu Z., Zhang Z., Liao C., and Quan Z. 'MIMO OTA testing on WiFi devices and 3D channel modeling in multi-probe anechoic chamber systems'. *The 2016 Progress in Electromagnetic Research Symposium (PIERS 2016)*, Shanghai, China, 2016.

[43] Wi-Fi Alliance *Certification* [online]. Available from: https://www.wi-fi.org/certification [Accessed February 2020].

[44] LoRa Alliance *Certification* [online]. Available from: https://lora-alliance.org/lorawan-certification [Accessed February 2020].

[45] Shen P., Qi Y., Yu W., Fan J., and Li F. 'OTA Measurement for IoT Wireless Device Performance Evaluation: Challenges and Solutions'. *IEEE Internet of Things Journal*, 2019, 6(1), pp. 1223–1237.

[46] IEEE Standards Association *Internet of Things Related Standards* [online]. Available from: https://standards.ieee.org/initiatives/iot/stds.html [Accessed February 2020].

[47] *Bluetooth Qualification Process* [online]. Available from: https://www.bluetooth.com/develop-with-bluetooth/qualification-listing/ [Accessed February 2020].

[48] Global Certification Forum (GCF) *Internet of Things (IoT) Solutions* [online]. Available from: https://www.globalcertificationforum.org/services/iot-devices.html [Accessed May 2020].

[49] IST-4-027756 WINNER II D1.1.2 V1.1 *WINNER II Channel Models* [online]. November 2006. Available from: http://www.ero.dk/6799B797-53CC-417E-B6FA-059A8E6AF350?frames=no& [Accessed February 2020].

[50] Sanchez D., Garciafernandez M. A., Arsalane N., Mouhamadou M., Carsenat D., and Decroze C., *Validation of 3GPP SCME Channel Models Emulated in Mode-Stirred Reverberation Chambers*, 3GPP TSG-Ran WG, 2012.

[51] 3GPP TR 37.842 V13.3.0 (2019-12) *Evolved Universal Terrestrial Radio Access (E-UTRA) and Universal Terrestrial Radio Access (UTRA); Radio Frequency (RF) Requirement Background for Active Antenna System (AAS) Base Station (BS)*, December 2019.

[52] Kyosti P., Hentila L., Fan W., Lehtomaki J., and Latva-Aho M. 'On Radiated Performance Evaluation of Massive MIMO Devices in Multiprobe Anechoic Chamber OTA Setups'. *IEEE Transactions on Antennas and Propagation*, 2018, 66(10), pp. 5485–5497.

[53] Fan W., Kyosti P., Rumney M., Chen X., and Pedersen G. F. 'Over-the-Air Radiated Testing of Millimeter-Wave Beam-Steerable Devices in a Cost-Effective Measurement Setup'. *IEEE Communications Magazine*, 2018, 56(7), pp. 64–71.

[54] Remley K. A., Gordon J. A., Novotny D., *et al.* 'Measurement Challenges for 5G and Beyond: An Update from the National Institute of Standards and Technology'. *IEEE Microwave Magazine*, 2017, 18(5), pp. 41–56.

[55] Li C., Loh T.-H., Humphreys D. A., Gui Y., Wang H., and Qin F. 'A LTE MIMO OTA test system using vector signal transceivers'. *The 25th European Conference on Networks and Communications (EuCNC 2016)*, Athens, Greece, 27–30 June 2016.

Chapter 10

Beamformer development challenges for 5G and beyond

Muhammad Ali Babar Abbasi[1] and Vincent F. Fusco[1]

Antenna's beamforming technology [1] is vital in the utilization of the microwave and millimeter-wave (mmWave) frequency bands for the fifth-generation (5G) communication technology, and beyond, applications that include the sixth generation (6G), Internet of Things and Industry 4.0 [2,3].

Antenna beamforming can be defined as the ability of an antenna array to steer the maximum radiation toward a prescribed direction, or, conversely, the ability of the antenna array to estimate the direction of arrival (DOA) of an impinging signal.

Beamforming is generally achieved by using multiple antennas, spatially colocated to either function as an independent radiator or as a unit cell in larger array arrangements. Recently, the term multiple-input–multiple-output (MIMO) has been used in conjugation with the beamforming technology since it also involves multiple antenna systems, thus linking it to the beamforming technology [4]. When the number of antennas used in the MIMO operation becomes very large, it is generally termed massive MIMO technology [5]. Different beamforming concepts are used for the two frequency spectrum classifications in cellular technology, i.e., sub-6 GHz and mmWave. The majority of beamformer challenges at sub-6 GHz are already resolved, and prototypes are now available [6]; however, there are still major fundamental challenges that engineers face while developing beamformers for use in the mmWave bands, i.e., >28 GHz. Challenges include, but are not limited to, high path loss, power generation, adaption to account for realistic channel estimation, hardware impairments, energy leakage in circuits, high development cost, spatial–temporal channel variations, signal blockages due to the presence of obstacles such as beamformer casing, user body and hand. Also, multiuser MIMO techniques are not easily implementable in mmWave frequencies. In addition to this, beam coordination at transmitter and receiver antenna systems especially in mmWave spectrum is very challenging. In this chapter, we will first classify the beamformers based on the system architecture, operational frequency

[1]Centre for Wireless Innovation (CWI), Institute of Electronics, Communications and Information Technology (ECIT), School of Electronics, Electrical Engineering and Computer Science (EEECS), Queen's University Belfast, Belfast, UK

band and use cases. We will then elaborate the hardware development challenges encountered in realizing beamformers for specialized purpose in 5G and beyond applications.

10.1 Introduction

The 5G communication technology is now ready to be deployed and efforts are now pushed forward to develop the technology for the 6G and beyond applications. As a first step of 5G deployment, communication base stations have been launched in many cities around the world and 5G-supported communication devices are available in the market since the mid-2019. The story does not end here since the 5G communication infrastructure launched so far has not fully satisfied user experience promises. There is still a long way before society fully harvests the actual benefits of research and development efforts that were pushed by the communication industry in 5G. The sub-6-GHz 5G technology currently available in marketplace performs better than previous communication technologies like the long-term evolution (LTE) and LTE Advanced (LTE-A). For full 5G evolution, we expect data rate of multiple gigabits per second per user and ten times better latency compared to 4G for all users in a cell [7]. It is expected to see this during the launch of mmWave 5G over the coming years.

Antenna array beamforming technology is now an integral part of the 5G infrastructure, as per the popular belief of researchers, communication engineers and major corporate telecom operators. It was envisioned in the early 2010s that the frequency spectrum used for the global system of mobile communication, LTE and LTE-A operating with other wireless technologies would quickly fill in the available sub-6-GHz spectrum. There will be less bandwidth available for further usage and moving to the higher frequencies will be the only option left. This gave rise to an immense research effort in engineering development at the higher frequencies into the millimeter range. This spectrum comprises mainly frequencies having wavelengths of millimeters and is generally referred to as mmWave spectrum. The migration of wireless technology from sub-6 GHz to mmWave has a high cost since the wireless devices operating at high frequencies face several problems that were solved due to technological advancements focused at sub-6-GHz spectrum. First problem is that the free-space path loss at mmWave frequencies is high compared to lower frequencies. The path loss of an electromagnetic wave is governed by the following expression:

$$path\ loss = \left(\frac{4\pi r \times f}{c}\right)^2 \tag{10.1}$$

where f is the frequency of an electromagnetic wave, r is the path length and c is the speed of light (all in SI units). This means that as we move higher along the frequency spectrum, the free-space path loss increases at the squared rate. A wireless link operating at 3 GHz suffering a path loss of around 82 dB will suffer a path loss of around 102 dB when operated at 30 GHz given all other variables are kept constant in the wireless link. Another way of looking at this is at the same distance and same transmitted power by the antenna in a wireless link, a 3-GHz carrier signal operating over a 1-km range, when replaced by 30-GHz carrier signal, will cover only 100 m and around 30 m at carrier signal of 100 GHz. To deal with this much path loss, we need compensation. This can be achieved by either enhancing the output signal of the power amplifier or directing the antenna energy into a more confined spatial direction. The second way is preferred because of the well-studied concept of antenna array beamforming [1] in which the radiated energy from an antenna array can be directed toward a predefined spatial sector and, if correctly designed, prevented from going in other directions where the radiated signal is not required. A system that performs this function is termed a beamformer.

A beamformer adds spatial "gain" by using antenna radiating elements in ensemble such that the resulting arrangement compensates path loss. Antenna gain is defined as the power radiated by the antenna per unit solid angle in a given direction, divided by the average power radiated per unit solid angle [8]. For example, assume a point source radiating electromagnetic energy equally in all directions, the power flux density can be written as

$$\Phi = \frac{P_T}{4\pi r^2} \tag{10.2}$$

where P_T is the transmitted power and r is the observation distance from the point source. A practical antenna cannot radiate equally in all directions, so its radiation is generally denoted in comparative forms of directivity (D) and gain (G) where directivity is defined as

$$D = \frac{4\pi\Phi(\theta,\phi)}{\int_0^{2\pi}\int_0^{\pi}\Phi(\theta,\phi)d\theta\, d\phi}, \tag{10.3}$$

and the gain is defined as

$$G = \eta D \tag{10.4}$$

where η is the antenna efficiency that is always less than 1 (for further details see [8]). Beamformer gain is normally tested in an anechoic environment using setups such as those in [9,10].

Most commonly used beamformers are arrays of antenna in which the gain of a single-antenna radiating element is enhanced by increasing the number of antennas radiating elements placed at approximately one-half wavelength spacing with respect to each other [8]. In this way, the magnitude of the antenna radiated by all the closely spaced antennas adds constructively or destructively in a spatial footprint defined by both the radiation characteristics of the antenna and the element separation. The directions where all the signals radiated by every antenna are added constructively, the array gain is the highest. The direction in which the maximum gain is achieved is generally referred to as boresight of the antenna array and the radiation along this direction is called a beam. A general rule of thumb is that as we increase the number of antennas in an array, we increase the array gain. If we add phase difference between the antenna elements in an array, we can steer the constructive interference between the signal radiated by each antenna away from the boresight direction. This concept is generally referred to as beam steering. By carefully managing the phases of each antenna element in an array using phase shifters, we can control the beam in multiple ways and achieve beam tilting, beam scanning, multi-beam radiation and interference rejection null formation [11]. An array that uses phase shifters to control the antenna array is called a phased array (Figure 10.1(b)). This is not a new concept and wealth of literature is available for design, synthesis and analysis of phased arrays.

According to the classical Shannon–Hartley theorem [12] that defines the tightest upper bound of the data rate, channel capacity (C) is defined by

$$C = B \times \log_2\left(1 + \frac{S}{N}\right) \tag{10.5}$$

where C is in bits per second. B is the bandwidth in Hertz and ideally is equal to the frequency passband used in signal transmission. S is the average signal power at the receiver end averaged over the bandwidth B whereas N is average noise/interference power averaged over B, and both S and N are in Watts. The ratio S/N is generally referred to as signal-to-noise ratio (SNR) and has a direct logarithmic (base 2) relationship with the channel capacity. Beamforming has a direct impact on increasing the signal power in (10.5), so by adding beamformer gain one can increase capacity.

Looking toward the example given previously, a wireless link suffering a path loss of 102 dB can use beamforming technology to reduce the path loss to 82 dB if an antenna array having a gain value of 20 dBi is used as a beamformer.

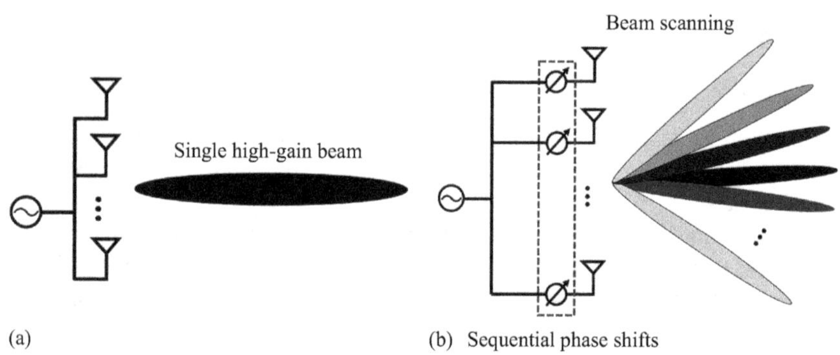

Figure 10.1 (a) Multiple antennas connected to a single radio-frequency source forming an antenna array capable of beamforming only along boresight direction. (b) Same antenna array with bank of phase shifters capable of beam scanning at an angle θ from the boresight direction

Beamformer gain is generally added to the communication system link budget directly.

As we move from sub-6 GHz toward the mmWave spectrum and beyond, the design of a standard phased array gets difficult because of the two main reasons. The first reason is that compensation of the path loss at mmWave requires high array gain, and in order to achieve this, increased numbers of antenna elements are required in an array. This requires additional hardware per antenna element (power amplifier, filter, mixer, etc.) and increases the overall system cost compared to a sub-6-GHz phased array where individual components are cheaper to realize at the present time due to volumes currently in use. The second difficulty lies with the technological limitations of mmWave electronics. Principally a phase shift is required for every radiating element in a phased array. At mmWave, this demands high-precision low-loss electronics. This trend of elevating cost and complexity generally increases as we move higher up in the frequency spectrum. Regardless of these problems, beamforming is inevitable if we want to use a high-frequency spectrum for communication technology, and innovative mmWave beamforming approaches are required now than ever before.

In this chapter, we will discuss the current state-of-the-art in beamforming technology. Our focus will be especially toward the frequencies higher than the sub-6-GHz bands. We will look at the beamformers in frequency spectrum bands centered at 26–32, 36–39, 50 and 60 GHz as these are frequencies currently allocated for 5G wireless [13]. The pressing need is to find beamforming architectures at these frequencies that are capable of meeting the high data rate and latency demands of the communication industry. The launch of mmWave 5G is expected soon and further technological migration from 5G to 6G will increase the requirement of high-efficiency beamforming system even more than at present [14].

Beamformer requirements and development challenges vary as we move from one communication system to another one. Due to this reason we will classify the types of beamformer based on system, operation and use cases in the following sections of the chapter. We will then discuss some state-of-the-art beamformer solutions as well as technical challenges that are yet to be solved.

10.2 Beamformer type classification

10.2.1 Classification based on architecture

10.2.1.1 Analog beamformer

Analog beamformer architectures generally consist of a matrix of fixed phase shifters. Conceptually, just by controlling phase of each transmitted signal from each antenna, we can form a beam in a particular direction. Analog beamformers normally contain radio-frequency (RF) switching matrices after the bank of phase shifters in order to facilitate the beam steering and to connect the required beams to a specific signal path. Some analog beamformers show capability of continuously scanning a radiated beam within a given sector (e.g., [15]). Analog beamformers can place nulls along interferences direction in order to cancel unwanted user

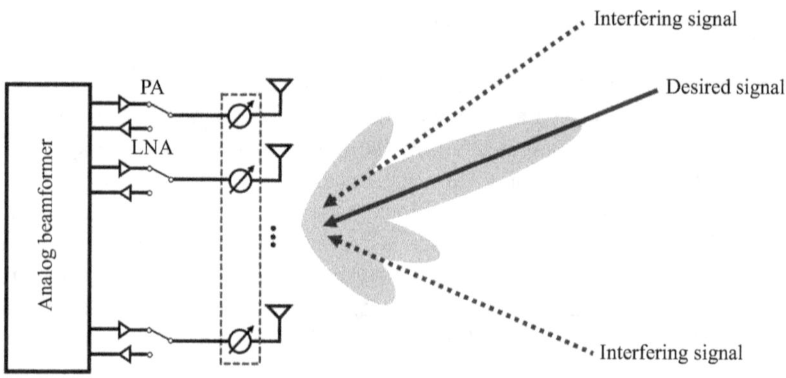

Figure 10.2 Analog beamformer when high beamforming gain is directed toward the desired signal and nulls are placed toward the direction of interfering or undesired signals. LNA, low-noise amplifier; PA, power amplifier

interference in real time. This concept is illustrated in Figure 10.2. Such a capability is required in almost every future multiuser communication standard where simultaneous signal is received by an antenna array from multiple users located along multiple directions.

10.2.1.2 Digital beamformer

In a digital beamformer, the radiated beam patterns of an array are constructed by processing the signal in the desired direction and cancelling the beams from interfering directions, all within the digital domain. This is generally accompanied using finite impulse response filters [1] that have the benefit of managing the weights of each antenna element adaptively. By optimizing the weights, theoretically perfect beamforming can be achieved with perfect nulls placed in the direction of interfering signal. There are several advantages of digital beamforming in large antenna array systems. These include high-resolution DOA estimation, high spectral efficiency, high system security and enhanced energy efficiency [16].

Beamforming is directly linked with the improvement of signal quality at the receiver end. When this attribute is coupled with high energy efficiency, the digital beamforming method can use the standards of massive MIMO technology to enhance spectral efficiency [17]. To elaborate on this, in a standard beamforming synthesis for massive MIMO, the power radiated by a single-antenna element is limited, which leads to energy efficiency [18,19]. Since massive MIMO technology can use digital beamforming, optimization of power consumption is possible through manipulation of the usage of number of antennas based on a specific criterion. For example, in a base station, the overall throughput of a MIMO system is optimized by utilizing the minimum number of radiating antennas and this is only possible by using the principles of weight management of antennas in digital domain [20]. The optimization goals are selected as per the required power control for a particular use case for example, at the mmWave base station in [21], analog phase shifters do not have a high resolution, so digital beamforming (pre-coding) is added. Digital beamforming works well at sub-6-GHz frequency spectrum and fully functional prototypes are already available, for example, see [22].

10.2.1.3 Hybrid or analog/digital beamformers

As discussed previously, analog beamforming is easy to implement using advancements in well-known phased array topology; however, the quality of interference cancellation is not as good as with a digital beamformer. However, to connect each antenna element to a digital domain requires complex hardware set-ups. These motivations lead to the introduction of a topology of a beamformer that attempts to yield the benefits of both digital and analog beamforming. Such a beamformer is classified as an analog/digital or hybrid beamformer [23,24]. The benefit of this topology is that digital beamforming methods generate digital signals that are good enough to allow good-quality DOA estimation, while the analog beamforming method enables a lower number of RF chains required to beamform. This eventually makes it possible, in principle, to develop a cost-effective hardware of the overall system. The analog parts of the beamformer include digital-to-analog conversion and power amplification on transmit and low-noise amplification on receive. Hybrid beamforming is usable not only at sub-6 GHz but also in the mmWave spectrum. It has been shown that in mmWave hybrid systems, with a given number of RF chains and analog-to-digital-converters (ADCs), the performance of a hybrid beamformer can bring closer to a purely digital beamformer with large number of RF chains [20]. This is done by inducing symbol multiplexing and comparatively smaller number of RF chains. In contrast to this, an increase in the number of RF chains and ADCs can improve the obtainable spectral efficiency [25,26]. This provides an attractive trade-off for communication hardware design engineers such that they can select the number of RF chains and radiating antennas as per cost cap and spectral efficiency requirements. Figure 10.3 shows a beamformer that can be purely analog or hybrid depending upon the number of antennas (N_{ant}) and the number of RF chains (L).

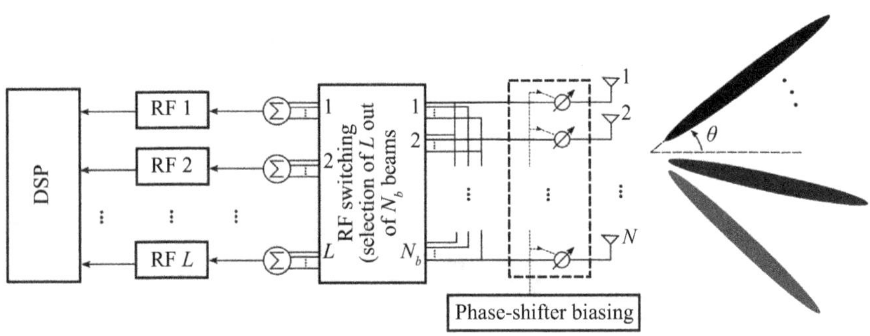

Figure 10.3 Phase-shifter-based analog or hybrid beamformer architecture. Hybrid beamforming when $N_{ant} > L$

A standard digital or hybrid beamformer relies on a weighting technique in which every antenna element's signal is independently multiplied by a weighting factor. This is done because when signal from one antenna is scaled compared to other antennas in an array, it can create a beamforming gain. Antenna weighting is difficult to manage accurately at mmWave and is generally done by programmable amplifier blocks that are located before the down conversion chain in the uplink. One such amplifier example is presented in [27] operating from 25 to 30 GHz and implemented on 64-nm CMOS (see Figure 10.4).

At the RF side, hybrid beamformers utilize analog beamforming hardware; however, it is a challenging task to select the optimum number of RF chains, switches, etc. for a given application scenario. In the current literature, this is decided by first analyzing hybrid beamforming as an optimization problem with a realistic formulation solvable in digital domain, e.g., as a minimum mean square problem. Techniques like [29] are then used to solve this problem, and as a result beam selection is assigned on the basis of the ratio between the number of users and the number of antenna elements required to serve the maximum number of users prescribed for a given scenario. Spectral efficiency in this case is reliant on the number of antennas, number of active RF chains, energy consumption and cell coverage area. The formulation of this kind is solved to find a "green point" that can be described as the maxima in the energy efficiency versus spectral efficiency

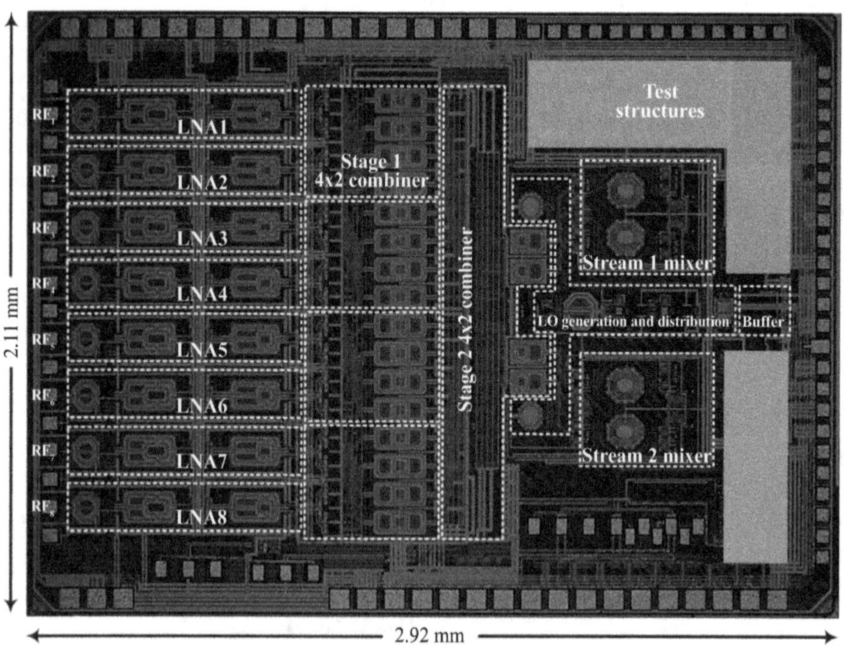

Figure 10.4 Die photograph of 25–30 GHz fully connected hybrid beamforming receiver. Reprinted, with permission, from [27]

graph. This approach is applicable for almost all hybrid beamformers and theoretically for any frequency [30]. Although accurate trade-offs between efficiency and energy consumption can be obtained, the solution does not consider the hardware-related challenges especially at mmWave spectrum where hardware impairments can be significant and should not be ignored [31,32].

Channel estimation using hybrid beamforming is also a challenging task, thanks to the amalgamation of digital and analog beamforming that delivers mixed signals to the digital signal processing (DSP) unit. Especially in mmWave systems, link availability and reliability are normally impaired due unfavorable propagation characteristics that can have a bad impact on the network service continuity. There is a requirement to identify the best beam pairs, and to do so, reference signal transmission schemes are required. Reference signals can serve several purposes, including beam alignment, channel state information exchange and initial access. Note that the purpose of reference signaling in mmWave channels is same as that of a cell-specific reference signal in Third-Generation Partnership Project (3GPP) LTE systems [33].

At sub-6-GHz spectrum, there is literature that shows that the reference signaling method can be used with hybrid beamformers for channel estimation, e.g., [35]. This method utilizes stochastic geometry; however, it is still not fully mature for use with mmWave hybrid beamformers. However, measurement campaigns in recent years are moving this line of research forward. One example is given in [36] where a reference signal is transmitted using highly directional antenna beamformers. The technique claimed to enable efficiency referencing in terms of angular coverage and yield the trade-off between channel information overheads and measurement accuracy. In a real-application scenario, this technique was used with a 28-GHz prototype depicting a high data rate in the downlink [37]. As per the description in [36], beam handover and switching at the base station end is simplified due to the directional nature of mmWave signals. Also, in addition to mitigating path loss, high-directivity reference signaling in mmWave beamforming can provide an assist with scattered signals such that the non-line-of-sight links can also be leveraged. An example of a high-directivity practical beamformer at 28 GHz, which provides high accuracy and precise phase shifting is given in [28] (see Figure 10.5(a)). While an evaluation study on a beamforming (receiver) array is shown in Figure 10.5(b) with digital baseband processing given in [34] wherein accurate phase shifting in a large-scale array is evaluated (Figure 10.6). Both systems in [28,34] are implementable with reference signaling [36] for mmWave channel estimation.

10.2.1.4 Lens-based hybrid beamformers

Most hybrid and analog beamforming strategies require a bank of phase shifters, power dividers and switches that are challenging to manufacture, especially in the mmWave spectrum. Lens structures are a plausible alternative since they inherently and simultaneously perform similar functions of power dividing, combining, phase shifting and even beam selection inside the lens structure.

A general architecture of lens-based beamformer is shown in Figure 10.7. There are a number of lens-based beamforming topologies and a comparative

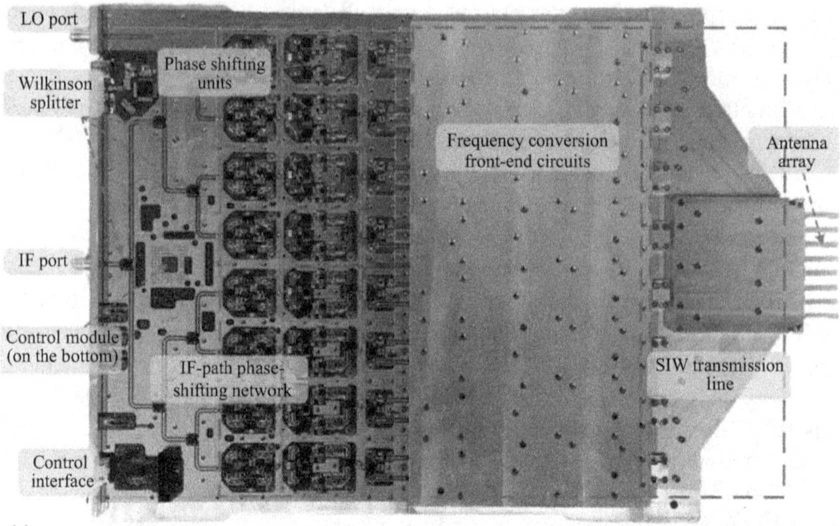

(a)

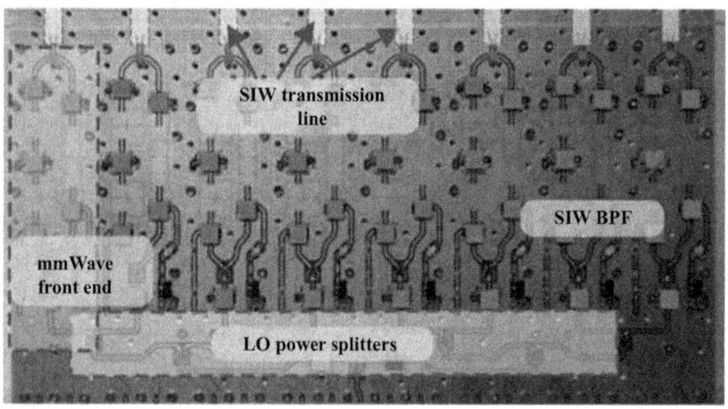

(b)

Figure 10.5 (a) Photograph of sub-array phase-shifting network and (b) Photograph of front-end circuits in one phased sub-array hybrid analog and digital beamforming transceiver system [28]

analysis is given in [38]. Lens-based beamformers rely on the lens-focusing property that is analogous to that of optical lenses. In spherical or semispherical optical lenses, an incoming beam of rays parallel to the lens's principal axis focuses at the so-called focal point [39]. In microwave and mmWave lens structures, a radiator placed at the focal point can create an outgoing electromagnetic signal traveling parallel to the principal axis. Theoretically, all sets of rays coming from infinity have an image at the focal point; however, practical lens structures have physical limitations. These limitations cause lens imperfections that end up creating spherical aberration, coma, chromatic aberration and astigmatism [40]. Regardless of the type of lens and the synthesis approach used in development, lens-based

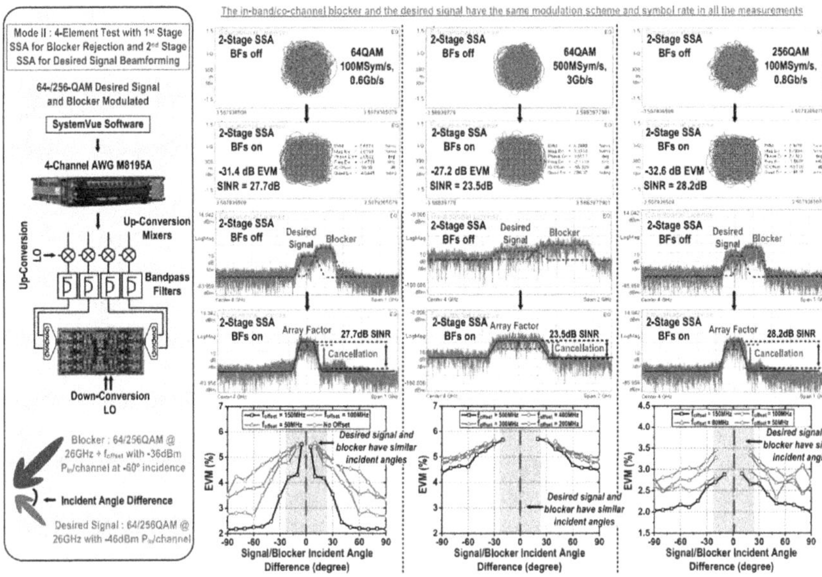

Figure 10.6 *Left-side hardware is a full field-of-view autonomous hybrid beamformer array chip and its scalability for phased array. The results in the right side demonstrate the blocker rejection and desired signal beamforming when both signals are broadband modulated at the same speed and scheme. Further information can be found in [34]*

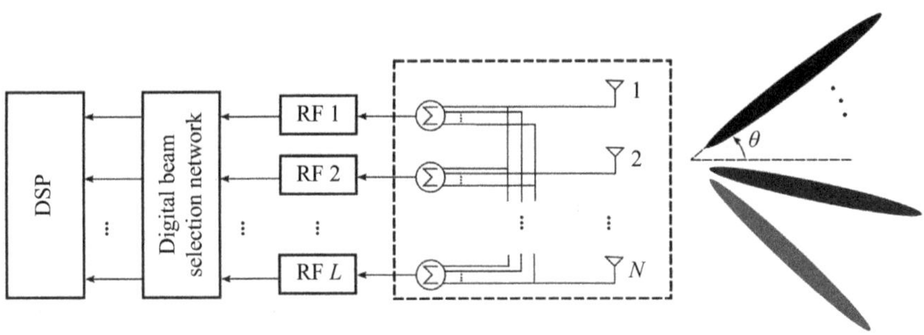

Figure 10.7 *Lens-based beamformer architecture when power dividing/ combining, and phase shifting happens at the lens structure*

beamformers are found to work well in the mmWave spectrum range when deployed for massive antenna array operation at the radio front end [39]. According to the study and conclusion in [41], a lens-based beamformer simplifies the signal processing effectively by exploiting the mmWave channel's angular sparsity [42].

This is extremely important in order to achieve high data rates with low latency required in "mmWave 5G and beyond" communication standards.

The most well-known lens-based beamformer operating in the microwave range was reported in 1964 and is known as Luneburg lens [43]. Since then, researchers worked on many lens variations, but the principle is the same as that of an antenna array beamformer, i.e., focus the radiated electromagnetic energy in one direction and reduce its leakage in directions where the signal is not required. Recently using transformation optics [43], researchers have formulated the lens refractive index grading in such a way that the grading principle enables the focusing of energy at the focal point. An example is a slab-based lens having properties of beamforming in azimuth direction given in [44]. A similar principle is shown in [45] where parallel plate spacing is varied along lens radius to achieve focusing. Another example is the conversion of 3D lens to 2D plane and achieving beamforming along azimuthal plane [46]. Transformation optics theory has been used to first test the lens focusing capabilities and then for its utility for sub-6 GHz and also for mmWave beamforming [38]. Antenna theory suggests that lens-based beamformers need to be electrically large to be able to ensure high-directivity beamforming so they are not preferred at low frequencies where wavelength is long.

However, since the lens size shrinks as we move higher along the frequency range, thanks to shorter wavelengths, lens-based beamformers work very well at mmWaves where size is no longer a core limitation. For example, a multilayer planar lens using metamaterial principles that use a planar substrate integrated waveguide (SIW) for specialized beamforming [47] has a practical size. Focusing capacity can also be achieved by arranging small-sized single unit cells in a lattice and replicating it along a body of lens. 3D monolithic periodic structures having low loss and uniform dielectric constant are shown to work well [48]. Nonperiodic drilling of holes changes the dielectric property of lens, and if controlled properly, this method has also been shown to work for lens-based beamformers [49]. Also, some classes of metasurface use subwavelength resonators (examples are given in Figure 10.8), and deployment of them over the lens profile can result in focused energy [50].

A special lens-based beamformer replicates the principle of phase shifters by using easy-to-develop Rotman lenses. The manufacturing cost of this type of lens-based beamformers is low, thanks to advancements in PCB development techniques and photolithography. For example, an SIW-based Rotman lens beamformer showing seven-beam selection capability was described in [51]. A more detailed investigation of a Rotman lens beamformer and its applicability in beamforming systems was given in [52], which showed that optimal beam selection can be done using this class of beamformer. Also, digital beamforming can further enhance interference reduction when deployed in conjunction with Rotman-lens-based beamformers [51]. A recent study has shown that a Rotman-lens-based beamformer with digital processing principles can work together to scan not only azimuth but also the elevation plane beamforming [42]. One example of Rotman-lens-based beamformer at higher frequencies is given in [53] where a 60-GHz module with

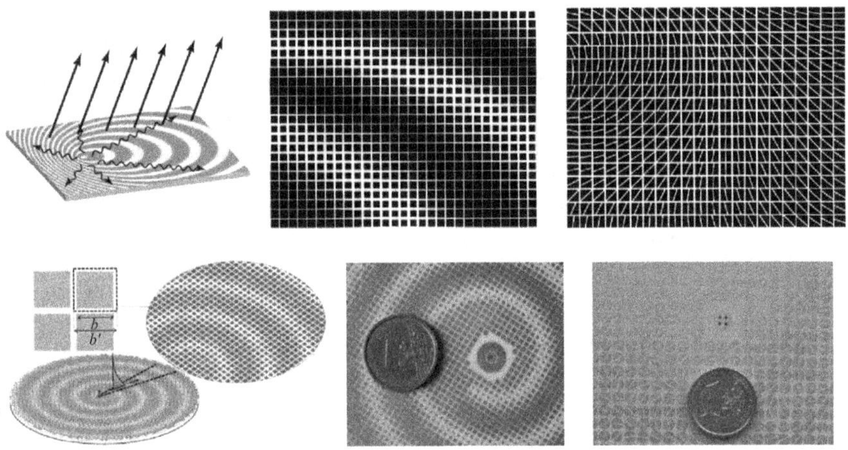

Figure 10.8 Multiple metasurfaces, sub-wavelength periodic and spiral structure modulating the required surface impedance to assist in beamforming [50]

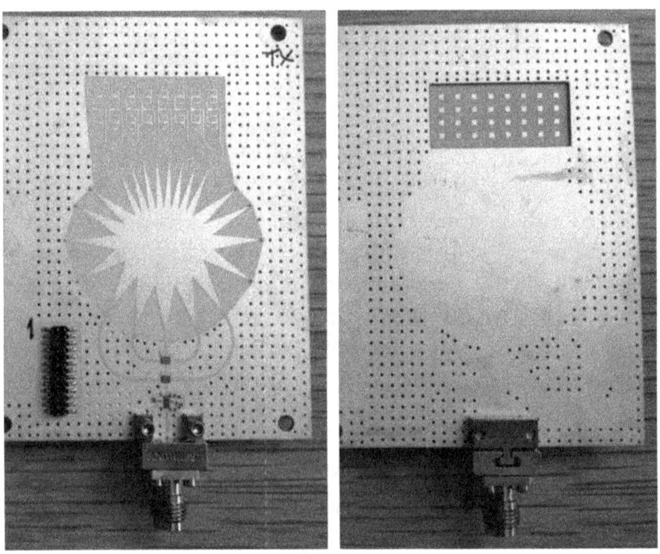

Figure 10.9 Photographs of the beam switching Rotman-lens-based beamformer with front and back view when back view has 4 × 8 antenna array [53]

integrated amplifier is shown to perform beamforming along the elevation and azimuth plane (see Figure 10.9).

Polarization control in lens-based beamformers is a challenging task since lens-based beamformers rely on geometrical positioning of feed antenna that normally supports only single polarization. Also, the frequency agility, multiband and multi-beam operation and signal amplification due to lens losses are some of the development challenges that need new innovative solutions. High complexity and sophisticated design approaches like the one in [54] have attempted to solve some of these challenges. The problem is that as we move from simple to complex development methods to achieve reliable beamforming, cost increases and may become even higher than those incurred when using standard antenna array analog beamforming methods.

10.2.2 Classification based on frequency

The second beamformer classification from an application point of view is based on the frequency of operation. In the previous section, we briefly touched on this classification also, but since the widely used MIMO and massive MIMO deployed systems are associated with both the sub-6-GHz and mmWave frequencies, it is important to highlight their differentiating characteristics with respect to beamformer development challenges. Sub-6-GHz frequencies are used for good network coverage since path loss is comparatively low. Also, the network coherence time is favorable at sub-6-GHz frequencies. The coherence time of the network is defined by an approximate time where the channel response is invariant [55]. It is defined by

$$T_0 \approx \frac{1}{f_d} \tag{10.6}$$

where f_d is the maximum Doppler frequency so we can say that there is an inverse relationship between the carrier frequency and the coherence time. In dense radio environments where the number of users is high, massive antenna arrays can, in principle, increase overall system efficiency by increasing the number of antennas [56]. Hence, already deployed base station infrastructure can often be uprated to implement sub-6-GHz massive antenna arrays. On the other hand, the efficiency of system at mmWave frequencies is increased by large available bandwidth but will only provide a shorter coverage area. The addition of large number of antennas in an array will add the beamformer gain, consequently helping the link budget as we discussed in the previous section. The major limiting factor for mmWave beamforming is noise power that is directly proportional to bandwidth so there is always a cap on the highest usable bandwidth at mmWave where more percentage bandwidth is available, which is not the case for sub-6-GHz usage. The following subsections will further detail the differences between sub-6-GHz and mmWave beamformers.

10.2.2.1 Beamformers at sub-6 GHz

Earlier investigations on massive MIMO were focused primarily sub-6-GHz frequencies and today, researchers associate the term massive MIMO technology largely only with the sub-6-GHz range. It is now well established that to achieve maximum theoretically achievable massive MIMO performance, digital beamforming is the preferred way forward for sub-6 GHz [57]. As mentioned previously, this requires each antenna element to be independently connected to the digital processing unit. The hardware required to support both digital and analog beamforming has advanced enough to handle such systems at sub-6 GHz. Here is an example in which hardware requirements for sub-6-GHz beamformers change as we move from digital to analog beamforming. Consider a 16-element antenna array performing digital beamforming at the receiver end. This requires each of the 16 element's received signal to have the capability to pass through a dedicated RF transmission line, and an ADC. Sixteen digital signal streams will then go into the DSP unit. Each signal stream will then be digitally added with a defined scaling factor or phase shift to get a resultant signal from the formed beam. In contrast to this, standard analog beamformers will require a scaling and phase shifting in the analog (RF) domain which will be done by 16 weighted phase shifters to get a resultant signal. Hence, the 16-element antenna array will require only a single ADC to convert this signal to digital data stream entering the DSP.

Since the mobility and number of users within a cell are high at sub-6 GHz, digital beamforming supports this operational scenario. Digital beamforming provides the necessary spatial multiplexing that can in turn enhance the spectral efficiency of the communication link. This multiplexing also compensates the low-bandwidth availability while keeping spectral efficiency as high as possible. Sub-6-GHz channels are rich in multipaths [6], which is another reason why multi-beam excitation in digital domain is preferred over the analog beamforming. Data rate in most propagation scenarios within the sub-6-GHz spectrum should be around 100 Mbits/s/user with reasonably uniform quality of service in a band of 40 MHz [6].

10.2.2.2 Beamformers at mmWave

The mobility at mmWave is also considered low and the number of users per cell is generally less compared to the number of users operating in sub-6-GHz cells [6]. It is generally recommended to have beamforming gain particularly for a signal or a small number of users so a portion of complexity of special multiplexing is moved to the analog domain suggesting the applicability of hybrid beamforming in the mmWave spectrum.

In recent years, measurement campaigns involving the mmWave beamforming reveal interesting information, i.e., the number of multipath required to serve users in mmWave cells is few [35]. Also, the multipath components are generally clustered in spatially independent sectors and a few beams radiating from the antenna array should be enough to serve a single user at a given time. Due to these reasons, in addition to large available bandwidth and a low number of users per cell, the spectral efficiency achieved by a mmWave beamformer is generally low [58].

Channel information is always important and becomes more difficult when fewer RF chains are used. Therefore techniques like power iterative adaptation [59] can be deployed, which when added to the benefit of mmWave sparsity allows channel estimation just by a few measurements. The disadvantage of this technique is the codebook burden that requires signaling and reconfiguration in every coherence time. As soon as the channel information is known to a beamformer, the sectoral mapping of the scatterers and receivers within a cell coverage area can further be optimized. This way, the known information will aid in increasing per user SNR values and interference signal. Load on the codebook is generally reduced by further micromanaging the mmWave channel into smaller clusters. An example of a user selection algorithm that is usable for multiuser cellular systems is reported in [60]. Additional difficulty is that by sub-clustering the channel, signal-to-noise-and-interference ratio will also start playing a role. Studies like [61] have tried to address this issue in which intercluster interferences are shared across all the clusters and algorithms manage the cluster allocation for a particular user dynamically. Further to this, algorithm 4 presented in [61] uses beam channel selection in multi-clusters in an attempt to reduce the complexity of channel estimation, and eventually the load on codebook. Another method of decreasing the codebook complexity of mmWave hybrid beamformer system is by using open-loop channel estimation given in [62] where decomposition of the clusters is defined using low-rank optimization problem. The channel estimation achieved using this method is suboptimal; however, it is stationary and usable for signal processing (refer to [62]). The technique is helpful at low SNR values compared to open-loop estimation techniques.

A summary here is that mmWave beamforming is not suitable for low data rate applications due to high network overheads, and mobility support in mmWave channels is also extremely challenging. Moreover, unlike sub-6-GHz beamforming that enjoys the spatial multiplexing of many users, mmWave beamforming can support high user density only when specialized forms of hybrid beamforming architectures are used.

10.2.3 Classification based on the use case

The overall purpose of beamforming is almost the same for all communication-related technologies; however, specific beamformer purpose varies from application to application. For instance, consider future mmWave 5G communication access points that can either be small-cell base stations, indoor wireless nodes or hotspots. The required beamformer in this case should be able serve a small number of users within a limited area so a high-gain multi-beam operation is desirable [39]. Conversely, an outdoor base station operating at sub-6-GHz 5G should be able to serve multiple users; here, a limited number of beams with high half-power beamwidth should suffice [63]. A beamformer that is a part of handheld mobile device (also referred to as "mobile user," and "user equipment" or simply "user" in literature) practically has a limited battery power. Since mmWave signal path loss is high, the beamformer gain should be enough to compensate for this path loss so that the transmitted signal form the handheld device is received by the mmWave beamformer with an acceptable SNR.

10.2.3.1 Fixed beamformers

A beamformer that is not a subject to mobility in a communication system can be classified as a fixed beamformer. Any classification of architecture described in

Section 10.2.1 can be used as a fixed beamformer to serve either single or multiple users. The type of antenna array to be used at the front end of a fixed beamformer purely depends upon the application and the frequency of operation. For low-frequency (sub-6 GHz) operations, generally a low number of antennas are enough to provide the service, so an easily realizable RF architecture should suffice. As we move higher in the frequencies, fixed beamformer architectures tend to demand an increased number of antennas, which consequently increase the complexity of the RF front end. Depending upon the application, in addition to beamformer gain, sometimes diversity gains like polarization and spatial are also required. There have always been concerns regarding the use of mmWave spectrum for fixed beamformers in a multiuser or a cellular communication system. According to the Friis transmission equation [64]:

$$P_r = P_t + G_t + G_r + 20\log\left(\frac{c}{4\pi f \times r}\right) \tag{10.7}$$

where P_r is the received power in dBm, P_t is the transmitted power, G_t and G_r are the transmitter and receiver antenna gains and r is the distance between the point of received power P_r and the radiating antenna.

The received power at r is inversely proportional to the frequency of operation squared, when an ideal isotropic radiator is considered, i.e., $G_t = 1$. At mmWave, antennas with G_t and G_r much greater than unity are typically used and the antenna gains are proportional to the square of the frequency of operation when a given aperture is to be used for the radiation. This allows the beamformer to yield gain to the antennas operating at mmWaves. To substantiate this, measurement campaigns are discussed in [65] in which same aperture area is used to host an antenna operating at 3 GHz and an antenna array operating at 30 GHz. In the experiment, a single microstrip patch antenna is used at 3 GHz while an array of patch antennas is designed on the same substrate size that was used to host a 3-GHz patch antenna.

Both antennas were placed in anechoic chamber such that the Friis transmission equation and the aforementioned argument can be verified. The experiment verified same amount of propagation loss regardless of the operating frequency when same physical aperture is used to host 3- and 30-GHz antennas. In addition to this, when a 30-GHz array antenna is used at both ends of the communication link, (consequently increasing the G_t and G_r), the measured received power was 20 dB higher than was achieved at 3 GHz. Other than these controlled measurement campaigns, outdoor measurements like the ones presented in [66,67] verify the applicability of mmWave frequency bands in fixed outdoor and indoor beamformers. The following two challenges still exist when it comes to development and deployment of such beamformers:

1. mmWave radio signals have propagation characteristics such as high penetration loss, foliage losses and losses due to precipitation. Although free-space path loss can be compensated by adding the beamformer gain at the fixed beamformer end, the channel characteristics and losses are uncontrollable.

2. Single-antenna radios are simple to design; however, for multiple antenna beamformers, when the number of antennas should be high enough to cover a comparatively large aperture area, the feed network and associated RF electronics require high precision. This can increase cost.

Consider again the example given in Figure 10.10. Although it has been shown that the beamformer gain is practically viable for a communication system implementation at mmWave spectrum, the beamforming array development all the way from antennas to the digital signal processing unit is again complex. Research efforts are focused on ways to reduce the cost and complexity of entire mmWave beamformer implementation. Classical phased arrays like the ones presented in [68] are one of the most commonly used antenna feed networks in fixed beamformers. A high-gain 16-element beamformer created using stacked patch antenna is given in [69]. This beamformer operates in 24.35–31.23 GHz band with a gain of 18.7 dBi. In addition to this, Butler matrix [70], Rotman-lens-based [42] and SIW-based feed network like the ones presented in [71] are commonly used fixed beamformer feed networks.

A frequency-selective surface (FSS) capable of assisting in beam tilt for a fixed beamformer with the frequency of operation 28–31 GHz is given in [72]. Generally, such FSS structures are scalable and higher frequency operation can also be achieved using the proof-of-concept module given in [72]. A positive–intrinsic–negative (p–i–n) diode-based beam switching system operating at 28 GHz is elaborated in [73] that can be used as a fixed beamformer. The maximum beam tilt achieved is shown to be around 45°. Similar to this, another fixed beamformer for the operation at 28-, 38- and 48-GHz 5G bands is discussed in [74]. An octagonal prism-like configuration of this array has been shown to achieve a radiation

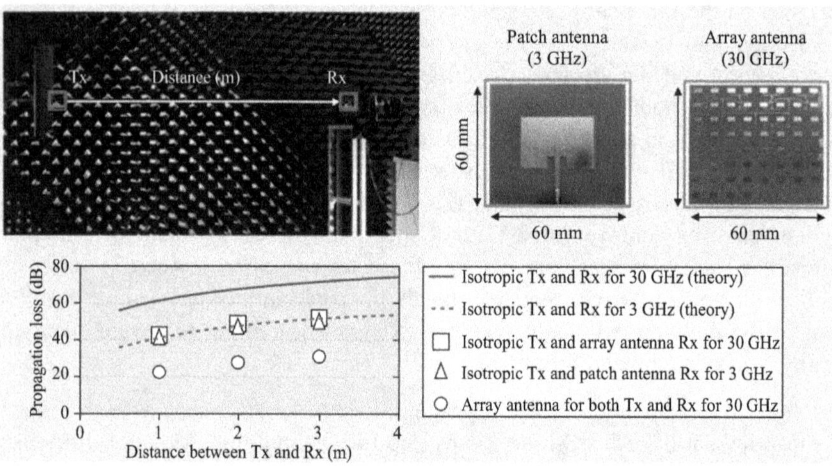

Figure 10.10 Verification of propagation losses using Friis equation and antenna apertures at 3 and 30 GHz [65]

efficiency of around 82%. This antenna array is specifically designed for 5G cellular base station. The antenna array gains in this system are around 7.8, 8.3 and 7.7 dBi at the centers of the three operating frequency bands.

Another subclassification of fixed beamformer is based on the ability of a beamformer to scan along either one or two planes. Fixed beamformers that are only capable of beam scanning along one plane generally use uniform linear arrays with phase shifters connected to each antenna. The illustrations in Figures 10.3 and 10.7 show uniform linear arrays in which beam can only scan along either horizontal or vertical planes. Also, the practical beamformer example given in [53] is capable of beam scanning along one plane. The drawback of single plane beamformer is that in a multipath channel, a significant amount of diversity information is lost since the fixed beamformer is operating along a single plane only. This is mitigated by using a fixed beamformer that utilizes multiple layers of uniform linear arrays forming a rectangular antenna array with a bank of phase-shifting network that can help in scanning two planes. This classification of beamformer is referred to as "full dimensional" in communication and multiple antenna system-related literature. The most important benefit of using full-dimensional fixed beamformers is when advanced beamforming technologies are used to eliminate interference. An example that compares the use of a uniform linear array and uniform rectangular array as a fixed beamformer is given in [42]. The end-of-end spectral efficiency of a fixed beamformer listening along two planes is shown to be better in terms of end-to-end spectral efficiency compared to the fixed beamformer listening to only along a single plane at uplink scenario. This is because by beam scanning along two planes, nonuniformly distributed users can be served using higher resolution. An example of full-dimensional precoding that yields the benefits of multiple antenna system is provided in [75]. This work also shows a two-layer phase shifter feed network architecture for mmWave use.

A significant amount of work is required to develop the mmWave RF hardware design required to facilitate complex fixed systems like full-dimensional beamformers and multilayer phase shifters. A pattern diversity fixed beamformer for cellular base station was discussed in [76] in which an antenna aperture was engineered for operation in the 27–30 GHz band and two fixed beams. The measured gain ranges from 7 to 8 dBi, while the mutual coupling is shown to be better than −18 dB at the center of the band of operation. Dual-polarization capability was added into beamformer by using zero-index metamaterial. By doing so, multiple-beam equalization is achieved. The end-fire antenna gain of polarization orthogonal ports was reported to stay between 9.2 and 9.6 dBi. Another way of beamforming in the fixed antenna array case is by using electromagnetic-bandgap-(EBG) backed structures. One example is provided in [77] where an EBG-backed structure is operating from 26 to 32 GHz. An example of a fixed beamformer that can be mounted onto a ceiling and is designed using tapered slot antennas is given in [78]. The antenna unit cell uses dielectric loading principle with metamaterial and tapered slot operates at the same frequency bands. The aperture and antenna efficiencies are shown to be 73%. This beamformer operates from 27 to 29.8 GHz. The beamformer radiations have pattern integrity and a gain of 9 dBi. The study shows an addition calibration after computing the path loss and updated version has beam-scanning capability of 45° with a higher gain of 12 dBi. A stacking topology is shown in [77] that uses the principles of artificial magnetic conductor and helps in the pattern diversity and antenna gain of 11.9 dBi. A planar Yagi–Uda antenna uses two parallel T-junctions form parallel antenna feeding resulting in circular polarization.

In recent years, photonic beamforming has also drawn significant attention especially for mmWave 5G transceivers. The main concept behind photonic beamforming is the use optical fiber for beamformer functions like beam scanning and beam switching. A wavelength-division multiplexing system is shown in [40] in which signals corresponding to multiple beams are routed within a multiantenna system. An RF signal of 15 GHz is utilized as a carrier frequency and a high-order modulation scheme is formulated. A similar system is shown in [79] where the wavelength of the operating frequency modifies the coupling coefficient of optical ring resonators and consequently multi-beam operation from 17.6 and 26 GHz is realized. A multicore fiber is used to feed multiple antennas with 128-QAM (quadrature amplitude modulation) signal using around 4 GHz of operational bandwidth and is shown to demonstrate experimentally as an mmWave beam-forming system. Beam steering is controlled by a central computer and fast steering is done by changing the wavelengths of the optical lasers and beam steering is shown to be 11.3°–23° (at 26 GHz the beam steering is 23°). The experiments are done for 5G NR standards in Europe. The beamformer can provide around 16 GB/s per user. Another study shows the use of phase shifter and Mach–Zehnder interferometer in the antenna array [80] capable of being used in fixed beamformer setups. This work shows that it is possible to successfully translate optical phase shifts to mmWave phase shift (at 28.5 GHz) using an amplitude-modulated optical signal.

Lens-based architectures as discussed in Section 10.2.1.4 are also viable for fixed beamformers. RF energy focusing structures like Luneburg, spherical, semi-spherical, gradient dielectric, constant dielectric and Rotman lenses are gaining a lot of attention as fixed beamformers, especially in mmWave spectrum. Primarily due to their focusing capability and ease in manufacturing, they are sometimes preferred over phased-array-based beamformers [39,81–83]. A simple 3D-printed Luneburg lens [82] has been shown to have a gain value of 21.2 dBi. The lens structure is built using rods that have gradient index and realized the required permittivity distribution as per Luneburg principle. The radiation efficiency of the lens is shown to be 75%. A magnetoelectric dipole antenna is used as a feeder of the lens structure, which can cover around 40% bandwidth at Ka-band. When different feeding points are excited on the same lens structure, measured mutual coupling as low as -17 dB is achievable. The lens is capable of beam scanning along $\pm 61°$ making it a good candidate for fixed beamformer application. Another example shows a different approach in which designing of a single-polarization Fresnel-zone plate lens antenna is discussed [84]. The operational band of the lens feed is 57–64 GHz. Antenna has characteristics, including trans-reflection and twist-reflection. By carefully yielding the benefits of these characteristics, the antenna is shown to have a maximum gain of around 32 dBi. Similar circular, semicircular and elliptical lens structures require multiple feeding points to be able to host multiple beams simultaneously. This requires multi-feed excitors like waveguide-based horn antennas [39], groove gap waveguides [85,86] and dipoles [82]. A specialized form of similar feed uses leaky-wave antenna operating at 21.9–23.9 GHz band. The feed structure comprises a metallic strip that has a grating structure and the end of this strip can be placed at the focal point of a lens-based beamformer [87].

In addition to 3D lens structures that require antenna feeds, Rotman and Fourier lenses are classified as 2D lenses in which the lens operation, or RF energy focusing action, happens in transmission-line domain [42]. Such lenses perform same operation of superimposing the radio signal in a transmission as it happens in a Butler matrix [70]. Recent example of a flexible Rotman lens that can be used for mmWave fixed beamformer application is shown in [83]. The passive lens structure can generate multiple beams at 28 GHz and a switching network performs the beam selection. Rotman and Fourier lenses typically have a very wide bandwidth; however, low coupling between multiple beams can be achieved at narrow bands [88]. A Rotman lens beamformer in [81] has a bandwidth from 26 to 40 GHz. The beam-scanning range in this beamformer is $\pm 39.5^\circ$ so it is usable in fixed beamforming scenarios. In [83], the Rotman is shown to have a bandwidth of 18–38 GHz. Another Rotman lens example at high frequencies is in [89] in which the lens operating frequency is from 25 to 31 GHz while generating seven beams.

It is expected that fixed beamformer operation may move to even higher frequencies post the successful implementation and deployment of 28 and 38 GHz 5G bands. Researchers have already pushed forward their efforts to facilitate these advancements. An example of parallel fed slot antenna beamformer is shown in [90], which has inherent mechanism of beam selection and uses low-temperature cofired ceramic technology for the array structure development. The operation band of this 11 beam beamformer is from 57 to 66 GHz, and experiments have verified high-accuracy beam scanning from -39° to $+39^\circ$. Another example of higher frequency beamformer is [84] operating at 60-GHz band. In another work, a novel approach of modeling and measuring channel characteristics using synthetic beamwidth at a fixed transceiver end is given [91]. This work is specific for indoor 60-GHz fixed beamformer.

Other than performing the beamforming to serve users, it is vital for practical fixed beamformers to understand and know the characteristics of the channel it is operating in. For this purpose, state-of-the-art algorithms and protocols are required [92]. As mentioned in previous sections, the classical channel models for sub-6 GHz may not be adequate for mmWave channels, so a new line of research has emerged in which novel methods are used at the fixed beamformer end for channel sounding and user location [92,93]. The channel information and user location are very important for successful beam alignment in order to yield the best data throughout. Localization algorithms designed for mmWave fixed beamformers should have fast initiation, and their runtime phase should be as short as possible [92].

Localization protocols, e.g., in [94], utilize in-band aiding in a standard mmWave geometric channel model. This protocol is shown to work well but imposes extra burden on the system. A new approach of user localization has shown to use the global positioning system (GPS) location data in conjugation to the protocols running purely at mmWaves for localization accuracy [75]. This approach simplifies localization problem by first reducing the size of beam numbers in beam training step by eliminating those beams that are not directly involved in user service. This helps reduce beam training overheads. In the next step, the protocol finds the beam that provides the best coverage, reliability and fastest link to the user. Further details about this protocol can be found in [75]. Another similar work shows an approach of radio-environment mapping of the wireless network, which helps in identifying the location of users, blockers and scatterers [95]. It is argued that even with successful radio-environment mapping, the likelihood of accurate localization is low. This is because of the requirement of processing of a likelihood function designed to find an unknown channel parameter vector can have many local maxima, making it difficult to find an optimum global solution. To achieve the required accuracy, a filter-based estimator at the fixed beamformer is proposed in this study [95].

Some of the channel estimation and user localization schemes that work well in low-density environments may not operate well for practical mmWave fixed beamformers. One way around this problem is increasing the number of fixed beamformers in an environment where the number of obstacles and users is high. In this way, the likelihood of hindrance between fixed beamformer and users can be decreased. There are two challenges associated with this approach:

1. Increasing the density of fixed beamformer or base stations for mmWave 5G cellular will increase network deployment cost.
2. The requirement of additional training overheads may overload the network.

One method as proposed in [96] tried to overcome these by introducing scalability into the high-density environment. It was shown that adaptive capabilities can be added into the localization protocol that can assist in functions like immediate reaction to changes in a channel and finding secondary propagation path for a blocked beam link. Also, the approach suggested the utilization of information of user location from out-of-band frequency applications such as GPS (like the ones presented in [75]) is also possible. This approach is scalable and is user-driven; hence, it scales well as the number of users increases in a dense environment. In addition to this, it is reported that any sensor information from a user mobile device is usable for localization purposes. The useful sensors in mobile device include all sensors that can share data with physical and MAC layers of the communication protocol running within the mobile device [96]. Although this approach is shown to improve the reliability of the mmWave communication standards, it slows down the beamformer functionality and reduces the capacity predictions that a theoretical mmWave antenna array system provides.

In summary, the out-of-band applications for the localization application in mmWave fixed beamformer have following advantages.

1. Network overheads required for operations like beam link training, handovers, load balancing and device tracking can be reduced.
2. Location-aided services can be included into the network without overloading the fixed beamformer hardware.

10.2.3.2 Variable beamwidth fixed beamformer

In an mmWave communication link, beamformer beamwidth has a vital role to play. Consider now a sequence of events that can explain the importance of beamwidth. Imagine a fixed beamformer mounted at the base station of an mmWave communication cell sector. The beamformer can generate high-directivity beams, sharp enough to serve users after mitigating the path loss and after surpassing multiple scatterings. Fixed beamformer theory associated to the large mmWave antenna array systems suggests that the first stage of mmWave network requires a beam training session during which fixed beamformer transmits the information that is received by a mobile terminal. The mobile terminal responses back with the information that contains the index of the best reception beam among all candidate beams. After receiving this information from all mobile users, the base station pairs beams with the corresponding users for the data transmission. During this procedure, mobility within the channel can be a deal breaker and every update in channel will require new beam pairing. This entire process requires a large network training overhead, and since training events require the same bandwidth allocated for signal transmission, these will reduce the achievable spectral efficiency in an mmWave channel. One way is to widen the beamwidth at the fixed beamformer end and serving a set of mobile users with the same beam rather than with individual beams serving individual users. This method can increase the spectral efficiency by reducing the training overhead. It can also add some relaxation in network state in terms of mobility. An example of a non-orthogonal multiple access system that can aid in enabling multiple mobile user service using same beam of a fixed beamformer is given in [97].

The trade-off between the beamwidth and network overheads depends upon the required number of service beams, cell size, number of mobile users, network mobility and coherence time. RF planning and engineering is required for specific use cases, area of coverage as well as the operational mmWave frequency band. One example that illustrates the aforementioned trade-off is given in [98] in which, 28- and 38-GHz fixed beamforming and planning is experimentally performed for an urban environment. It is shown that beamwidth compromise can be more useful when the feasibility study of propagation loss in mmWave 5G communication system is done beforehand. After this quantification, beam number and beamwidth figures can be optimized until the time-targeted spectral efficiency is achieved. In this study, the importance of beam misalignment due to beamwidth reduction and its impact on the power loss is also shown. It is also shown that the signal bandwidth of around 1 GHz is required for fixed beamformer to be able to match the theoretical capacity [99].

The beamwidth of the radiating mmWave beamformer depends upon the coverage area and specific location of mobile users. For instance, consider a fixed beamformer at base station serving in a dense urban environment. The number of mobile users required to be served on roads, streets, vehicles and shops will be higher as compared to the number of users in high-rise buildings. If the same beamformer with a universal fixed beamwidth is used to serve all mobile users, there will be wastage of spectral resources, which is expensive at mmWaves. Such challenging scenario requires fixed beamformer to be flexible in terms of beam-width, and research efforts like [75] have shown a possible solution. In this study a variable (nonuniform) beamwidth full-dimensional beamforming system uses narrow beamwidth directive beams to serve high-density mobile user areas like streets and uses wider beamwidth beams to serve multiple mobile users in buildings, etc. Such a scenario is illustrated in Figure 10.11.

The study in [91] reveals interesting information about the beamwidth of a fixed beamformer. The signal delay spread, and the number of multipath components reduces with the reduction in beamwidth of a beamformer. It also experimentally shows that the reduction in beamwidth required to make the beams narrower can provide up to 20-dB improvement in gain, yielding a directionality benefit to the link budget at mmWave. It is importation to note that low beamwidth in a fixed array does not necessarily imply a high number of antenna elements with associated RF electronics. Using array synthesis techniques, including compressive sensing algorithms (like the ones presented in [11,100]), a considerably fewer number of antenna elements with predefined complex weights can be used to achieve a beamwidth in a sparse antenna array format that is comparable to the beamwidth achievable by a fully populated large antenna array.

10.2.3.3 Mobile beamformers

A beamformer that is mounted to a device that is mobile can be classified as a mobile beamformer. This includes all wireless devices that are not a fixed beamformer and require an inbuilt beamformers to establish communication.

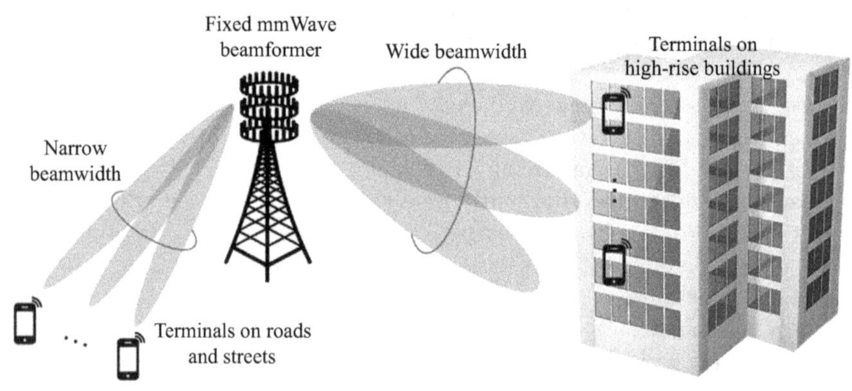

Figure 10.11 Illustration of a full-dimensional nonuniform fixed beamformer at mmWave base station array serving users in a dense urban environment

A quasi-omnidirectional single-antenna transmitter/receiver is not considered as a mobile beamformer since it cannot yield a high-directivity beamforming gain. With this description, mmWave devices (like handheld phones, laptops, vehicles and wearable gadgets) that may contain antenna arrays or lens antennas capable of establishing single of multiple-beam-based communication with other beamformers can loosely be classified as mobile beamformers. The 3GPP has identified the spherical coverage area requirements for mmWave mobile beamformers that are based on the cumulative distribution function on the effective isotropic radiated power [101]. At sub-6-GHz frequency bands, antenna radiators with low directivity are enough to provide a complete 360° coverage, while this is not necessarily the case when it comes to mmWave mobile beamformers. The isotropic radiated power from a signal antenna at mmWave frequencies is typically not enough to establish reliable communication with other mobile beamformers or fixed beamformers, so beamformer gain is a requisite. According to [98], the directivity of beams generated by the mmWave beamformer can decrease the capturable number of multipath components, so there is always a trade-off between multipath resolution and free-space penetration when it comes to designing a reliable mobile beamformer.

In the mmWave spectrum, mobile beamformers are generally made reconfigurable in which the beam selection is controlled separately. In doing so, it is possible to select of the best beam covering suitable direction around the mobile beamformer for a reliable communication. In sub-6-GHz bands, a low number of antennas operating at the same frequency band are enough to provide the required coverage around handheld mobile devices. In mmWave frequencies, energy focusing and directive radiation are requirements due to several reasons discussed in previous section; hence, an antenna array is required. Sometimes, a single-antenna array can only cover a portion of 360° coverage area so more than one antenna arrays are required, placed at multiple locations on handheld device (Figure 10.12). An example of reconfigurable beam switching network is given in [102] where 16-element high-directivity antenna array having 1.5-GHz bandwidth is shown. The phase control network in this array is experimentally validated and as a result five beam states are achieved. In addition to this, beam splitting is also shown that can assist in realizing multipath signal reception. Another example of a phased array for 5G handset using 40-nm CMOS technology is shown in [103]. The proof-of-concept module operating from 27 to 30 GHz is developed with standard block-level specifications and the hardware performance is thoroughly investigated. A carrier-aggregated 64-QAM OFDM is tested on a 1P6M bulk CMOS-based prototype to verify the reliability of the communication link. Experiments show a noise figure of around 5.5–6.0 dB and a third-order intercept point of −6 to −8.5 dB.

Some of the recent works on mobile beamformers show applications of multi-stream transmission in which multiple frequencies are used for multiple data streams. This is done by multiband operation that uses cartesian-combining-based complex weighting factors. It is shown in [104] that with the image-rejection scheme, a heterodyne beamformer can be realized. Also, reconfigurability can also

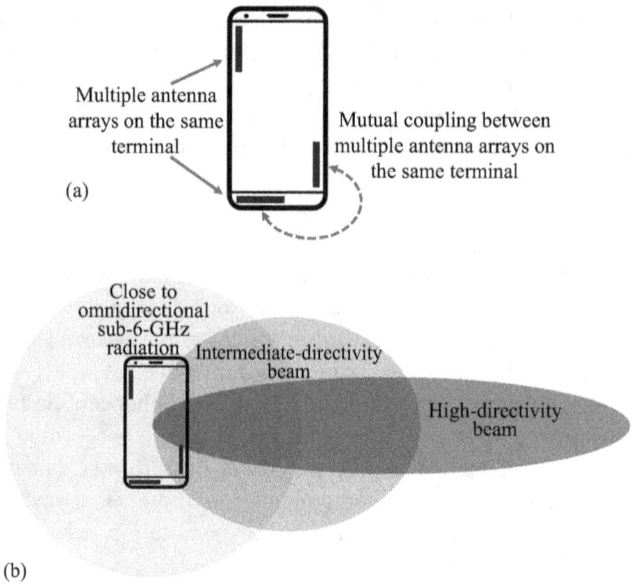

Figure 10.12 (a) Illustration of a handheld mobile device supporting multiple mmWave mobile beamforming arrays while trying to achieve complete 360° coverage. (b) Comparison of a low-directivity radiation requirement at sub-6 GHz bands with intermediate- and high-directivity requirement at mmWave frequencies

be added by which two bands of 28 and 37 GHz can be used. The technology used in receiver side is 65-nm CMOS. The conversion gains at the receiver end are shown to be 5.7 and 8.5 dB at 28 and 37 GHz.

It is now believed that large antenna array systems in conjugation with the digital beamformers are best suited for the sub-6-GHz frequency bands. This conclusion is yet to be further explored when it comes to mmWave spectrum. A lot of research is going on in finding the best beamformer architecture at mmWave spectrum, e.g., [21], where a fully digital 64-channel radio is shown to work at 28-GHz carrier frequency. Another fully digital beamforming scheme implemented on a software-defined radio is given in [105]. This work uses slot antennas with a bank of low-noise amplifiers connected to eight RF chains. This hardware has been shown to achieve a phase shifter resolution of around 0.72°. The in-band signal that has no utility in communication is utilized to correct the errors in phase and amplitudes, hence calibrating each RF chain. As a result, the amplitude and phase errors are confined to below 0.5 dB and 0.9°, respectively. At higher frequencies, a comparative study between hybrid and fully digital beamforming schemes is given in [20] wherein the 60-GHz band is discussed.

In mobile beamformers, phase errors can easily occur; this can happen because of placement of multiple mobile beamformers on the same handheld platform, tempting undesirable mutual coupling (Figure 10.12(a)). An innovative beamforming technique is shown in [106] where frequency-modulated diverse array is shown. In such an array, small increments in frequency in a time-dependent signal can generate multiple beams without the necessity of phase shifter network. Handheld mobile devices with mmWave beamformers are likely to face energy leakages due to the presence of human body (hand, head, etc.). Some studies have suggested the confinement of electromagnetic energy within the structures like coplanar waveguides, dielectric slabs and plasmonic structures [107]. These methods are helpful in transmitting the energy from mmWave feeds to the antenna front ends and then to the required mobile direction; however, the signal quality becomes weaker because of the slow-wave propagations and spoof surface plasmon polarization effects. A novel approach is given in [108] by which bounded spoof surface modes are translated to radiation modes in mobile beamformer. A comparison with a horn antenna is shown in this study, which proves the proposed approach. Moreover, it is shown that a radiation efficiency of close to 90% and beamformer gain of 15 dB is achievable. Further to this, the applications of mobile beamformers for beyond 5G are presented in [109].

A detailed comparison of beamformer performance on multiple frequencies is given in [37]. The comparison is focused on 2.9, 29 and 61 GHz and several real-life scenarios are discussed, e.g., shopping center, indoor office and outdoor. The propagation losses because of the materials, buildings and walls are thoroughly investigated for each carrier frequency. This study is very helpful for beamformer gain understanding and about how the channel interacts with the signal radiated from a fixed or mobile beamformer operating at multiple carrier frequencies from 2.9 to 61 GHz.

10.3 Conclusion

In this chapter, we first discussed the definitions that are vital to understand the working of a beamformer. We then classified the beamformer based on the architecture, frequency of operation and a use case. These clarifications are done keeping in mind the future technological advancements in the communication industry. Beamformer architectures are further divided into purely analog, digital and hybrid types, when each one of them has a specific need in specialized communication standards. Frequency bands of operations are divided into the well-known 5G subbands that are sub-6-GHz and mmWave bands. We further discussed the ways in which a beamformer function differs when they are operating at different frequency bands. Lastly, we classified beamformers in terms of their utility as a fixed or mobile radio in a communication system. State-of-the-art beamformer examples are comparatively analyzed to better predict the most suitable choice for a given classification of beamformers in the 5G and beyond applications.

References

[1] C. A. Balanis, *Antenna theory: Analysis and design*. John Wiley & Sons, 2016.

[2] D. Jiang and G. Liu, "An overview of 5G requirements," in *5G Mobile Communications*, Springer, 2017, pp. 3–26.

[3] J. Gozalvez, "New 3GPP standard for IoT [mobile radio]," *IEEE Veh. Technol. Mag.*, vol. 11, no. 1, pp. 14–20, 2016.

[4] M. Passoja, "5G NR: Massive MIMO and Beamforming–What Does It Mean and How Can I Measure It in the Field," https//www.rcrwireless.Com/20180912/5g/5g-nr-massive-mimo-and-beamformingwhat-does-it-mean-and-how-can-i-measure-it-in-the-f., 2018.

[5] D. Qiao, X. Yang, W. Tan, C. Wen, and S. Jin, "Low-cost massive MIMO: Pilot length and ADC resolution." *J. Commun. Inf. Netw.*, vol. 3, no. 2, pp. 23–36, 2018.

[6] E. Bjornson, L. Van der Perre, S. Buzzi, and E. G. Larsson, "Massive MIMO in sub-6 GHz and mmWave: Physical, practical, and use-case differences," *IEEE Wirel. Commun.*, vol. 26, no. 2, pp. 100–108, 2019.

[7] C.-G. Gheorghe, D. A. Stoichescu, and R. Dragomir, "Latency requirement for 5G mobile communications," in *2018 10th International Conference on Electronics, Computers and Artificial Intelligence (ECAI)*, 2018, pp. 1–4.

[8] V. F. Fusco, *Foundations of antenna theory and techniques*. Pearson Education, 2005.

[9] J. Fridén, A. Razavi, and A. Stjernman, "Angular sampling, test signal, and near-field aspects for over-the-air total radiated power assessment in anechoic chambers," *IEEE Access*, vol. 6, pp. 57826–57839, 2018.

[10] W. Fan, P. Kyosti, M. Rumney, X. Chen, and G. F. Pedersen, "Over-the-air radiated testing of millimeter-wave beam-steerable devices in a cost-effective measurement setup," *IEEE Commun. Mag.*, vol. 56, no. 7, pp. 64–71, 2018.

[11] M. A. B. Abbasi, V. Fusco, and D. E. Zelenchuk, "Compressive sensing multiplicative antenna array," *IEEE Trans. Antennas Propag.*, vol. 66, no. 11, pp. 5918–5925, 2018.

[12] C. E. Shannon, "A mathematical theory of communication," *Bell Syst. Tech. J.*, vol. 27, no. 3, pp. 379–423, 1948.

[13] J. Lee, E. Tejedor, K. Ranta-aho, *et al.*, "Spectrum for 5G: Global status, challenges, and enabling technologies," *IEEE Commun. Mag.*, vol. 56, no. 3, pp. 12–18, 2018.

[14] Z. Zhang, Y. Xiao, Z. Ma, *et al.*, "6G wireless networks: Vision, requirements, architecture, and key technologies," *IEEE Veh. Technol. Mag.*, vol. 14, no. 3, pp. 28–41, 2019.

[15] V. Venkateswaran and A.-J. van der Veen, "Analog beamforming in MIMO communications with phase shift networks and online channel estimation," *IEEE Trans. Signal Process.*, vol. 58, no. 8, pp. 4131–4143, 2010.

[16] S. Darzi, T. S. Kiong, M. T. Islam, H. R. Soleymanpour, and S. Kibria, "A memory-based gravitational search algorithm for enhancing minimum variance distortionless response beamforming," *Appl. Soft Comput.*, vol. 47, pp. 103–118, 2016.

[17] M. Behjati, M. Ismail, and R. Nordin, "Limited feedback MU-MIMO in LTE-A system: Concepts, performance, and future works," *Wirel. Pers. Commun.*, vol. 84, no. 2, pp. 935–957, 2015.

[18] J. Gozalvez, "Fifth-generation technologies trials [mobile radio]," *IEEE Veh. Technol. Mag.*, vol. 11, no. 2, pp. 5–13, 2016.

[19] E. Björnson, L. Sanguinetti, J. Hoydis, and M. Debbah, "Designing multi-user MIMO for energy efficiency: When is massive MIMO the answer?," in *2014 IEEE wireless communications and networking conference (WCNC)*, 2014, pp. 242–247.

[20] K. Roth, H. Pirzadeh, A. L. Swindlehurst, and J. A. Nossek, "A comparison of hybrid beamforming and digital beamforming with low-resolution ADCs for multiple users and imperfect CSI," *IEEE J. Sel. Top. Signal Process.*, vol. 12, no. 3, pp. 484–498, 2018.

[21] B. Yang, Z. Yu, J. Lan, R. Zhang, J. Zhou, and W. Hong, "Digital beamforming-based massive MIMO transceiver for 5G millimeter-wave communications," *IEEE Trans. Microw. Theory Tech.*, vol. 66, no. 7, pp. 3403–3418, 2018.

[22] R. M. Vaghefi, G. Miranda, R. Srirambhatla, *et al.*, "First commercial hybrid massive MIMO system for sub-6Hz bands," in *2018 IEEE 5G World Forum (5GWF)*, 2018, pp. 357–362.

[23] S. Noh, M. D. Zoltowski, and D. J. Love, "Training sequence design for feedback assisted hybrid beamforming in massive MIMO systems," *IEEE Trans. Commun.*, vol. 64, no. 1, pp. 187–200, 2015.

[24] D. Ying, F. W. Vook, T. A. Thomas, and D. J. Love, "Hybrid structure in massive MIMO: Achieving large sum rate with fewer RF chains," in *2015 IEEE International Conference on Communications (ICC)*, 2015, pp. 2344–2349.

[25] S. Hur, T. Kim, D. J. Love, J. V Krogmeier, T. A. Thomas, and A. Ghosh, "Millimeter wave beamforming for wireless backhaul and access in small cell networks," *IEEE Trans. Commun.*, vol. 61, no. 10, pp. 4391–4403, 2013.

[26] E. Vlachos, J. Thompson, M. A. B. Abbasi, V. F. Fusco, and M. Matthaiou, "Robust estimator for lens-based hybrid MIMO with low-resolution sampling," in *2019 IEEE 20th International Workshop on Signal Processing Advances in Wireless Communications (SPAWC)*, 2019, pp. 1–5.

[27] S. Mondal, R. Singh, A. I. Hussein, and J. Paramesh, "A 25–30 GHz fully-connected hybrid beamforming receiver for MIMO communication," *IEEE J. Solid-State Circuits*, vol. 53, no. 5, pp. 1275–1287, 2018.

[28] R. Zhang, J. Zhou, J. Lan, B. Yang, and Z. Yu, "A high-precision hybrid analog and digital beamforming transceiver system for 5G millimeter-wave communication," *IEEE Access*, vol. 7, pp. 83012–83023, 2019.

[29] R. Guo, Y. Cai, M. Zhao, Q. Shi, B. Champagne, and L. Hanzo, "Joint design of beam selection and precoding matrices for mmWave MU-MIMO systems relying on lens antenna arrays," *IEEE J. Sel. Top. Signal Process.*, vol. 12, no. 2, pp. 313–325, 2018.

[30] S. Han, I. Chih-Lin, Z. Xu, and C. Rowell, "Large-scale antenna systems with hybrid analog and digital beamforming for millimeter wave 5G," *IEEE Commun. Mag.*, vol. 53, no. 1, pp. 186–194, 2015.

[31] M. A. B. Abbasi, V. F. Fusco, and M. Matthaiou, "Millimeter wave hybrid beamforming with Rotman lens: Performance with hardware imperfections,"

in *2019 16th International Symposium on Wireless Communication Systems (ISWCS)*, 2019, pp. 203–207.

[32] M. A. B. Abbasi, H. Tataria, V. F. Fusco, M. Matthaiou, and G. C. Alexandropoulos, “Aggressive RF circuit reduction techniques in millimeter wave cellular systems,” in *2019 16th International Symposium on Wireless Communication Systems (ISWCS)*, 2019, pp. 500–504.

[33] S. Parkvall, A. Furuskär, and E. Dahlman, “Evolution of LTE toward IMT-advanced,” *IEEE Commun. Mag.*, vol. 49, no. 2, pp. 84–91, 2011.

[34] M.-Y. Huang, T. Chi, F. Wang, T.-W. Li, and H. Wang, “A full-FoV autonomous hybrid beamformer array with unknown blockers rejection and signals tracking for low-latency 5G mm-Wave links,” *IEEE Trans. Microw. Theory Tech.*, 2019.

[35] M. Rebato, J. Park, P. Popovski, E. De Carvalho, and M. Zorzi, “Stochastic geometric coverage analysis in mmWave cellular networks with realistic channel and antenna radiation models,” *IEEE Trans. Commun.*, vol. 67, no. 5, pp. 3736–3752, 2019.

[36] C.-W. Weng, B. P. S. Sahoo, H.-Y. Wei, and C.-H. Yu, “Directional reference signal design for 5G millimeter wave cellular systems,” *IEEE Trans. Veh. Technol.*, vol. 67, no. 11, pp. 10740–10751, 2018.

[37] V. Raghavan, A. Partyka, A. Sampath, *et al.*, “Millimeter-wave MIMO prototype: Measurements and experimental results,” *IEEE Commun. Mag.*, vol. 56, no. 1, pp. 202–209, 2018.

[38] W. Hong, Z. H. Jiang, C. Yu, *et al.*, “Multibeam antenna technologies for 5G wireless communications,” *IEEE Trans. Antennas Propag.*, vol. 65, no. 12, pp. 6231–6249, 2017.

[39] M. A. B. Abbasi, V. F. Fusco, H. Tataria, and M. Matthaiou, “Constant-ϵ_r lens beamformer for low-complexity millimeter-wave hybrid MIMO,” *IEEE Trans. Microw. Theory Tech.*, 2019.

[40] C. Tsokos, E. Mylonas, P. Groumas, L. Gounaridis, H. Avramopoulos, and C. Kouloumentas, “Optical beamforming network for multi-beam operation with continuous angle selection,” *IEEE Photonics Technol. Lett.*, vol. 31, no. 2, pp. 177–180, 2018.

[41] R. G. Stephen and R. Zhang, “Millimeter-wave CRAN with lens antenna arrays,” in *GLOBECOM 2017-2017 IEEE Global Communications Conference*, 2017, pp. 1–6.

[42] M. A. B. Abbasi, H. Tataria, V. F. Fusco, and M. Matthaiou, “Performance of a 28 GHz two-stage Rotman lens beamformer for millimeter wave cellular systems,” in *2019 13th European Conference on Antennas and Propagation (EuCAP)*, 2019, pp. 1–4.

[43] R. K. Luneburg and M. Herzberger, *Mathematical theory of optics*. University of California Press, 1964.

[44] Q.-W. Lin and H. Wong, “A low-profile and wideband lens antenna based on high-refractive-index metasurface,” *IEEE Trans. Antennas Propag.*, vol. 66, no. 11, pp. 5764–5772, 2018.

[45] C. Hua, X. Wu, N. Yang, and W. Wu, "Air-filled parallel-plate cylindrical modified Luneberg lens antenna for multiple-beam scanning at millimeter-wave frequencies," *IEEE Trans. Microw. Theory Tech.*, vol. 61, no. 1, pp. 436–443, 2013.
[46] H.-T. Chou and Z.-D. Yan, "Parallel-plate Luneburg lens antenna for broadband multibeam radiation at millimeter-wave frequencies with design optimization," *IEEE Trans. Antennas Propag.*, vol. 66, no. 11, pp. 5794–5804, 2018.
[47] M. Jiang, Z. N. Chen, Y. Zhang, W. Hong, and X. Xuan, "Metamaterial-based thin planar lens antenna for spatial beamforming and multibeam massive MIMO," *IEEE Trans. Antennas Propag.*, vol. 65, no. 2, pp. 464–472, 2016.
[48] K. F. Brakora, J. Halloran, and K. Sarabandi, "Design of 3-D monolithic MMW antennas using ceramic stereolithography," *IEEE Trans. Antennas Propag.*, vol. 55, no. 3, pp. 790–797, 2007.
[49] H. F. Ma and T. J. Cui, "Three-dimensional broadband ground-plane cloak made of metamaterials," *Nat. Commun.*, vol. 1, p. 21, Jun. 2010.
[50] S. B. Glybovski, S. A. Tretyakov, P. A. Belov, Y. S. Kivshar, and C. R. Simovski, "Metasurfaces: From microwaves to visible," *Phys. Rep.*, vol. 634, pp. 1–72, 2016.
[51] Y. J. Cheng, W. Hong, K. Wu, *et al.*, "Substrate integrated waveguide (SIW) Rotman lens and its Ka-band multibeam array antenna applications," *IEEE Trans. Antennas Propag.*, vol. 56, no. 8, pp. 2504–2513, 2008.
[52] Y. Gao, M. Khaliel, F. Zheng, and T. Kaiser, "Rotman lens based hybrid analog–digital beamforming in massive MIMO systems: Array architectures, beam selection algorithms and experiments," *IEEE Trans. Veh. Technol.*, vol. 66, no. 10, pp. 9134–9148, 2017.
[53] A. Lamminen, J. Säily, M. Kaunisto, M. Pokorný, J. Aurinsalo, and Z. Raida, "Gain enhanced millimetre-wave beam-switching Rotman lens antenna designs on LCP," in *2017 11th European Conference on Antennas and Propagation (EUCAP)*, 2017, pp. 2781–2785.
[54] S. V. Hum and J. Perruisseau-Carrier, "Reconfigurable reflectarrays and array lenses for dynamic antenna beam control: A review," *IEEE Trans. Antennas Propag.*, vol. 62, no. 1, pp. 183–198, 2013.
[55] A. Goldsmith, *Wireless communications*. Cambridge University Press, 2005.
[56] E. Björnson, E. G. Larsson, and T. L. Marzetta, "Massive MIMO: Ten myths and one critical question," *IEEE Commun. Mag.*, vol. 54, no. 2, pp. 114–123, 2016.
[57] E. Ali, M. Ismail, R. Nordin, and N. F. Abdulah, "Beamforming techniques for massive MIMO systems in 5G: Overview, classification, and trends for future research," *Front. Inf. Technol. Electron. Eng.*, vol. 18, no. 6, pp. 753–772, 2017.
[58] C.-R. Tsai and A.-Y. Wu, "Structured random compressed channel sensing for millimeter-wave large-scale antenna systems," *IEEE Trans. Signal Process.*, vol. 66, no. 19, pp. 5096–5110, 2018.

[59] A. M. A. Abdo, X. Zhao, R. Zhang, *et al.*, "MU-MIMO downlink capacity analysis and optimum code weight vector design for 5G big data massive antenna millimeter wave communication," *Wirel. Commun. Mob. Comput.*, vol. 2018, 2018.

[60] J. Wang, J. Weitzen, O. Bayat, V. Sevindik, and M. Li, "Interference coordination for millimeter wave communications in 5G networks for performance optimization," *EURASIP J. Wirel. Commun. Netw.*, vol. 2019, no. 1, p. 46, 2019.

[61] J.-S. Sheu, W.-H. Sheen, and T.-X. Guo, "On the design of downlink multiuser transmission for a beam-group division 5G system," *IEEE Trans. Veh. Technol.*, vol. 67, no. 8, pp. 7191–7203, 2018.

[62] W. Zhang, T. Kim, D. J. Love, and E. Perrins, "Leveraging the restricted isometry property: Improved low-rank subspace decomposition for hybrid millimeter-wave systems," *IEEE Trans. Commun.*, vol. 66, no. 11, pp. 5814–5827, 2018.

[63] W. El-Halwagy, R. Mirzavand, J. Melzer, M. Hossain, and P. Mousavi, "Investigation of wideband substrate-integrated vertically-polarized electric dipole antenna and arrays for mm-Wave 5G mobile devices," *IEEE Access*, vol. 6, pp. 2145–2157, 2017.

[64] H. T. Friis, "A note on a simple transmission formula," *Proc. IRE*, vol. 34, no. 5, pp. 254–256, 1946.

[65] W. Roh, J. Seol, J. Park, *et al.*, "Millimeter-wave beamforming as an enabling technology for 5G cellular communications: Theoretical feasibility and prototype results," *IEEE Commun. Mag.*, vol. 52, no. 2, pp. 106–113, 2014.

[66] T. S. Rappaport, F. Gutierrez, E. Ben-Dor, J. N. Murdock, Y. Qiao, and J. I. Tamir, "Broadband millimeter-wave propagation measurements and models using adaptive-beam antennas for outdoor urban cellular communications," *IEEE Trans. Antennas Propag.*, vol. 61, no. 4, pp. 1850–1859, 2012.

[67] H. Zhao, R. Mayzus, S. Sun, *et al.*, "28 GHz millimeter wave cellular communication measurements for reflection and penetration loss in and around buildings in New York city," in *2013 IEEE International Conference on Communications (ICC)*, 2013, pp. 5163–5167.

[68] T. A. Hill and J. R. Kelly, "28 GHz Taylor feed network for sidelobe level reduction in 5G phased array antennas," *Microw. Opt. Technol. Lett.*, vol. 61, no. 1, pp. 37–43, 2019.

[69] M. Khalily, R. Tafazolli, P. Xiao, and A. A. Kishk, "Broadband mm-Wave microstrip array antenna with improved radiation characteristics for different 5G applications," *IEEE Trans. Antennas Propag.*, vol. 66, no. 9, pp. 4641–4647, 2018.

[70] M. Ikram, M. S. Sharawi, K. Klionovski, and A. Shamim, "A switched-beam millimeter-wave array with MIMO configuration for 5G applications," *Microw. Opt. Technol. Lett.*, vol. 60, no. 4, pp. 915–920, 2018.

[71] J.-W. Lian, Y.-L. Ban, Q.-L. Yang, B. Fu, Z.-F. Yu, and L.-K. Sun, "Planar millimeter-wave 2-D beam-scanning multibeam array antenna fed by

compact SIW beam-forming network," *IEEE Trans. Antennas Propag.*, vol. 66, no. 3, pp. 1299–1310, 2018.

[72] M. Mantash, A. Kesavan, and T. A. Denidni, "Beam-tilting endfire antenna using a single-layer FSS for 5G communication networks," *IEEE Antennas Wirel. Propag. Lett.*, vol. 17, no. 1, pp. 29–33, 2017.

[73] Y. Yashchyshyn, K. Derzakowski, G. Bogdan, *et al.*, "28 GHz switched-beam antenna based on S-PIN diodes for 5G mobile communications," *IEEE Antennas Wirel. Propag. Lett.*, vol. 17, no. 2, pp. 225–228, 2017.

[74] K. R. Mahmoud and A. M. Montaser, "Performance of tri-band multi-polarized array antenna for 5G mobile base station adopting polarization and directivity control," *IEEE Access*, vol. 6, pp. 8682–8694, 2018.

[75] W. Liu and Z. Wang, "Non-uniform full-dimension MIMO: New topologies and opportunities," *IEEE Wirel. Commun.*, vol. 26, no. 2, pp. 124–132, 2019.

[76] K. G. Sadananda, M. P. Abegaonkar, and S. K. Koul, "Gain equalized shared-aperture antenna using dual-polarized ZIM for mmWave 5G base stations," *IEEE Antennas Wirel. Propag. Lett.*, vol. 18, no. 6, pp. 1100–1104, 2019.

[77] M. Mantash and T. A. Denidni, "CP antenna array with switching-beam capability using electromagnetic periodic structures for 5G applications," *IEEE Access*, vol. 7, pp. 26192–26199, 2019.

[78] G. S. Karthikeya, M. P. Abegaonkar, and S. K. Koul, "Pattern diversity of path loss compensated antennas for 5G base stations," *Int. J. RF Microw. Comput. Eng.*, vol. 29, no. 8, p. e21800, 2019.

[79] M. Morant, A. Trinidad, E. Tangdiongga, T. Koonen, and R. Llorente, "Experimental demonstration of mm-Wave 5G NR photonic beamforming based on ORRs and multicore fiber," *IEEE Trans. Microw. Theory Tech.*, 2019.

[80] C. Tsokos, E. Mylonas, P. Groumas, *et al.*, "Analysis of a multibeam optical beamforming network based on Blass matrix architecture," *J. Light. Technol.*, vol. 36, no. 16, pp. 3354–3372, 2018.

[81] E. H. Mujammami, I. Afifi, and A. B. Sebak, "Optimum wideband high gain analog beamforming network for 5G applications," *IEEE Access*, vol. 7, pp. 52226–52237, 2019.

[82] Y. Li, L. Ge, M. Chen, Z. Zhang, Z. Li, and J. Wang, "Multibeam 3-D-printed Luneburg lens fed by magnetoelectric dipole antennas for millimeter-wave MIMO applications," *IEEE Trans. Antennas Propag.*, vol. 67, no. 5, pp. 2923–2933, 2019.

[83] A. Rahimian, Q. H. Abbasi, A. Alomainy, and Y. Alfadhl, "A low-profile 28-GHz Rotman lens-fed array beamformer for 5G conformal subsystems," *Microw. Opt. Technol. Lett.*, vol. 61, no. 3, pp. 671–675, 2019.

[84] M. R. D. Kodnoeih, Y. Letestu, R. Sauleau, E. M. Cruz, and A. Doll, "Compact folded Fresnel zone plate lens antenna for mm-Wave communications," *IEEE Antennas Wirel. Propag. Lett.*, vol. 17, no. 5, pp. 873–876, 2018.

[85] E. Rajo-Iglesias, M. Ferrando-Rocher, and A. U. Zaman, "Gap waveguide technology for millimeter-wave antenna systems," *IEEE Commun. Mag.*, vol. 56, no. 7, pp. 14–20, 2018.

[86] A. Tamayo-Dominguez, J.-M. Fernandez-Gonzalez, and M. S. Castaner, "Low-cost millimeter-wave antenna with simultaneous sum and difference patterns for 5G point-to-point communications," *IEEE Commun. Mag.*, vol. 56, no. 7, pp. 28–34, 2018.

[87] D. Comite, S. K. Podilchak, P. Baccarelli, *et al.*, "Analysis and design of a compact leaky-wave antenna for wide-band broadside radiation," *Sci. Rep.*, vol. 8, no. 1, p. 17741, 2018.

[88] A. Rahimian, Y. Alfadhl, and A. Alomainy, "Design and performance analysis of millimetre-wave Rotman lens-based array beamforming networks for large-scale antenna subsystems," *Prog. Electromagn. Res.*, vol. 78, pp. 159–171, 2017.

[89] S. E. Ershadi, A. Keshtkar, A. Bayat, A. H. Abdelrahman, and H. Xin, "Rotman lens design and optimization for 5G applications," *Int. J. Microw. Wirel. Technol.*, vol. 10, no. 9, pp. 1048–1057, 2018.

[90] F. F. Manzillo, M. Śmierzchalski, L. L. Coq, *et al.*, "A wide-angle scanning switched-beam antenna system in LTCC technology with high beam crossing levels for V-band communications," *IEEE Trans. Antennas Propag.*, vol. 67, no. 1, pp. 541–553, 2018.

[91] R. Sun, C. A. Gentile, J. Senic, *et al.*, "Millimeter-wave radio channels vs. synthetic beamwidth," *IEEE Commun. Mag.*, vol. 56, no. 12, pp. 53–59, 2018.

[92] R. W. Heath, N. Gonzalez-Prelcic, S. Rangan, W. Roh, and A. M. Sayeed, "An overview of signal processing techniques for millimeter wave MIMO systems," *IEEE J. Sel. Top. Signal Process.*, vol. 10, no. 3, pp. 436–453, 2016.

[93] M. Giordani, M. Polese, A. Roy, D. Castor, and M. Zorzi, "A tutorial on beam management for 3GPP NR at mmWave frequencies," *IEEE Commun. Surv. Tutorials*, vol. 21, no. 1, pp. 173–196, 2019.

[94] G. E. Garcia, G. Seco-Granados, E. Karipidis, and H. Wymeersch, "Transmitter beam selection in millimeter-wave MIMO with in-band position-aiding," *IEEE Trans. Wirel. Commun.*, vol. 17, no. 9, pp. 6082–6092, 2018.

[95] M. Ruble and I. Güvenç, "Wireless localization for mmWave networks in urban environments," *EURASIP J. Adv. Signal Process.*, vol. 2018, no. 1, p. 35, 2018.

[96] C. Fiandrino, H. Assasa, P. Casari, and J. Widmer, "Scaling millimeter-wave networks to dense deployments and dynamic environments," *Proc. IEEE*, vol. 107, no. 4, pp. 732–745, 2019.

[97] N. Rupasinghe, Y. Yapıcı, I. Güvenç, and Y. Kakishima, "Non-orthogonal multiple access for mmWave drone networks with limited feedback," *IEEE Trans. Commun.*, vol. 67, no. 1, pp. 762–777, 2018.

[98] J. Lee, M. Kim, J. Park, and Y. J. Chong, "Field-measurement-based received power analysis for directional beamforming millimeter-wave systems: Effects of beamwidth and beam misalignment," *ETRI J.*, vol. 40, no. 1, pp. 26–38, 2018.

[99] V. Ariyarathna, A. Madanayake, X. Tang, *et al.*, "Analog approximate-FFT 8/16-beam algorithms, architectures and CMOS circuits for 5G beamforming MIMO transceivers," *IEEE J. Emerg. Sel. Top. Circuits Syst.*, vol. 8, no. 3, pp. 466–479, 2018.

[100] M. Abbasi and V. F. Fusco, "Hardware constraints in compressive sensing based antenna array," *arXiv Prepr. arXiv1907.08821*, 2019.

[101] A. Alammouri, J. Mo, B. L. Ng, J. C. Zhang, and J. G. Andrews, "Hand grip impact on 5G mmWave mobile devices," *IEEE Access*, vol. 7, pp. 60532–60544, 2019.

[102] V. Basavarajappa, B. B. Exposito, L. Cabria, and J. Basterrechea, "Binary phase-controlled multi-beam-switching antenna array for reconfigurable 5G applications," *EURASIP J. Wirel. Commun. Netw.*, vol. 2019, no. 1, pp. 1–16, 2019.

[103] S. Shakib, M. Elkholy, J. Dunworth, V. Aparin, and K. Entesari, "A wideband 28-GHz transmit-receive front-end for 5G handset phased arrays in 40-nm CMOS," *IEEE Trans. Microw. Theory Tech.*, 2019.

[104] S. Mondal and J. Paramesh, "A reconfigurable 28-/37-GHz MMSE-adaptive hybrid-beamforming receiver for carrier aggregation and multi-standard MIMO communication," *IEEE J. Solid-State Circuits*, vol. 54, no. 5, pp. 1391–1406, 2019.

[105] D.-C. Kim, S.-J. Park, T.-W. Kim, L. Minz, and S.-O. Park, "Fully digital beamforming receiver with a real-time calibration for 5G mobile communication," *IEEE Trans. Antennas Propag.*, vol. 67, no. 6, pp. 3809–3819, 2019.

[106] S. Y. Nusenu, "Development of frequency modulated array antennas for millimeter-wave communications," *Wirel. Commun. Mob. Comput.*, vol. 2019, 2019.

[107] X.-F. Zhang, W.-J. Sun, and J.-X. Chen, "Millimeter-wave ATS antenna with wideband-enhanced endfire gain based on coplanar plasmonic structures," *IEEE Antennas Wirel. Propag. Lett.*, vol. 18, no. 5, pp. 826–830, 2019.

[108] X.-F. Zhang, J. Fan, and J.-X. Chen, "High gain and high-efficiency millimeter-wave antenna based on spoof surface plasmon polaritons," *IEEE Trans. Antennas Propag.*, vol. 67, no. 1, pp. 687–691, 2018.

[109] S. Ghosh and D. Sen, "An inclusive survey on array antenna design for millimeter-wave communications," *IEEE Access*, vol. 7, pp. 83137–83161, 2019.

Chapter 11

Massive MIMO channels

Trinh Van Chien[1,2] *and Hien Quoc Ngo*[3]

Massive multiple-input multiple-output (MIMO) technology offers significant improvements in both spectral and energy efficiency by utilizing the same time and frequency resource as current radio networks to simultaneously serve multiple users with the use of channel state information measurements and linear processing schemes at the base station. These remarkable gains are obtained by equipping each base station with an array of many (hundreds or thousands) antennas to enable spatial multiplexing of many user terminals. The performance analysis of massive MIMO relies on its fundamental properties comprising the channel hardening, the favorable propagation, and the sparsity. This chapter presents some well-known channel models used in massive MIMO and evaluates their corresponding fundamental properties.

11.1 Introduction

In the conventional multiuser MIMO systems, each base station is only equipped with a few antennas to serve several users, and hence, the spectral efficiency is notably limited by mutual interference [1,2]. One potential solution is using orthogonal multiplexing techniques where the base station communicates with each user in separate time–frequency resources so that the mutual interference is removed. However, these techniques reduce the degree of freedoms (a.k.a. the multiplexing gain) and, hence, may reduce the spectral efficiency. It has been proved via information theory that orthogonal multiplexing techniques are not optimal, and therefore, higher spectral efficiency can be definitely obtained when the system allows multiple users to share the same time–frequency resources [3]. Unfortunately, the advanced signal processing techniques with the very high computational complexity as maximum-likelihood multiuser detection in the uplink or dirty-paper coding in the

[1]Department of Electrical Engineering (ISY), Linköping University, Linköping, Sweden
[2]School of Electronics and Telecommunications, Hanoi University of Science and Technology, Hanoi, Vietnam
[3]School of Electronics, Electrical Engineering and Computer Science, Queen's University Belfast, Belfast, UK

downlink should be used to mitigate mutual interference effectively. One of the disruptive technologies for 5G-and-beyond wireless communications called massive MIMO overcomes these drawbacks in the current multiuser MIMO systems by allowing the base station to be equipped with many antennas simultaneously serving many users in the same time–frequency resource [4–6]. Massive MIMO systems inherit all the benefits of what traditional MIMO has, but with a significantly larger scale as a function of the number of base station antennas [7–9]. In particular, compared with the conventional MIMO systems, massive MIMO systems with linear processing can improve the spectral and energy efficiency together with communication reliability by several orders of magnitude.

Such the abovementioned attractive benefits of massive MIMO systems are obtained by the following key assumptions. First, the propagation channel vectors between a base station and users should be nearly orthogonal, which is defined as the *favorable propagation* [10]. As a consequence, the use of a simple linear processing technique can provide the spectral efficiency close to the optimal one in many scenarios. More precisely, on the uplink data communication, a linear detection such as maximum ratio combining or zero forcing can cancel out both noise and interference effectively [11,12]. On the downlink data communication, the base station only uses a linear beamforming technique to simultaneously steer multiple data streams to different users with small interference to each other. Second, the so-called *channel hardening* allows to approximating an instantaneous channel gain by its mean value with minor influences on the spectral efficiency [13]. This important property indicates that the effective channel gain becomes deterministic and close to its mean value when there is a sufficiently large number of antennas installed at the base station. Consequently, the use of this mean value as the true effective channel gain for signal detection works well in many scenarios. This simplifies the computational complexity and channel estimation overhead [3,14]. Third, it is well known that the channel estimation is crucial in massive MIMO communication, and the *channel sparsity* can be commonly considered in the literature as prior information to improve the channel estimation quality and reduce the signaling overhead [15]. A propagation channel offers a sparsity structure if the number of nonzero coefficients (e.g., the impulse responses or the eigenvalues of covariance matrices) is less than the total available coefficients. More importantly, the measurements [16,17] have observed that the channels very often have a sparsity structure, especially when propagation environments lack scatterers.

All the above fundamental properties of massive MIMO channels may vary and depend a lot on the propagation environments. In other words, a channel model may not have all these properties simultaneously. For instance, the favorable propagation and channel hardening properties are clearly observed in Rayleigh fading. However, the channel sparsity is preferable in the opposite way, where the propagation environment only has a few scatterers. This typical observation motivates us to investigate these fundamental properties in different channel models, especially those which are well used in massive MIMO literature. In particular, we give an overview of these properties and interpret how it contributes to the performance (e.g., spectral efficiency and channel estimation quality) of massive MIMO systems. We then define useful metrics to evaluate and compare these properties in different well-known channel models.

The remaining of this chapter is organized as follows: Section 11.2 introduces a massive MIMO system. Section 11.3 gives the popular channel models that are widely used in massive MIMO literature. The favorable propagation property is presented in Section 11.4. We also provide a method to check if the channel is favorable or not. Meanwhile, Section 11.5 describes the channel hardening and the use of this property to formulate a lower bound on the ergodic channel capacity. The method to measure the channel hardening property is also presented. Section 11.6 briefly discusses about the channel sparsity and how to evaluate the sparsity structure from the covariance matrices. Finally, some main conclusions are given in Section 11.7.

Notations: The bold lower and uppercase letters are used for vectors and matrices, respectively. The expectation of the random variable is denoted by $\mathbb{E}\{\cdot\}$, whereas its variance is $\text{Var}\{\cdot\}$. We use $\mathcal{CN}(\cdot,\cdot)$ to denote a circularly symmetric Gaussian distribution. The Euclidean norm is $\|\cdot\|$ and the Frobenius norm is $\|\cdot\|_F$. The transpose and Hermitian of a matrix are denoted by $(\cdot)^T$ and $(\cdot)^H$, respectively. Finally, $\mathcal{O}(\cdot)$ denotes the big $-\ \mathcal{O}$ notation.

11.2 Massive MIMO system

We consider the uplink transmission of a massive MIMO system that comprises a base station equipped with M antennas serving K single-antenna users as illustrated in Figure 11.1. Though the propagation channels vary over time and frequency, we consider the classical block fading channels. The channel responses are approximately constant within a coherence time and a coherence bandwidth, which is defined as a coherence block. We study the channel models in massive MIMO communications, so it is sufficient to investigate only the uplink data transmission. The extension of our framework to the downlink transmission is straightforward.

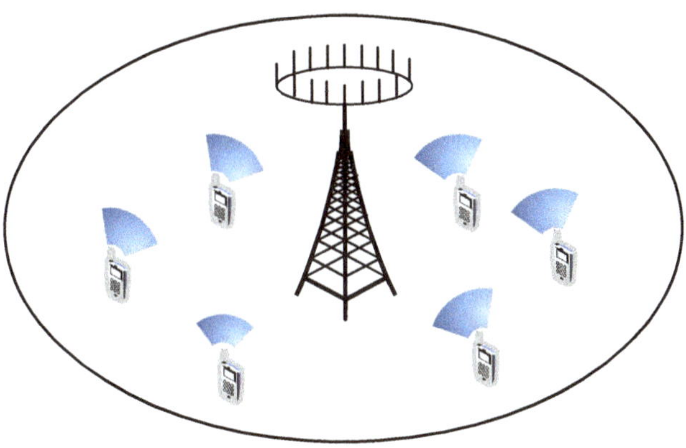

Figure 11.1 The uplink transmission of a massive MIMO system where multiple users communicate with the base station by utilizing the same time and frequency resource

Let us denote the channel between user k and the base station as a vector with M elements

$$\mathbf{h}_k = [h_{k1}, \ldots, h_{kM}]^T \in \mathbb{C}^M, \tag{11.1}$$

where h_{kM} is the channel response from user k to the base station antenna M.

In the uplink data transmission,* we assume that all K users send the complex signal symbols $x_1, \ldots, x_K$ to the base station. Here, x_k, where $\mathbb{E}\{|x_k|^2\} = 1$, is the symbol sent by user k. The received signal at the base station, $\mathbf{y} \in \mathbb{C}^M$, is

$$\mathbf{y} = \sum_{k=1}^{K} \sqrt{p_k} \mathbf{h}_k x_k + \mathbf{w}, \tag{11.2}$$

where p_k is the transmit power of user k and $\widetilde{w} \sim \mathcal{CN}(\mathbf{0}, \sigma_{\mathrm{UL}}^2 \mathbf{I}_M)$. In a compact form, the received signal in (11.2) is rewritten as

$$\mathbf{y} = \mathbf{HPx} + \mathbf{w}, \tag{11.3}$$

where the signal vector is denoted as $\mathbf{x} = [x_1, \ldots, x_K]^T \in \mathbb{C}^K$, the channel matrix is $\mathbf{H} = [\mathbf{h}_1, \ldots, \mathbf{h}_K] \in \mathbb{C}^{M \times K}$, and the diagonal matrix $\mathbf{P} = \mathrm{diag}([\sqrt{p_1}, \ldots, \sqrt{p_K}]) \in \mathbb{C}^{K \times K}$ contains the transmit powers in the diagonal elements. We now present several channel models that are popularly used to evaluate the performance of massive MIMO systems.

11.3 Channel models

This section provides some well-known channel models that are widely used in massive MIMO literature. We notice that the channel in (11.1) comprises both large- and small-scale fading effects. We assume that the channels are ergodic random variables in the sense that new independent realizations in each coherence block are assumed.

The vector containing all the mean values, $\bar{h}_{k1}, \ldots, \bar{h}_{kM}$, of the original channel $\mathbf{h}_k$ is denoted by $\bar{\mathbf{h}}_k \in \mathbb{C}^M$, which is mathematically given as

$$\bar{\mathbf{h}}_k = \mathbb{E}\{\mathbf{h}_k\} = \left[\bar{h}_{k1}, \ldots, \bar{h}_{kM}\right]^T, \tag{11.4}$$

where $\bar{h}_{kM}$ denotes the mean value of the channel response h_{kM}. The nonzero mean value $\bar{\mathbf{h}}_k \neq 0$ indicates there is a direct path (called light-of-sight (LoS) path) from user k to the base station. In contrast, $\bar{\mathbf{h}}_k = \mathbf{0}$ indicates that there is not LoS paths due to obstacles, for example, buildings or large objects, between user k and the base station.

The large-scale fading coefficient of the channel h_{kM} is denoted by

$$\beta_k = \mathbb{E}\{|h_{kM}|^2\}, \tag{11.5}$$

*In the massive MIMO literature, there is also the uplink pilot training which supports the base station to estimate instantaneous channel state information.

which is independent of the antenna index. The large-scale fading β_k can model the attenuation of signals by diffraction around large objects such as high towers and by the propagation over a long distance between the base station and the users. We stress that the large-scale fading comprises the contributions from both the LoS and non-LoS (NLoS) paths. The 3GPP LTE standard [18] suggests the large-scale fading defined in decibel (dB), for example, using a single-slope model, as

$$\beta_k = \xi_k - 10\alpha\log_{10}\left(\frac{d_k}{d_0}\right) + z_k, \tag{11.6}$$

where ξ_k denotes the median channel gain at the reference distance d_0 and α is the path loss exponent. The parameter d_k represents the distance from user k to the base station. The random parameter z_k denotes the *shadow fading* effect, which is usually modeled by a lognormal distribution as $\tilde{z}_k \sim \mathcal{CN}(0, \sigma^2)$.

11.3.1 Uncorrelated Rician channels

This model is used in many works on massive MIMO since it reasonable for many practical scenarios and simple for the performance analysis. The channel between user k and the base station is defined as

$$\mathbf{h}_k \sim \mathcal{CN}\left(\bar{\mathbf{h}}_k, \beta_k^{\text{NLoS}}\mathbf{I}_M\right), \tag{11.7}$$

where β_k^{NLoS} is the large-scale fading coefficient corresponding to the NLoS path. The large-scale fading coefficient corresponding to the NLoS path is denoted by β_k^{LoS} that is equal to $|\bar{h}_{kM}|^2$. We must have

$$\beta_k = \beta_k^{\text{LoS}} + \beta_k^{\text{NLoS}}.$$

By introducing a Rician factor $\kappa_k \geq 0$ that stands for the ratio between power of the LoS and the NLoS paths, we can decompose β_k (in linear scale) into the LoS and NLoS paths as

$$\beta_k^{\text{LoS}} = \frac{\kappa_k}{\kappa_k + 1}\beta_k, \text{ and } \beta_k^{\text{NLoS}} = \frac{1}{\kappa_k + 1}\beta_k. \tag{11.8}$$

As an example, when a uniform linear array is exploited at the base station, the mean vector in (11.7) is modeled as [7, Section 1.3]

$$\bar{\mathbf{h}}_k = \sqrt{\beta_k^{\text{LoS}}}\left[1, e^{j2\pi d \sin\theta_k}, \ldots, e^{j2\pi d(M-1)\sin\theta_k}\right]^T, \tag{11.9}$$

where d is the antenna spacing at the base station, and θ_k is the angle of arrival. The mean vector in (11.9) produces

$$\|\bar{\mathbf{h}}_k\|^2 = \beta_k^{\text{LoS}} \sum_{m=1}^{M} e^{j2\pi d(m-1)\sin\theta_k} e^{-j2\pi d(m-1)\sin\theta_k} = M\beta_k^{\text{LoS}}, \tag{11.10}$$

which is proportional to the number of base station antennas and independent of the antenna space and the angle of arrival.

As reported in [19] for the collocated antenna arrays, uncorrelated Rician channels match well with the propagation environments that have many scatterers. Moreover, we stress that when $\bar{\mathbf{h}}_k = \mathbf{0}$, this channel model is called *uncorrelated Rayleigh fading* channel model and follows a zero-mean circularly symmetric Gaussian distribution $\mathscr{CN}(\mathbf{0}, \beta_k^{\mathrm{NLoS}} \mathbf{I}_M)$.

11.3.2 Spatial correlated channels

With more likely in practice, the correlated Rician channels have attracted a lot of interest recently. Mathematically, the channel of user k and the base station is modeled as

$$\mathbf{h}_k \sim \mathscr{CN}(\bar{\mathbf{h}}_k, \mathbf{R}_k), \tag{11.11}$$

where $\mathbf{R}_k \in \mathbb{C}^{M \times M}$ is the covariance matrix, that is defined by

$$\mathbf{R}_k = \mathbb{E}\left\{ (\mathbf{h}_k - \bar{\mathbf{h}}_k)(\mathbf{h}_k - \bar{\mathbf{h}}_k)^H \right\}.$$

One simple example for the covariance matrix is the one using the exponential correlation model for the uniform linear arrays, given by [14]:

$$\mathbf{R}_k = \beta_k^{\mathrm{NLoS}} \begin{bmatrix} 1 & \xi e^{-j\theta_k} & \dots & \xi^{M-1} e^{-j\theta_k(M-1)} \\ \xi e^{j\theta_k} & 1 & \dots & \xi^{M-2} e^{-j\theta_k(M-2)} \\ \vdots & \vdots & \ddots & \vdots \\ \xi^{M-1} e^{j\theta_k(M-1)} & \xi^{M-2} e^{j\theta_k(M-2)} & \dots & 1 \end{bmatrix}, \tag{11.12}$$

where $\xi \in [0, 1]$ is the correlation for which $\xi = 0$ becomes the uncorrelated Rician channel model. The strength of spatial correlation is achieved by increasing the value of ξ, i.e., $\xi \to 1$. In addition, θ_k is the user incidence angle to the array boresight. It notices that $\mathbf{R}_k$ is Toeplitz matrix. An alternative to model the covariance matrix is to use the local scattering channel model [7]:

$$[\mathbf{R}_k]_{lm} = \beta_k^{\mathrm{NLoS}} \int e^{2\pi j d_H (l-m) \sin \hat{\theta}_k} f(\hat{\theta}_k) d\hat{\theta}_k, \tag{11.13}$$

where $\hat{\theta}_k$ is the nominal angle of arrival, and $f(\hat{\theta}_k)$ is the probability density function of $\hat{\theta}_k$ [20]. In (11.13), we assume that it lacks scattering around the base station, thus we can assume that a rich scattering environment with an infinite number of scatterers is around an arbitrary user k [7]. When there are more than one scattering cluster, say $\hat{S}$ clusters, (11.13) is defined as

$$[\mathbf{R}_k]_{lm} = \beta_k^{\mathrm{NLoS}} \sum_{s=1}^{\hat{S}} \int e^{2\pi j d_H (l-m) \sin \hat{\theta}_{ks}} f(\hat{\theta}_{ks}) d\hat{\theta}_{ks}, \tag{11.14}$$

and the illustration of this finite scattered channel model is in Figure 11.2. In (11.14), $\hat{\theta}_{ks}$ is a random variable standing for the nominal angle of arrival for the

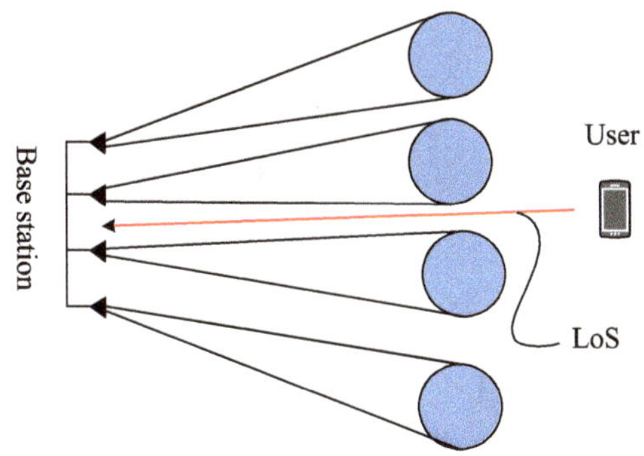

Figure 11.2 An illustration of the local scattering channel model, where the scattering clusters are located around the user (each big circle stands for a scattering cluster)

sth scatterer. Note that the multipath components of a scattering cluster follows, for example, a Gaussian distribution with the angular standard deviation $\widehat{\sigma}$. When $\widehat{\sigma}$ is sufficiently small, (11.14) is approximated to as

$$[\mathbf{R}_k]_{lm} = \frac{\beta_k^{\text{NLoS}}}{\widehat{S}} \sum_{s=1}^{\hat{S}} e^{j\pi(l-m)\sin\hat{\theta}_{ks}} e^{-\left(\hat{\sigma}^2/2\right)(\pi(l-m)\cos\hat{\theta}_{ks})^2}, \tag{11.15}$$

which simplifies the computational complexity [7, Section 2.6].

11.3.3 Double-scattering channels

More complicated than the channel models have been presented so far, the real propagation channels may be non-Gaussian. The authors in [21,22] studied the information rate of massive MIMO systems under one of the non-Gaussian channels, named the double-scattering channel model. A schematic of the double-scattering channels between user k and the base station is illustrated in Figure 11.3 with several geometric parameters. Roughly speaking, several scatterers are located around the base station and every user k. Mathematically, the channel between user k and the base station is formulated as

$$\mathbf{h}_k = \bar{\mathbf{h}}_k + \sqrt{\frac{\beta_k^{\text{NLoS}}}{S}} \bar{\mathbf{R}}_k^{1/2} \mathbf{G}_k \widehat{\mathbf{R}}_k^{1/2} \mathbf{g}_k, \tag{11.16}$$

where S is the number of scatterers located around the base station and user sides. $\mathbf{G}_k \in \mathbb{C}^{M \times S}$ is the small-scale fading matrix between the base station and its scattering cluster, while $\mathbf{g}_k \in \mathbb{C}^S$ is the small scale-fading vector between the user and its scattering cluster. As demonstrated in [23], the elements of $\mathbf{G}_k$ and $\mathbf{g}_k$ follow a normal complex Gaussian distribution $\mathcal{CN}(0, 1)$. The (m, m')th element of the

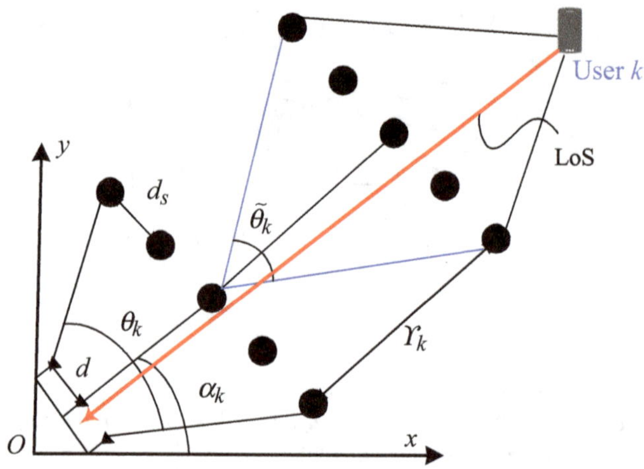

Figure 11.3 An illustration of the double-scattering channel model, where the scatterers (the black circles) are located around both the base station and the user sides

correlation matrix $\overline{\mathbf{R}}_k \in \mathbb{C}^{M\times M}$ between the base station antenna array and its S scatterers, $m, m' = 1, \ldots, M$, is defined as

$$[\overline{\mathbf{R}}_k]_{mm'} = \frac{1}{S}\sum_{n=(1-S)/2}^{(S-1)/2} e^{-2\pi j(m-m')d\cos((\pi/2)+\alpha_k+\theta_{kn})}, \tag{11.17}$$

where $d > 0$ is the antenna spacing at the base station, θ_{kn} represents the angle spread between scatterer n and the base station antenna array, which is computed by using the same method as in [24]:

$$\theta_{kn} = \frac{n\theta_k}{S-1}, \forall n = \frac{1-S}{2}, \ldots, \frac{S-1}{2}. \tag{11.18}$$

In (11.17) and (11.18), we explicitly assume that S is an odd number and $S > 1$. A channel model with only one scatterer (i.e., $S = 1$) is discussed hereafter. We stress that the correlation matrix $\overline{\mathbf{R}}_k$ models the space among the antennas installed at the base station and the effects of the scattering distribution as well. Furthermore, the covariance matrix between the transmit and receive scatterers is defined by assuming that there are virtual antenna arrays, so the angle spread $\widetilde{\theta}_k$ is given as

$$\tan\left(\frac{\widetilde{\theta}_k}{2}\right) = \frac{d_S(S-1)}{2r_k}, \tag{11.19}$$

where $d_S > 0$ denotes the distance between two scatterers. Moreover, the (m, m')th element of the covariance matrix $\hat{\mathbf{R}}_k$, $\forall, m, m' = 1, \ldots, S$, is computed as

$$[\hat{\mathbf{R}}_k]_{mm'} = \frac{1}{S} \sum_{n=(1-S)/2}^{(S-1)/2} e^{-2\pi j(m-m')d_S \cos\left((\pi/2)+\alpha_k+\tilde{\theta}_{kn}\right)}, \tag{11.20}$$

where $\widetilde{\theta}_{kn}$ denotes the angle between the scatterers and is defined as

$$\widetilde{\theta}_{kn} = \frac{n\widetilde{\theta}_k}{S-1}, \forall n = \frac{1-S}{2}, \ldots, \frac{S-1}{2}. \tag{11.21}$$

By controlling the number of scatterers, the large-scale fading coefficients, and the correlation matrices, the double-scattering channel model is capable of containing several key perspectives of massive MIMO channels comprising the spatial correlation, the rank deficiency, and the signal attenuation. Additionally, this channel model spans from uncorrelated Rician to the keyhole channels: by setting $\overline{\mathbf{R}}_k = \mathbf{I}_M$,, $\hat{\mathbf{R}}_k = \mathbf{I}_S$, and $S \to \infty$, the uncorrelated Rician fading is obtained. This aligns with the rich scattering environments with high-rank correlation matrix. For the worst case of rank deficiency with rank $(\widehat{\mathbf{R}}_k) = 1$, for example, all entries of $\hat{\mathbf{R}}_k$ equal to 0.5. In addition, $S = 1$ reduces the double-scattering channel model to a single keyhole channel model.

Figure 11.4 visualizes the covariance matrices of the considered channel models with color mapping. The base station is located at the origin. The NLoS large-scale fading coefficients are equal to 0.5, i.e., $\beta_k^{\text{NLoS}} = 0.5$. The number of antennas at the base station is 100, and user k is located at the position having the coordinate (500,500) with the distance to the base station of about 707 m. The correlation factor of the exponential channel model is 0.9. For the local scattering channel model, there have six scattering clusters located in the range of $[-2\pi/9, 2\pi/9]$ with the distance between the two scatterers being $4\pi/45$. The antenna space is normalized to 1/2. The local scattering channel model has 21 scatterers, and the angular spread is $\pi/3$. The uncorrelated Rician fading gives the covariance matrix with the diagonal elements equal to 1 and the off-diagonal elements equal to 0. It demonstrates the uncorrelated property of this channel model. The exponential correlation has nonzero values not only on the diagonal elements but also on the off-diagonal ones. In each row, the magnitudes get smaller when the elements are farther than the diagonal ones. Alternatively, the correlation reduces. With this setting, the local scattering and double-scattering channel models have a stronger spatial correlation than the exponential correlation model does.

11.4 Favorable propagation

This section discusses the favorable propagation property in massive MIMO, which is interpreted as the quasi-orthogonality among the channel vectors from the base

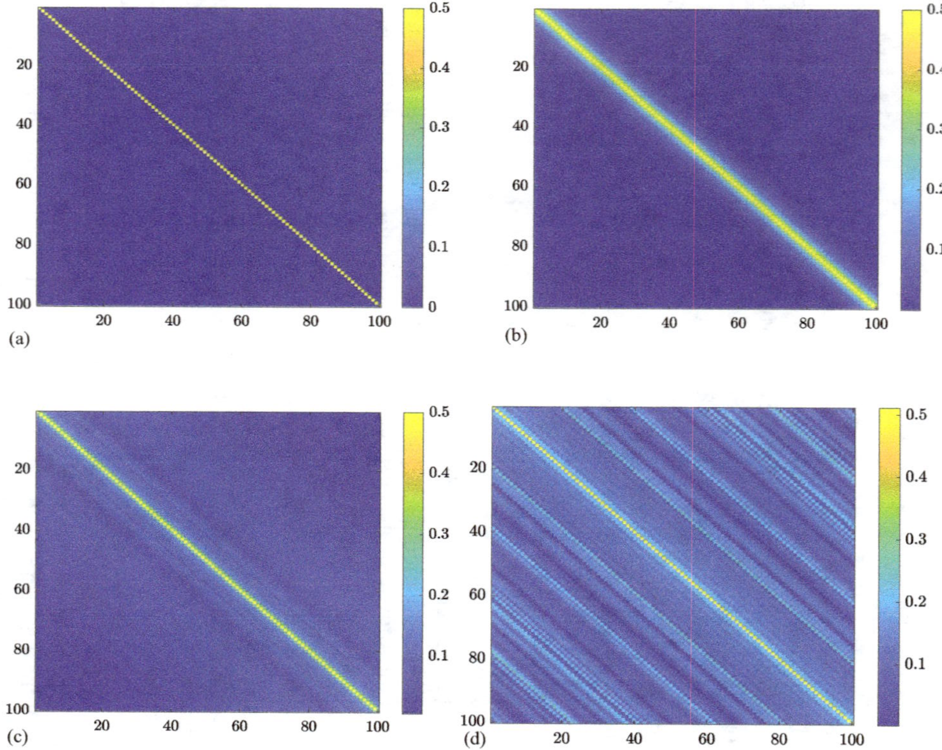

Figure 11.4 The covariance matrix for the different channel models: (a) uncorrelated Rician, (b) exponential correlation, (c) local scattering, and (d) double-scattering

station to the users. Under favorable propagation, simple linear processing techniques can obtain the (nearly) optimal performance.

The favorable propagation is defined as the mutual orthogonality among the channels to the users [10], as shown in the following definition.

Definition 11.1 *A channel offers the favorable propagation property if the inner product of the two arbitrary channel vectors* $\mathbf{h}_k$ and $\mathbf{h}_{k'}, k \neq k'$, satisfies

$$\mathbf{h}_k^H \mathbf{h}_{k'} = 0. \tag{11.22}$$

The key observation from (11.22) is that the base station can completely decode the signal sent by user k without inter-user interference when it observes the received signal by using the channel $\mathbf{h}_k$ as

$$\begin{aligned} \widehat{s}_k &= \mathbf{h}_k^H \mathbf{y} & (11.23) \\ &= \sqrt{p_k} \mathbf{h}_k^H \mathbf{h}_k s_k + \sum_{k'=1, k' \neq k}^{K} \sqrt{p_{k'}} \mathbf{h}_k^H \mathbf{h}_{k'} s_{k'} + \mathbf{h}_k^H \mathbf{w} & (11.24) \\ &= \sqrt{p_k} \|\mathbf{h}_k\|^2 s_k + \mathbf{h}_k^H \mathbf{w}. & (11.25) \end{aligned}$$

The same approach can be applied to the other users, for example, the base station decodes the desired signal from an arbitrary user k' by utilizing $\mathbf{h}_{k'}^H \mathbf{y}$. The favorable propagation property (11.22) offers the optimal performance with a simple linear processing, which represents the most desirable scenario from the perspective of maximizing the information throughput. However, we should seek for an approximation of favorable propagation since the condition (11.22) is never exactly satisfied in reality. We define this as *asymptotically favorable propagation* as presented in Definition 11.2.

Definition 11.2 A channel offers the asymptotically favorable propagation if

$$\frac{\mathbf{h}_k^H \mathbf{h}_{k'}}{\sqrt{\mathbb{E}\{\|\mathbf{h}_k\|^2\}\mathbb{E}\{\|\mathbf{h}_{k'}\|^2\}}} \rightarrow 0, \text{ as } M \rightarrow \infty \tag{11.26}$$

holds for any $k \neq k'$.

Different from (11.22), (11.26) shows that it does not require the strict condition in which the inner product $\mathbf{h}_k^H \mathbf{h}_{k'}$ must go to zero. For practical massive MIMO systems, the base station is equipped with a large but not infinite number of antennas. Therefore, if the channel offers the asymptotically favorable propagation, then it is interesting to measure how close to the asymptotically favorable propagation we are with a given finite value M. To do this, we define the following deterministic favorable propagation metric.

Definition 11.3 *Given channel vectors* $\mathbf{h}_k$ *and* $\mathbf{h}_{k'}$, *their corresponding deterministic favorable propagation metric is defined as*

$$\mathrm{FP}_{kk'} = \frac{\mathrm{Var}\{\mathbf{h}_k^H \mathbf{h}_{k'}\}}{\mathbb{E}\{\|\mathbf{h}_k\|^2\}\mathbb{E}\{\|\mathbf{h}_{k'}\|^2\}}. \tag{11.27}$$

The deterministic favorable propagation metric in (11.27) measures how fast a channel model offers the favorable propagation. This metric depends only on the statistical information instead of the instantaneous channels as in the previous definitions. We next provide some common assumptions used in massive MIMO literature. Then, we exploit the favorable propagation property of different channel models by using (11.26) and (11.27).

Assumption 11.1 *The LoS components* $\bar{\mathbf{h}}_k$ *and* $\bar{\mathbf{h}}_{k'}$ *of two users* k *and* k' $(k \neq k')$ *satisfy*

$$\frac{1}{M}|\bar{\mathbf{h}}_k^H \bar{\mathbf{h}}_{k'}| \to 0, \text{ as } M \to \infty. \tag{11.28}$$

Assumption 11.2 *The LoS component* $\bar{\mathbf{h}}_k$ *of user* k *satisfies*

$$\limsup_{M\to\infty} \frac{1}{M}\|\bar{\mathbf{h}}_k\|^2 < \infty. \tag{11.29}$$

Assumption 11.3 *In the case of correlated fading channels, the covariance matrix* $\mathbf{R}_k$ *of user* k *satisfies*

$$\limsup_{M\to\infty} \frac{1}{M}\|\mathbf{R}_k\|_2 < \infty, \tag{11.30}$$

$$\liminf_{M\to\infty} \frac{1}{M}\mathrm{tr}(\mathbf{R}_k) > 0, \tag{11.31}$$

where $\|\cdot\|_2$ *is the spectral norm of a matrix.*

The first assumption implies that the LoS components of the two separate users should be asymptotically orthogonal. The second assumption implies that the energy of each LoS component is bounded at the asymptotic regime. While the last assumption implies that the antenna array functions properly in the sense that it gathers signal energy from different directions and proportionally to the number of antennas.

Lemma 11.1 *If the propagation channels are uncorrelated Rician fading channels defined by (11.7), then they offer asymptotically favorable propagation, and the deterministic favorable propagation metric can be computed in the closed form as*

$$\mathrm{FP}_{kk'}^{\mathrm{uR}} = \frac{M\beta_k^{\mathrm{NLoS}}\beta_{k'}^{\mathrm{NLoS}} + \beta_{k'}^{\mathrm{NLoS}}\|\bar{\mathbf{h}}_k\|^2 + \beta_k^{\mathrm{NLoS}}\|\bar{\mathbf{h}}_{k'}\|^2}{\left(\|\bar{\mathbf{h}}_k\|^2 + M\beta_k^{\mathrm{NLoS}}\right)\left(\|\bar{\mathbf{h}}_{k'}\|^2 + M\beta_{k'}^{\mathrm{NLoS}}\right)}. \tag{11.32}$$

Proof. The proof is completed by computing the statistical information in the numerator and denominator of (11.27) by using the channel model in (11.7). The detailed proof is available in Appendix A.2.

It emphasizes that (11.32) visualizes the contributions of the channel statistics, including the large-scale fading coefficients and the mean values to the speed of convergence. The numerator of (11.32) scales as $\mathcal{O}(M)$, while the denominator of (11.32) scales as $\mathcal{O}(M^2)$, and hence, $\mathcal{O}(M)/\mathcal{O}(M^2) \to 0$ as $M \to \infty$. Therefore, in uncorrelated Rician propagation, the convergence of (11.26) is very fast. If the LoS path is modeled as in (11.9), then we can simplify the deterministic favorable propagation unit in (11.32) to as

$$\mathrm{FP}_{kk'}^{\mathrm{uR}} = \frac{\beta_k^{\mathrm{NLoS}}\beta_{k'}^{\mathrm{NLoS}} + \beta_{k'}^{\mathrm{NLoS}}\beta_k^{\mathrm{LoS}} + \beta_k^{\mathrm{NLoS}}\beta_{k'}^{\mathrm{LoS}}}{M\left(\beta_k^{\mathrm{LoS}} + \beta_k^{\mathrm{NLoS}}\right)\left(\beta_{k'}^{\mathrm{LoS}} + \beta_{k'}^{\mathrm{NLoS}}\right)} \to 0, \text{ as } M \to \infty, \tag{11.33}$$

by using the fact in (11.10) that $\|\bar{\mathbf{h}}_k\|^2 = \beta_k^{\mathrm{LoS}}M$ and $\|\bar{\mathbf{h}}_{k'}\|^2 = \beta_{k'}^{\mathrm{LoS}}M$. When there is non-line-of-sight channels (i.e., $\bar{\mathbf{h}}_k = \bar{\mathbf{h}}_{k'} = \mathbf{0}$), (11.32) is simplified to $\mathrm{FP}_{kk'}^{\mathrm{uR}} = 1/M$ which was reported, for example, in [10].

We now consider the favorable propagation for correlated Rician fading channels by computing the closed-form expression of (11.27) as shown in Lemma 11.2.

Lemma 11.2 *If the propagation channels are correlated Rician fading channels defined by (11.11), then they offer asymptotically favorable propagation, and the deterministic favorable propagation metric can be computed in the closed form as*

$$\mathrm{FP}_{kk'}^{\mathrm{cR}} = \frac{\mathrm{tr}(\mathbf{R}_k\mathbf{R}_{k'}) + \bar{\mathbf{h}}_k^H\mathbf{R}_{k'}\bar{\mathbf{h}}_k + \bar{\mathbf{h}}_{k'}^H\mathbf{R}_k\bar{\mathbf{h}}_{k'}}{(\|\bar{\mathbf{h}}_k\|^2 + \mathrm{tr}(\mathbf{R}_k))(\|\bar{\mathbf{h}}_{k'}\|^2 + \mathrm{tr}(\mathbf{R}_{k'})).} \tag{11.34}$$

Proof. The proof is completed by computing the statistical information in the numerator and denominator of (11.27) by using the channel model in (11.11). The detailed proof is available in Appendix A.3.

Lemma 11.2 reveals the spatial correlations among antennas contribute to the speed of the convergence that is one of the main differences between (11.34) and (11.32). The first part in the numerator of (11.34) shows the correlation of the two variance matrices. The remaining parts demonstrate the effects of the LoS paths in the relationship with the covariance matrices. For the correlated Rayleigh channels, the deterministic favorable propagation in (11.34) is simplified to as $\mathrm{FP}_{kk'}^{\mathrm{cR}} = \mathrm{tr}(\mathbf{R}_k\mathbf{R}_{k'})/(\mathrm{tr}(\mathbf{R}_k)\mathrm{tr}(\mathbf{R}_{k'}))$.

For the double-scattering channel model, we can compute the closed-form expression of the deterministic favorable propagation unit as in Lemma 11.3.

Lemma 11.3 *If the propagation channels are the double-scattering channels defined by (11.16), then they offer favorable propagation, and the deterministic favorable propagation metric can be computed in the closed-form expression as*

$$\mathrm{FP}_{kk'}^{\mathrm{dsC}} = \frac{\beta_k^{\mathrm{NLoS}}\bar{\mathbf{h}}_{k'}^H\overline{\mathbf{R}}_k\bar{\mathbf{h}}_{k'} + \beta_{k'}^{\mathrm{NLoS}}\bar{\mathbf{h}}_k^H\overline{\mathbf{R}}_{k'}\bar{\mathbf{h}}_k + \beta_{k'}^{\mathrm{NLoS}}\beta_k^{\mathrm{NLoS}}\mathrm{tr}(\overline{\mathbf{R}}_{k'}\overline{\mathbf{R}}_k)}{\left(\|\bar{\mathbf{h}}_{k'}\|^2 + M\beta_{k'}^{\mathrm{NLoS}}\right)\left(\|\bar{\mathbf{h}}_k\|^2 + M\beta_k^{\mathrm{NLoS}}\right)}. \tag{11.35}$$

Proof. The proof is completed by computing the statistical information in the numerator and denominator of (11.27) by using the channel model in (11.16). The detailed proof is available in Appendix A.4.

Lemma 11.3 shows the effects of statistic information as the mean values and spatial correlation to the deterministic favorable propagation unit. Interestingly, the number of scatterers does not have any contributions to the convergence speed of the asymptotically favorable propagation.

Figure 11.5 displays the real part of the asymptotically favorable propagation of the different channel models by utilizing 10,000 realizations of the small-scale fading coefficients. The network topology, the location of user k, and the channel models have the same setup as that of Figure 11.4. User k' is located at the coordinate $(500, -500)$ with the distance to the base station is of about 707 m. This simulation setup confirms that all the channel models offer the favorable propagation. Interestingly, the exponential correlation has the best asymptotically favorable propagation. Moreover, Figure 11.6 shows the deterministic favorable propagation unit as a function of the base station antenna number with the different channel models. The local scattering model uses 6 scattering clusters, while the number of scatterers in the double-scattering model is 21. It confirms the significant reductions of the deterministic favorable propagation when increasing the number of antennas equipped at the base station. This general trend is applied for all the considered channel models. For this network setting, the double-scattering channel model has the highest deterministic favorable propagation unit when increasing the number of base station antennas, leading to the fact that this channel model has the stronger non-orthogonality of the channel vectors compared with the other models.

11.5 Channel hardening

This section investigates the hardening property that indicates that one random variable observed from a communication system can be converted to a

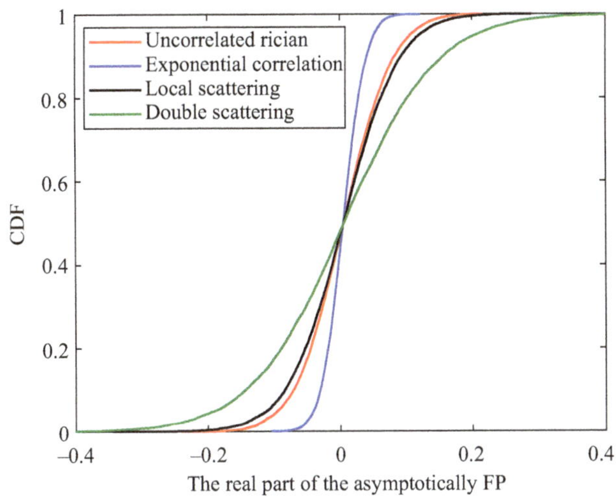

Figure 11.5 The real part of the metric (11.26) for the asymptotically favorable propagation property

deterministic value at the asymptotic regime [25,26]. Therefore, the base station can treat an instantaneous gain as a deterministic value, which reduces computational complexity.

11.5.1 Channel hardening for different channel models

In massive MIMO communications, channel hardening makes the effective channel gain after some linear processing becomes very close to its mean value. This property reduces the need for compensating the fluctuation of small-scale fading such as performing a power allocation scheme for each small-scale fading realization. A propagation channel gives channel hardening if the condition in Definition 11.4 is satisfied.

Definition 11.4 *A channel offers the channel hardening property if*

$$\frac{\|\mathbf{h}_k\|^2}{\mathbb{E}\{\|\mathbf{h}_k\|^2\}} \to 1, \text{ as } M \to \infty, \tag{11.36}$$

where convergence is the almost sure convergence.

The condition (11.36) indicates that the effective channel gain $\|\mathbf{h}_k\|^2$ converges to its mean value thanks to many antennas equipped at the base station. However, since $\mathbf{h}_k$ is a random variable, it is hard to use the above condition to

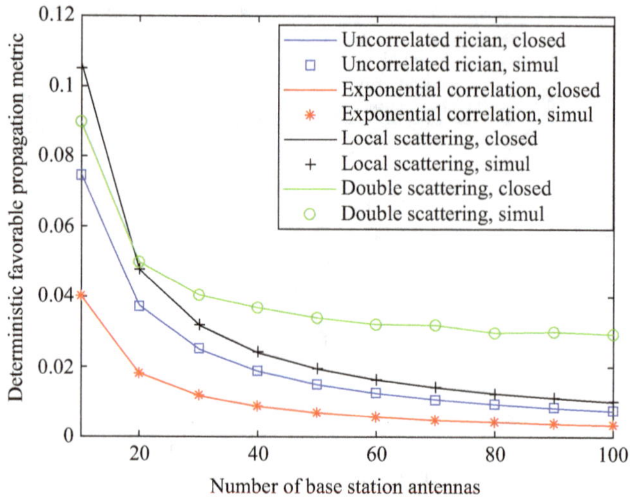

Figure 11.6 The deterministic favorable propagation unit versus the number of base station antennas with the different channel models

check the hardening property of the channels. In order to measure the channel hardening quality, we use the Chebyshev inequality [13,27], for which[†]

$$\Pr\left\{\left|\frac{\|\mathbf{h}_k\|^2}{\mathbb{E}\{\|\mathbf{h}_k\|^2\}}-1\right|^2\leq\varepsilon\right\}=1-\Pr\left\{\left|\frac{\|\mathbf{h}_k\|^2}{\mathbb{E}\{\|\mathbf{h}_k\|^2\}}-1\right|^2\geq\varepsilon\right\} \tag{11.37}$$

$$\geq 1-\frac{1}{\varepsilon^2}\times\frac{\mathrm{Var}\left\{\|\mathbf{h}_k\|^2\right\}}{\left(\mathbb{E}\left\{\|\mathbf{h}_k\|^2\right\}\right)^2},\forall\varepsilon\geq 0. \tag{11.38}$$

From (11.38), we can see that if

$$\frac{\mathrm{Var}\left\{\|\mathbf{h}_k\|^2\right\}}{\left(\mathbb{E}\left\{\|\mathbf{h}_k\|^2\right\}\right)^2}\to 0,\quad \text{as } M\to\infty, \tag{11.39}$$

we have channel hardening. Otherwise, we do not have channel hardening. Therefore, we can use the entity $\mathrm{Var}\{\|\mathbf{h}_k\|^2\}/(\mathbb{E}\{\|\mathbf{h}_k\|^2\})^2$ to measure where or not a channel model has the hardening property.

[†]Let x be a random variable with the mean $\bar{x}$ and the variance σ_x^2, then the Chebyshev inequality holds that $\Pr(|x-\bar{x}|\geq\varepsilon)\leq\sigma_x^2/\varepsilon^2$.

We definite a deterministic metric to estimate the channel hardening property of an arbitrary channel model as in Definition 11.5.

Definition 11.5 *The deterministic channel hardening metric is defined as follows:*

$$\mathrm{CH}_k = \frac{\mathrm{Var}\left\{\|\mathbf{h}_k\|^2\right\}}{\left(\mathbb{E}\left\{\|\mathbf{h}_k\|^2\right\}\right)^2}. \tag{11.40}$$

The deterministic channel hardening metric defined in (11.40) can be used to determine whether or not the channel hardening holds for a particular propagation environment. For uncorrelated Rayleigh fading, Lemma 11.4 shows the deterministic channel hardening metric in the closed-form expression.

Lemma 11.4 *If the propagation channels are uncorrelated Rician fading channels defined by (11.7), the deterministic channel hardening metric is computed in the closed form as*

$$\mathrm{CH}_k^{\mathrm{uR}} = \frac{2\beta_k^{\mathrm{NLoS}}\|\bar{\mathbf{h}}_k\|^2 + M\left(\beta_k^{\mathrm{NLoS}}\right)^2}{\left(\|\bar{\mathbf{h}}_k\|^2 + M\beta_k^{\mathrm{NLoS}}\right)^2}. \tag{11.41}$$

Proof. The proof is completed by computing the statistical information in the numerator and denominator of (11.40) by using the channel model in (11.7). The detailed proof is available in Appendix A.5.

From Assumption 11.2, the numerator of (11.41) scales as $\mathcal{O}(M)$ as M increases, while the denominator scales as $\mathcal{O}(M^2)$. Consequently, $\mathrm{CH}_k^{\mathrm{uR}} \to 0$ as $M \to \infty$ that implies the uncorrelated Rician channel offers the channel hardening. If the mean vector is modeled as in (11.9), the deterministic channel hardening unit is simplified to as

$$\mathrm{CH}_k^{\mathrm{uR}} = \frac{2\beta_k^{\mathrm{NLoS}}\beta_k^{\mathrm{LoS}} + \left(\beta_k^{\mathrm{NLoS}}\right)^2}{M\left(\beta_k^{\mathrm{LoS}} + \beta_k^{\mathrm{NLoS}}\right)^2}. \tag{11.42}$$

When $\bar{\mathbf{h}}_k = \mathbf{0}$, we have uncorrelated Rayleigh fading scenarios where $\mathrm{CH}_k^{\mathrm{uR}} = 1/M$.

For correlated Rician fading given in (11.11), the deterministic channel hardening metric is computed in the closed-form expression as shown in Lemma 11.5.

Lemma 11.5 *If the propagation channels are correlated Rician fading channels defined by (11.11), the deterministic channel hardening unit is computed in the* *closed form as*

$$\mathrm{CH}_k^{\mathrm{cR}} = \frac{2\bar{\mathbf{h}}_k^H \mathbf{R}_k \bar{\mathbf{h}}_k + \mathrm{tr}\left(\mathbf{R}_k^2\right)}{\left(\|\bar{\mathbf{h}}_k\|^2 + \mathrm{tr}(\mathbf{R}_k)\right)^2}. \tag{11.43}$$

Proof. The proof is completed by computing the statistical information in the numerator and denominator of (11.40) by using the channel model in (11.7). The detailed proof is available in Appendix A.6.

Lemma 11.5 demonstrates the contributions of channel statistic information as mean vector and covariance matrix to evaluate the channel hardening property. Interestingly, depending on the channel statistic property, the channel can harden or not. For example, if $\mathbf{R}_k = \mathrm{diag}([M, 0, \ldots 0])$, then $\mathrm{CH}_k^{\mathrm{cR}}$ does not converge to 0 when $M \to \infty$, and hence, we do not have channel hardening. But if $\mathbf{R}_k = \mathbf{I}_M$, then $\mathrm{CH}_k^{\mathrm{cR}} \to 0$, as $M \to \infty$, and hence, we have channel hardening.

When the covariance matrix is defined by the exponential correlation channel model or the local scattering channel model, (11.43) can further reformulated as

$$\mathrm{CH}_k^{\mathrm{cR}} = \frac{2\bar{\mathbf{h}}_k^H \mathbf{R}_k \bar{\mathbf{h}}_k + \mathrm{tr}\left(\mathbf{R}_k^2\right)}{\left(\|\bar{\mathbf{h}}_k\|^2 + M\beta_k^{\mathrm{NLoS}}\right)^2}, \tag{11.44}$$

by the fact that $\mathrm{tr}(\mathbf{R}_k) = M\beta_k^{\mathrm{NLoS}}$. Furthermore, if we set $\mathbf{R}_k = \beta_k^{\mathrm{NLoS}}\mathbf{I}_M$, then (11.43) and (11.42) are identical.

For the double-scattering fading channels, we can compute the closed-form expression of (11.40) as shown in Lemma 11.6.

Lemma 11.6 *If the propagation channels are the double-scattering channels defined by (11.16), the deterministic channel hardening unit is computed in the closed form expression as*

$$\mathrm{CH}_k^{\mathrm{dsC}} = \frac{\mathrm{Num}_k}{\left(\|\bar{\mathbf{h}}_k\|^2 + M\beta_k^{\mathrm{NLoS}}\right)^2}, \tag{11.45}$$

where the numerator of (11.45) is given as

$$\mathrm{Num}_k = 2\beta_k^{\mathrm{NLoS}} \bar{\mathbf{h}}_k^H \bar{\mathbf{R}}_k \bar{\mathbf{h}}_k + \left(\beta_k^{\mathrm{NLoS}}\right)^2 \left(1 + \frac{\mathrm{tr}\left(\hat{\mathbf{R}}_k^2\right)}{S^2}\right) \mathrm{tr}\left(\bar{\mathbf{R}}_k^2\right) + \left(\beta_k^{\mathrm{NLoS}}\right)^2 \frac{M^2 \mathrm{tr}\left(\hat{\mathbf{R}}_k^2\right)}{S^2}. \tag{11.46}$$

Proof. The proof is completed by computing the statistical information in the numerator and denominator of (11.40) by using the channel model in (11.16). The detailed proof is available in Appendix A.7.

Lemma 11.6 reveals the effects of the mean vector and covariance matrix to the deterministic channel hardening metric. Similarly to the correlated Rician channels, the double-scattering channel can harden or not depending on its statistic properties. It shows that the second covariance matrix $\hat{\mathbf{R}}_k$ does not have any contributions to harden the channels. Additionally, increasing the number of scatterers always provides better the channel hardening since

$$\mathrm{CH}_k^{\mathrm{dsC}} \geq \frac{2\beta_k^{\mathrm{NLoS}} \bar{\mathbf{h}}_k^H \overline{\mathbf{R}}_k \bar{\mathbf{h}}_k + \left(\beta_k^{\mathrm{NLoS}}\right)^2 \mathrm{tr}\left(\overline{\mathbf{R}}_k^2\right)}{\left(\|\bar{\mathbf{h}}_k\|^2 + M\beta_k^{\mathrm{NLoS}}\right)^2}, \tag{11.47}$$

which the bound, for a given M, is tight when the number of scatterers is sufficiently large such that

$$\frac{\left(\beta_k^{\mathrm{NLoS}}\right)^2 \mathrm{tr}\left(\hat{\mathbf{R}}_k^2\right) \mathrm{tr}\left(\bar{\mathbf{R}}_k^2\right)}{S^2} \approx 0 \text{ and } \frac{\left(\beta_k^{\mathrm{NLoS}}\right)^2 M^2 \mathrm{tr}\left(\hat{\mathbf{R}}_k^2\right) s}{S^2} \approx 0, \tag{11.48}$$

since $\mathrm{tr}\left(\hat{\mathbf{R}}_k^2\right)$ only scales up as $\mathcal{O}(S)$, and especially the equality in (11.48) holds as $S \to \infty$. Moreover, for a given number of scatterers S, $\mathrm{CH}_k^{\mathrm{dsC}} \to \mathrm{tr}\left(\hat{\mathbf{R}}_k^2\right)/S$ as $M \to \infty$, and therefore, the channel hardening property is the worst in the case of the keyhole channels ($S = 1, \mathrm{CH}_k^{\mathrm{dsC}} = 1$, as $M \to \infty$) compared to the other propagation channels as reported in [13].

Figure 11.7 plots $\|\mathbf{h}_k\|^2/\mathbb{E}\{\|\mathbf{h}_k\|^2\}$ provided in (11.36) over 10,000 different realizations of the small-scale fading coefficients for the considered channel models. The system setup is the same as that of Figure 11.4. The uncorrelated Rician fading channel offers the best asymptotically channel hardening with less fluctuation, following by the local scattering channel model. Among them, the double-scattering channel has the highest fluctuation. Figure 11.8 shows the deterministic channel hardening metric as the function of the base station antennas. Equipping one or more antennas at the base station always improves the channel hardening. The closed-form expressions have very close results as the Monte Carlo simulations do. It is because of the limited small number of scatterers (only 21 scatterers), we also observe that, for this simulation setting, the double-scattering model yields the worst hardening property with the highest deterministic channel hardening metric.

Figure 11.9 plots the deterministic channel hardening metric as the function of scatterers, and there are 50 antennas equipped at the base station. The closed-form expression curves generated from (11.43) match well with the Monte Carlo simulation curves. As expected, the double-scattering channel model always gives the highest deterministic channel hardening metric, and this metric reduces significantly when we increase the number of scatterers.

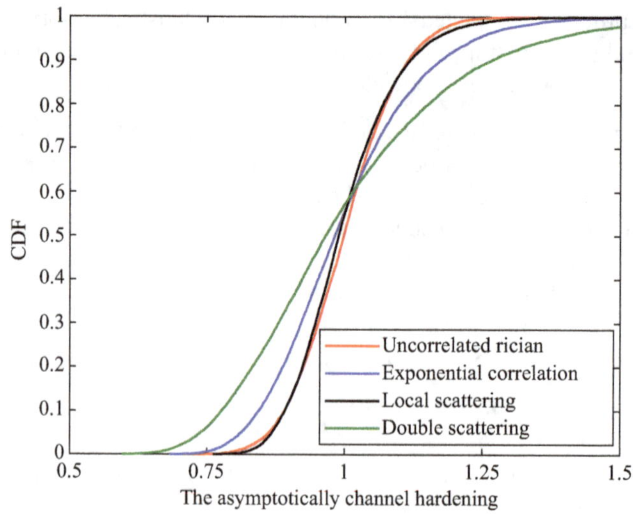

Figure 11.7 The asymptotically channel hardening metric in (11.36) for different channel models

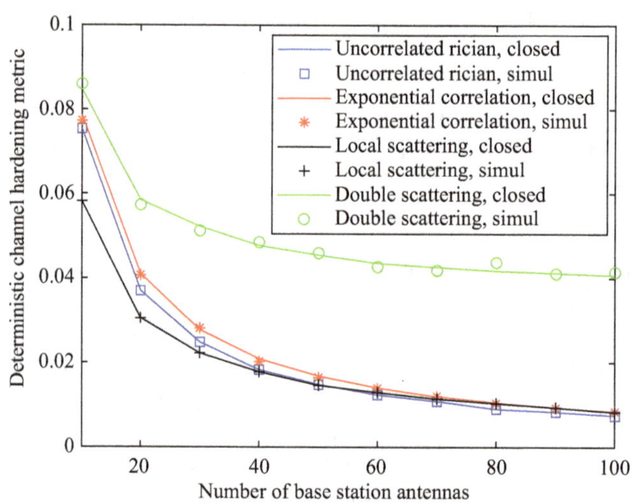

Figure 11.8 The deterministic hardening metric versus the number of base station antennas with the different channel models

11.5.2 Channel hardening and spectral efficiency

In the uplink data transmission, when user k is transmitting a signal x_k with $\mathbb{E}\{|x_k|^2\} = 1$ as in (11.2), by exploiting a combining vector $\mathbf{v}_k \in \mathbb{C}^M$, the

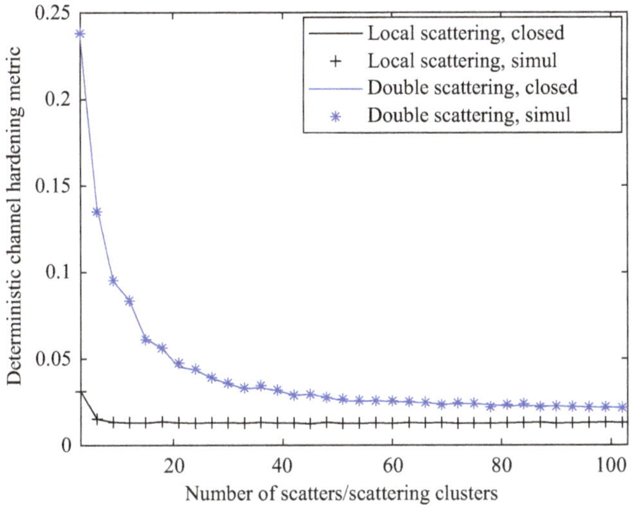

Figure 11.9 The deterministic hardening metric versus the number of scatterers (for double-scattering channels) or the number of scattering clusters (for local scattering channels) with the different channel models

transmitted signal from user k is detected by using

$$\mathbf{v}_k^H \mathbf{y} = \sum_{k'=1}^{K} \sqrt{p_{k'}} \mathbf{v}_k^H \mathbf{h}_{k'} x_{k'} + \mathbf{v}_k^H \mathbf{w} \tag{11.49}$$

$$= \sqrt{p_k} \mathbf{v}_k^H \mathbf{h}_k x_k + \sum_{\substack{k'=1,\\ k' \neq k}}^{K} \sqrt{p_{k'}} \mathbf{v}_k^H \mathbf{h}_{k'} x_{k'} + \mathbf{v}_k^H \mathbf{w}. \tag{11.50}$$

If the channel hardens (i.e., the effective channel gain is very close to its mean with high probability), then we can use $\mathbb{E}\{\mathbf{v}_k^H \mathbf{h}_k\}$ as the true effective channel gain $\mathbf{v}_k^H \mathbf{h}_k$ to detect x_k. More precisely, (11.50) can be rewritten as

$$\mathbf{v}_k^H \mathbf{y} = \sqrt{p_k} \mathbb{E}\{\mathbf{v}_k^H \mathbf{h}_k\} x_k + \sqrt{p_k} \left(\mathbf{v}_k^H \mathbf{h}_k - \mathbb{E}\{\mathbf{v}_k^H \mathbf{h}_k\} \right) x_k + \sum_{\substack{k'=1,\\ k' \neq k}}^{K} \sqrt{p_{k'}} \mathbf{v}_k^H \mathbf{h}_{k'} x_{k'} + \mathbf{v}_k^H \mathbf{w}, \tag{11.51}$$

where the first term of (11.51) is treated as the desired signal, and the remaining parts (including the second, third, and fourth terms) are considered as the effective channel gain. It is straightforward to show that the effective noise is uncorrelated with the desired signal. Thus, by using the fact that uncorrelated Gaussian noise is the worst case regarding system rate, we can obtain the following achievable rate

$$R_k = \frac{\tau_d}{\tau_c} \log_2(1 + \mathrm{SINR}_k), \tag{11.52}$$

where τ_c is the total number of symbols in each coherence block, $\tau_d < \tau_c$ is the number of symbols dedicated to the uplink data transmission phase, and SINR_k is the effective signal-to-interference-and-noise ratio (SINR), given by

$$\mathrm{SINR}_k = \frac{p_k \left| \mathbb{E}\{\mathbf{v}_k^H \mathbf{h}_k\} \right|^2}{\Sigma_{k'=1,k' \neq k}^{K} p_{k'} \mathbb{E}\left\{ \left| \mathbf{v}_k^H \mathbf{h}_{k'} \right|^2 \right\} - p_k \left| \mathbb{E}\{\mathbf{v}_k^H h_k\} \right|^2 + \mathbb{E}\left\{ \left| \mathbf{v}_k^H \mathbf{n} \right|^2 \right\}} \tag{11.53}$$

Note that the achievable rate (11.52) is obtained following the *use-and-then-forget* capacity bounding technique in [25]. This result can be applied for an arbitrary channel propagation model and an arbitrary combining technique. For example, if we denote $\hat{\mathbf{H}} = [\hat{\mathbf{h}}_1, \ldots, \hat{\mathbf{h}}_K]^T \in \mathbb{C}^{M \times K}$, where $\hat{\mathbf{h}}_k$ is the channel estimate of the original version $\mathbf{h}_k$, then the linear combining matrix $\hat{\mathbf{V}} = [\hat{\mathbf{v}}_1, \ldots, \hat{\mathbf{v}}_K]^T \in \mathbb{C}^{M \times K}$ is formulated as

$$\hat{\mathbf{V}} = \begin{cases} \hat{\mathbf{H}} & \text{Maximum–ratio combining}, \\ \hat{\mathbf{H}}\left(\hat{\mathbf{H}}^H \hat{\mathbf{H}}\right)^{-1}, & \text{Zero–forcing combining}. \end{cases} \tag{11.54}$$

In many scenarios, such as uncorrelated Rayleigh fading with the use of a linear combining technique, for example, maximum-ratio or zero-forcing, the closed-form expressions, which are independent of small-scale fading coefficients, are obtained [7]. The closed-form expressions do not only give insights on the array gain, mutual interference, and noise but also have significant contributions on the resource allocation problems that are simple and implemented based on the large-scale fading coefficients [28,29].

11.6 Channel sparsity

In massive MIMO, channel estimation plays an essential role [14,30]. The linear minimum mean-squared error (MMSE) estimator or element-wise linear MMSE estimator can be used in many phenomena if the covariance matrix (for the linear MMSE estimator) or channel gain (for element-wise linear MMSE estimator) is available. Such prior information may be burdensome as the base station is equipped with a massive number of antennas. Without any prior information, one can use the least squares channel estimator to perform the channel estimation from the uplink pilot training phase with a tolerable reduction of the spectral efficiency. Alternatively, channel sparsity is of interest to improve the quality of a channel estimation process.

The propagation channels may be themselves sparse in the temporal domain, which is the so-called temporal sparsity. In more detail, the channel impulse response has much less nonzero coefficients than their ambient dimension suggests. It contains many pulses with zero values in between. In multipath propagation environments, these nonzero pulses represent different paths from different time delays [31]. It worth emphasizing that the sparse channel impulse response does not offer by propagation channels with the rich of scatterers, but it may happen in mmwave communications with a limited of reflections and propagation paths [15].

In general, the original channels may not be sparse at all. However, some of their properties, such as covariance matrices or angular spreads, could be sparse thanks to an over-complete dictionary, which is defined by a linear combination of basic elements (atoms) [32,33]. Given hints on the selected channel models, we can select the over-complete dictionary that projects a proper channel property onto the sparse domain. For example, by exploiting a discrete Fourier transform, the angle of arrival and the angle of departure information can be transformed into the sparse domain. On the other hand, the propagation channel between a user and the base station only involves paths in a limited subset of all angular directions.

Here is often the case that it is difficult to define an over-complete dictionary. Therefore, in general, projecting the channel property onto a proper sparsity domain is room for many applications in 5G-and-beyond networks. One way to verify the existence of spatial sparsity is projecting the spatial covariance matrix using the eigen-decomposition and analyzing the eigenvalues. Since each eigenvalue represents the average squared amplitude in a subset of angular directions, the spatial sparsity results in zero eigenvalues. Mathematically, for a given covariance matrix $\mathbf{R}_k \in \mathbb{C}^{M\times M}$, we can project it onto eigen domain with the eigen-decomposition as

$$\mathbf{R}_k = \mathbf{U}_k \mathbf{\Sigma}_k \mathbf{U}_k^H, \tag{11.55}$$

where $\mathbf{U}_k \in \mathbb{C}^{M\times M}$ are the unitary matrices, and $\mathbf{\Sigma}_k = \mathrm{diag}([\sigma_1,\ldots,\sigma_M]) \in \mathbb{C}^{M\times M}$ is the diagonal matrix whose diagonal elements, $\sigma_1,\ldots,\sigma_M$, are the eigenvalues.

Figure 11.10 plots the sorted eigenvalues of the covariance matrices of all the considered channel models. For the uncorrelated Rician fading model, all the eigenvalues are equal so that there is no sparsity. The exponential correlation model has different eigenvalues with many small values, but all of them are nonzero. For a tolerable level, these eigenvalues are viewed as having a compressed structure [33]. Hence, by defining a threshold, we can only keep the significantly large eigenvalues and force the small ones to zeros for a sparsity structure. The local scattering and double-scattering channel models have sparsity structure with many zero eigenvalues. Therefore, we observe that not all channel models provide the sparsity, and the eigen-decomposition is a good choice to testify if the spatial correlation is sparse in the eigen-domain.

11.7 Conclusion

This chapter has presented three fundamental properties of massive MIMO channels: *favorable propagation*, *channel hardening*, and *channel sparsity*. We provided the definitions of these properties together with some measures (deterministic metrics) to evaluate these properties when there is a limited number of antenna equipped at the base station. The closed-form expressions of the deterministic

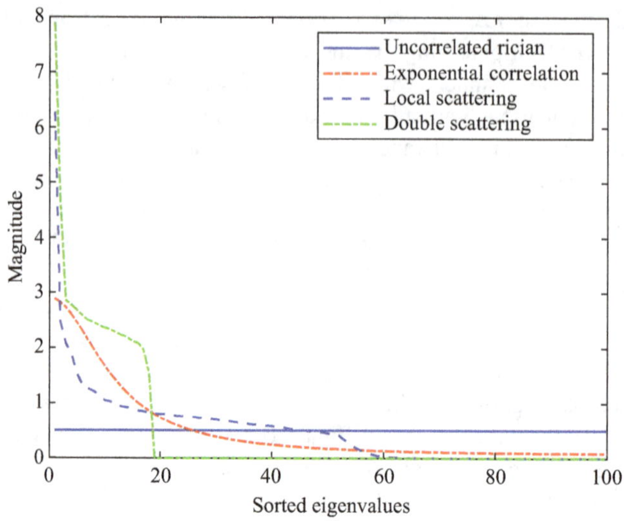

Figure 11.10 The eigenvalues of the spatial covariance matrices with the different channel models

metrics corresponding to the channel hardening and favorable propagation for the different channel models in the massive MIMO literature were derived. Furthermore, we provided some insightful observations on the channel sparsity. Even though the channel vectors are not sparse in general, there exists an over-complete dictionary where the projected channel properties are sparse in some aspects such as the covariance matrices. By utilizing a proper projection, it manifests that the selected channel models provide the sparse structure.

Appendix A

A.1 Useful lemmas

This appendix provides the three useful lemmas that are widely used to prove all main results of this chapter.

Lemma 11.7 *Let us consider the two independently random vectors* $\mathbf{u} \sim \mathcal{CN}(\bar{\mathbf{u}}, \mathbf{R}_u)$ *and* $\mathbf{v} \sim \mathcal{CN}(\bar{\mathbf{v}}, \mathbf{R}_v)$ *with the mean values* $\bar{\mathbf{u}}, \bar{\mathbf{v}} \in \mathbb{C}^M$ *and the covariance matrices* $\mathbf{R}_u, \mathbf{R}_v \in \mathbb{C}^{M \times M}$. *It holds that*

$$\mathbb{E}\{|\mathbf{u}^H \mathbf{v}|^2\} = \mathrm{tr}(\mathbf{R}_v \mathbf{R}_u) + \bar{\mathbf{u}}^H \mathbf{R}_v \bar{\mathbf{u}} + \bar{\mathbf{v}}^H \mathbf{R}_u \bar{\mathbf{v}} + |\bar{\mathbf{u}}^H \bar{\mathbf{v}}|^2. \tag{A.1}$$

Proof. The proof is completed by computing the moments of Gaussian distributions [34]. More precisely, we first represent $\mathbf{u}, \mathbf{v}$ as

$$\mathbf{u} = \bar{\mathbf{u}} + \widetilde{\mathbf{u}}, \quad \text{where } \widetilde{\mathbf{u}} \sim \mathcal{CN}(\mathbf{0}, \mathbf{R}_u), \tag{A.2}$$

$$\mathbf{v} = \bar{\mathbf{v}} + \widetilde{\mathbf{v}}, \text{ where } \widetilde{\mathbf{v}} \sim \mathcal{CN}(\mathbf{0}, \mathbf{R}_v), \tag{A.3}$$

then, we can express $\mathbf{u}^H\mathbf{v} = (\bar{\mathbf{u}}^H + \widetilde{\mathbf{u}}^H)(\bar{\mathbf{v}} + \widetilde{\mathbf{v}})$ and the expectation in (A.1) is computed as

$$\mathbb{E}\{|\mathbf{u}^H\mathbf{v}|^2|\} = \mathbb{E}\{(\bar{\mathbf{u}}^H + \widetilde{\mathbf{u}}^H)(\bar{\mathbf{v}} + \widetilde{\mathbf{v}})(\bar{\mathbf{v}}^H + \widetilde{\mathbf{v}}^H)(\bar{\mathbf{u}} + \widetilde{\mathbf{u}})\} \tag{A.4}$$

$$= |\bar{\mathbf{u}}^H\bar{\mathbf{v}}|^2 + \mathbb{E}\left\{|\bar{\mathbf{u}}^H\widetilde{\mathbf{v}}|^2\right\} + \mathbb{E}\left\{|\widetilde{\mathbf{u}}^H\bar{\mathbf{v}}|^2\right\} + \mathbb{E}\{|\widetilde{\mathbf{u}}^H\widetilde{\mathbf{v}}|\}. \tag{A.5}$$

The three expectations in (A.5) are further computed as

$$\mathbb{E}\{|\bar{\mathbf{u}}^H\widetilde{\mathbf{v}}|^2\} = \bar{\mathbf{u}}^H\mathbf{R}_v\bar{\mathbf{u}}, \quad \mathbb{E}\{|\widetilde{\mathbf{u}}^H\bar{\mathbf{v}}|^2\} = \bar{\mathbf{v}}^H\mathbf{R}_u\bar{\mathbf{v}}, \quad \mathbb{E}\{|\widetilde{\mathbf{u}}^H\widetilde{\mathbf{v}}|^2\} = \mathrm{tr}(\mathbf{R}_v\mathbf{R}_u), \tag{A.6}$$

and we obtain the result as shown in the lemma.

Lemma 11.8 *Let us consider a random matrix* $\mathbf{G} \in \mathbb{C}^{M\times S}$ *with each element distributed as* $\mathcal{CN}(0, 1)$*. For a deterministic positive semi-definite matrix* $\mathbf{M} \in \mathbb{C}^{S\times S}$*, it holds that*

$$\mathbb{E}\{\mathbf{GMG}^H\} = \mathrm{tr}(\mathbf{M})\mathbf{I}_M. \tag{A.7}$$

Proof. By using the eigen-decomposition, we can reformulate $\mathbf{M} = \mathbf{U\Sigma U}^H$, where $\mathbf{U} \in \mathbb{C}^{S\times S}$ is a complex unitary matrix, and $\mathbf{\Sigma} \in \mathbb{C}^{S\times S}$ contains the eigenvalues $\sigma_1, \ldots, \sigma_S$ of matrix $\mathbf{M}$, we can compute the expectation in (A.7) as

$$\mathbb{E}\{\mathbf{GMG}^H\} = \mathbb{E}\{\mathbf{GU\Sigma U}^H\mathbf{G}^H\} = \mathbb{E}\left\{\hat{\mathbf{G}}\mathbf{\Sigma}\hat{\mathbf{G}}^H\right\}, \tag{A.8}$$

where $\hat{\mathbf{G}} = \mathbf{GU}$ with each element distributed as $\mathcal{CN}(0, 1)$ due to the fact that the unitary matrix $\mathbf{U}$ preserves the distribution of matrix $\mathbf{G}$. Let $\hat{\mathbf{G}} = [\hat{\mathbf{g}}_1, \ldots, \hat{\mathbf{g}}_S]$, where $\hat{\mathbf{g}}_s \in \mathbb{C}^M$ is the sth column of matrix $\hat{\mathbf{G}}$, (A.8) is further computed as

$$\mathbb{E}\{\mathbf{GMG}^H\} = \sum_{s=1}^{S}\sigma_s\mathbb{E}\{\hat{\mathbf{g}}_s\hat{\mathbf{g}}_s^H\} = \sum_{s=1}^{S}\sigma_s\mathbf{I}_M = \mathrm{tr}(\mathbf{M})\mathbf{I}_M, \tag{A.9}$$

where the last equality of (A.9) is obtained by the fact $\mathrm{tr}(\mathbf{M}) = \sum_{s=1}^{S}\sigma_s$ and we complete the proof.

Lemma 11.9 *Let us consider a random vector* $\mathbf{u} \in \mathbb{C}^M$ *distributed as* $\mathcal{CN}(\bar{\mathbf{u}}, \mathbf{R}_u)$*. For a given deterministic matrix* $\mathbf{M} \in \mathbb{C}^{M\times M}$*, it holds that*

$$\begin{aligned}\mathbb{E}\{|\mathbf{u}^H\mathbf{Mu}|^2\} \;=\;& |\bar{\mathbf{u}}^H\mathbf{M}\bar{\mathbf{u}}|^2 + 2\bar{\mathbf{u}}^H\mathbf{M}\bar{\mathbf{u}}\mathrm{tr}(\mathbf{MR}_u) + \bar{\mathbf{u}}^H\mathbf{MR}_u\mathbf{M}^H\bar{\mathbf{u}} \\ &+ \bar{\mathbf{u}}^H\mathbf{M}^H\mathbf{R}_u\mathbf{M}\bar{\mathbf{u}} + |\mathrm{tr}(\mathbf{R}_u\mathbf{M})|^2 + \mathrm{tr}(\mathbf{R}_u\mathbf{MR}_u\mathbf{M}^H).\end{aligned} \tag{A.10}$$

Proof. We first decompose $\mathbf{u} = \bar{\mathbf{u}}+\tilde{\mathbf{u}}$, with $\tilde{\mathbf{u}} \sim \mathcal{CN}(\mathbf{0}, \mathbf{R}_u)$. Then, we obtain

$$\mathbb{E}\{|\mathbf{u}^H\mathbf{M}\mathbf{u}|^2\} = \mathbb{E}\{|(\bar{\mathbf{u}}^H + \tilde{\mathbf{u}}^H)\mathbf{M}(\bar{\mathbf{u}} + \tilde{\mathbf{u}})|^2\} \tag{A.11}$$

$$= \mathbb{E}\{|\bar{\mathbf{u}}^H\mathbf{M}\bar{\mathbf{u}} + \bar{\mathbf{u}}^H\mathbf{M}\tilde{\mathbf{u}} + \tilde{\mathbf{u}}^H\mathbf{M}\bar{\mathbf{u}} + \tilde{\mathbf{u}}^H\mathbf{M}\tilde{\mathbf{u}}|^2\} \tag{A.12}$$

$$= |\bar{\mathbf{u}}^H\mathbf{M}\bar{\mathbf{u}}|^2 + \mathbb{E}\{|\tilde{\mathbf{u}}^H\mathbf{M}\tilde{\mathbf{u}}|^2\}$$

$$+ \mathbb{E}\{|\bar{\mathbf{u}}^H\mathbf{M}\tilde{\mathbf{u}}|^2\} + \mathbb{E}\{|\tilde{\mathbf{u}}^H\mathbf{M}\bar{\mathbf{u}}|^2\} + 2\bar{\mathbf{u}}^H\mathbf{M}\bar{\mathbf{u}}\mathbb{E}\{\tilde{\mathbf{u}}^H\mathbf{M}\tilde{\mathbf{u}}\} \tag{A.13}$$

$$= |\bar{\mathbf{u}}^H\mathbf{M}\bar{\mathbf{u}}|^2 + 2\bar{\mathbf{u}}^H\mathbf{M}\bar{\mathbf{u}}\mathrm{tr}(\mathbf{M}\mathbf{R}_u) + \bar{\mathbf{u}}^H\mathbf{M}\mathbf{R}_u\mathbf{M}^H\bar{\mathbf{u}} + \bar{\mathbf{u}}^H\mathbf{M}^H\mathbf{R}_u\mathbf{M}\bar{\mathbf{u}}$$

$$+|\mathrm{tr}(\mathbf{R}_u\mathbf{M})|^2 + \mathrm{tr}(\mathbf{R}_u\mathbf{M}\mathbf{R}_u\mathbf{M}^H). \tag{A.14}$$

where (A.13) contains the nonzero values of (A.12), while the other terms of (A.12) are zero due to the circular symmetry property. In (A.14), the expectation $\mathbb{E}\{\|\tilde{\mathbf{u}}\|^4\}$ is computed by utilizing [14, Lemma 1].[‡] We complete the proof.

A.2 Proof of Lemma 11.1

We first decompose the channel between user k and the base station into the two components as

$$\mathbf{h}_k = \bar{\mathbf{h}}_k + \tilde{\mathbf{h}}_k,\ \forall k = 1, \ldots, K, \tag{A.15}$$

where $\tilde{\mathbf{h}}_k \sim \mathcal{CN}(\mathbf{0}, \beta_{k'}^{\mathrm{NLoS}}\mathbf{I}_M)$, then the two expectations in the denominator of (11.27) are computed as

$$\mathbb{E}\{\|\mathbf{h}_k\|^2\} = \mathbb{E}\{\|\bar{\mathbf{h}}_k + \tilde{\mathbf{h}}_k\|^2\} = \|\bar{\mathbf{h}}_k\|^2 + M\beta_k^{\mathrm{NLoS}}, \tag{A.16}$$

$$\mathbb{E}\{\|\mathbf{h}_{k'}\|^2\} = \mathbb{E}\{\|\bar{\mathbf{h}}_{k'} + \tilde{\mathbf{h}}_{k'}\|^2\} = \|\bar{\mathbf{h}}_{k'}\|^2 + M\beta_{k'}^{\mathrm{NLoS}}, \tag{A.17}$$

Substituting (A.15)–(A.17) into (11.26), we obtain

$$\frac{\mathbf{h}_k^H\mathbf{h}_{k'}}{\sqrt{\mathbb{E}\{\|\mathbf{h}_k\|^2\}\mathbb{E}\{\|\mathbf{h}_{k'}\|^2\}}} = \frac{\bar{\mathbf{h}}_k^H\bar{\mathbf{h}}_{k'} + \bar{\mathbf{h}}_k^H\tilde{\mathbf{h}}_{k'} + \tilde{\mathbf{h}}_{k'}^H\bar{\mathbf{h}}_k + \tilde{\mathbf{h}}_k^H\tilde{\mathbf{h}}_{k'}}{\sqrt{\left(\|\bar{\mathbf{h}}_k\|^2 + M\beta_k^{\mathrm{NLoS}}\right)\left(\|\bar{\mathbf{h}}_{k'}\|^2 + M\beta_{k'}^{\mathrm{NLoS}}\right)}} \tag{A.18}$$

$$\leq \frac{1}{\sqrt{\beta_k^{\mathrm{NLoS}}\beta_{k'}^{\mathrm{NLoS}}}}\left(\frac{\bar{\mathbf{h}}_k^H\bar{\mathbf{h}}_{k'}}{M} + \frac{\bar{\mathbf{h}}_k^H\tilde{\mathbf{h}}_{k'}}{M} + \frac{\tilde{\mathbf{h}}_{k'}{}^H\bar{\mathbf{h}}_k}{M} + \frac{\tilde{\mathbf{h}}_k^H\tilde{\mathbf{h}}_{k'}}{M}\right) \to 0,\ \text{as}\ M \to \infty, \tag{A.19}$$

[‡]Let us consider a random variable $\tilde{\mathbf{u}} \sim \mathcal{CN}(\mathbf{0}, \mathbf{R}_u)$ and a deterministic matrix $\mathbf{M}$, it holds that $\mathbb{E}\{|\tilde{\mathbf{u}}^H\mathbf{M}\tilde{\mathbf{u}}|^2\} = |\mathrm{tr}(\mathbf{R}_u\mathbf{M})|^2 + \mathrm{tr}(\mathbf{R}_u\mathbf{M}\mathbf{R}_u\mathbf{M}^H)$.

thanks to Assumption 11.1, and the law of large numbers. Consequently, the uncorrelated Rician channels offer the asymptotically favorable propagation.

We now compute the numerator of (11.27) as

$$\mathrm{Var}\{\mathbf{h}_k^H\mathbf{h}_{k'}\} = \mathbb{E}\left\{|\mathbf{h}_k^H\mathbf{h}_{k'}|^2\right\} - |\mathbb{E}\{\mathbf{h}_k^H\mathbf{h}_{k'}\}|^2 \tag{A.20}$$

$$= M\beta_k^{\mathrm{NLoS}}\beta_{k'}^{\mathrm{NLoS}} + \beta_{k'}^{\mathrm{NLoS}}\|\bar{\mathbf{h}}_k\|^2 + \beta_k^{\mathrm{NLoS}}\|\bar{\mathbf{h}}_{k'}\|^2 + |\bar{\mathbf{h}}_k^H\bar{\mathbf{h}}_{k'}|^2, \tag{A.21}$$

where (A.21) is obtained from (A.20) by computing $\mathbb{E}\{|\mathbf{h}_k^H\mathbf{h}_{k'}|^2\}$ utilizing Lemma 11.7. Finally, plugging (A.16), (A.17), and (A.21) into (11.27), we obtain the result as in (11.32).

A.3 Proof of Lemma 11.2

Exploiting the decomposition in (A.15) with taking note that $\widetilde{\mathbf{h}}_k \sim \mathcal{CN}(\mathbf{0}, \mathbf{R}_k)$, the two expectations in the denominator of (11.27) are computed as

$$\mathbb{E}\{\|\mathbf{h}_k\|^2\} = \mathbb{E}\{\|\bar{\mathbf{h}}_k + \widetilde{\mathbf{h}}_k\|^2\} = \|\bar{\mathbf{h}}_k\|^2 + \mathrm{tr}(\mathbf{R}_k), \tag{A.22}$$

$$\mathbb{E}\{\|\mathbf{h}_{k'}\|^2\} = \mathbb{E}\{\|\bar{\mathbf{h}}_{k'} + \widetilde{\mathbf{h}}_{k'}\|^2\} = \|\bar{\mathbf{h}}_{k'}\|^2 + \mathrm{tr}(\mathbf{R}_{k'}). \tag{A.23}$$

The substitution of (A.15), (A.22), and (A.23) into (11.26) yields

$$\frac{\mathbf{h}_k^H\mathbf{h}_{k'}}{\sqrt{\mathbb{E}\{\|\mathbf{h}_k\|^2\}\mathbb{E}\{\|\mathbf{h}_{k'}\|^2\}}} = \frac{\bar{\mathbf{h}}_k^H\bar{\mathbf{h}}_{k'} + \bar{\mathbf{h}}_k^H\widetilde{\mathbf{h}}_{k'} + \widetilde{\mathbf{h}}_{k'}^H\bar{\mathbf{h}}_k + \widetilde{\mathbf{h}}_k^H\widetilde{\mathbf{h}}_{k'}}{\sqrt{(\|\bar{\mathbf{h}}_k\|^2 + \mathrm{tr}(\mathbf{R}_k))(\|\bar{\mathbf{h}}_{k'}\|^2 + \mathrm{tr}(\mathbf{R}_{k'}))}} \tag{A.24}$$

$$\leq \frac{\bar{\mathbf{h}}_k^H\bar{\mathbf{h}}_{k'} + \bar{\mathbf{h}}_k^H\widetilde{\mathbf{h}}_{k'} + \widetilde{\mathbf{h}}_{k'}^H\bar{\mathbf{h}}_k + \widetilde{\mathbf{h}}_k^H\widetilde{\mathbf{h}}_{k'}}{\sqrt{\mathrm{tr}(\mathbf{R}_k)\mathrm{tr}(\mathbf{R}_{k'})}} \leq \frac{1}{a_k a_{k'}}\left(\frac{\bar{\mathbf{h}}_k^H\bar{\mathbf{h}}_{k'}}{M} + \frac{\bar{\mathbf{h}}_k^H\widetilde{\mathbf{h}}_{k'}}{M} + \frac{\widetilde{\mathbf{h}}_{k'}{}^H\bar{\mathbf{h}}_k}{M} + \frac{\widetilde{\mathbf{h}}_k^H\widetilde{\mathbf{h}}_{k'}}{M}\right), \tag{A.25}$$

where the last inequality is obtained by Assumption 11.3 with $\mathrm{tr}(\mathbf{R}_k) = a_k M$ and $\mathrm{tr}(\mathbf{R}_{k'}) = a_{k'} M$, for $a_k, a_{k'} > 0$ at the asymptotic regime. Consequently, the spatially correlated channels offer the asymptotically favorable propagation.

By utilizing Lemma 11.7, the numerator of (11.27) is now computed as

$$\mathrm{Var}\{\mathbf{h}_k^H\mathbf{h}_{k'}\} = \mathrm{tr}(\mathbf{R}_k\mathbf{R}_{k'}) + \bar{\mathbf{h}}_k^H\mathbf{R}_{k'}\bar{\mathbf{h}}_k + \bar{\mathbf{h}}_{k'}{}^H\mathbf{R}_k\bar{\mathbf{h}}_{k'} + \left|\bar{\mathbf{h}}_k^H\bar{\mathbf{h}}_{k'}\right|^2. \tag{A.26}$$

Substituting (A.22), (A.23), and (A.26) into (11.27), we obtain the result as in (11.34).

A.4 Proof of Lemma 11.3

By letting $\mathbf{h}_{k'} = \bar{\mathbf{h}}_{k'} + \widetilde{\mathbf{h}}_{k'}$ where $\widetilde{\mathbf{h}}_{k'} = \sqrt{\beta_{k'}^{\text{NLoS}}}\bar{\mathbf{R}}_{k'}{}^{1/2}\mathbf{G}_{k'}\hat{\mathbf{R}}_{k'}^{1/2}\mathbf{g}_{k'}/\sqrt{S}$, we have

$$\mathbb{E}\{\mathbf{h}_{k'}\mathbf{h}_{k'}^H\} = \mathbb{E}\Big\{(\bar{\mathbf{h}}_{k'} + \widetilde{\mathbf{h}}_{k'})\Big(\bar{\mathbf{h}}_{k'}{}^H + \widetilde{\mathbf{h}}_{k'}{}^H\Big)\Big\} \tag{A.27}$$

$$= \bar{\mathbf{h}}_{k'}\bar{\mathbf{h}}_{k'}{}^H + \mathbb{E}\Big\{\widetilde{\mathbf{h}}_{k'}\widetilde{\mathbf{h}}_{k'}^H\Big\}. \tag{A.28}$$

We now compute $\mathbb{E}\{\widetilde{\mathbf{h}}_{k'}\widetilde{\mathbf{h}}_{k'}^H\}$ by utilizing the property of the double-scattering channel structure as

$$\mathbb{E}\{\widetilde{\mathbf{h}}_{k'}\widetilde{\mathbf{h}}_{k'}^h\} = \frac{\beta_{k'}^{\text{NLoS}}}{S}\overline{\mathbf{R}}_{k'}^{1/2}\mathbb{E}\Big\{\mathbf{G}_{k'}\hat{\mathbf{R}}_{k'}^{1/2}\mathbf{g}_{k'}\mathbf{g}_{k'}^H\hat{\mathbf{R}}_{k'}^{1/2}\mathbf{G}_{k'}^H\Big\}\overline{\mathbf{R}}_{k'}^{1/2} \tag{A.29}$$

$$= \frac{\beta_{k'}^{\text{NLoS}}}{S}\overline{\mathbf{R}}_{k'}^{1/2}\mathbb{E}\{\mathbf{G}_{k'}\hat{\mathbf{R}}_{k'}\mathbf{G}_{k'}^H\}\bar{\mathbf{R}}_{k'}{}^{1/2} \tag{A.30}$$

$$= \frac{\beta_{k'}^{\text{NLoS}}\text{tr}(\hat{\mathbf{R}}_{k'})}{S}\overline{\mathbf{R}}_{k'} \tag{A.31}$$

$$= \beta_{k'}^{\text{NLoS}}\overline{\mathbf{R}}_{k'}, \tag{A.32}$$

where (A.30) is obtained by the fact $\mathbb{E}\{\mathbf{g}_{k'}\mathbf{g}_{k'}^H\} = \mathbf{I}_S$, (A.31) is obtained by utilizing Lemma 11.8, and (A.32) is based on the fact that $\text{tr}(\hat{\mathbf{R}}_{k'}) = S$. Consequently, we obtain

$$\mathbb{E}\{\mathbf{h}_{k'}\mathbf{h}_{k'}^H\} = \bar{\mathbf{h}}_{k'}\bar{\mathbf{h}}_{k'}{}^H + \beta_{k'}^{\text{NLoS}}\overline{\mathbf{R}}_{k'}. \tag{A.33}$$

Similarly, the expectation $\mathbb{E}\{\mathbf{h}_k\mathbf{h}_k^H\}$ is computed in the closed-form expression as

$$\mathbb{E}\{\mathbf{h}_k\mathbf{h}_k^H\} = \bar{\mathbf{h}}_k\bar{\mathbf{h}}_k^H + \beta_k^{\text{NLoS}}\overline{\mathbf{R}}_k. \tag{A.34}$$

Thus, the asymptotically favorable propagation is

$$\frac{\mathbf{h}_k^H\mathbf{h}_{k'}}{\sqrt{\mathbb{E}\{\|\mathbf{h}_k\|^2\}\mathbb{E}\{\|\mathbf{h}_{k'}\|^2\}}} = \frac{\bar{\mathbf{h}}_k^H\bar{\mathbf{h}}_{k'} + \bar{\mathbf{h}}_k^H\widetilde{\mathbf{h}}_{k'} + \widetilde{\mathbf{h}}_{k'}^H\bar{\mathbf{h}}_k + \widetilde{\mathbf{h}}_k^H\widetilde{\mathbf{h}}_{k'}}{\sqrt{\Big(\|\bar{\mathbf{h}}_k\|^2 + \beta_k^{\text{NLoS}}\text{tr}(\bar{\mathbf{R}}_k)\Big)\Big(\|\bar{\mathbf{h}}_{k'}\|^2 + \beta_k^{\text{NLoS}}\text{tr}(\bar{\mathbf{R}}_{k'})\Big)}}$$

$$\leq \frac{1}{\sqrt{\beta_k^{\text{NLoS}}\beta_k^{\text{NLoS}}}}\left(\frac{\bar{\mathbf{h}}_k^H\bar{\mathbf{h}}_{k'}}{M} + \frac{\bar{\mathbf{h}}_k^H\widetilde{\mathbf{h}}_{k'}}{M} + \frac{\widetilde{\mathbf{h}}_{k'}{}^H\bar{\mathbf{h}}_k}{M} + \frac{\widetilde{\mathbf{h}}_k^H\widetilde{\mathbf{h}}_{k'}}{M}\right) \to 0, \quad \text{as } M \to \infty. \tag{A.35}$$

which manifests that the double-scattering channel models offer the asymptotically favorable propagation.

The numerator of (11.27) is first reformulated by $\text{Var}\{\mathbf{h}_k^H \mathbf{h}_{k'}\} = \mathbb{E}\{|\mathbf{h}_k^H \mathbf{h}_{k'}|^2\} - |\bar{\mathbf{h}}_k^H \bar{\mathbf{h}}_{k'}|^2$, then the expectation $\mathbb{E}\{|\mathbf{h}_k^H \mathbf{h}_{k'}|^2\}$ is computed as

$$\mathbb{E}\{|\mathbf{h}_k^H \mathbf{h}_{k'}|^2\} = \text{tr}(\mathbb{E}\{\mathbf{h}_{k'} \mathbf{h}_{k'}^H\}\mathbb{E}\{\mathbf{h}_k \mathbf{h}_k^H\}). \tag{A.36}$$

Combining (A.33), (A.34), and (A.36), the numerator of (11.27) is derived in closed form for the case of double-scattering channel as

$$\text{Var}\{\mathbf{h}_k^H \mathbf{h}_{k'}\} = \beta_k^{\text{NLoS}} \bar{\mathbf{h}}_{k'}^H \overline{\mathbf{R}}_k \bar{\mathbf{h}}_{k'} + \beta_{k'}^{\text{NLoS}} \bar{\mathbf{h}}_k^H \mathbf{R}_{k'} \bar{\mathbf{h}}_k + \beta_{k'}^{\text{NLoS}} \beta_k^{\text{NLoS}} \text{tr}(\overline{\mathbf{R}}_{k'} \overline{\mathbf{R}}_k). \tag{A.37}$$

As a consequence of (A.33) and (A.34), the two expectations in the denominator of (11.27) are computed in closed form as

$$\mathbb{E}\{\|\mathbf{h}_{k'}\|^2\} = \|\bar{\mathbf{h}}_{k'}\|^2 + \beta_{k'}^{\text{NLoS}} \text{tr}(\overline{\mathbf{R}}_{k'}) = \|\bar{\mathbf{h}}_{k'}\|^2 + M\beta_{k'}^{\text{NLoS}}, \tag{A.38}$$

$$\mathbb{E}\{\|\mathbf{h}_k\|^2\} = \|\bar{\mathbf{h}}_k\|^2 + \beta_k^{\text{NLoS}} \text{tr}(\overline{\mathbf{R}}_k) = \|\bar{\mathbf{h}}_k\|^2 + M\beta_k^{\text{NLoS}}, \tag{A.39}$$

with noting that $\text{tr}(\overline{\mathbf{R}}_k) = M, \forall k$. Using (A.37), (A.38), and (A.39) into (11.27) then doing some algebra, we complete the proof.

A.5 Proof of Lemma 11.4

We first compute the numerator of (11.40) as

$$\text{Var}\{\|\mathbf{h}_k\|^2\} = \mathbb{E}\{\|\mathbf{h}_k\|^4\} - (\mathbb{E}\{\|\mathbf{h}_k\|^2\})^2 \tag{A.40}$$

$$= \mathbb{E}\{\|\mathbf{h}_k\|^4\} - \left(\|\bar{\mathbf{h}}_k\|^2 + M\beta_k^{\text{NLoS}}\right)^2 \tag{A.41}$$

$$= 2\beta_k^{\text{NLoS}} \|\bar{\mathbf{h}}_k\|^2 + M(\beta_k^{\text{NLoS}})^2, \tag{A.42}$$

where $\mathbb{E}\{\|\mathbf{h}_k\|^4\}$ is computed by utilizing Lemma 11.9 with the mean $\bar{\mathbf{h}}_k$ and the variance $\beta_k^{\text{NLoS}} \mathbf{I}_M$. After that, plugging (A.42) and (A.16) into (11.40), we complete the proof.

A.6 Proof of Lemma 11.5

The numerator of (11.40) is computed as

$$\text{Var}\{\|\mathbf{h}_k\|^2\} = \mathbb{E}\{\|\mathbf{h}_k\|^4\} - (\|\bar{\mathbf{h}}_k\|^2 + \text{tr}(\mathbf{R}_k))^2 \tag{A.43}$$

$$= 2\bar{\mathbf{h}}_k^H \mathbf{R}_k \bar{\mathbf{h}}_k + \text{tr}(\mathbf{R}_k^2), \tag{A.44}$$

where $\mathbb{E}\{\|\mathbf{h}_k\|^4\}$ is computed by utilizing Lemma 11.9 with the mean $\bar{\mathbf{h}}_k$ and the variance $\mathbf{R}_k$. After that, plugging (A.44) and (A.22) into (11.40), we complete the proof.

A.7 Proof of Lemma 11.6

We compute the numerator of (11.40) by decomposing $\mathbf{h}_k = \bar{\mathbf{h}}_k + \widetilde{\mathbf{h}}_k$ where $\sqrt{\beta_k^{\text{NLoS}}}\overline{\mathbf{R}}_k^{1/-2}\mathbf{G}_k\hat{\mathbf{R}}_k^{1/2}\mathbf{g}_k/\sqrt{S}$, similar to what has been done in Appendix A.5 as

$$\text{Var}\{\|\mathbf{h}_k\|^2\} = \mathbb{E}\{\|\mathbf{h}_k\|^4\} - (\mathbb{E}\{\|\mathbf{h}_k\|^2\})^2. \tag{A.45}$$

The expectation $\mathbb{E}\{\|\mathbf{h}_k\|^4\}$ in (A.45) is computed as

$$\begin{aligned}\mathbb{E}\{\|\mathbf{h}_k\|^4\} &= \mathbb{E}\left\{\left|\left(\bar{\mathbf{h}}_k^H + \widetilde{\mathbf{h}}_k^H\right)(\bar{\mathbf{h}}_k + \widetilde{\mathbf{h}}_k)\right|^2\right\}\\ &= \|\bar{\mathbf{h}}_k\|^4 + \mathbb{E}\left\{\left|\bar{\mathbf{h}}_k^H\widetilde{\mathbf{h}}_k\right|^2\right\} + \mathbb{E}\left\{\left|\widetilde{\mathbf{h}}_k^H\bar{\mathbf{h}}_k\right|^2\right\} + 2\|\bar{\mathbf{h}}_k\|^2\mathbb{E}\{\|\widetilde{\mathbf{h}}_k\|^2\} + \mathbb{E}\{\|\widetilde{\mathbf{h}}_k\|^4\}.\end{aligned} \tag{A.46}$$

The first three expectations in (A.46) are computed by using the statistic information of $\widetilde{\mathbf{h}}_k$ in (A.32) as

$$\mathbb{E}\left\{\left|\bar{\mathbf{h}}_k^H\widetilde{\mathbf{h}}_k\right|^2\right\} = \bar{\mathbf{h}}_k^H\mathbb{E}\left\{\widetilde{\mathbf{h}}_k\widetilde{\mathbf{h}}_k^H\right\}\bar{\mathbf{h}}_k = \beta_k^{\text{NLoS}}\bar{\mathbf{h}}_k^H\overline{\mathbf{R}}_k\bar{\mathbf{h}}_k, \tag{A.47}$$

$$\mathbb{E}\left\{\left|\bar{\mathbf{h}}_k^H\widetilde{\mathbf{h}}_k\right|^2\right\} = \text{tr}\left(\bar{\mathbf{h}}_k\bar{\mathbf{h}}_k^H\mathbb{E}\left\{\widetilde{\mathbf{h}}_k\widetilde{\mathbf{h}}_k^H\right\}\right) = \beta_k^{\text{NLoS}}\bar{\mathbf{h}}_k^H\overline{\mathbf{R}}_k\bar{\mathbf{h}}_k, \tag{A.48}$$

$$\mathbb{E}\{\|\widetilde{\mathbf{h}}_k\|^2\} = \beta_k^{\text{NLoS}}\text{tr}\left(\overline{\mathbf{R}}_k\right) = M\beta_k^{\text{NLoS}}. \tag{A.49}$$

The expectation $\mathbb{E}\{\|\widetilde{\mathbf{h}}_k\|^4\}$ is computed by utilizing the structure of double-scattering channel model as

$$\mathbb{E}\{\|\widetilde{\mathbf{h}}_k\|^4\} = \frac{\left(\beta_k^{\text{NLoS}}\right)^2}{S^2}\mathbb{E}\left\{\left|\mathbf{g}_k^H\hat{\mathbf{R}}_k^{1/2}\mathbf{G}_k^H\overline{\mathbf{R}}_k\mathbf{G}_k\hat{\mathbf{R}}_k^{1/2}\mathbf{g}_k\right|^2\right\} \tag{A.50}$$

$$= \frac{\left(\beta_k^{\text{NLoS}}\right)^2}{S^2}\mathbb{E}\left\{\|\hat{\mathbf{R}}_k^{1/2}\mathbf{g}_k\|^4\left|\frac{\mathbf{g}_k^H\hat{\mathbf{R}}_k^{1/2}\mathbf{G}_k^H}{\|\hat{\mathbf{R}}_k^{1/2}\mathbf{g}_k\|}\overline{\mathbf{R}}_k\frac{\mathbf{G}_k\hat{\mathbf{R}}_k^{1/2}\mathbf{g}_k}{\|\hat{\mathbf{R}}_k^{1/2}\mathbf{g}_k\|}\right|^2\right\}. \tag{A.51}$$

By setting $\mathbf{z}_k = \mathbf{G}_k\hat{\mathbf{R}}_k^{1/2}\mathbf{g}_k/\|\hat{\mathbf{R}}_k^{1/2}\mathbf{g}_k\|$, then $\mathbf{z}_k \sim \mathcal{CN}(\mathbf{0}, \mathbf{I}_M)$ noticing that $\mathbf{g}_k$ and $\mathbf{z}_k$ are independent Gaussian random vectors. We now reformulate (A.51) to as

$$\mathbb{E}\{\|\widetilde{\mathbf{h}}_k\|^4\} = \frac{\left(\beta_k^{\text{NLoS}}\right)^2}{S^2}\mathbb{E}\left\{\|\hat{\mathbf{R}}_k^{1/2}\mathbf{g}_k\|^4\right\}\mathbb{E}\{|\mathbf{z}_k^H\overline{\mathbf{R}}_k\mathbf{z}_k|^2\} \tag{A.52}$$

$$= \left(\beta_k^{\text{NLoS}}\right)^2\left(1 + \frac{\text{tr}\left(\hat{\mathbf{R}}_k^2\right)}{S^2}\right)\left(M^2 + \text{tr}\left(\overline{\mathbf{R}}_k^2\right)\right), \tag{A.53}$$

where the forth moments of Gaussian random vectors in (A.52) are computed by using Lemma 11.9 and noticing that $\hat{\mathbf{R}}_k^{1/2}\mathbf{g}_k \sim \mathcal{CN}(\mathbf{0}, \hat{\mathbf{R}}_k)$. Plugging (A.47)–(A.49), and (A.53) into (A.52), we obtain

$$\mathbb{E}\left\{\|\mathbf{h}_k\|^4\right\} = \|\bar{\mathbf{h}}_k\|^4 + 2\beta_k^{\mathrm{NLoS}}\bar{\mathbf{h}}_k^H\bar{\mathbf{R}}_k\bar{\mathbf{h}}_k + 2M\beta_k^{\mathrm{NLoS}}\|\bar{\mathbf{h}}_k\|^2 + \left(\beta_k^{\mathrm{NLoS}}\right)^2\left(1+\frac{\mathrm{tr}\left(\hat{\mathbf{R}}_k^2\right)}{S^2}\right)\left(M^2+\mathrm{tr}\left(\bar{\mathbf{R}}_k^2\right)\right). \tag{A.54}$$

Subtracting (A.39) from (A.54), we obtain the $\mathrm{Var}\{\|\mathbf{h}_k\|^2\}$ for the double-scattering channel model as

$$\mathrm{Var}\{\|\mathbf{h}_k\|^2\} = 2\beta_k^{\mathrm{NLoS}}\bar{\mathbf{h}}_k^H\mathbf{R}_k\bar{\mathbf{h}}_k + \left(1+\frac{\mathrm{tr}\left(\hat{\mathbf{R}}_k^2\right)}{S^2}\right)\mathrm{tr}\left(\bar{\mathbf{R}}_k^2\right) + \left(\beta_k^{\mathrm{NLoS}}\right)^2\frac{M^2\mathrm{tr}\left(\hat{\mathbf{R}}_k^2\right)}{S^2}. \tag{A.55}$$

Utilizing (A.39) and (A.55) into (11.40), we complete the proof.

References

[1] Mochaourab R, and Jorswieck E. Optimal beamforming in interference networks with perfect local channel information. IEEE Trans Signal Process. 2011;59(3):1128–1141.

[2] Chae CB, Hwang I, Heath RW, *et al.* Interference aware-coordinated beamforming system in a multi-cell environment. IEEE Trans Wireless Commun. 2012;11(10):3692–3703.

[3] Shariati N, Björnson E, Bengtsson M, *et al.* Low-complexity polynomial channel estimation in large-scale MIMO with arbitrary statistics. IEEE J Sel Top Signal Process. 2014;8(5):815–830.

[4] Marzetta TL. Noncooperative cellular wireless with unlimited numbers of base station antennas. IEEE Trans Wireless Commun. 2010;9(11):3590–3600.

[5] Van Chien T, and Björnson E. Massive MIMO communications. In: 5G Mobile Communications. Springer; 2017. p. 77–116.

[6] Al-Salihi H, Van Chien T, Le TA, *et al.* A successive optimization approach to pilot design for multi-cell massive MIMO systems. IEEE Commun Lett. 2018;22(5):1086–1089.

[7] Björnson E, Hoydis J, and Sanguinetti L. Massive MIMO networks: spectral, energy, and hardware efficiency. Found Trends Signal Process. 2017; 11(3–4):154–655. Available from: http://dx.doi.org/10.1561/2000000093.

[8] Van Chien T, Canh TN, Björnson E, *et al.* Power control in cellular massive MIMO with varying user activity: a deep learning solution. arXiv preprint arXiv:190103620. 2019.

[9] Atzeni I, Arnau J, and Debbah M. Fractional pilot reuse in massive MIMO systems. In: Proc. IEEE ICC; 2015.

[10] Ngo H, Larsson E, and Marzetta T. Aspects of favorable propagation in massive MIMO. In: Proc. EUSIPCO; 2014.

[11] Ngo H, Larsson E, and Marzetta T. Energy and spectral efficiency of very large multiuser MIMO systems. IEEE Trans Commun. 2013;61(4):1436–1449.

[12] Yang H, and Marzetta TL. Total energy efficiency of cellular large scale antenna system multiple access mobile networks. In: Proc. IEEE Online GreenComm; 2013.

[13] Ngo HQ, and Larsson EG. No downlink pilots are needed in TDD massive MIMO. IEEE Trans Wireless Commun. 2017;16(5):2921–2935.

[14] Van Chien T, Mollén C, and Björnson E. Large-scale-fading decoding in cellular massive MIMO systems with spatially correlated channels. IEEE Trans Commun. 2019;67(4):2746–2762.

[15] Kim T, and Love DJ. Virtual AoA and AoD estimation for sparse millimeter wave MIMO channels. In: Proc. IEEE SPAWC. IEEE; 2015. p. 146–150.

[16] Matthaiou M, Sayeed AM, and Nossek JA. Sparse multipath MIMO channels: performance implications based on measurement data. In: Proc. IEEE SPAWC. IEEE; 2009. p. 364–368.

[17] Kolomvakis N, Matthaiou M, and Coldrey M. Massive MIMO in sparse channels. In: Proc. IEEE SPAWC. IEEE; 2014. p. 21–25.

[18] 3GPP. Further advancements for E-UTRA physical layer aspects (Release 9). 3GPP TS 36.814; 2010.

[19] Gao X, Tufvesson F, and Edfors O. Massive MIMO channels—measurements and models. In: Proc. ASILOMAR. IEEE; 2013. p. 280–284.

[20] Matthaiou M, Ngo HQ, Smith PJ, *et al.* Massive MIMO with a generalized channel model: fundamental aspects. In: Proc. IEEE SPAWC. IEEE; 2019. p. 1–5.

[21] Van Chien T, Björnson E, and Larsson EG. Multi-cell massive MIMO performance with double scattering channels. In: Proc. IEEE CAMAD. IEEE; 2016. p. 231–236.

[22] Kammoun A, Debbah M, Alouini MS, *et al.* Asymptotic analysis of RZF over double scattering channels with MMSE estimation. IEEE Trans Wireless Commun. 2019;18(5):2509–2526.

[23] Gesbert D, Bölcskei H, Gore DA, *et al.* Outdoor MIMO wireless channels: models and performance prediction. IEEE Trans Commun. 2002;50(12): 1926–1934.

[24] Ertel R, Cardieri P, Sowerby K, *et al.* Overview of spatial channel models for antenna array communication systems. IEEE Pers Commun Mag. 1998;5 (1):10–22.

[25] Marzetta TL. Fundamentals of Massive MIMO. Cambridge University Press; 2016.

[26] Van Chien T. Resource Allocation for Max-Min Fairness in Multi-Cell Massive MIMO. vol. 1797. Linköping University Electronic Press; 2018.

[27] Lim YG, Chae CB, and Caire G. Performance analysis of massive MIMO for cell-boundary users. IEEE Trans Wireless Commun. 2015;14(12):6827–6842.

[28] Van Chien T, Björnson E, and Larsson EG. Joint power allocation and user association optimization for massive MIMO systems. IEEE Trans Wireless Commun. 2016;15(9):6384–6399.

[29] Guo K, Guo Y, Ascheid G, *et al.* Uplink power control with MMSE receiver in multi-cell MU-massive-MIMO systems. In: Proc. IEEE ICC. IEEE; 2014. p. 5184–5190.

[30] Björnson E, Larsson EG, and Marzetta TL. Massive MIMO: ten myths and one critical question. IEEE Commun Mag. 2016;54(2):114–123.

[31] Gerlach D, and Paulraj A. Base station transmitting antenna arrays for multipath environments. Signal Process. 1996;54(1):59–73.

[32] Van Chien T, Dinh KQ, Jeon B, *et al.* Block compressive sensing of image and video with nonlocal Lagrangian multiplier and patch-based sparse representation. Signal Process: Image Commun. 2017;54:93–106.

[33] Eldar YC, and Kutyniok G. Compressed Sensing: Theory and Applications. Cambridge University Press; 2012.

[34] Van Chien T, Björnson E, and Larsson EG. Joint pilot design and uplink power allocation in multi-cell massive MIMO systems. IEEE Trans Wireless Commun. 2018;17(3):2000–2015.

Chapter 12

Novel aspects in terahertz wireless communications

Fawad Sheikh[1], *Muath Al-Hasan*[2] *and Thomas Kaiser*[1]

The new 'Tera-Era' visions extreme-high-speed wireless data transfer of 1 Tb/s and beyond, machines computing at rates of teraflops, and electronic devices performing operations on the femtosecond timescale. The key to break this 1-Tb/s barrier by wireless means incites numerous applications in wireless communications, material characterization, spectroscopy, imaging, sensing, and screening. In other words, 1 Tb/s implies that one may transfer 1 GB in 0.001 s. For instance, some of the most promising 5G applications such as augmented reality/mixed reality in the near future require a wireless data transmission capacity exceeding several hundred Gb/s. Meanwhile, with rapid prevalence of smartphones, more data-hungry applications are to emerge in 5G era. In addition, a forthcoming next-generation video format, namely super hi-vision with a resolution 7,680 $\times$ 4,320, 16 times larger than the current 1080p format requires more than 24-Gb/s data rate depending on its frame rate and colour depth [1]. Likewise, cloud desktop and online gaming relying on Wi-Fi hotspot might accelerate this trend in data rates' traffic beyond the 1-Tb/s limit. Honestly speaking, the current wireless networks are not yet on the foreseeable horizon of 'wireless everything' or 'everything wireless'. *Wireless* stands here for point-to-point radio systems that propagate inside rooms as will be reasoned later. However, by exploiting the power and peril of terahertz (THz) band frequencies, i.e. in the frequency range of 250 GHz to 4 THz, numerous applications for high-speed data links such as 5G cellular network, terabit wireless local area networks, terabit wireless personal area networks, cloud servers, and ultrafast kiosk downloads are practically realizable.

Yet despite the various benefits offered, the venture of employing the THz frequency range in real-world applications still sounds quite elusive. However, it seems clear that to achieve 1-Tb/s wireless links, one requires either larger swaths of spectrum or very high spectral efficiencies. In the microwave regime, wireless data capacity is exploited by increasing spectral efficiency with advanced modulation schemes and robust signal-processing techniques [2–5]. The realization of

[1]Fachgebiet Digitale Signalverarbeitung, Universität Duisburg-Essen, Duisburg, Germany
[2]College of Engineering, Al Ain University (AAU), Al Ain, UAE

such high data rates although becomes quite challenging because of natural fundamental limitation of current technologies: limited bandwidth. Likewise, the data rate provided by the *biologically friendly* technologies with frequencies above 10 THz such as infrared and visible light communications are restricted due to several technical and safety limitations [6].

A promising approach to overcome this bottleneck is to transmit in higher frequency bands instead where large swaths of spectrum are available. An incarnation of this approach is transmission at millimetre-wave (mm-wave) frequencies (20–100 GHz), which has gained much attention in the context of Wi-Fi (802.11ad) [7] and 5G [8]. However, even the large bandwidths available there (up to several GHz) may not be sufficient for many aforementioned future applications, so a move to even higher frequencies becomes desirable, where extreme bandwidths (i.e. 20 GHz and beyond) are available as listed in Table 12.1.

Fortunately, the two atmospheric spectral windows in lower THz band centred at 300 and 350-GHz band offer 47 GHz of continuous bandwidth [10,11] with relatively low atmospheric attenuation. The wireless data transmission of 100 Gb/s and beyond in the THz range is referred to as *THz communications*. For instance, to transmit at 300 GHz, this carrier frequency is the choice of fortune for five reasons: (i) it is five times higher than the highest frequency of 60 GHz used in wireless communications today; (ii) the atmospheric attenuation is less extreme that amounts to no more than 2.8 dB/km, hence falling within the spectral windows; (iii) it procures a 47 GHz of continuous bandwidth which is larger than the unified spectral resources globally available for all kinds of wireless systems; (iv) the wide choice of indoor material groups such as wood, glass, and plastic is still considered to be optically (but not ideally) smooth and, hence, obey geometrical optics rules; and (v) this former barely accessible spectral window due to technological limitations allows a 100-Gb/s high throughput even with a simple modulation scheme for short-range wireless communications. Helping this technological development is the rapid improvement in silicon germanium [12], gallium nitride [13], and graphene technology [14], so that THz science and technology becomes a highly investigated research topic.

Table 12.1 List of five potential spectral windows at THz frequencies with corresponding minimum attenuations

Window number (No.)	Centre frequency (GHz)	Bandwidth (GHz)	FSPL (at 5 m) (dB)	Atmospheric Attn. (dB/km)
I	300	47	~96	>3
II	350	47	~97	>3
III	410	51	~99	12
IV	670	87	~103	38
V	850	111	105	51

Source: [9].

The wireless data transmission of 100 Gb/s and beyond in the THz band frequencies, i.e. in the frequency range of 250 GHz to 4 THz, is referred to as THz communications.

In this chapter, the primary contribution is the comprehensive study of the distinct THz propagation channel and using advanced modelling approaches to overcome the scattering problem of random rough surfaces. A few details about the approval of IEEE Std 802.15.3d-2017 Standards for the lower THz frequency range are also revealed. Prior to this, an analysis of the candle flame impact on ultra-broadband THz communication links across a spectrum of interest from 300 to 310 GHz is performed experimentally to elaborate the propagation loss factor.

12.1 Terahertz wave propagation characteristics

The THz communication systems promise to transfer huge files in the blink of an eye by offering wireless data transfer of 1 Tb/s and beyond for indoor networks. Now, it is of interest to understand the distinct THz propagation mechanisms which form the basis for this reliable high-speed and ultra-broadband THz system design. The individual propagation mechanisms are briefly described with the notion to provide an overall THz multi-ray model of the indoor THz wireless channel. Note that only the essential propagation mechanisms neglecting transmission are mentioned since indoor material thicknesses in the order of few centimetres cause high transmission attenuations up to several 10 dBs for the indoor material groups. At THz frequencies, all surfaces of the common building materials reflect radio waves rather allowing transmission of the waves. Thereupon, transmission never contributes to the multipath propagation in indoor scenarios and can most often be disregarded at THz frequencies. While two major barriers, free-space path loss (FSPL) and transmission attenuation (room entry or exit loss through walls), limit the signal transmission distance and coverage range to (inroom) the scenarios. As mentioned earlier, the term *wireless* cites to the point-to-point radio systems propagating inside rooms. The THz multi-ray model for line-of-sight (LoS) propagation (direct path), reflected (specular), and diffracted and diffuse reflection (non-specular) paths is expressed in the following equation:

$$\boldsymbol{E}_{\mathrm{RX}} = \boldsymbol{E}_{\mathrm{LoS}} + \boldsymbol{E}_{\mathrm{Ref}} + \boldsymbol{E}_{\mathrm{Diff}} + \boldsymbol{E}_{\mathrm{Scatt}} \tag{12.1}$$

12.2 Free-space propagation

LoS propagation is the most simple mode in which the propagation path between a transmitter (TX) and receiver (RX) does not consider any interaction with the absorbing or reflecting materials. It is also referred to as free-space propagation. The LoS electric field $\boldsymbol{E}_{\mathrm{LoS}}$ received by an antenna RX accounting for the polarization vector as well as corresponding antenna gain $\boldsymbol{G}^{\mathrm{RX}}$ and related to the emitted field $\boldsymbol{E}_{\mathrm{TX}}$ radiated by the TX antenna with polarization and antenna gain $\boldsymbol{G}^{\mathrm{TX}}$ is obtained from [15] as

$$\boldsymbol{E}_{\text{LoS}} = \left[\boldsymbol{G}_{\text{LoS}}^{\text{RX}}\right]^{*}\boldsymbol{G}_{\text{LoS}}^{\text{TX}}\ L_{\text{LoS}}\ \boldsymbol{E}_{\text{TX}} \tag{12.2}$$

Furthermore, the FSPL can be calculated based on the Friis equation as a function of transmission distance d and signal wavelength λ with the following definition in decibel (dB) [16]

$$\text{FSPL} = \left(\frac{4\pi d}{\lambda}\right)^2 \ \text{(dB)} \tag{12.3}$$

Equation (12.3) is only valid when the distance is greater enough to fulfil the far-field condition. At THz regime, there is much more FSPL than at lower frequencies since it increases quadratically with the frequency assuming gain of the antennas is constant, as illustrated in Figure 12.1. This FSPL is the fundamental cause that even without any additional attenuation in the transmission path of the THz wave, the received power is much lower than the transmitted power. It poses a major constraint on the communication distance and makes THz regime a promising candidate in delivering 1 Tb/s and beyond indoors rather than in outdoor applications.

12.2.1 Propagation loss factor of atmospheric attenuation

At THz frequencies, the LoS propagation is not only affected by FSPL, i.e. spreading of energy, the power is also lost along a transmission path due to the phenomena called atmospheric attenuation or molecular absorption. Therefore, the term L_{LoS} from (12.2) includes both of these losses. It is worth mentioning that

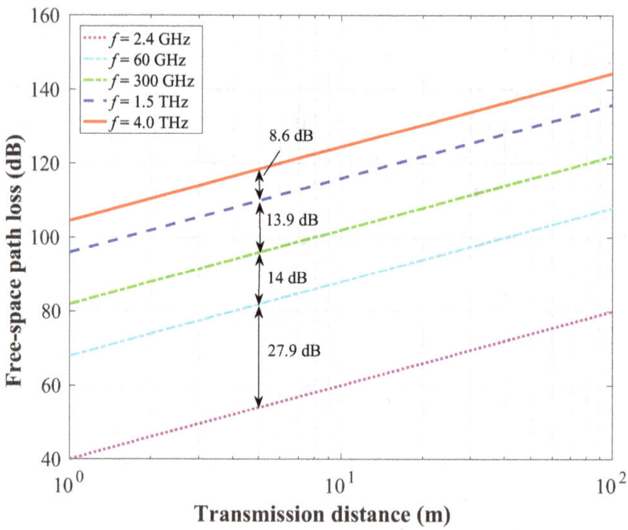

Figure 12.1 Free-space path loss vs. transmission distance d

the resonances for frequencies below 100 GHz occur at 24 GHz for water molecules and 60 GHz for oxygen, while the frequency range from 100 GHz to 1 THz is dominated by only water molecules as evident in Figure 12.2.

Conclusively an approximation of total free-space loss for the traveling wave in THz regime thereupon can be defined as the addition in dB of the FSPL and atmospheric attenuation loss:

$$\text{Total free} - \text{space loss (dB)} = \text{FSPL (dB)} + \text{atmospheric attenuation loss (dB)}$$

Despite the atmospheric attenuation, there are certain spectral windows available for *THz communication*. Atmospheric attenuation values can vary hugely depending on the transmission distance, frequency, and the medium composition (i.e. how many molecules it interacts with). Between 250 GHz and 1 THz, five windows can be identified (cf. Figure 12.2) where the worst case attenuation is below 100 dB/km based on the method described in the Recommendation P.676-10 from the International Telecommunications Union [9].

Meanwhile, in Figure 12.3, the channel transfer functions (CTFs) are depicted by ignoring and considering the atmospheric attenuation for LoS and non-LoS (NLoS) scenarios. The geometry is given in Figure 12.22 where the TX and RX are at heights 2 and 0.75 m, respectively, and separated by a distance of 5 m. In order to show the impact of atmospheric attenuation only, the plaster walls and ceiling are modelled as ideally smooth surfaces ($\sigma_h = 0$). The details about smooth and rough surfaces are given in Sections 12.3 and 12.4. As deduced, the peak-to-peak frequency-dependent variation of up to 87.5 dB at $f_c = 557$ GHz (water vapour

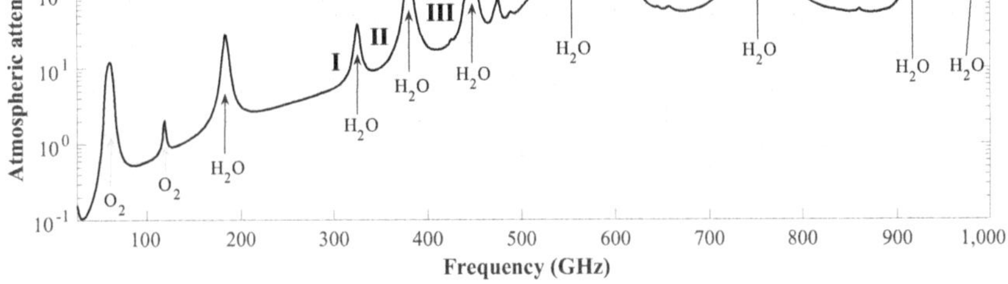

Figure 12.2 Atmospheric attenuation resulting from oxygen absorption and water vapour attenuation (water content 7.5 g/m³, pressure 1013 hPa, and atmospheric temperature 20°C)

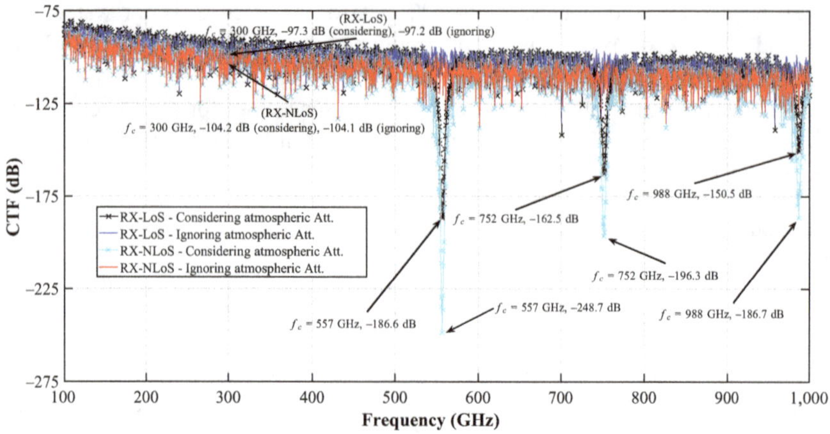

Figure 12.3 CTFs with and without considering the atmospheric attenuation for receivers RX-LoS and RX-NLoS

resonances) is well evident. Conversely, it is 0.1 dB at $f_c = 300$ GHz (spectral window). The atmospheric attenuation for short-range indoor wireless communications in LoS transmission links can thus be neglected at lower THz band frequencies.

For NLoS case, the results depict as expected that the multipaths are comparatively more attenuated. Moreover, in NLoS scenario, most of the total received power contribution comes from reflections. For LoS scenario, however, the contribution of reflection is less poignant and the total received power depends highly on direct path. Unlike LoS, the peak-to-peak frequency-dependent variation of up to 87.5 dB at $f_c = 557$ GHz, 87.5 dB at $f_c = 752$ GHz, and 87.5 dB at $f_c = 988$ GHz is ascertained. Nevertheless, at $f_c = 300$ GHz (spectral window), the atmospheric attenuation for both LoS and NLoS cases is identical.

12.2.2 Novel findings from candle flame analysis

To exploit the power and peril of THz frequencies in the presence of fire, a preliminary analysis of candle flame impact on the ultra-broadband THz communication links across the spectrum of interest from 300 to 310 GHz is investigated. The measurement setup is schematically illustrated in Figure 12.4. This approach is based on recording the frequency-dependent complex scattering parameter S_{21} extracted from channel measurements using a Rohde & Schwarz vector network analyzer ZNA76 to study the variations in total received power and phase. All S_{21}, in addition to the system losses, include the respective antenna gains (i.e. 26 dBi for horn antenna and 9.7 dBi for open-ended waveguide antenna).

In Figure 12.5, the open-ended waveguide and horn antenna are depicted. The channel measurements are performed in a lecture room under LoS environment. The chosen distances d between TX and RX are 0.5 and 1 m, respectively. Perfect

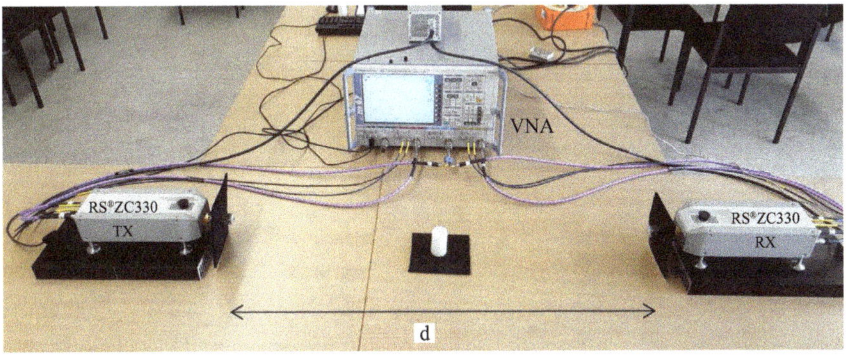

Figure 12.4 300-GHz VNA-based channel sounder

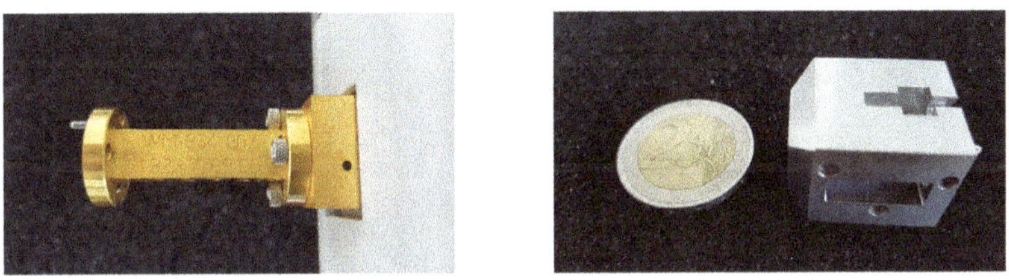

Figure 12.5 Measurement setup: waveguide WM-864 (left) and horn antenna (right)

phase coherence is mandatory to get correct measurements for experimental analysis. Measurements are taken in different scenarios with respect to the distance between the two RSZC330 extenders as well as the three different antenna configurations: horn–horn, horn–open, and open–open. The number of chosen sweep points is 801, and the output power level from the test port is -10 dBm. A two-port TOSM (through, open, short, match) calibration is executed on the converter's waveguide ports considering our experiment's parameters. A detailed description of the scenario is omitted here because of its minor relevance in the results presented in the author's former works [17,18].

The transfer functions for the three scenarios are illustrated in Figure 12.6 for $d = 0.5$ m and in Figure 12.7 for $d = 1$ m. The respective phase shift is also illustrated for the above-mentioned distances in Figures 12.8 and 12.9. In the measurement process, the S_{21} is calculated by averaging ten snapshots taken at different time instances as

$$PL[dB] = -10\log_{10}\left[\frac{1}{N}\sum_{i=1}^{10}|S_{21}(f_i)|^2\right] \tag{12.4}$$

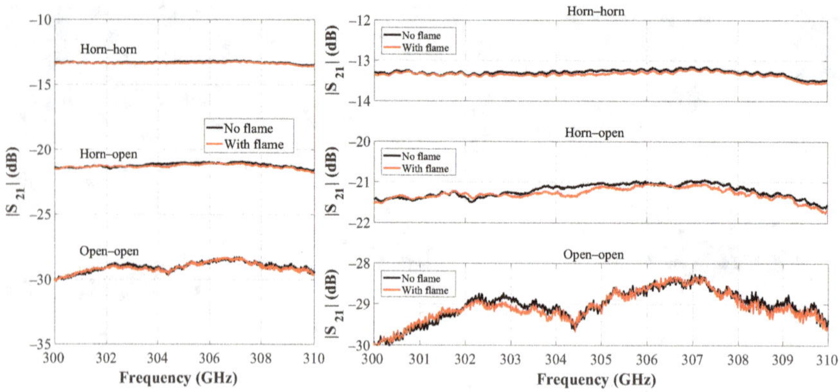

Figure 12.6 Transfer functions for three channel scenarios: horn–horn, horn–open, and open–open for d = 0.5 m

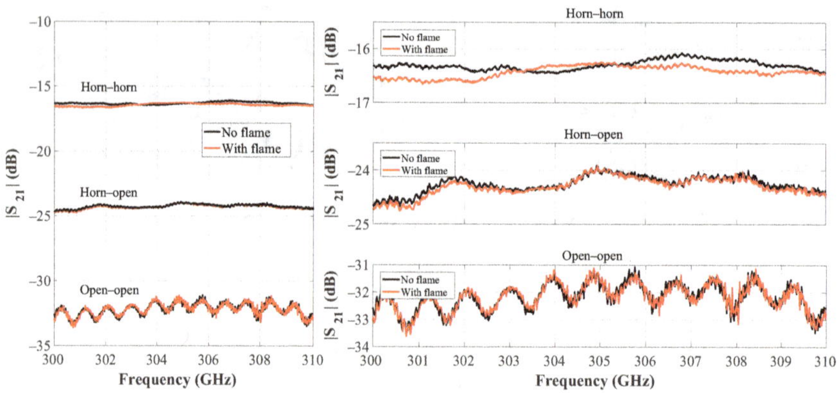

Figure 12.7 Transfer functions for three channel scenarios: horn–horn, horn–open, and open–open for d = 1 m

The conclusions drawn from this work are as follows:

- For both scenarios (i.e. d = 0.5 m and d = 1 m) in all the three channel configurations (i.e. horn–horn, horn–open, and open–open), the experimental results demonstrate that the THz electromagnetic (EM) wave propagation is exposed to slight attenuation (i.e. amplitude) in the course of free-space propagation through candle flame and this is mainly caused by water vapour attenuation. Hence, this slight attenuation does not pose any serious challenge to the precise characterization and localization of materials at THz frequencies.
- For both distances and all of the three channel configurations in the considered scenarios, we experimentally investigated the THz EM wave propagation concluding that the wave propagation is delayed (phase shifted). However, this phase shift is more prominent in both of the Open–Open antenna configurations.

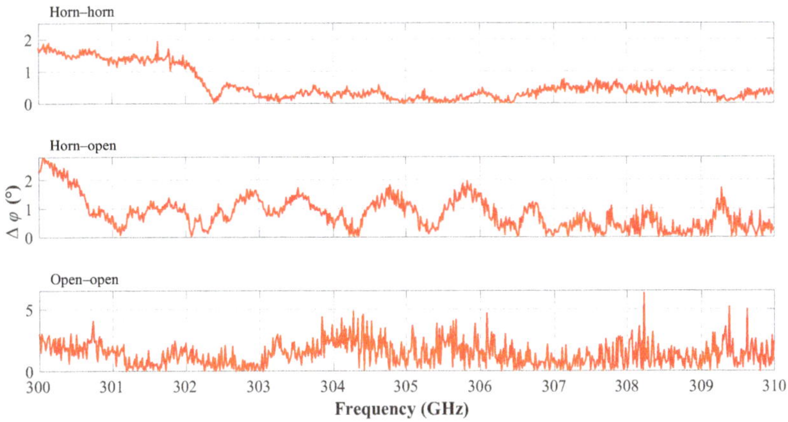

Figure 12.8 Phase difference with and without candle flame for three channel scenarios: horn–horn, horn–open, and open–open for d = 0.5 m at 801 frequency points for $f = 300 \ldots 310$ GHz *in free-space measurements*

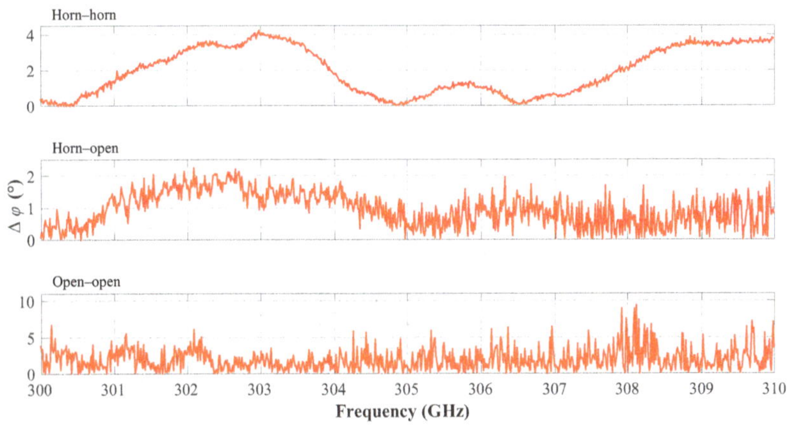

Figure 12.9 Phase difference with and without candle flame for three channel scenarios: horn–horn, horn–open, and open–open for d = 1 m at 801 frequency points for $f = 300 \ldots 310$ GHz *in free-space measurements*

- The previous results provide useful information for designing the ultra-broadband THz system. Meanwhile, adaptive beamformers on the TX as well as RX ends play a crucial role for precise localization of materials and to adequately address the time-invariant nature of the THz wireless channel.

The beam spot size for free-space propagation (after the beam exits from the horn antenna) is bypassed. As the beam spot size for the 1-m distance is about 16.6 cm at

$f = 300$ GHz horizontally and 12.5 cm vertically, the measurements for analysis cannot be performed in the lab due to safety reasons.

12.3 Reflection by a smooth surface

A reliable communication link can be guaranteed only if a THz communication system utilizes indirect paths (i.e. reflections) as a backup. The direct path can be blocked completely by moving objects in the LoS propagation, such schemes have to rely on NLoS propagation paths by making use of multiple reflections from the walls, floor, and the ceiling. The indoor propagation environments usually have dielectric and conducting materials in their surroundings. The surfaces of these materials could be either smooth or rough. When an EM wave is incident on a smooth surface, a portion of the wave is reflected and part of the energy is lost due to the transmission in the material. This reflection can be perceived as specular in a single direction where the angle of incidence (grazing angle) and reflection angle are equal. If the reflection point belongs to the dimension of the surface, the first-order reflected field at the receiving point O can be expressed from [15] as

$$\boldsymbol{E}_{\mathrm{Ref}}(O) = \left[\mathrm{G}_{\mathrm{Ref}}^{\mathrm{RX}}\right]^{*}\mathbf{R}\ G_{\mathrm{Ref}}^{\mathrm{TX}}\ L_{\mathrm{Ref}}\ \boldsymbol{E}_{\mathrm{TX}} \tag{12.5}$$

where

$$\mathbf{R} = \mathbf{R}_{\mathrm{RX}}\begin{pmatrix} \Gamma_{\mathrm{TE}} & \zeta_1 \\ \zeta_2 & \Gamma_{\mathrm{TM}} \end{pmatrix}\mathbf{R}_{\mathrm{TX}} \tag{12.6}$$

Please note that the path length of reflections in L_{Ref} is the path length from TX to reflection point and then further from reflection point to point O. The ζ_1 and ζ_2 are cross-polarization coupling coefficients. For specular reflection paths, these coefficients are equal to 0 [7]. Note that $\mathbf{R}_{\mathrm{RX}}$ and $\mathbf{R}_{\mathrm{TX}}$ are the geometrical depolarization vectors from RX and TX, respectively [19], explained in more detail later in Section 12.4.

At THz frequencies, in the case of ideally smooth and homogeneous surfaces such as glass and wood, the characteristics of the reflection can be described by the well-known Fresnel reflection coefficient Γ. The smooth surface Fresnel reflection coefficient depends not only on the electrical properties (permittivity, permeability, and conductivity) of the reflecting material, but also on the polarization, frequency, and grazing angle briefly explained next.

12.3.1 Dependence of the reflection coefficient on polarization

The Fresnel reflection coefficient expressions for perpendicular (Γ_{TE}) and parallel (Γ_{TM}) polarizations for smooth surfaces can be found in [20, p. 21], as

$$\Gamma_{\mathrm{TE}}(f, \Theta_i) = \frac{Z \cos \Theta_i - Z_0 \cos \Theta_t}{Z \cos \Theta_i + Z_0 \cos \Theta_t} \tag{12.7}$$

and

$$\Gamma_{\mathrm{TM}}(f,\Theta_i) = \frac{Z\cos\Theta_t - Z_0\cos\Theta_i}{Z\cos\Theta_t + Z_0\cos\Theta_i} \tag{12.8}$$

where Θ_i is the grazing angle (incident angle), $\Theta_t = \arcsin(\sin(\Theta_i)Z/Z_0)$ is the angle of refraction, $Z_0(=377\,\Omega)$ is the free-space impedance, and the wave impedance Z of the reflecting material is calculated as

$$Z = \sqrt{\frac{\mu_o}{\varepsilon_o(n^2 - (\alpha c/4\pi f)^2 - j(2n\alpha\, c/4\pi f))}} \tag{12.9}$$

where μ_o, ε_o, c, and f are free-space permeability, permittivity, velocity, and the frequency of the incident wave, respectively. α is the absorption coefficient and the frequency-dependent complex refractive index n of the smooth surface is given by

$$n = \sqrt{\varepsilon_r}, \quad \text{and} \quad \varepsilon_r = \widetilde{\varepsilon}_r - j\widetilde{\widetilde{\varepsilon}}_r = \widetilde{\varepsilon}_r - j60\lambda\sigma \tag{12.10}$$

where $\widetilde{\varepsilon}_r$, $\widetilde{\widetilde{\varepsilon}}_r$, λ, and σ are the real part of the relative complex permittivity, imaginary part of the relative complex permittivity, the wavelength at hand, and the medium conductivity in mhos/metre, respectively. The empirically computed complex dielectric constants (ε_r) of the indoor materials at 300-GHz carrier frequency used in our modelling approaches are described in Table 12.2.

In the case of smooth surface, Γ_{TM} corresponds to vertical linear polarization, while Γ_{TE} corresponds to horizontal polarization. Comparing these two reflection coefficients, one can see that the response of horizontal polarization increases by increasing the grazing angle, as indicated in Figure 12.10.

12.3.2 Dependence on the grazing angle

The minimum in Γ_{TM} occurs at the Brewster's polarization angle (cf. Figure 12.10), the angle where the transmitted wave is completely absorbed by the surface but a sudden increase in the response of vertical polarization is witnessed again with further increase of the grazing angle. In other words, the grazing angle marks the

Table 12.2 Materials and their parameters used in ray-tracing (from [21,22,46])

Material	$\widetilde{\varepsilon}_r$	$\widetilde{\widetilde{\varepsilon}}_r$	ℓ_{cr} (mm)	σ_{h} (mm)
Plaster x (perf. smooth)	3.691	0.217	0.00	0.00
Plaster x1 (slightly rough)	3.691	0.217	1.30	0.05
Plaster x2 (slightly rough)	3.691	0.217	1.70	0.15
Plaster x3 (very rough)	3.691	0.217	1.80	0.25
Wood (perf. smooth)	1.734	0.073	0.00	0.00
PVC (perf. smooth)	2.788	0.069	0.00	0.00
Polycarbonate (perf. smooth)	2.750	0.011	0.00	0.00
Glass (perf. smooth)	6.656	0.539	0.00	0.00

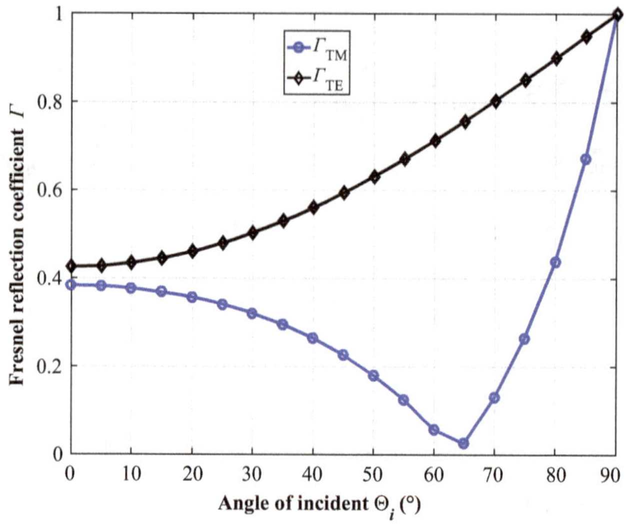

Figure 12.10 Magnitude of reflection coefficients Γ_{TE} *and* Γ_{TM} *from smooth plaster* ($\tilde{\varepsilon}_r = 3.691, \tilde{\tilde{\varepsilon}}_r = 0.217$) *at* $f = 350$ GHz *for angle of incidence between 0° and 90°*

sign change in the real part of the reflection coefficient for parallel polarization. Furthermore, in the case of very small grazing angle ($\Theta_i \approx 0°$), we have

$$\Gamma_{\mathrm{TE}} = \Gamma_{\mathrm{TM}} \tag{12.11}$$

Conversely, when ($\Theta_i \approx 90°$), we get

$$\Gamma_{\mathrm{TE}} = -\Gamma_{\mathrm{TM}} \tag{12.12}$$

12.3.3 Dependence on frequency

The Fresnel equations have the frequency-dependent refractive index and absorption coefficient in the equation of wave impedance Z, which means the reflectivity is also frequency dependent. At THz frequencies, however, the refractive indices do not vary considerably with frequency for common indoor building materials [21]. In contrast, a significant increase of the absorption coefficient with the frequency is conspicuous.

12.4 Reflection by a rough surface

Reflection by a rough surface or diffuse reflection is the most critical and important propagation phenomenon at THz frequencies. The diffuse reflection of EM waves is merely dependent on the surface roughness, incident angles, complex refractive index of materials, and wavelength at hand. At THz frequencies, diffuse reflection tends to be particularly strong due to the increased surface roughness and this causes an additional attenuation even in the specular direction of reflection (cf. Fig. 6a–b in [22]). Generally speaking, when an EM wave is incident on a rough surface, the phenomenon of diffuse scattering splits the wave into a specular and many non-specular paths. When a wave is scattered, the resulting reflections occur in many directions, though certain privileged directions may receive more energy as depicted in Figure 12.11. It is noteworthy to mention that we deal with *random rough surfaces* throughout the chapter. The reason is straightforward that the shape of the indoor rough surfaces met in practice can merely be described by a random function of coordinates; hence, it can be described statistically. Henceforth, the diffuse scattering from indoor surfaces should also be dealt as a statistical problem to determine the statistical characteristics of the scattered field (distribution functions, moments, correlation functions, etc.) from the statistical properties of the real physical surface [23].

Additionally, the surface topography of any real indoor material reveals that periodic surfaces are actually never encountered. Likewise, even little violations of periodicity lead to qualitatively unique phenomena, namely disappearance of specular lobe and distortion of the side lobes. Besides, it is nonessential to investigate the fine structure of the scattered field and knowledge of certain parameters of the reflected signal averaged over a whole class of surfaces and reflecting objects proves sufficient [23]. As a matter of fact, the statistical approach for the solution of scattering problems from a rough surface received broad acceptance in the past [20], [24–27].

12.4.1 Basic geometry of scattering

Let us assume the incident field $\boldsymbol{E}_{\mathrm{TX}}$, a harmonic plane wave of unit amplitude, is incident on the random rough surface element dS. The elevation angle of incidence is represented as Θ_i and the elevation and azimuth reflected angles in the scattered region as Θ_r and Θ_s, respectively. To avoid confusion while representing single diffuse reflection in a basic geometry, we actually need two reflection angles to define it. One angle (Θ_r) which goes from the normal (cf. Figure 12.12, it is z-axis)

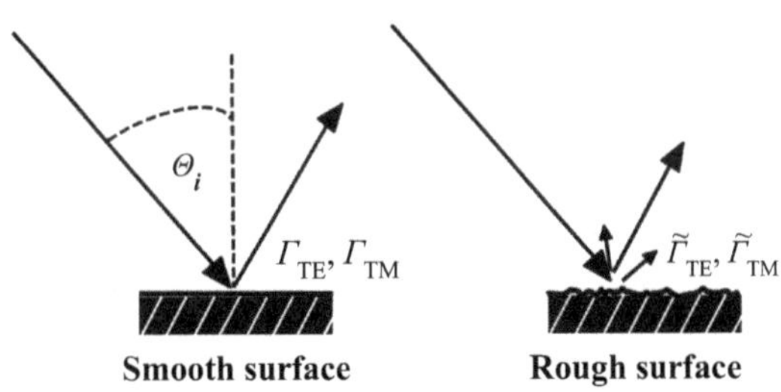

Figure 12.11 Specular reflection by a smooth (left) and diffuse reflections by a rough (right) surface

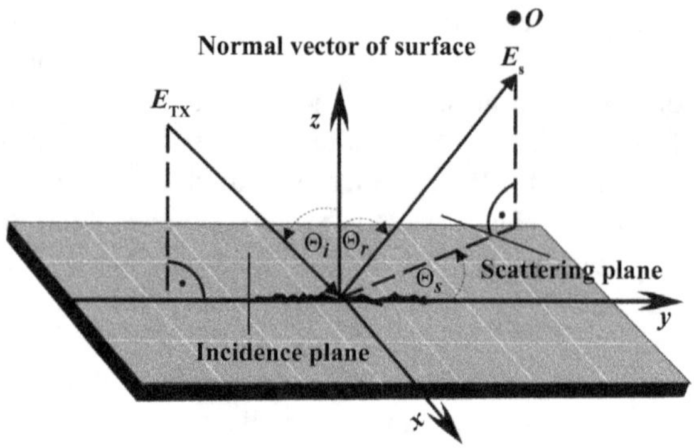

Figure 12.12 Illustration of the basic scattering geometry

and the other angle (Θ_s) which goes from the incident plane (it is y-axis). However, there are various other diffuse reflections also, not displayed in the figure. The angles are measured with respect to the z-axis as schematically shown in Figure 12.12.

Assuming the observation point O is far above the selected rough surface to fulfil the far-field condition. The scattered field at the receiving point O has the following expression:

$$\boldsymbol{E}_{\text{Scatt}}(O) = \left[\mathrm{G}_{\text{Scatt}}^{\text{RX}}\right]^* \mathbf{R}\, \boldsymbol{G}_{\text{Scatt}}^{\text{TX}}\, L_{\text{Scatt}}\, \boldsymbol{E}_{\text{TX}} \tag{12.13}$$

Notably, for diffuse scattering from rough surfaces, the terms ζ_1 and ζ_2 in $\mathbf{R}$ are now not equal to 0. Rather, they are determined by the perturbation theory [28]. Moreover, due to the rough surface, the Γ_{TE} and Γ_{TM} are also changed in $\mathbf{R}$. Please refer to the modified reflection coefficients in Section 12.4.3.

12.4.2 Statistical description of rough surface

A surface categorized as rough in macroscale might be smooth in microscale or vice versa [29]. The surfaces can perhaps be perceived as rough in various ways, i.e. they can be periodically rough like sinusoidal, saw-tooth, or rectangular. Nevertheless, the surfaces just mentioned have one common thing – *they produce a scattered field*. In consequence, there is a need to describe a rough surface accurately and concisely. In more general terms, the physical characteristics of a random rough surface are described by its simplest statistical functions, like the height and slope distribution. These can be computed as non-cumulative or cumulative. Basically, these functions are computed as normalized histograms of the height or slope values. Note that the height and slope distribution quantities belong to the first-order statistical quantities, thus describing only the statistical properties of the individual points. To simplify the discussion, a rough surface can be defined as the surface height having a non-zero variance $\sigma_h \neq 0$ with respect to the mean height $\bar{h}$, which constitutes an absolutely smooth reference surface.

In Figure 12.13, a rough surface with standard deviation height of the surface roughness $\sigma_h = 0.10$ mm is shown, while the corresponding surface height histogram is illustrated in Figure 12.14. The average surface height $\bar{h}$ or the first moment of the random variable $h(\mathbf{r})$ at a position $\mathbf{r}$ is assumed 0. The $\bar{h} = 0$ mm for any position on the surface warrants that the random field $h(\mathbf{r})$ is homogeneous.

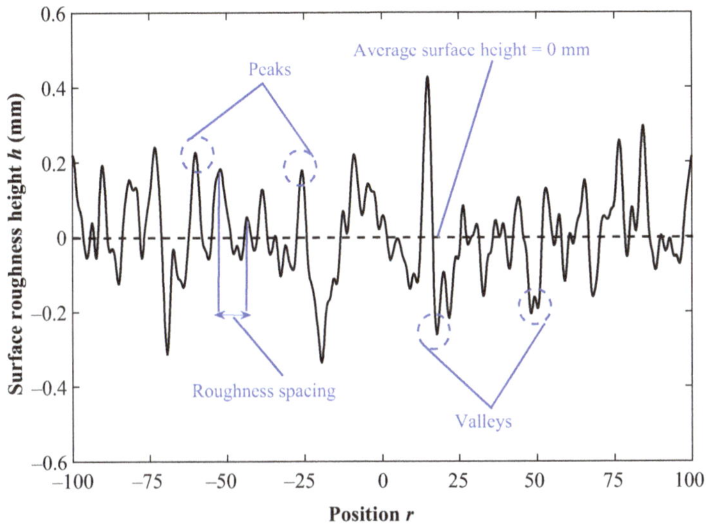

Figure 12.13 *Rough surface with* $\sigma_h = 0.10$ mm. *The dotted lines in the figure indicate the average surface height positions*

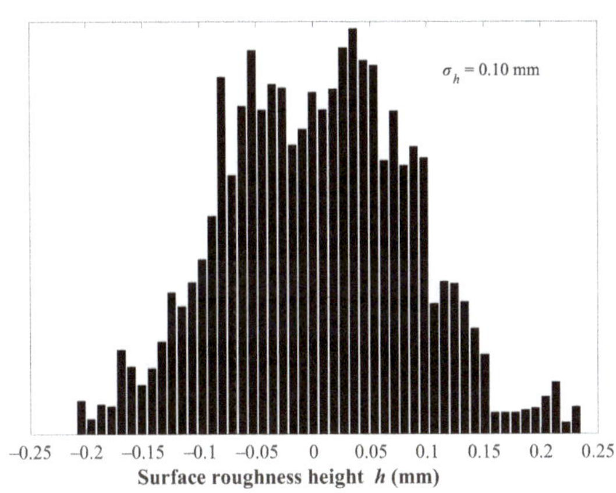

Figure 12.14 *Histogram of rough surface with* $\sigma_h = 0.10$ mm *from Figure 12.13*

12.4.3 The Rayleigh method

To account for the decrease in the reflected energy in specular direction due to the roughness of materials in the THz range, Fresnel reflection coefficients must be multiplied by the Rayleigh roughness factor [20]. Namely, the field reflected from the rough surface is reduced from that of a smooth surface by this factor. Thus, the scattering losses for the specular reflections are equal to the Rayleigh roughness factor (ρ_{spec}) expressed as

$$\rho_{\text{spec}} = e^{-g/2} = \exp\left(-\frac{8\pi^2 f^2 \sigma_h^2 \cos^2\Theta_i}{c^2}\right) \tag{12.14}$$

where

$$g = \sigma_h^2\left(\frac{2\pi f}{c}\right)^2 (\cos(\Theta_i))^2 \tag{12.15}$$

or

$$g = k^2\sigma_h^2(\cos(\Theta_i) + (\cos(\Theta_r))^2 \tag{12.16}$$

where g is the roughness parameter of a material, f the frequency of incident wave, σ_h the standard deviation height of surface roughness, Θ_i the angle of incidence and reflection relative to surface normal, and c the velocity of light. The parameter g, an interesting and simple means of estimating the degree of EM roughness of a surface, i.e. if a surface can be qualified as smooth ($g = 0$), slightly rough ($g \ll 1$), moderately rough ($g \geq 1$), or very rough ($g \gg 1$), is depicted with its reflection and scattering pattern in Figure 12.15.

From (12.14), it is apparent that the roughness parameter g of a material depends on standard deviation height σ_h, angle of incidence Θ_i, and the carrier frequency f. Further, the consequence of cosine function in the g formula indicates that the surface roughness impact for an impinging wave is lower towards higher angles of incidence. The corresponding g values of the rough materials, i.e. plaster x1, plaster x2, and plaster x3 used in our scenario are illustrated in Figure 12.16 for

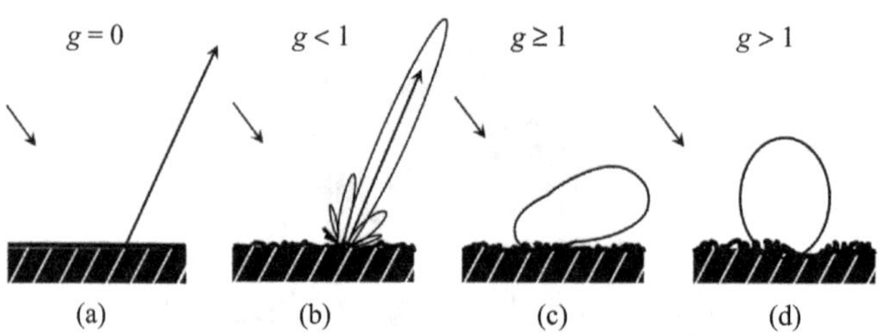

Figure 12.15 Transition from specular reflection to diffuse scattering. The surfaces are (a) smooth, (b) slightly rough, (c) moderately rough, and (d) very rough. Figure adapted from Fig. 5.1 in [20]

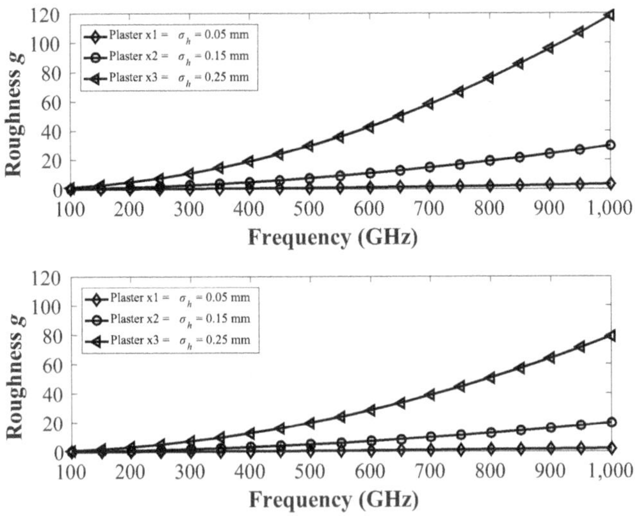

Figure 12.16 Calculated roughness parameter g of three different plasters $(\widetilde{\varepsilon}_r = 3.691, \widetilde{\varepsilon}_r = 0.217)$ *for angles of incidence 30° (top) and 45° (bottom)*

the frequency range between 100 GHz and 1 THz for two angles of incidence: 30° and 45°.

The modified reflection coefficients for parallel $\widetilde{\Gamma}_{\mathrm{TE}}$ and perpendicular $\widetilde{\Gamma}_{\mathrm{TM}}$ polarizations in the presence of rough surfaces taking into account the different standard deviation heights are then

$$\widetilde{\Gamma}_{\mathrm{TE}}(f, \Theta_i) = \rho_{\mathrm{spec}} \Gamma_{\mathrm{TE}}(f, \Theta_i) \tag{12.17}$$

also

$$\widetilde{\Gamma}_{\mathrm{TM}}(f, \Theta_i) = \rho_{\mathrm{spec}} \Gamma_{\mathrm{TM}}(f, \Theta_i) \tag{12.18}$$

In [30], the THz time-domain spectroscopy (THz-TDS) is used to measure the absorption coefficient and refractive index of rough surfaces (i.e. granular wall-paper and concrete plaster). Then, the reflection coefficients for the specular reflection are calculated using the conventional Fresnel equations. Next, the modified reflection coefficients $\widetilde{\Gamma}$ are calculated from (12.17) and (12.18) owing to the roughness $(g \neq 0)$ of the investigated samples. By using the aforementioned procedure, the coverage simulation results in [30] show a good agreement with measurements upon comparison. Please note that the conventional reflection coefficients do conform with the measurements for long wavelengths (i.e. at lower frequencies) but deviate for shorter wavelengths (i.e. at THz frequencies). Thus, in [30], the deviation is rectified by including the Rayleigh roughness factor. Consequently, this entails channel models with surface roughness at THz

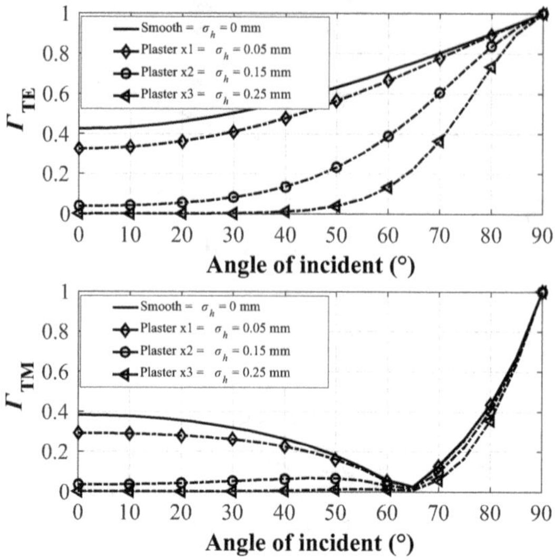

Figure 12.17 Magnitude of (modified) reflection coefficient for angle of incidence between 0° and 90° at a frequency of 350 GHz for TE and TM polarized wave (material: plaster)

frequencies. It is now employed to calculate magnitude of the reflection coefficients Γ_{TE} and Γ_{TM}, shown in Figure 12.17 as a function of angle of incidence at the centre frequency 350 GHz for the rough plasters used in the scenario depicted in Figure 12.18. The solid line (cf. Figure 12.17) is used as a benchmark for comparing smooth (i.e. $\sigma_h = 0$) and rough plasters.

12.4.3.1 Coverage simulations attributive of roughness

The diffuse reflection of EM waves is merely dependent on the surface roughness, incident angles, complex refractive index of materials, and wavelength at hand. At THz frequencies, diffuse reflection tends to be particularly strong due to the increased surface roughness and this causes an additional attenuation even in the specular direction of reflection. In the above-mentioned figures, the influence of standard deviation height σ_h is theoretically evaluated in the case of smooth and rough plasters. However, in order to investigate the influence of different materials along with their roughness on the 300-GHz propagation channel, simulations in a realistic environment are indispensable. This expedites the perception of plausible TX/RX constellations and assists in bringing out the materials that affect the quality of the THz links as a result of scattering.

These simulations are carried out in an empty office room of 7 m (length) × 7 m (width) × 3 m (height) which is an LoS cubic scenario. The TX is placed at the points $x = 6$ m, $y = 1$ m, $z = 2$ m. There are once again 729 receiver points (RXPs). Taking into account the tabletop terminals, RXPs are chosen to possess a *height* = 0.75 m. All walls and the ceiling are made of plaster except for the floor. The floor is covered by smooth polyvinyl chloride (PVC) with $\widetilde{\varepsilon}_r = 2.788, \widetilde{\widetilde{\varepsilon}}_r = 0.069$. Due to the smooth flooring, only specular reflections from the floor are included. Up to twice-reflected paths (i.e. second order) are considered in the modelling. To maximize scattering contributions and lower scattering losses, vertically polarized omnidirectional antennas are chosen at both TX and RXPs. The diffraction (i.e. no objects) and transmission are neglected. RX sensitivity level is set to -140 dBm.

The results in Figure 12.18 show the impact of surface roughness on the THz propagation channel in terms of coverage maps obtained from ray-tracing simulations under LoS scenario. As deduced, the roughness (i.e. variable roughness σ_h ranging from 0.05 to 0.25 mm) peculiarly persists without any noticeable influence on the relative received power where an average attenuation of around 91.72 dB is recorded ignoring the surface roughness. Regardless of the direct path, the scattering is quite significant in the points near to the walls and distant from the TX. Nevertheless, the average attenuations for $\sigma_h = 0.05,\ 0.10,\ 0.15,\ 0.20,$ and 0.25 mm are recorded as 92.08, 92.72, 93.13, 93.30, and 93.38 dB, respectively. Intuitively, these almost identical average attenuations are perhaps highlighting the effect of a dominant LoS path components (direct path) compared to the scattered channel paths with almost negligible effect. In other words, the relative scattered power is negligible.

Furthermore, we calculate the modified reflection coefficients $\widetilde{\Gamma}_{\mathrm{TE}}$ and $\widetilde{\Gamma}_{\mathrm{TM}}$ of the corresponding material (rough plaster) used in all walls and ceiling taking into

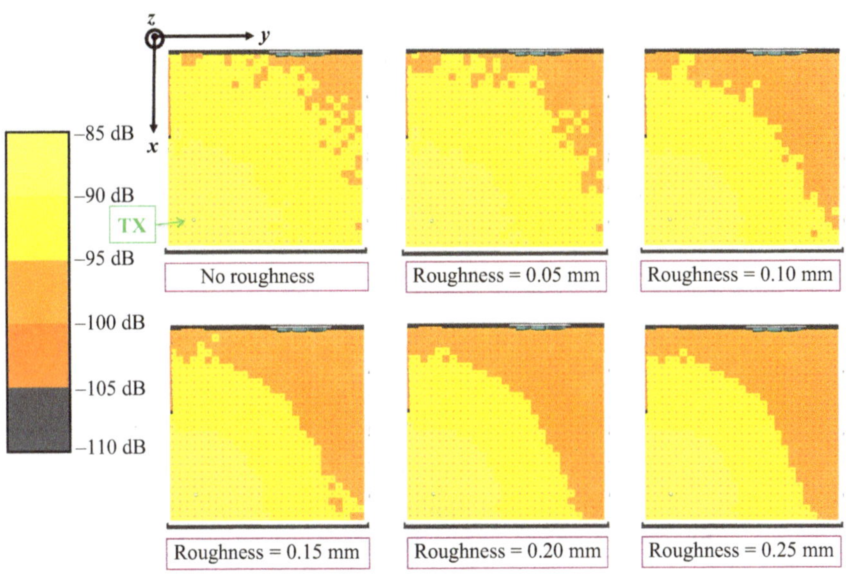

Figure 12.18 Coverage maps attributive of roughness dependence on the total received power at 300 GHz for LoS scenario. TX at the position $x = 6$ m, $y = 1$ m, $z = 2$ m, and 729 RXPs are 0.75 m in height

account the different standard deviations of surface height to identify exactly how the material is impairing the quality of THz wireless links for first-order reflections (Figure 12.19).

12.4.4 Depolarization

Reflection by a rough surface not only scatters the field towards non-specular directions but also depolarizes it. This change in polarization of the incident wave on being scattered at the surface is proportional to the surface roughness of the material in a distinctive manner. Thus, one should distinguish between depolarization that occurs due to scattering from a rough surface and geometrical depolarization. Both of these polarization effects can be measured in terms of the cross-polar discrimination (XPD) ratio, where an orthomode transducer enables the vertically polarized transmitting signal to capture both horizontal (cross-polar) and vertical (co-polar) signal components at the RX end. The XPD is defined in [31] by the following expression:

$$XPD = 10\log_{10}\left(\frac{P_V}{P_H}\right) \tag{12.19}$$

where P_V and P_H correspond to the received power of the vertical and the horizontal component, respectively. In [31], it is shown that the depolarization is more dominant in the case of rough surfaces than that caused by smooth surfaces.

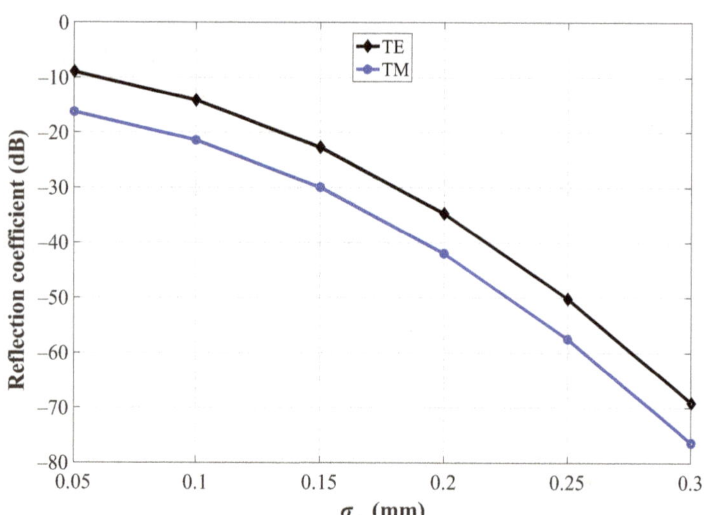

Figure 12.19 Calculated magnitude of reflection coefficients $\widetilde{\Gamma}_{TE}$ and $\widetilde{\Gamma}_{TM}$ for specular direction on rough plaster ($\widetilde{\varepsilon}_r = 3.691$, $\widetilde{\widetilde{\varepsilon}}_r = 0.217$) at $f = 300$ GHz for incidence angle $\Theta_i = 45°$ and standard deviation heights σ_h from 0.05 to 0.30 mm

Besides, the XPD becomes bigger for higher incident angles but no dependency on distance is observed. Similarly, in [32], it is shown that depolarization arises in backscattering from dielectric rough surface. However, in [33] it is concluded that neither depolarization nor polarization dependence is found in scattering from a perfectly conducting rough surface. Note that a rough surface is often assumed to be perfectly conducting such that the boundary condition is satisfied. Perhaps, the scattering from a perfectly rough surface is also a convenient starting point as an assumption for the development of mathematical models.

12.5 Diffraction

Diffraction occurs when an obstacle with sharp irregularities (i.e. wedges, edges, and circular cylinders) and large dimensions as compared to the propagated wavelength comes between the TX and RX. Even a moving person blocking the path of an EM wave can be considered as a diffracting object at THz frequencies [34]. Apart from persons, the diffracting objects in an indoor environment are metallic or wooden edges of file cabinets, bookshelves, cupboards, tables, workplace-privacy panels, and open windows and doors. Due to diffraction phenomena, the propagated wave bends around the obstacle and continues propagating in the geometrical shadow region (or incident shadow boundary) behind the obstacles, as evident from Figure 12.20.

The well-known models with analytical description of the diffraction behaviour of many different objects are the uniform theory of diffraction (UTD) [35] and the knife-edge diffraction [28]. The first approach not only holds valid for

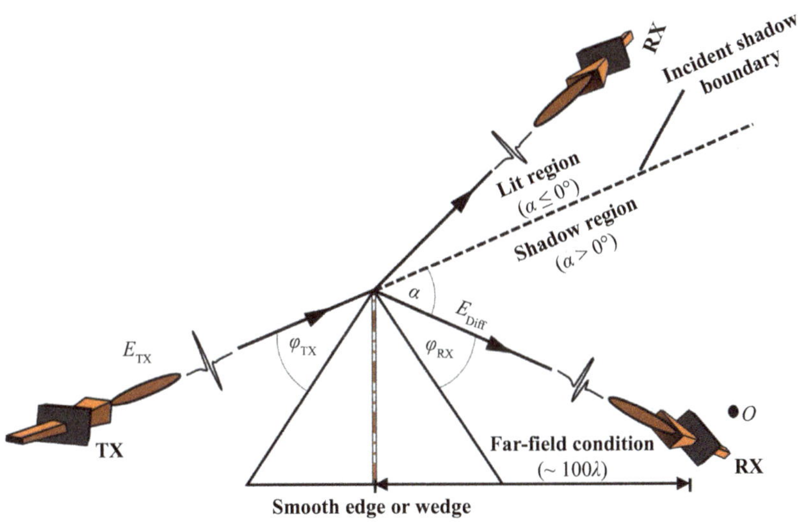

Figure 12.20 Geometry for smooth edge/wedge diffraction

various objects with arbitrary dielectric parameters but also accounts for polarization in contrast to the latter approach. However, the complexity of UTD is very high, and it requires numerical solutions for path loss [36]. The first-order diffracted field at the receiving point O can be expressed from [15] as

$$\boldsymbol{E}_{\mathrm{Diff}}(O) = \left[\boldsymbol{G}_{\mathrm{Diff}}^{\mathrm{RX}}\right]^{*} \boldsymbol{G}_{\mathrm{Diff}}^{\mathrm{TX}}\, D(\varphi_{\mathrm{TX}}, \varphi_{\mathrm{RX}})\, Q(d_{\mathrm{TX}}, d_{\mathrm{RX}})\, L_{\mathrm{Diff}}\, \boldsymbol{E}_{\mathrm{TX}} \tag{12.20}$$

Based on (12.20), $\boldsymbol{E}_{\mathrm{Diff}}$ depends on D and Q. The diffraction coefficient D is determined by diffraction angles at objects and the boundary conditions. Likewise, the geometry factor Q considers the distance from the TX and RX to the edge or wedge. So far, a very few angular-dependent diffraction measurements at THz frequencies are reported [34,37–39]. In [39], the analysis of the frequency dependency on wedges (metallic cuboids) between 275 and 325 GHz is experimentally studied, and no significant dispersion of the diffracted wave is noticed. Moreover, the measurements show a good agreement with the calculations based on the UTD regardless of the polarization and angular range. To summarize, the UTD proves well suited for the diffraction modelling at 300 GHz from smooth surfaces. *However, the diffraction from rough surfaces at THz frequencies has not been studied yet.*

12.6 Scenario environments

As far as propagation in a real environment is concerned, the office room BB121 in our Institute of Digital Signal Processing is modelled. This typical office room of 7 m (length) × 7 m (width) × 3 m (height) covers both LoS and NLoS scenarios. In general, NLoS links can either take the form of specular reflection, diffuse scattering (non-specular reflection) or diffracted paths. Note that the diffraction effect can be neglected particularly for the indoor environments considered and shall be reasoned later within each environment in the following subsections. Up to twice-reflected paths (i.e. second order) are considered in the modelling process unless otherwise stated. The multipaths that bounce back and have a power level lower than −200 dBm are also neglected by using the RX threshold function of ray tracer [40]. Note that the minimum path amplitude applied to the simulations is selected keeping in mind that the maximum influence of diffuse reflection is to be investigated, and any paths with very small power levels are to be omitted from the latter stages of computations.

An access point serving as the TX is positioned in a corner of the office just below the ceiling at the points $x = 6$ m, $y = 1$ m, $z = 2$ m. The transmitting antenna is kept away from any obstructions in order to provide maximum coverage. The transmit output power is 0 dBm. Taking into account the tabletop terminals (i.e. laptops, tablets, and cellular phones), the chosen RXs are 0.75 m in height, unless otherwise indicated. Moreover, the simulations are performed with isotropic omnidirectional antennas at both TX and RX ends with radiation pattern in the azimuth plane to realize an estimation of the maximum occurring multipath influence, despite the demand of the THz frequencies for directional antennas of high gains. It is noteworthy to mention that the directional antenna though supports the multipath suppression and the temporal dispersion but simultaneously shows a high sensitivity towards mispointing. Please refer to [41] for antenna directivity impact and [42] for misalignment impact on THz indoor channel characteristics.

The ceiling and walls are covered with rough plasters given in Table 12.2 possessing identical electrical parameters but with different statistical surface parameters of standard deviation height and correlation length. The key idea is to use realistic statistical surface parameters measured by the authors in [30]; and by varying these statistical parameters of rough plasters, our focus is to investigate and demonstrate the influence of degree of roughness on the 300-GHz propagation characteristics. The floor is made of concrete and covered with ideally smooth PVC, unless otherwise indicated. Hence, due to the smooth flooring, only specular reflections are included from the floor. In order to attain coverage of the whole office, maximize scattering contributions, and lower scattering losses, vertically polarized antennas are chosen at both TX and RXs' ends. The three glass windows $1.65 \times 1.7\ m^2$ each are modelled. These windows are 1.05 m above the floor and separated by a distance of 0.45 m from each other. The white board on the wall is an inactive object in the simulation environment. In Figure 12.22, the pine wood wardrobe with the dimensions $2.25 \times 2.6 \times 0.1\ m^3$ is clearly depicted. The wooden door towards the corridor is 2.1-m high and 1.05-m wide. The other comparatively smaller door is 1.9-m high and 0.8-m wide.

12.6.1 First environment

At THz frequencies, the environment where channel modelling and measurements are of great interest are offices, residential structures, conference rooms, corridors, and libraries with distance up to 10 m. The simulated model of an empty office shown in Figure 12.21 is an LoS cubic scenario.

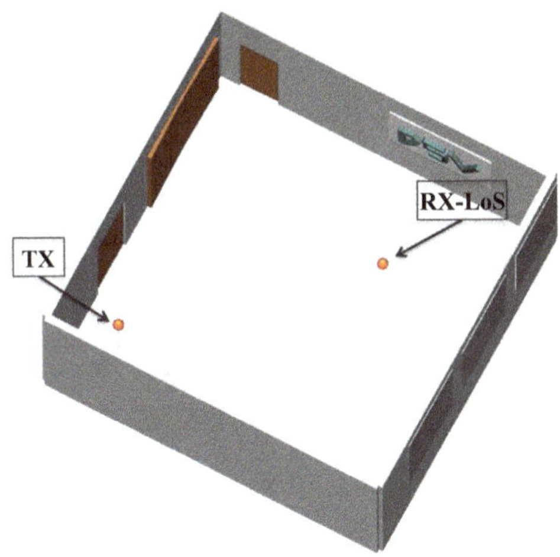

Figure 12.21 3D view of simulated empty office room BB121 (first environment)

The ultimate goal of this empty office model is to efficiently yield accurate predictions on radio propagation key statistics, including complex impulse responses (CxIRs), received signal strengths for mobile locations, angular spreads, delays of direct/indirect rays, electric and magnetic field strengths, and interference measures. Besides, this empty environment is also the choice of fortune to study the propagation channels when only the impact of roughness on THz ultra-broadband channel is the focus of investigation. The RX-LoS is chosen to have the 0.75-m height at the points $x = 2$ m, $y = 4.4$ m, $z = 0.75$ m. Hereupon, the TX and the RX have a separation distance of 5 m. The diffraction in this empty office scenario is not considered (i.e. no furniture).

12.6.2 Second environment

Figure 12.22 shows a simple environment which ascertains both LoS and NLoS scenarios. The RX-LoS and the RX-NLoS are chosen to have 0.75-m height at the points $x = 3.5$ m, $y = 4.45$ m, $z = 0.75$ m and $x = 1.7$ m, $y = 2.75$ m, $z = 0.75$ m, respectively. The empty workplace has a height of 0.7 m, width of 1.8 m, and length of 0.8 m. This workplace is separated by a metallic panel 1.8×0.55 m^2 in dimensions and pine wood tables with metallic legs. The TX and the two RXs are 5 m apart.

Please note that no diffraction is observed when conducting simulations in the presence of an empty work space, the reason being the absence of any reflection points that occur near sharp edges and wedges.

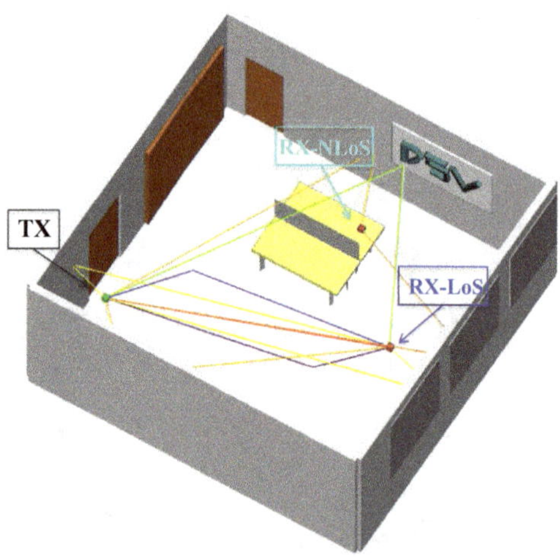

Figure 12.22 3D view of office room BB121 with an empty work place (second environment)

12.7 Frequency dependence of material properties

In the lower microwave region (below 6 GHz) with a fairly long wavelength (e.g. the wavelength is 60 mm at 5 GHz), reflections from common building materials hardly consider surface roughness and simply treat surfaces as smooth, flat, and reflective. However, multiple works at mm-wave frequencies such as [31,43,44] investigated the impact of surface roughness (diffuse scattering) experimentally. The experiments in most of these works are setup for outdoor rough surfaces with standard deviation heights of few centimetres. The authors in [22] measured the statistical parameters of rough surfaces of indoor common building materials such as wallpaper and plaster using commercially available equipment [45]. *Please note that the lateral and vertical scan resolutions are the most important characterization parameters for surface roughness measurements but not the carrier frequencies.* The calculated relative standard deviation of the surface height is 130 μm for wallpaper and 50 μm for plaster x1. These standard deviations are comparatively small even at mm-wave frequencies and thus are usually neglected. On the contrary, the THz frequencies have a wavelength on the order of several hundred micrometres (e.g. $\lambda_{300\text{GHz}} = 1$ mm) which is comparable to the surface deviation heights. Hence, the impact of roughness is to be investigated quantitatively. Thereupon, the empirical parameters of the materials used in our modelling approach along with their standard deviation heights and correlation lengths are tabulated in Table 12.2.

Figure 12.23 shows the reflection loss for common building materials at 300 GHz for different incident angles corresponding to the experimentally calculated Fresnel reflection coefficient in [46]. It is evident that the roughness of a surface causes far more reflection loss than done by its electrical properties. Apparently, this higher reflection loss is also due to shorter wavelength at THz frequencies. In contrast, for larger wavelengths, the reflection loss is comparatively less. Now, if rough plasters are assumed to be smooth (i.e. plaster x), the resulting reflection losses are illustrated in Figure 12.24, which clearly depicts the dominance of material electrical properties.

12.8 Development of THz standards

In this section, we briefly discuss the current status of the regulation and standardization effort of the physical layer (PHY) for the lower THz frequency range. THz technology has come of age, and the standards of IEEE 802.15.3d-2017 operating from 252 to 325 GHz, designed for data rates of up to 100 Gb/s for intra-device communication (e.g. board-to-board communication), close proximity communication, wireless data centres, and backhaul/fronthaul links, are already approved [47]. Two PHY modes are defined in the specification for the THz PHY: the single carrier (THz-SC PHY) and the on–off keying mode (THz-OOK PHY). The two

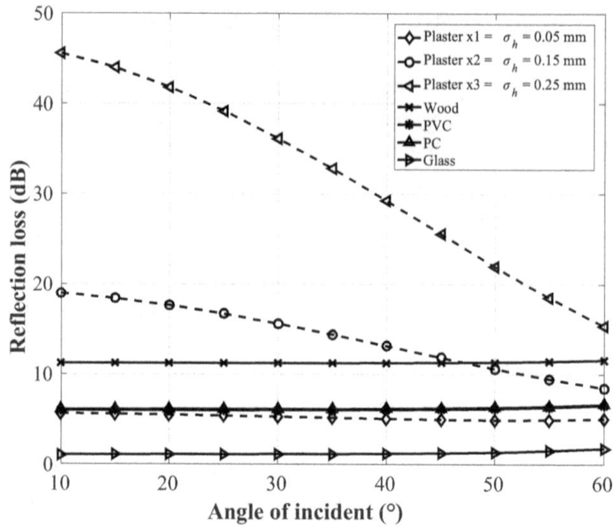

Figure 12.23 Reflection losses from combined Fresnel at 300 GHz for different incident angles for materials listed in Table 12.2

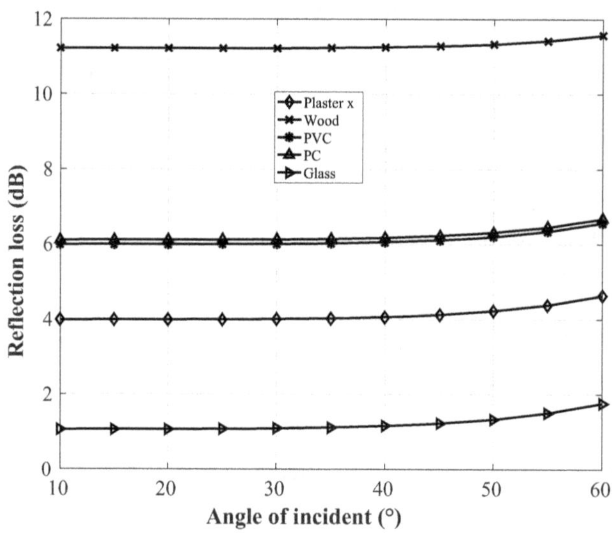

Figure 12.24 Reflection losses from combined Fresnel at 300 GHz for different incident angles (smooth materials)

PHYs are mapped to the distinctive advantages in supporting specific applications. For instance, the THz-SC PHY is aimed for extremely high data rates up to 100 Gb/s. Besides, it is designed to support a wide range of modulations, namely $\pi/2$ BPSK, $\pi/2$ QPSK, $\pi/2$ 8-PSK, $\pi/2$ 8-APSK, 16-QAM, and 64-QAM. The modulations of

$\pi/2$ BPSK and $\pi/2$ QPSK are mandatory for THz-SC PHY, while the other modulations are optional. On the other hand, THz-OOK PHY is designed to support low-cost and simple design. However, it supports a single modulation scheme, OOK, and three forward error correction schemes. The Reed–Solomon code is mandatory here. The standard IEEE 802.15.3d-2017 provides 69 channels with eight ranges of channel bandwidths from 2.16 to 69.12 GHz.

12.9 Rough surfaces at THz frequencies

We interrogate in this section the statistical description of random rough surfaces encountered in indoor environments which may have implications on the throughput of THz communication systems. The earlier discussion (cf. Section 12.4.2) encompasses surface profiles to be functions of coordinate x only, and their physical characteristics are further described by one-dimensional (1D) distribution function, namely height distribution function. These surfaces are termed 1D. In fact, most common random surfaces met in practice are two-dimensional (2D), interpreted as functions of two spatial directions, i.e. the surface height h is expressed as a random function of the two coordinates x and y. Basically, the surface height distribution function describes the 2D variation in surface elevation and azimuth above an arbitrary plane. Hereupon, to approximate the scattering pattern of random rough surfaces, the roughness parameters and characteristic functions are absolutely imperative. They are categorized as Gaussian and non-Gaussian with several approaches being used to describe the statistical nature of the surface. The considered random rough surface is homogeneous, single layered, isotropic, and Gaussian distributed.

12.9.1 Gaussian rough surfaces

The behaviour of diffuse scattering of an incoming wave from typical indoor rough surfaces inside rooms is ascertained by the statistical surface characteristics. The surface height h as already mentioned is a random function of the two coordinates x and y. The probability distribution of h then facilitates to figure out the shape of that surface. The surface heights of indoor materials used in our simulated environments (cf. Section 12.6) are assumed to be Gaussian distributed (i.e. the surface heights are normally distributed). In Gaussian height distribution, surface height fluctuates systematically around the average surface height [29]. Considering h normally distributed with average surface height $\bar{h} = 0$ and standard deviation σ_h, then the probability distribution of h is given by

$$p_h(h) = \frac{1}{\sqrt{2\pi}\sigma_h} e^{-\left(h^2/2\sigma_h^2\right)} \tag{12.21}$$

The standard deviation σ_h represents the surface roughness. This is among the most important physical parameters but insufficient to express anything about the lateral distance between the peaks and valleys of the surface profile. For instance, though apparently non-identical, the rough surfaces shown in Figure 12.25 exhibit identical

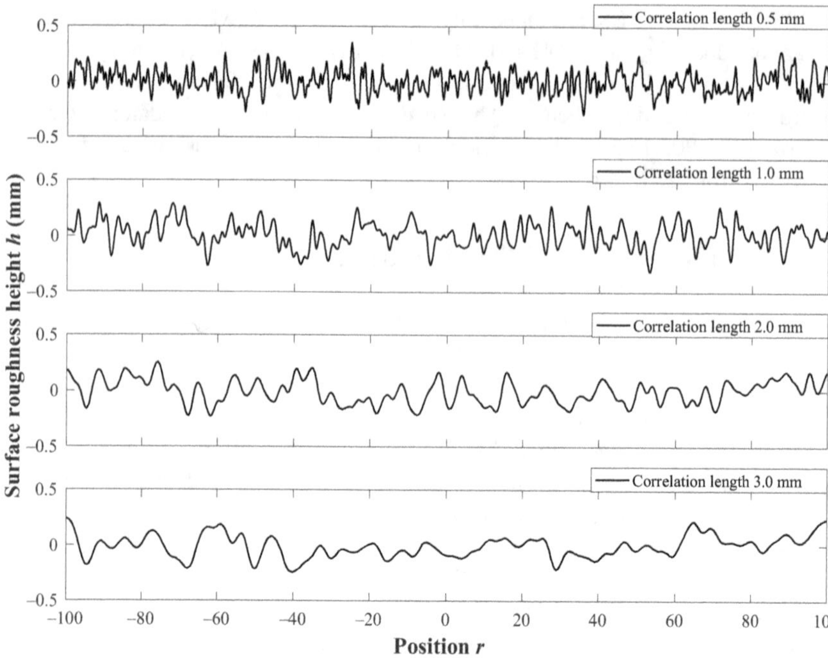

Figure 12.25 Random rough surface with the unique standard deviation height h but different correlation lengths ℓ_{cr}

Gaussian distribution (i.e. $\bar{h}$ and σ_h are same for both surfaces). The reason for this difference in appearance is the surface correlation length ℓ_{cr}. The correlation length gives us the typical distance between two irregularities (peaks) on the surface profile. It represents the lateral dimension of a rough surface. In other words, the correlation length provides information whether the surface consists of densely packed irregularities or slowly varying features. In brief, the two statistical parameters necessary for characterizing indoor rough surfaces are (i) the standard deviation height σ_h and (ii) the correlation length or correlation distance ℓ_{cr} [48]. In fact, by varying these two statistical parameters, one can generate surfaces that match in appearance to almost any rough surface met in practice. It is noteworthy to mention here the use of an autocorrelation function $C(\tau)$ of the surface to determine the correlation length. The autocorrelation function is of fundamental importance in describing the surface morphology of random rough surfaces. In particular, it gives the distance beyond which two surface points can be considered independent of each other. The autocorrelation function widely used and assumed mostly is *Gaussian*. This normalized Gaussian autocorrelation function for the computation of the correlation length in a single dimension is expressed from [20, p. 81] as

$$C(\tau) = e^{-\left(\tau^2/l_{cr}^2\right)} \tag{12.22}$$

where τ is the distance between two surface points (x_1, y_1) and (x_2, y_2) randomly selected at the respective surface heights $h_1 = h(x_1, y_1)$ and $h_2 = h(x_2, y_2)$. Hence, the correlation length is the value of distance τ for which the autocorrelation function $C(\tau)$ drops to e^{-1} of its value. Most noteworthy in (12.22) is the assumption of an isotropically rough surface. However, different rough surfaces may have different correlation functions, and it is quite likely that other auto-correlation functions may give a better fit to the measured surface data.

12.9.2 Non-Gaussian rough surfaces

In practice, the surface height distribution is not always Gaussian. A non-Gaussian rough surface requires additional statistical parameters such as skewness and kurtosis in addition to the standard deviation height σ_h and correlation length ℓ_{cr} [29]. Thus, the fundamental question arises as to how this non-Gaussian height distribution affects the total received and scattered power, and what scattering models may be adopted for resolving these problems at THz frequencies? Not dwelling deep into that and rather proceeding with the efforts mainly on Gaussian rough surfaces keeps this research gap open for the time.

12.10 Novel solution of the scattering problem in THz

Among the major challenges at THz frequencies is the modelling of the most significant propagation phenomenon of diffuse scattering by which an incident ray may split into a specular and several non-specular rays after bouncing off from rough materials. Notice that the scattering problem of EM waves is still not completely solved and no exact closed-form solutions exist as of yet. However, numerous approximate methods are developed for wave scattering at rough surfaces in order to predict and interpret experimental data. Among these, three popular surface scattering models are (i) Rayleigh–Rice (R–R), (ii) classical Beckmann–Kirchhoff (cB–K), and (iii) modified Beckmann–Kirchhoff (mB–K). Pertaining to the roughness characteristic, R–R model depends only on standard deviation height σ_h which is irregular and arbitrary in distribution, whereas B–K model [20] considers not only the standard deviation height which takes Gaussian hypothesis on the height distributions into consideration rather the irregularity or aperiodicity of surface roughness in spatial direction also. This second statistical parameter is called correlation length l_{cr}. In fact, by varying these two statistical parameters, one can generate surfaces that match in appearance to almost any rough surface met in practice. The main characteristics of these aforementioned models are summarized in more detail in the preceding subsections.

12.10.1 Rayleigh–Rice (R–R) model

The R–R approach can be seen as the most rigorous analytical solution of Maxwell's equations for the limiting case of optically smooth surface (i.e. slightly rough surface). Rayleigh expressed optical smoothness by following the accurate criterion from [49] as

$$\left(\frac{4\pi\sigma_h \cos\Theta_i}{\lambda}\right)^2 \ll 1 \tag{12.23}$$

Notice that (12.23) does not contain the scattering angle as an argument. The reason is straightforward, no assumptions are made regarding this angle when developing the theory. Thus, when employing the R–R model, the incoherent diffuse scattering is considered as multipath not directed towards RX and is therefore neglected. Besides, the optically smooth surface check can be a formidable figure depending on the composite surface or wavelength at hand. It is usually formulated as $\sigma_h/\lambda \ll 1$. Unfortunately, there is no explicit number to decide whether the criterion is fulfilled or not. For example, in [49,50], it is stated that σ_h/λ should be smaller than 0.01 and 0.05, respectively. However, in the case of THz frequencies, the optically smooth surface requirement is possibly reinstated owing to the extremely short wavelength. Therefore, in our simulation model, we hypothetically assume that slightly rough surfaces are $\sigma_h = 0.15$ and 0.30 mm. We consider $\sigma_h = 0$ mm as a benchmark for conceptualizing ideally smooth surfaces as opposed to rough surfaces.

The R–R theory is developed on the basis of boundary conditions for a perfectly conducting surface. Besides, this theory takes polarization of the incident and scattered wave into account. A small parameter of this theory is the Rayleigh roughness parameter (ρ_{spec}). For a Gaussian height probability density function, this term is equal to (cf. 12.14). At THz frequencies, diffuse reflection tends to be higher due to the increased surface roughness, and this surface roughness causes an additional attenuation even in a specular direction of reflection (by the amount that is scattered into non-specular directions). The surface scattering process for diffuse reflection is analysed based on R–R theory (i.e. the specular losses occur due to the diffuse reflection). Simulations are conducted in order to study the indoor multipath propagation and its impact on the ultra-broadband THz channel by considering two different degrees of surface roughness (i.e. $\sigma_h = 0.15$ and 0.30 mm), respectively, and as a benchmark compared to the ideally smooth wall (i.e. $\sigma_h = 0$ mm).

The simulation setup is depicted in Figure 12.21 (first environment, cf. Section 12.6). All walls, the ceiling, and floor are made of plasters ($\tilde{\varepsilon}_r = 3.691,\ \tilde{\tilde{\varepsilon}}_r = 0.217$) with focus to investigate and demonstrate the influence of degree of roughness on the ultra-broadband 300-GHz propagation characteristics. The multipaths between TX and RX are computed in terms of the frequency-domain CTF according to Kirchhoff approximation (KA), which accounts for specular losses by introducing Rayleigh roughness factor calculated from the surface height distribution. The time-domain channel impulse response or $h(t)$ is obtained via inverse fast Fourier transform (IFFT). Here it shall be noted that at each frequency point, the frequency-dependent CxIR of the channel is computed with the corresponding number of multipaths, angle of arrival, angle of departure, and time of arrival of individual paths. The complete CTF is then obtained through the coherent addition of the individual CxIR. We employ the tangent plane

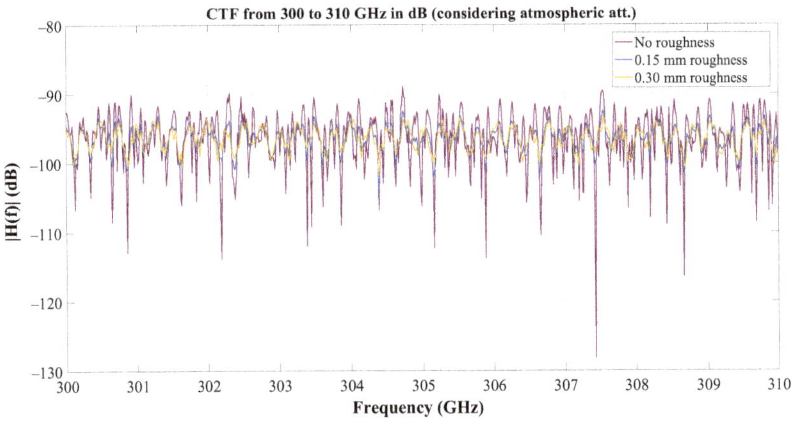

Figure 12.26 CTF of ray-tracing simulation of twice-reflected TE polarized plane wave when considering smooth and rough surfaces

approximation to scattering properties of common building materials in our propagation models which take Gaussian hypothesis on the height distributions into consideration.

At THz frequencies the multipath propagation characteristics are likely to vary significantly over the band of interest in case the environment is surrounded by rough surfaces. This is actually the case in our simulation model. Figures 12.26 and 12.27 depict the LoS results with and without scattering. Upon dual analysis (cf. Figures 12.26 and 12.27), in the case of no scattering (or no roughness), the highest frequency selectivity of the channel is observed. In contrast, the impact of surface roughness on the THz propagation channel leads to hardening of the THz channel. However, the average attenuation over the whole bandwidth in Figure 12.26 is found to be 96.59, 97.02, and 97.04 dB for $\sigma_h = 0,\ 0.15,$ and 0.30 mm, respectively. Apparently, the relatively identical average attenuation observed for $\sigma_h = 0.15$ and 030 mm is most probably due to the presence of a dominant LoS path (direct path) and the scattered channel paths may exert little or no impact on the total received power. In other words, varying σ_h does not affect the relative scattered power in LoS case. Furthermore, if the surface roughness is not properly taken into account in the propagation modelling, the received power levels for the worst scenario (i.e. frequency point $f = 307.4$ GHz) are miscalculated up to 28.3 and 30.5 dB for $\sigma_h = 0.15$ and 30 mm, respectively. This approves the necessity to include scattering in THz propagation modelling.

Next, Figure 12.27 presents quite interesting results. Varying the σ_h from 0.15 to 0.30 mm it is evident that the longer paths are crossing the threshold limit because their amplitudes are impacted. Thus, the surface roughness on the one hand increases the scattering richness but on the other hand decreases the channel strength.

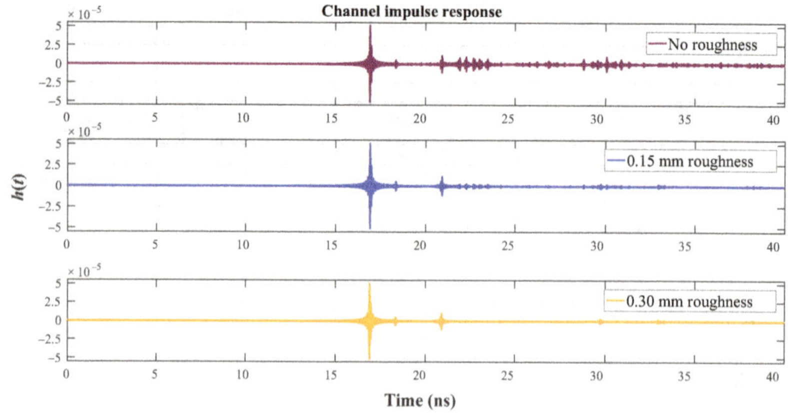

Figure 12.27 CIR of ray-tracing simulation of twice-reflected TE polarized plane wave when considering smooth and rough surfaces

12.10.2 Classical Beckmann–Kirchhoff (cB–K) model

The cB–K model is derived from the analytical model. Its theory is more realistic and provides more insight of the physical processes involved in the surface scattering. More realistic in contrast to R–R, cB–K accounts for the diffuse reflection impact from rough surfaces in both specular and non-specular directions. The cB–K model is validated against ultra-broadband measurements [51], and it is applicable to both dielectric and metallic surfaces. However, the cB–K theory is a scalar treatment, i.e. the wave scattering theory accounts only for the distribution of energy and not for more complex effects such as *polarization*. The B–K theory is derived from the Helmholz integral [48], and it predicts a symmetrical scattered field distribution around the specular direction under some assumptions for *slightly* and *very rough surfaces* [20]. The details of the theory along with its derivation are described comprehensively in the monograph by Beckmann and Spizzichino [20], and the key steps involved in the derivation of the model are only mentioned. But let us summarize the assumptions under which cB–K model is derived. Perhaps being forthright and honest, these assumptions ensure the validity of B–K model, and it is not fair to withdraw any of the following listed assumptions at the cost of mathematical simplicity.

12.10.3 Assumptions

The assumptions are as follows:

1. The surface is perfectly conducting.
2. The radius of curvature of surface irregularities is large compared with the wavelength of the incident field.
3. Shadowing and multiple scattering (cf. Figure 12.28(b)) are neglected.
4. The reflection coefficient of the surface has unity magnitude.

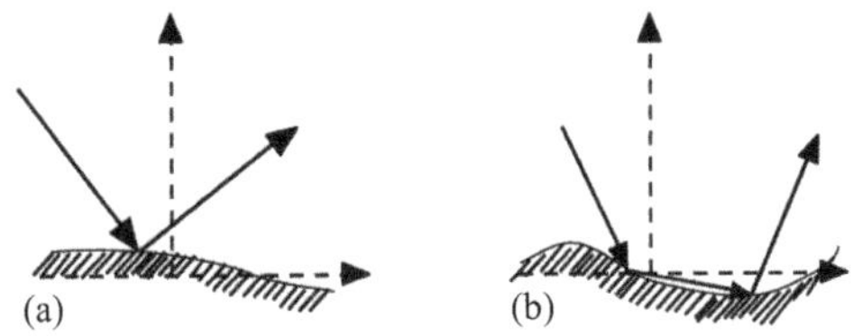

Figure 12.28 *Schematic view of surface scattering: (a) single and (b) multiple*

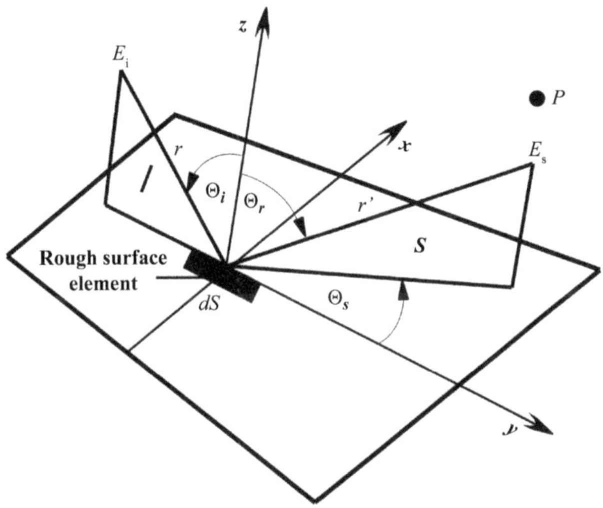

Figure 12.29 *Geometry of scattered angles from a rough surface element dS, I is the incident plane and S is the scattering plane explaining B–K model*

5. The incident wave is plane and linearly polarized with the electric field vector in the plane of incidence or perpendicular to it.
6. The observation point (i.e. **P**) is sufficiently far from surface for the scattered waves to be exactly planar.
7. The rough surface has a Gaussian probability density of height and a Gaussian correlation function.

The field scattered by a surface in any direction can be determined from the field present at the surface. Let us assume the incident field E_i, a harmonic plane wave of unit amplitude, is incident at the rough surface element dS. The elevation angle of incidence is represented as Θ_i and the elevation and azimuth reflected angles in the scattered region as Θ_r and Θ_s, respectively. The angles are measured with respect to the z-axis as shown in Figure 12.29. The total scattered field $\mathbf{E_s}$ at a point P far above the rough surface is given by

$$\mathbf{E_s}(\mathrm{P}) = \frac{1}{4\pi}\int\int_S \left(E_i\frac{\partial\psi}{\partial n} - \psi\frac{\partial E_i}{\partial n}\right)dS \tag{12.24}$$

This is called the *Helmholtz integral*. Its complete derivation is provided in [20]. Here S is the reflection surface, $\mathbf{n}$ is the normal to the rough surface element at the considered point, and ψ is a continuous scalar function given by

$$\psi = \frac{e^{jkr'}}{r'} \tag{12.25}$$

where $j = \sqrt{-1}$ and $k = 2\pi/\lambda$ is the propagation constant of the reflected wave. $\mathbf{E_i}$ and $\partial\mathbf{E_i}/\partial\mathbf{n}$ are the field and its normal derivative on dS. The exact values of these two quantities in general are unknown. The KA or *tangent plane approximation* may be used in approximating the values of the field and its normal derivative at each point on the surface and then evaluating the integral in (12.24). This approximation breaks down if the roughness includes sharp edges or sharp points compared to the wavelength of the incident wave. Thus, the field at a point on the rough surface is equal to the field that would be present on a tangent plane at that point.

Moreover, to calculate the local scattered fields E_s, the KA treats the facets (rough surface elements) of a rough surface as tangent planes and applies the Fresnel reflection coefficient Γ. Within this approximation, the field on the surface S based on [20, p. 20] is defined as

$$E_s = (1 + \Gamma)E_i \tag{12.26}$$

In general, the characteristics of the reflection can be described by the well-known Fresnel reflection coefficient only in the case of an ideal smooth surface. Although, we have considered a rough surface, the KA is still applicable due to its assumption of a rough surface that is however locally smooth.

In the case of very rough surfaces, the reflection behaviour of the wave is dominated by the scattering phenomena. We introduce a scattering coefficient from [20, p. 22]

$$\rho = \frac{E_s}{E_r} \tag{12.27}$$

where E_r is the field specularly reflected ($\Theta_i = \Theta_r$) by a smooth and perfect conducting surface of similar dimensions as the rough one, and under the same angle of incidence, at the same distance, by considering vertical polarization. From [20, Appendix A], we have

$$E_r = \frac{jkA\cos(\Theta_i)e^{jkr_0}}{\pi r_0} \tag{12.28}$$

Area $A = l_x l_y$ is the rectangular surface area. The Kirchhoff solution requires the lateral dimension l_x and l_y of the area A to be much greater than the wavelength, i.e. $l_x \gg \lambda$ and $l_y \gg \lambda$.

It is worth remarking that to attain ρ, in general an integration of the scattered field over the whole surface needs to be performed. However, usually rough surfaces found in indoor environments with common building materials (i.e. wallpaper and plaster) have randomly rough Gaussian height distributions at THz frequencies [22]. From a complex quantity ρ in (12.27), we determine the mean value of $|\rho|^2$

$$\langle \rho\rho^* \rangle = \langle |\rho|^2 \rangle = \left| \frac{E_s^2}{E_r^2} \right| \tag{12.29}$$

where the operator $\langle . \rangle$ is an ensemble mean which represents a statistical average and $(.)^*$ denotes the complex conjugate.

Now, assume a rectangular surface of area $A = l_x l_y$ with infinite conductivity. In general, the average scattering coefficient of an incident wave on a rough surface of angle Θ_i, scattered at angles Θ_r and Θ_s, respectively, is determined by the following expressions [20, p. 88]

$$\langle \rho\rho^* \rangle_\infty = \left(\rho_0^2 + \frac{\pi l_{cr}^2 F^2}{A} \sum_{m=1}^{\infty} \frac{g^m}{m!m} e^{-\left(v_{xy}^2 l_{cr}^2/4m\right)} \right) e^{-g} \tag{12.30}$$

where ρ_0 the scattering coefficient of a plane surface with area $A = l_x l_y$ is given by

$$\rho_0 = \operatorname{sinc}(v_x l_x) \operatorname{sinc}(v_y l_y) \tag{12.31}$$

From trigonometry, it follows:

$$v_x = k(\sin(\Theta_i) - \sin(\Theta_r)\cos(\Theta_s)) \tag{12.32}$$

$$v_y = k(-\sin(\Theta_r)\sin(\Theta_s)) \tag{12.33}$$

$$v_{xy} = \sqrt{v_x^2 + v_y^2} \tag{12.34}$$

The geometrical factor, a function of incident and scattered angles, is given as

$$F = \frac{1 + \cos(\Theta_i)\cos(\Theta_r) - \sin(\Theta_i)\sin(\Theta_r)\cos(\Theta_s)}{\cos(\Theta_i)(\cos(\Theta_i) + \cos(\Theta_r))} \tag{12.35}$$

The quantity g, a measure of phase variation introduced by surface roughness σ_h, is expressed as

$$g = (v_z \sigma_h)^2 = \sigma_h^2 \left(\frac{2\pi f}{c} \right)^2 (\cos(\Theta_i) + (\cos(\Theta_r))^2$$

From expression in (12.30), it is apparent that the average scattering coefficient consists of two terms. The first term $\rho_0^2 e^{-g}$ describes the influence of the scattering in the direction of specular reflection (i.e. specular spike component) only and the succeeding (second) term corresponds to the diffusely scattered field (i.e. side lobes'

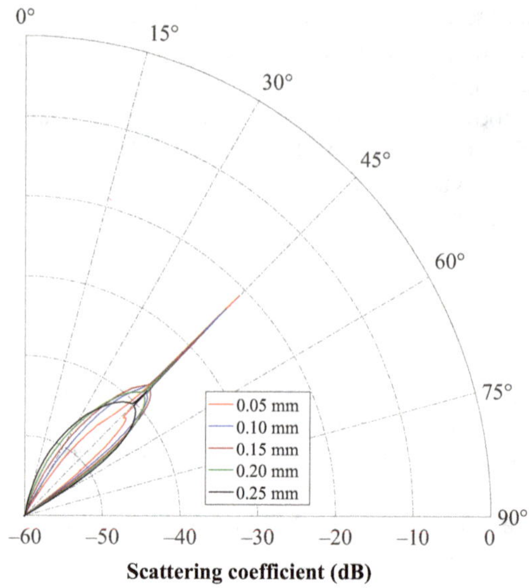

Figure 12.30 Simulated scattering coefficient at 300 GHz of the plaster x_i, ($r' = 5$ m)

components). The average scattering coefficient with different surface height distributions is schematically depicted in Figure 12.30. Further, in Figure 12.31, it is apparent that the incident power is reflected and scattered in and around the specular direction with an attenuation of 20.93 and 37.03 dB, respectively.

Owing to (12.30), B–K model uses two statistical parameters to characterize rough surfaces: (i) the standard deviation height σ_h and (ii) the correlation length ℓ_{cr}. In general, the height values of the topographic surface features about the mean surface level are measured at equally spaced digitized data points. On the other hand, the correlation length ℓ_{cr} is defined as the lag length at which the Gaussian correlation function drops to $1/e$ of its maximum [48].

The exponential series given by the summation in the lobe component can be approximated for slightly rough ($0 < g \ll 1$) and very rough surfaces ($g \gg 1$). The approximation results in simpler expressions of the scattering coefficient for these two extreme surface conditions are

$$\langle \rho\rho^* \rangle_{\text{slightly rough}} = \left(\rho_0^2 + \frac{\pi \ell_{cr}^2 F^2 g}{A} e^{-\left(v_{xy}^2 \ell_{cr}^2/4\right)} \right) e^{-g} \tag{12.36}$$

$$\langle \rho\rho^* \rangle_{\text{very rough}} = \frac{\pi l_{cr}^2 F^2}{A v_z^2 \sigma_h^2} e^{-\left(v_{xy}^2 l_{cr}^2 / 4 v_z^2 \sigma_h^2\right)} \tag{12.37}$$

Importantly, the above approximation in (12.30) assumed the surface to be a perfect conductor. Indeed, the materials used in our study are not perfect conductors. Next, in order to approximate the average scattering coefficient for finite conductors, we

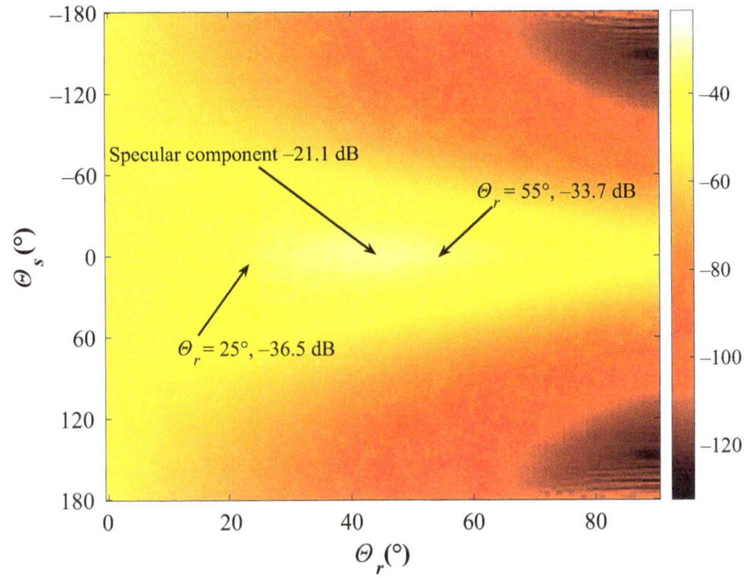

Figure 12.31 Angular-dependent power reflection factor of scattering coefficient (B–K model) at 300 GHz of the plaster sample x1 ($r' = 5$ m)

average the conventional Fresnel reflection coefficient (Γ) over the entire surface area and use the resultant value ($\langle \Gamma \rangle$) as a constant in the Helmholtz integral [48]. Finally, for finite conducting surfaces, the average scattering coefficient becomes

$$\langle \rho \rho^* \rangle_{\text{finite}} = \langle \Gamma \Gamma^* \rangle \langle \rho \rho^* \rangle_{\infty} \tag{12.38}$$

Therefore, the mean scattered power is given by

$$\langle P_s \rangle = \frac{|E_s^2|}{2Z_0} = \frac{|E_r^2|}{2Z_0} \langle \rho \rho^* \rangle_{\text{finite}} \tag{12.39}$$

Likewise, the incident power is

$$\langle P_i \rangle = \frac{|E_i^2|}{2Z_0} \tag{12.40}$$

Thus, the average power reflection coefficient of a surface area A, specifying the scattered field in a distance r' from surface to the observation point P relative to the incident power P_i, leads to [20, p. 89] and let $|E_i^2| = 1$

$$\langle R_{\text{power}} \rangle_{\text{scattered}} = \frac{\langle P_s \rangle}{\langle P_i \rangle} = \langle |E_r^2| \rangle \langle \rho \rho^* \rangle_{\text{finite}} \tag{12.41}$$

The significance of this result can be seen once we substitute the value of E_r from (12.28), and we find

$$\langle R_{\text{power}} \rangle_{\text{scattered}} = \frac{4A^2 \cos^2(\Theta_i)}{\lambda^2 r'2} \langle \rho\rho^* \rangle_{\text{finite}} \tag{12.42}$$

Similarly, for the direction of specular reflection $\Theta_r = \Theta_i, \Theta_s = 0$, we have

$$\langle R_{\text{power}} \rangle_{\text{specular}} = \langle \rho\rho^* \rangle_{\text{finite}} = \langle \Gamma\Gamma^* \rangle \langle \rho\rho^* \rangle_{\infty} \tag{12.43}$$

since $\rho_0 = 1$ for $v_{xy} = 0$, (12.30) is rewritten as [20, p. 93]

$$\langle \rho\rho^* \rangle_{\infty} = e^{-g} \tag{12.44}$$

Substituting (12.44) in (12.43), we obtain

$$\langle R_{\text{power}} \rangle_{\text{specular}} = \langle \rho\rho^* \rangle_{\text{finite}} = \langle \Gamma\Gamma^* \rangle e^{-g} \tag{12.45}$$

12.10.4 Modified Beckmann–Kirchhoff (mB–K) model

A modified B–K theory is attained by replacing the geometrical factor (F-factor squared) used by Beckmann in (12.30) with the cos (Θ_i) in Lambert's cosine law. We can rewrite this for slightly and very rough surfaces as

$$\langle \rho\rho^* \rangle_{\text{slightly rough}} = \left(\rho_0^2 + \frac{\pi \ell_{cr}^2 K g}{A} e^{-\left(v_{xy}^2 \ell_{cr}^2/4\right)} \right) e^{-g} \tag{12.46}$$

$$\langle \rho\rho^* \rangle_{\text{very rough}} = \frac{\pi \ell_{cr}^2 K}{A v_z^2 \sigma_h^2} e^{-\left(v_{xy}^2 \ell_{cr}^2/4 v_z^2 \sigma_h^2\right)} \tag{12.47}$$

The renormalization constant K in this reformulation of scalar diffraction theory is given by the following expression:

$$K = \frac{\int_{\alpha=-\infty}^{\infty} \int_{\beta=-\infty}^{\infty} L(\alpha, \beta - \beta_0) d\alpha \; d\beta}{\int_{\alpha=-1}^{1} \int_{\beta=-(1-\alpha^2)^{1/2}}^{(1-\alpha^2)^{1/2}} L(\alpha, \beta - \beta_0) d\alpha \; d\beta} \tag{12.48}$$

The details about the variables used in (12.48) are given in [52]. This model can be worthy of use for the solution of scattering problems in volume scattering while it accommodates larger scattering angles.

12.10.4.1 Comparison between the surface scattering models

In order to demonstrate a comparison between the surface scattering models along with their specific advantages and limitations, the ultra-broadband channel behaviour by using our self-developed THz 3D ray-tracing algorithm tool [53] in terms of the frequency-domain CTF dynamics at 3,201 frequency points for $f = 300 \ldots 310$ GHz in LoS and NLoS scenarios is simulated. Figures 12.32 and 12.33 depict the CTF results for the respective R–R, cB–K and mB–K models for

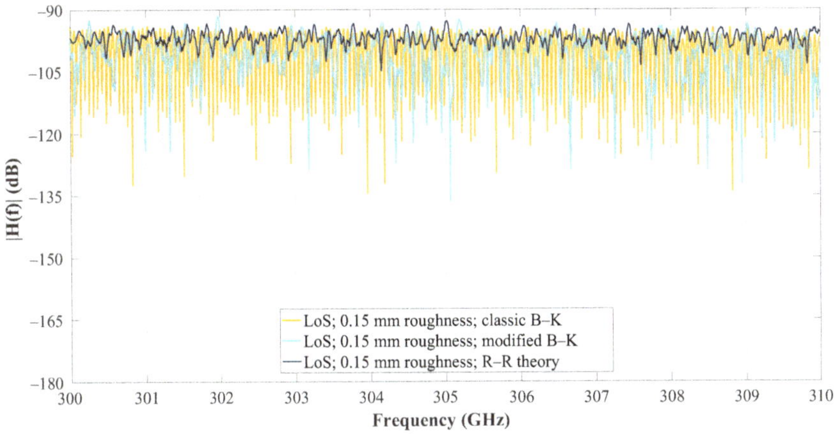

Figure 12.32 *Comparison of CTFs between classical B–K, modified B–K, and R–R models in the presence of slightly rough surfaces at RX-LoS ($x = 3.5$ m, $y = 4.45$ m, $z = 0.75$ m). The surface correlation length is $\ell_{cr} = 1.7$ mm, $l_x = l_y = 10\ \ell_{cr}$, and $\sigma_h = 0.15$ mm*

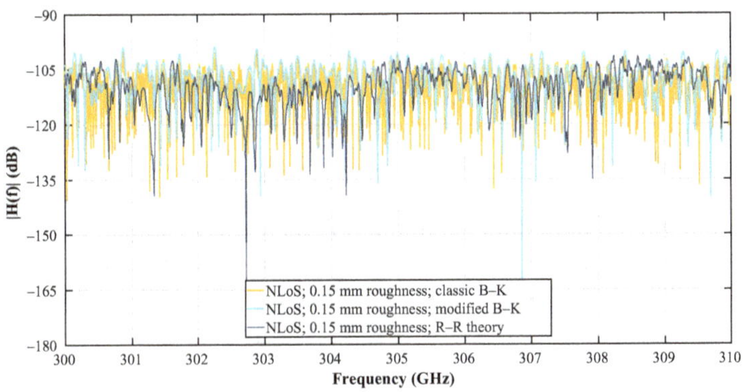

Figure 12.33 *Comparison of CTFs between classical B–K, modified B–K, and R–R models in the presence of slightly rough surfaces at RX-NLoS ($x = 1.7$ m, $y = 2.75$ m, and $z = 0.75$ m). The surface correlation length is $\ell_{cr} = 1.7$ mm, $l_x = l_y = 10\ \ell_{cr}$, and $\sigma_h = 0.15$ mm*

Figure 12.22 scenario. For LoS case, the respective average attenuation over the whole bandwidth is 97.02, 100.86, and 101.59 dB. Meanwhile, the standard deviation for the respective models is 1.70, 5.96, and 5.29. However, for NLoS case, the respective average attenuation is 110.09, 109.90, and 107.95 dB with the standard deviation being 5.51, 5.46, and 5.20, respectively.

12.11 Summary

The 1-Tb/s wireless links anywhere and anytime can only be achieved by first studying thoroughly the basic propagation mechanisms which further necessitate the detailed typology of the surfaces that enables one to combat the problems posed by diffuse scattering from these surfaces.

In this chapter, three different antenna configurations (horn–horn, horn–open, and open–open) are employed to study the impact of candle flame on the ultra-broadband THz communication across the spectrum of interest from 300 to 310 GHz. The channel measurements are performed for two different distances, i.e. 0.5 and 1 m. The findings demonstrate that the THz EM wave is exposed to slight attenuation mostly caused by the water vapour attenuation. The wave propagation is delayed (phase shifted) and comparatively more prominent in the open–open antenna configuration. However, this slight attenuation does not cause any hindrance in the characterization and localization of materials at THz frequencies.

A peculiar scattering pattern is revealed from the coverage map results of the ray-tracing simulations in an LoS environment. It is further perceived that the scattering phenomenon significantly persists in the points near to the walls and distant from the TX. The almost identical average attenuations for varied degrees of roughness recorded from our simulations in LoS scenario highlight the effect of a dominant direct path as opposed to the diffuse reflection paths with almost negligible effect.

The diffuse scattering is among the major challenges at THz frequencies when modelling the THz propagation channels. Hence, a thorough study on the four widely accepted scattering models (R–R, ER, cB–K, and mB–K) along with their comparison is presented in the chapter to assist one in resolving this scattering problem. A noteworthy finding from this comparison and further affirmed by the simulation results as well is that the ER model can be considered for modelling THz channels, though originally proposed for channel modelling only at 60 GHz.

References

[1] H. J. Song and T. Nagatsuma, "Handbook of Terahertz Technologies: Devices and Applications", *Pan Standford Pub.*, Singapore, 2015.

[2] G. J. Foschini and M. J. Gans, "On limits of wireless communications in a fading environment when using multiple antennas", *Wireless Personal Communications*, vol. 6, no. 3, pp. 311–335, 1998.

[3] A. Goldsmith, S. A. Jafar, N. Jindal, and S. Vishwanath, "Capacity limits of MIMO channels", *IEEE Journal on Selected Areas in Communications*, vol. 21, no. 5, pp. 684–702, 2003.

[4] T. L. Marzetta and B. M. Hochwald, "Capacity of a mobile multiple-antenna communication link in Rayleigh flat fading", *IEEE Transactions on Information Theory*, vol. 45, no. 1, pp. 139–157, 1999.

[5] T. L. Marzetta, "Noncooperative cellular wireless with unlimited numbers of base station antennas", *IEEE Transactions on Wireless Communications*, vol. 9, no. 11, pp. 3590–3600, 2010.
[6] S. Berman, D. S. Greenhouse, I. Bailey, R. D. Clear, and T. Raasch, "Human electroretinogram responses to video displays, fluorescent lighting, and other high frequency sources", *Optometry and Vision Science*, vol. 68, no. 11, pp. 645–662, 1991.
[7] A. Maltsev, A. Sadri, R. Maslennikov, A. Davydov, and A. Khoryaev, "Channel Models for 60 GHz WLAN Systems", *IEEE 802.11-08/0811r1*, Jul. 2008.
[8] M. Shafi, A. F. Molisch, P. J. Smith, *et al.*, "5G: A tutorial overview of standards, trials, challenges, deployment and practice", *IEEE Journal on Selected Areas in Communications*, vol. 35, no. 6, pp. 1201–1221, 2017.
[9] International Telecommunication Union, "Attenuation by atmospheric gases", P Series Radiowave propagation, Recommendation ITU-R P.676-10, 2013.
[10] F. Sheikh, N. Zarifeh, and T. Kaiser, "Terahertz band: Channel modelling for short-range wireless communications in the spectral windows", *IET Microwaves, Antennas & Propagation*, vol. 10, no. 13, pp. 1435–1444, 2016.
[11] R. Piesiewicz, T. K. Ostmann, N. Krumbholz, *et al.*, "Short-range ultra-broadband terahertz communications: Concepts and perspectives", *IEEE Antennas and Propagation Magazine*, vol. 49, no. 6, pp. 24–39, 2007.
[12] A. Fox, B. Heinemann, H. Rücker, *et al.*, "Advanced heterojunction bipolar transistor for half-THz SiGe BiCMOS technology", *IEEE Electron Device Letters*, vol. 36, no. 7, pp. 642–644, 2015.
[13] K. Shinohara, A. C. Regan, Y. Tang, *et al.*, "Scaling of GaN HEMTs and Schottky diodes for submillimeter-wave MMIC applications", *IEEE Transactions on Electron Devices*, vol. 60, no. 10, pp. 2982–2996, 2013.
[14] S. Lara-Avila, D. Danilov, D. Golubev, *et al.*, "Towards quantum-limited coherent detection of terahertz waves in charge-neutral graphene", *Nature Astronomy*, vol. 3, no. 11, pp. 983–988, 2019.
[15] C. Oestges, B. Clerckx, L. Raynaud, and D. V. Janvier, "Deterministic channel modelling and performance simulation of microcellular wide-band communication systems", *IEEE Transactions on Vehicular Technology*, vol. 51, no. 6, pp. 1422–1430, 2002.
[16] H. T. Friis, "A note on a simple transmission formula", *Proceedings of the IRE*, vol. 34, no. 5, pp. 254–256, 1946.
[17] Y. Zantah, F. Sheikh, A. Abbas, M. Alissa, and T. Kaiser, "Channel Measurements in Lecture Room Environment at 300 GHz", *IEEE 2nd Int. Workshop on Mobile THz Systems (IWMTS 2019)*, pp. 1–5, Jul. 2019.
[18] A. Ghwaji, F. Sheikh, T. Schulze, I. Willms, and T. Kaiser, "Preliminary Analysis of Candle Flame Impact on THz Electromagnetic Wave Propagation", *IEEE 2nd Int. Workshop on Mobile THz Systems (IWMTS 2019)*, pp. 1–5, Jul. 2019.
[19] A. Maltsev, R. Maslennikov, A. Lomayev, A. Sevastyanov, and A. Khoryaev, *IEEE 802.11-09/0431r0*, Apr. 2009.

[20] P. Beckmann and A. Spizzichino, "The Scattering of Electromagnetic Waves from Rough Surfaces", *Artech House Radar Library*, Boston, MA, 1963.
[21] R. Piesiewicz, C. Jansen, S. Wietzke, D. Mittleman, M. Koch, and T. Kürner, "Properties of building and plastic materials in the THz range", *International Journal of Infrared and Millimeter Waves*, vol. 28, no. 5, pp. 363–371, 2007.
[22] C. Jansen, S. Priebe, C. Möller, *et al.*, "Diffuse scattering from rough surfaces in THz communication channels", *IEEE Transactions on Terahertz Science and Technology*, vol. 1, no. 2, pp. 462–472, 2011.
[23] F. G. Bass and I. M. Fuks, "Wave Scattering From Statistically Rough Surfaces", *International Series in Natural Philosophy* (Volume 93), Pergamon Press, New York, NY, 1979.
[24] E. L. Feinberg, "The propagation of radio waves along the surface of the earth", AF Systems Command, 1967.
[25] I. M. Fuks, "Radio wave scattering theory on the distributed surface of the sea", *Institute of Radio Physics and Electronics of the Ukrainian Academy of Sciences.*, vol. 9, no. 5, pp. 876–887, 1966.
[26] R. N. Gurzhi and S. M. Shevchenko, "Thermal conductivity theory of thin dielectric models", *Soviet Physics - Journal of Experimental and Theoretical Physics*, vol. 25, no. 3, pp. 534–536, 1967.
[27] G. T. Ruck, "Radar Cross Section Handbook", *Plenum Press*, New York, 1970.
[28] R. Vaughan and J. Andersen, "Channels, Propagation and Antennas for Mobile Communications", *IET*, London, 2003.
[29] Y. Zhao, G. Wang, and T. Lu, "Characterization of Amorphous and Crystalline Rough Surface: Principles and Applications", *1st Edition (Volume 37)*, Academic Press, San Diego, CA, 2000.
[30] R. Piesiewicz, C. Jansen, D. Mittleman, T. K. Ostmann, M. Koch, and T. Kürner, "Scattering analysis for the modeling of THz communication systems", *IEEE Transactions on Antennas and Propagation*, vol. 55, no. 11, pp. 3002–3009, 2007.
[31] A. Goulianos, A. Freire, T. Barratt, *et al.*, "Measurements and Characterisation of Surface Scattering at 60 GHz", *2017 IEEE 86th Vehicular Technology Conference (VTC-Fall)*, pp. 1–5, Sep. 2017.
[32] A. K. Fung, "Scattering and depolarization of EM waves from a rough surface", *Proceedings of the IEEE (Communications)*, vol. 54, pp. 395–396, 1966.
[33] P. Beckmann, "The Depolarization of Electromagnetic Waves Scattered From Rough Surfaces", *Proc. Symp. Electromagnetic Theory and Antennas*, E. C. Jordan, Ed. New York: Pergamon, pp. 717–726, 1963.
[34] M. Jacob, S. Priebe, R. Dickhoff, T. Ostmann, T. Schrader, and T. Kürner, "Diffraction in mm and sub-mm wave indoor propagation channels", *IEEE Transactions on Microwave Theory and Technique*, vol. 60, no. 3, pp. 833–844, 2012.
[35] R. Kouyoumjian and P. Pathak, "Uniform geometrical theory of diffraction for an edge in a perfectly conducting surface", *Proceedings of the IEEE*, vol. 62, no. 11, pp. 1448–1461, 1974.

[36] C. Han, A. O. Bicen, and I. F. Akyildiz, "Multi-ray channel modeling and wideband characterization for wireless communications in the terahertz band", *IEEE Transactions on Wireless Communications*, vol. 14, no. 5, pp. 2402–2412, 2015.

[37] J. Kokkoniemi, P. Rintanen, J. Lehtomäki, and M. Juntti, "Diffraction Effects in Terahertz Band – Measurements and Analysis", *2016 IEEE Global Communications Conference (GLOBECOM)*, pp. 1–6, Dec. 2016.

[38] C. Cheng, S. Kim, and A. Zajic "Study of Diffraction at 30 GHz, 140 GHz, and 300 GHz", *IEEE AP-S Symposium on Antennas and Propagation and USNC-URSI*, pp. 1–2, Jul. 2018.

[39] T. Ostmann, M. Jacob, S. Priebe, R. Dickhoff, T. Schrader, and T. Kürner, "Diffraction Measurements at 60 GHz and 300 GHz for Modeling of Future THz Communication Systems", *Proc. 37th Int. Conf. Infr. Millim. Terahertz Waves (IRMMW-THz)*, pp. 1–2, Sep. 2012.

[40] Remcom, Inc. "Wireless InSite Users Manual version 2.7", Electromagnetic Simulation Software by Remcom, 2020. www.remcom.com.

[41] S. Priebe, M. Jacob, and T. Kürner, "The Impact of Antenna Directivities on THz Indoor Channel Characteristics", *Proceedings of the 6th European Conference on Antennas and Propagation (EUCAP)*, pp. 478–482, May 2012.

[42] S. Priebe, M. Jacob, and T. Kürner, "Affection of THz Indoor Communication Links by Antenna Misalignment", *Proceedings of the 6th European Conference on Antennas and Propagation (EUCAP)*, pp. 483–487, May 2012.

[43] Y. Cocheril, R. Vauzelle, L. Aveneau, and M. Khoudeir, "Rough Surfaces Influence on an Indoor Propagation Simulation at 60 GHz", *ECPS*, Mar. 2005.

[44] S. Tariq, C. Despins, S. Affes, and C. Nerguizian, "Rough Surface Scattering Analysis at 60 GHz in an Underground Mine Gallery", *2014 IEEE International Conference on Communications Workshops (ICC)*, pp. 724–729, Jun. 2014.

[45] Alicona Imaging InfiniteFocus, "Alicona Imaging GmbH", *Online*, Available: www.alicona.com, accessed in August 2019.

[46] R. Piesiewicz, T. Kleine-Ostmann, N. Krumbholz, D. Mittleman, M. Koch, and T. Kürner, "Terahertz characterisation of building materials", *Electronics Letters*, vol. 41, no. 18, pp. 1002–1004, 2005.

[47] IEEE Standard for High Data Rate Wireless Multi-Media Networks, "100 Gb/s Wireless Switched Point-to-Point Physical Layer", *IEEE Std 802.15.3d-2017*, pp. 1–55, Oct. 2017.

[48] S. Nayar, K. Ikeuchi, and T. Kanade, "Surface reflection: Physical and geometrical perspectives", *IEEE Transactions on Pattern Analysis and Machine Intelligence*, vol. 13, no. 7, pp. 611–634, 1991.

[49] J. Stover, "Optical Scattering, Measurement and Analysis", *2nd Edition*, SPIE Press, Washington, DC, 1995.

[50] T. Vorburger, E. Marx, and T. Lettieri, "Regimes of surface roughness measurable with light scattering", *Applied Optics*, vol. 32, no. 19, pp. 3401–3408, 1993.

[51] S. Priebe, M. Jacob, C. Jansen, and T. Kürner, "Non-Specular Scattering Modeling for THz Propagation Simulations", *Proceedings of the 5th EUCAP*, pp. 15, Apr. 2011.

[52] J. Harvey, C. Vernold, A. Krywonos, and P. Thompson, "Diffracted radiance: a fundamental quantity in nonparaxial scalar diffraction theory", *Applied Optics*, vol. 38, no. 31, pp. 6469–6481, 1999.

[53] F. Sheikh, D. Lessy, and T. Kaiser, "A Novel Ray-Tracing Algorithm for Non-specular Diffuse Scattered Rays at Terahertz Frequencies", *in IEEE 1st. International Workshop on Mobile Terahertz Systems (IWMTS)*, pp. 1–6, Jul. 2018.

Chapter 13
Conclusion and future perspectives

Qammer H. Abbasi[1], *Syeda F. Jilani*[2], *Akram Alomainy*[2] *and Muhammad A. Imran*[1]

This book presents a comprehensive overview of the communication transition towards the fifth generation (5G) networks and the role of advanced antenna solutions in this advancement. It is anticipated that 5G will build the foundation of many emerging technologies such as Internet of Things (IoT), device-to-device (D2D) communication and smart applications with cognitive control that can support industries to increase efficiency, open new productivity domains and effectively improve consumers' lifestyles. Faster connectivity, high data rates, low latency, security, energy efficiency and high mobility are the defining features of 5G to realise a real-wireless experience. The vision of 5G demands highly efficient antenna front ends and sophisticated algorithms to define the high-frequency wireless propagation. Many novel antenna design techniques such as metamaterial antennas, beamformers comprising reflectarray, phased-array antennas and massive multiple-input–multiple-output (MIMO) antennas, and state-of-the-art fabrication techniques such as three-dimensional (3D) printed antennas, antennas-on-chip and novel compact antenna designs for wideband operation have gained huge attention in 5G networks. The progress towards 5G has secured success in the complete redesigning of the mobile network framework, establishing the new protocols for data throughputs and network security, dealing with the fronthaul and backhaul issues and ensuring an efficient spectrum utilisation by enabling smart algorithms and channel-modelling techniques.

13.1 Conclusion

In this book, the spectrum goals for 5G in terms of bandwidth, potential of millimetre-wave (mm-wave) bands, efficient choices of the antenna arrays for the beamforming networks, performance analysis and channel modelling of the MIMO configurations, reliable antenna topologies and effective fabrication and integration

[1]James Watt School of Engineering, University of Glasgow, Glasgow, UK
[2]School of Electronic Engineering and Computer Science, Queen Mary University of London, London, UK

techniques are thoroughly discussed. The initial deployment of 5G wireless networks on commercial grounds has begun in 2019 in many countries with the release of the first set of 5G standards and soon will be available worldwide. The antenna demands are highly challenging for 5G; thus, more compact designs and efficient performance attributes are needed for their suitable integration in 5G wireless devices designed for many industrial, clinical, biomedical, personal and communication purposes. In addition, the transition towards mm-waves has originated many efficient fabrication techniques for bulk manufacturing such as 3D fabrication and antennas-on-chip. Many advanced artificially designed materials like metamaterials have found their place in the design of antennas and propagation links for 5G and beyond. In order to enhance the capacity, large panels of massive MIMO antennas have gained attention at many forums. Reflectarrays and phased arrays have been anticipated as efficient solutions to enable line-of-sight links specifically to handle propagation losses due to high attenuations at mm-wave spectrum.

The trends of future wireless networks integrated in the 5G architecture are not limited to this recent progress but have the potential to be utilised in the communication networks beyond 5G. For instance, advanced antenna arrays and beamforming networks, complex multiplexing techniques for massive MIMO, integration of metamaterials to manipulate wave-propagation, antenna design techniques to achieve wideband, complex integration of antenna-on-chip and many other emerging techniques will be useful in 5G as well as beyond. In addition, the parameters of the network's physical layer radio frame proposed for the 5G cellular systems are expected to be flexibly configured to cope with the diversity of numerous applications and services. This book has also elaborated on the design of channel-modelling parameters enabling massive connection requests, wireless Xhaul and associate impact on network provisioning, over-the-air testing, including both antenna and propagation characteristics at the same time, and massive MIMO communication channels in wireless propagation scenarios.

13.2 Future perspectives

This has been established that the 5G wireless network marks the beginning of a true digital society, accomplishes substantial advancements in its path towards maturation and sets even bigger goals for the next-generation networks beyond 5G. It is important to highlight the promising features in the current wireless communication and further prospects for the antennas and propagation for the next generation (i.e. 6G). The initial study on the design architecture of 6G networks and vision beyond 5G has already been established at various research forums. It has been predicted that 6G will provide the real user experience to many novel applications recently introduced in the 5G networks. The ever-increasing size, complexity, number of users and applications and high-performance needs of the future wireless devices necessitate envisioning innovative techniques for enabling and harmonising the forecasted highly densified heterogeneous networks. In this section, some recent advanced trends for the future wireless networks beyond 5G have been discussed.

13.2.1 Artificial intelligence and cognitive radio

The artificial intelligence (AI) has gained huge attention in recent years, which has motivated the research trends to improvising the essential intelligence, cognitive control and autonomy in 5G cellular networks and beyond. The AI and cognitive radio are intelligent technologies with the capabilities of sensing, analysis and decision-making for dynamic resource allocation and spectrum management, though this intelligence at this stage is only limited to perform certain basic tasks related to optimisation, logic, control and management. It is anticipated that the AI and cognitive radio will fully establish their place in the future networks beyond 5G. Thus, the future AI system is expected to establish smart communication networks, IoT, the Internet of Vehicles' intelligence and many other applications of a smart city to optimise the benefits of individuals.

13.2.2 Wireless system requirements beyond 5G

It is essential to ensure high yield, low delay time, low payment energy, high battery efficiency and compactness, improved connectivity, high scalability, spectral efficiency and performance reliability in the future wireless networks. In beyond 5G, hundreds of Gbps data rate (up to 1 Tbps), ultra-low latency of microseconds and 1,000 times higher capacity than 5G networks are estimated and according to the forecast by International Telecommunication Union, global mobile data traffic will reach 5 zettabytes by 2030. Multi-antenna systems such as beamforming and MIMO are expected to be highly involved in 5G and beyond 5G communications. In addition, following the trends of 5G networks, the energy efficiency will be one of the key considerations in the future networks. Increasing the mobile network's energy efficiency facilitates to minimise both its operating expenditures and its carbon emissions. Potential implementations of applications like IoT and D2D communication in the imminent future raise the bars for high spectral efficiency in the wireless network, where heterogeneous services and devices should be integrated in a unified mesh of smart networks. Novel approaches for the spatial efficiency, resource allocation and frequency reuse for the ultra-dense small cell deployments will be further improved in the future networks.

13.2.3 Terahertz frequency spectrum

Another interesting feature proposed for the wireless networks beyond 5G is terahertz (THz) (i.e. 0.1–10 THz) communication. When compared to mm-wave communication, THz signals provide higher link directionality and higher security. However, the limitation of increased attenuation is more critical at THz than mm-waves, which can be handled by designing shorter distance and ultra-dense networks probably residing within a building area for beyond 5G systems. THz communication permits implementation of wireless nano-networks comprising a network of wirelessly connected nano-machines. Given the progress of nanotechnology and recent developments in advanced materials like carbon nanotubes and graphene, nanoscale antennas operating on the THz frequency band can be manufactured on nano-machines in future. Nano-networks are also anticipated to facilitate Internet-of-Nano-Things for numerous important areas such as environmental monitoring, diagnosis services in healthcare, industrial manufacturing, lab-on-a-chip, connected vehicles and many smart city applications. Also, research has been in progress on the scope of optical communication for future wireless networks that allow even higher data rates and better physical security.

13.2.4 Intelligent reflective surfaces and machine learning

Intelligent reflecting surfaces (IRSs) have found a great interest as potential solutions for the future wireless propagation beyond 5G for boosting the spectrum and energy efficiency. The IRS constitutes artificially intelligent and programmable meta-surfaces exhibiting numerous exotic electromagnetic functionalities such as perfect and controllable absorption, beam steering, beam splitting, wavefront shaping, polarisation control, broadband pulse delay or harmonic generation. Due to these exceptional properties, IRSs have been investigated in wireless power transfer, as beamformers and beam splitters in multi-user communication, in phase modulation scenarios and as relays. These advantages make the IRS a promising candidate for 6G wireless communications especially when integrated in highly advanced applications of IoT such as machine learning (ML) which is envisioned to be revolutionising a wide variety of future applications. The incorporation of IRS in ML is anticipated to drive the wireless environment into a programmable and partially deterministic space that can be used as a parameter subject to optimisation in the network design, thus, to obtain the desired performance characteristics, such as latency, data rate, energy efficiency, secrecy and reliability.

13.2.5 Energy harvesting

Future wireless communication is also focusing on the harnessing energy from ambient sources such as sunlight, mechanical, thermal and even electromagnetic sources. Energy harvesting and wireless power transfer are of major interests to make possible powering of wireless networks via clean energy and renewable sources. This enables energy sharing among wireless network nodes increasing the lifetime of nodes and also has the potential to get rid of large and bulky batteries and can facilitate the flexibility and conformity of handheld devices and wearable gadgets. Many efficient approaches are in the process of development and maturation to make energy harvesting and wireless power transfer more feasible in the future networks.

These advantages can impact heavily on applications related to medical, remote sensing and industrial automation. Based on the 5G development and future prospects, it can be concluded that 5G has engraved the foundation for many innovative technologies which will flourish further, and in future the evolution of mobile communication will undoubtedly continue beyond 5G to facilitate the evergrowing number of devices and services.

Index

www.ingramcontent.com/pod-product-compliance
Lightning Source LLC
LaVergne TN
LVHW011833200926
848484LV00002B/350
* 9 7 8 1 8 3 9 5 3 0 9 7 5 *